BOUNDED GAPS BETWEEN PRIMES

Searching for small gaps between consecutive primes is one way to approach the twin primes conjecture, one of the most celebrated unsolved problems in number theory. This book documents the remarkable developments of recent decades, whereby an upper bound on the known gap length between infinite numbers of consecutive primes has been reduced to a tractable finite size. The text is both introductory and comprehensive: the detailed way in which results are proved is fully set out and plenty of background material is included. The reader journeys from selected historical theorems to the latest best result, exploring the contributions of a vast array of mathematicians, including Bombieri, Goldston, Motohashi, Pintz, Yildirim, Zhang, Maynard, Tao and Polymath8. The book is supported by a linked and freely available package of computer programs. The material is suitable for graduate students and of interest to any mathematician curious about recent breakthroughs in the field.

KEVIN BROUGHAN is Emeritus Professor at the University of Waikato, New Zealand. He cofounded and is a fellow of the New Zealand Mathematical Society. Broughan brings a unique set of knowledge and skills to this project, including number theory, analysis, topology, dynamical systems and computational mathematics. He previously authored the two-volume work *Equivalents of the Riemann Hypothesis* (Cambridge University Press, 2017) and wrote a software package which is part of Goldfeld's *Automorphic Forms and L-Functions or the Group* GL(n,R) (Cambridge University Press, 2006).

BOUNDED GAPS BETWEEN PRIMES

The Epic Breakthroughs of the Early Twenty-First Century

KEVIN BROUGHAN

University of Waikato

CAMBRIDGE
UNIVERSITY PRESS

CAMBRIDGE
UNIVERSITY PRESS

University Printing House, Cambridge CB2 8BS, United Kingdom

One Liberty Plaza, 20th Floor, New York, NY 10006, USA

477 Williamstown Road, Port Melbourne, VIC 3207, Australia

314–321, 3rd Floor, Plot 3, Splendor Forum, Jasola District Centre, New Delhi – 110025, India

79 Anson Road, #06–04/06, Singapore 079906

Cambridge University Press is part of the University of Cambridge.

It furthers the University's mission by disseminating knowledge in the pursuit
of education, learning, and research at the highest international levels of excellence.

www.cambridge.org
Information on this title: www.cambridge.org/9781108836746
DOI: 10.1017/9781108872201

First published 2021

Printed in the United Kingdom by TJ Books Limited, Padstow Cornwall

A catalogue record for this publication is available from the British Library.

Library of Congress Catalouging-in-Publication Data
Names: Broughan, Kevin A. (Kevin Alfred), 1943– author.
Title: Bounded gaps between primes : the epic breakthroughs of the early
twenty-first century / Kevin Broughan.
Description: New York : Cambridge University Press, 2021. |
Includes bibliographical references and index.
Identifiers: LCCN 2020043768 (print) | LCCN 2020043769 (ebook) |
ISBN 9781108836746 (hardback) | ISBN 9781108799201 (paperback) | ISBN 9781108872201 (epub)
Subjects: LCSH: Numbers, Prime. | Number theory.
Classification: LCC QA246 .B743 2021 (print) | LCC QA246 (ebook) | DDC 512.7/3–dc23
LC record available at https://lccn.loc.gov/2020043768
LC ebook record available at https://lccn.loc.gov/2020043769

ISBN 978-1-108-83674-6 Hardback
ISBN 978-1-108-79920-1 Paperback

Dedicated to Jackie, Jude and Beck

Pintz took a closer look at the flawed proof and came up with a key insight for the ultimate fix. He contacted Goldston in December 2004, and the three number theorists, Goldston, Pintz and Yildirim, had a complete proof by early February. They circulated the manuscript to a handful of experts . . . One of these, Motohashi, found a shortcut that led to a surprising short proof of the basic qualitative result.

– Science Magazine

In April 2013, a lecturer at the University of New Hampshire submitted a paper to the Annals of Mathematics. Within weeks word spread – a little-known mathematician, with no permanent job. The rest is history.

– IMDb

In November 2013, inspired by Zhang's extraordinary breakthrough, James Maynard dramatically slashed the bound (for an infinite number of prime pairs) to 600, by a substantially easier method. Both Maynard and Terry Tao, who had independently developed the same idea, were able to extend their proofs to show that for any given integer $m \geq 1$ there exists a bound such that there are infinitely many intervals of that length containing at least m distinct primes. If Zhang's method is combined with the Maynard–Tao setup, then it appears that the bound can be further reduced to 246. If all of these techniques could be pushed to their limit, then we would obtain a bound of 12 (or arguably 6), so new ideas are still needed to have a feasible plan for proving the twin prime conjecture.

– Andrew Granville

The conquest of the bounded gaps is an historical event that will continue to attract attention long into the future.

– Yoichi Motohashi

Contents

Preface

This book has been written to mark, celebrate and detail the remarkable developments which occurred in the number theory area of gaps between primes in the first two decades of the twenty-first century and were published in the main between 2006 and 2015. Many mathematicians contributed to these developments, and most of the main participants are acknowledged and their work set out in Chapters 4 through 8.

In addition, there is an introductory chapter giving background material including an overview of the developments, their historical context, technical prerequisites and a brief guide to introductory material. The second chapter is also introductory, with details of the Brun and Selberg sieves, and a third gives details of a selection of early work relating to prime gaps.

The main part of the book, Chapters 4 through 7, is time sequential, reporting successively the work of Goldston, Motohashi, Pintz and Yildirim, then Motohashi, Pintz and Zhang, then Maynard, and finally Polymath8b. Each chapter is roughly self-contained, even if ideas from one advance sparked discoveries in the next, and the reader might wish to dwell especially on Chapter 6 (Maynard) and/or its "completion" Chapter 7 (Polymath8b), where the prime gap bound 246 is derived.

Chapter 8 is devoted in part to an exposition of Zhang's variation, but mostly to Polymath8a's variations of the Bombieri–Vinogradov theorem. An extensive set of appendices support the work of that chapter, and even this is insufficient to fully report on the work, which would take a similar sized book to this. The final chapter takes a different course than the others, giving a summary of further developments which have resulted already from the developments in the short space of time since 2015.

To aid the reader, definitions are repeated in various places, and major steps in proofs numbered to give a clear indication of the main parts and allow for easy proof internal referencing. For the most part, we have kept to the methods and logic of the original authors, while clarifying the argument when possible and providing background results when this seemed useful for readers.

The reader should be warned this is not an introductory number theory text. The expected background for a beginner would be a study of elementary analytic number theory, especially that relating to the distribution of the primes, and complex calculus, especially for Chapter 4. The Fourier transform is used intensively in Chapter 7. There are excellent texts from which a selection might be made. For example, those by Apostol [4], Hardy and Wright [90], Nathanson [152, 153], De Koninck and Luca [121] and especially Montgomery and Vaughan [144].

We use the arithmetic functions $\mu(n)$, $\varphi(n)$, $\Omega(n)$, $\omega(n)$ and the Landau–Vinogradov expressions

$$f(x) \ll g(x),\ f(x) = O(g(x)),\ f(x) = o(g(x)),$$

each with its usual meaning.

In writing a book of this nature, it is somewhat tempting to dwell on particular results and see if they could be improved. Too much of that ensures the work would never be finished and has been resisted. However, there are some small improvements. For example, proofs of some lemmas have been provided, such as Lemma 2.21, the sum of $\log p/p$ over the divisors of an integer n in Chapter 2; the derivation of a particularly nasty binomial sum identity in Lemma 4.3 in Chapter 4; a clearer presentation of the derivation of Zhang's lower bound given comments in his introduction which are apparently misleading, showing that Maynard's choices of parameters in two of his explicit forms is optimal for Theorem 6.17 and Lemma 6.11; and the derivation of a bound for smaller values of ϵ to extend Theorem 7.17.

Many computations have been repeated, so numerical results may at times look a little different from published results. There is a website for errata and notes, and readers are encouraged to communicate with the author in this regard at *kab@waikato.ac.nz*. The website is linked to:

www.math.waikato.ac.nz/∼kab

Also linked to this website is the suite of Mathematica™ programs, PGpack, related to the material in this volume, which is available for free download. Instructions on how to download the software are given in Appendix I. The functions `RayleighQuotientMaynard` and `RayleighQuotientPolymath` are essentially those of Maynard and Polymath respectively. The application of these functions is an essential part of the derivation of their results. Readers should be aware that the result along the lines of that of Maynard was shown by Polymath to be optimal, and the author was not able to improve the result of the Polymath, even though, on the face of it, this should be possible. Consulting the epilogue in Chapter 9 could be useful in this regard.

Even though I have done my best to properly acknowledge the contributions of those whose results have made this work possible, a range of limitations mean there will undoubtedly be omissions, for which I am regretful. One of the functions of the book web page "Errata and Notes" is to provide a remedy in such situations, and readers are encouraged to contact the author at *kab@waikato.ac.nz* if necessary.

It is not the purpose of this book to detail the story of bounded gaps between primes as a human endeavour, just the technical details and then only part of these. The human side, the "backstory", is of great interest, and, according to this author, quite unique in the history of mathematics, and even more generally of science. It should be told before too many years have elapsed and memory blurs the edges of details. It already has begun to do that – it is a natural process.

The human aspect of the prime gaps discoveries has many components – group and individual creativity, generous leadership, courage in the face of error, determination to solve problems over decades, rivalry and disappointment, deserved and undeserved fame, excellent and poor communication, generosity in giving credit and obstinacy when it comes to attribution. There is a link established on the book home page, entitled Backstory, giving some summary information and further links. I have started collecting these and they should grow in number: for example, articles in *Science* and *Quanta* magazines, the Notices and Bulletin of the American Mathematical Society (AMS), in the Newsletter of the European Math Union (EMU), and in *Ideas of the IAS* (Institute of Advanced Studies, Princeton). In addition, links to a Csicsery film, to the book by Vicky Neal, and to Polymath8's home page with of the order of thousand further links. Finally, this web-based information would not in any way take the place of a coherent write-up, preparation for which would involve extensive interviewing as well as investigating primary and secondary sources.

Kevin Broughan

Acknowledgements

Many people have assisted with the development and production of this book. Without their help and support, the work would not have been possible, and certainly not completed in a reasonable period of time. They include Dave Baird, Enrico Bombieri, Jackie Broughan, Nick Cavenagh, Daniel Delbourgo, Pat Gallagher, Dan Goldston, Roger Heath-Brown, Annika Heinz, Geoff Holmes, Henric Iwaniec, Stephen Joe, Yoichi Motohashi, Wei Minn Phee, Neil Quigley, Terence Tao, Cem Yildirim, Yitang Zhang and the hard-working folk of Cambridge University Press, including Roger Astley, Clare Dennison and Andy Saff.

I need to emphasize that the greatest contribution to this work belongs to those responsible for the remarkable breakthroughs which have been reported: in the main (in chronological order) to Bombieri, Davenport, Goldston, Pintz, Yildirim, Motohashi, Zhang, Polymath8a, Maynard, Tao and Polymath8b. Note that in Section 1.4 there is a list of some of the principal contributors to the two Polymath8 projects.

Associated software PGpack is freely available from the author's website under the GNU general public license. It was written by the author with two exceptions which are kindly acknowledged. The functions RayleighQuotientMaynard and Rayleigh QuotientPolymath are adaptations of the software written and used by James Maynard and Polymath8b respectively.

Cover image: Doubtful Sound is a fjord on the south coast of the South Island of New Zealand. It was named by the explorer James Cook in 1770 as Doubtful Harbour. (Photo: Terry Latson / 500px / Getty Images.)

1

Introduction

1.1 Why This Study?

This book provides an account of the remarkable progress which has been made during the first two decades of the twenty-first century, reducing the size of the known gap length between infinite numbers of consecutive primes. Parts of the story, and the mathematicians who have been responsible for this progress, are well known. Here, in one book, are the mathematical details needed to follow the developments and, it is hoped, extend them. In addition, since computation is an important component of some of the methods which are used, a suite of Mathematica programs is provided for checking results and for further experimentation.

Most mathematicians believe there are an infinite number of twin primes. However, by 2013 it was not known for sure whether or not the minimum separation for pairs of consecutive primes tended to infinity. That is to say, if p_n is the nth prime with $p_1 = 2$, then

$$\liminf_{n \to \infty} p_{n+1} - p_n = \infty \, ?$$

In 2014, all this changed. Two groups and three individuals, Yitang Zhang, Polymath8a, James Maynard, Terence Tao and Polymath8b, both separately, together and progressively, first showed that $p_{n+1} - p_n \leq 7 \times 10^7$ for an infinite number of p_n, and then, in many, many steps, lowered the upper bound to 246. This incredible progress on the question of prime gaps then ceased. As far as prime gaps is concerned, at the time of writing the last several years have not witnessed any further reductions in the proven minimal gap width.

It is the purpose of this book to explain how this progress came about. In particular, the book describes how it is rooted in the conjecture of Dickson, Hardy and Littlewood on finite patterns of primes, uses variations on the Selberg sieve and would not have happened at all were it not for the earlier breakthrough of Goldston, Motohashi, Pintz and Yildirim (known collectively as GMPY with the subset GPY also appearing quite often).

There is considerable technical detail and much computational support needed to derive the best results. Just because an idea has been used to derive a prime gap which is bigger than the current best, its deeper understanding and combination with other ideas and methods could bring about further improvements. In any case, the early ideas are an essential part of this story.

1.2 Summary of This Chapter

Section 1.3 gives an overview of the contents of the book. Section 1.4 describes Timothy Gowers' idea of a polymath project, and lists named contributors to Polymath8. Section 1.5 gives a timeline of the developments which are covered, in whole or in part, or which were a prelude to the breakthroughs. Section 1.6 discusses the twin primes constant and the Dickson–Hardy–Littlewood conjecture, which relates to the essential underlying concept of "admissible tuples" of integers. Section 1.7 delves into the nature of the prime gap distribution by discussing the issue of which prime gap up to increasing positive x is most common, and reports on the recent work on "jumping champions". Section 1.8 gives the derivation of some useful properties of the von Mangoldt function which are needed in the following sections and chapters. Section 1.9 is devoted to a discussion of the Bombieri–Vinogradov theorem, its statement, history and references to proofs. Section 1.10 introduces admissible tuples, which describe patterns of primes which are expected to repeat infinitely often, and the intriguing relationship between the Dickson–Hardy–Littlewood conjecture and the so-called second Hardy–Littlewood conjecture. Section 1.11 is a brief guide to the literature, including a film, some introductory secondary sources and the primary sources, with a reader's guide. This latter includes a note on which parts of the book could be skipped to get to the best result in minimum time.

In an end note, Section 1.12, in contrast to the small gaps between primes that are the principal focus of the book, successive results on large gaps between consecutive primes are summarized in a table with references.

1.3 History and Overview of These Developments

Searching for small gaps between consecutive primes is of course one way to approach the twin primes conjecture, which is one of the most celebrated unsolved problems in number theory. Some believe the conjecture was known to Euclid, but there is no evidence to support this. The first known mention is in Polignac in 1849 [170], where it is included in his more general conjecture that for each even integer $2k$ there is an infinite number of prime pairs (p, q), which do not need to be consecutive, such that $q = p + 2k$. A comment attributed to Dan Goldston from PBS's *Nova*:

No one really knows if Euclid made the twin primes conjecture. He does have a proof that there are infinitely many primes, and he or other Greeks could easily have thought of this problem, but the first published statement seems to be due to de Polignac in 1849 [170]. Strangely enough, the Goldbach conjecture, that every even number is a sum of two primes, seems less natural but was conjectured about 100 years before this.

Part of the context for this work was already established by Dickson in 1904 in [39]. He conjectures that infinite sets of primes occur in patterns, unless there is a modular condition outlawing a pattern. In detail, given a finite set of distinct integers $\mathcal{H} = \{h_1, h_2, \ldots, h_k\}$, then there exists an infinite number of translates $n + \mathcal{H}$ where all of the $n + h_j$ are prime, unless for some fixed prime p the size of $\mathcal{H}$ modulo p is p – in that case, for each n there is a j_n such that $p \mid n + h_{j_n}$ so for all n sufficiently large an element of the translated tuple cannot be prime. See Section 1.10 in this chapter. Twin primes have the pattern $\{0, 2\}$, and de Polignac's conjecture is the pattern $\{0, 2k\}$ for all $k \in \mathbb{N}$.

Hardy and Littlewood made a precise asymptotic form of Dickson's conjecture, which form is now called the Dickson–Hardy–Littwood conjecture, or DHL, and is described in Section 1.6. Work on this problem and on another conjecture Hardy and Littlewood made in the same paper [89], namely that for all $x, y \geq 2$ we have $\pi(x + y) \leq \pi(x) + \pi(y)$, added significantly to our knowledge of admissible k-tuples, i.e., those for which the number of distinct congruence classes modulo p is less than p for all primes p. See Section 1.10.3 in this chapter. It is an interesting fact, demonstrated in that section (but not used subsequently), that DHL and this second Hardy–Littlewood conjecture, cannot both be true at the same time.

The early work on gaps between primes focused on showing that there were gaps significantly smaller than the average. Up to real x, by the prime number theorem, this average is asymptotically $\log x$ of course. The first unconditional result is due to Erdős and was published in 1940. He proved that there were an infinite number of prime pairs strictly closer than this average by a small but unspecified amount, which was expected. His proof is given in Chapter 3, which includes also the reduction in gap size demonstrated by Bombieri and Davenport in 1966. Table 1.1 sets out most of the main other published work along these lines until 1988.

The breakthrough, signifying the start of the exciting modern era for work on the prime gap problem, was by Dan Goldston, Yoichi Motohashi, Janos Pintz and Cem Yildirim (known colloquially as GMPY), essentially done by 2006 but published in 2009. They showed that for any given $\epsilon > 0$ there are an infinite number of prime pairs $p < q \leq x$ with $(q - p) < \epsilon \log x$. See the introductory section of Chapter 4 for an overview of their work, or read Kannan Soundararajan's excellent paper [198]. Their principal tool was the sieve developed by Atle Selberg. Used for a wide variety of number theory problems, this tool was not only used in some of the early work (see for example Chapter 3), but also sufficiently flexible to be the main underpinning device for each of the later breakthroughs. See Chapter 2 for the elements of Selberg's method and an application which is used subsequently. There is also a brief overview of other commonly used sieves.

Then Yitang Zhang read the works of GPY, carefully and in detail, and thought long and hard about how to improve the methods. In particular, he found a way of improving the unconditional Bombieri–Vinogradov theorem, which is roughly speaking the generalized Riemann hypothesis for primes in arithmetic progressions on average, and used this improvement, and other very advanced techniques, to show that there were an infinite number of prime pairs $p < q$ such that $q - p \leq 7 \times 10^7$. Written in 2013, this work was published in 2014. His work was unusual, in that the results were based on explicit constants, but he made no claim that his upper bound was optimal. Indeed, in [215] we find a remark at the end of his section 1, which we paraphrase:

This result is of course not optimal. To replace the constant 7×10^7 by a value as small as possible is an open problem that will not be discussed in this paper.

For some biographical background on Zhang, and an overview of his approach, see the introduction to Chapter 5. The reader could with benefit consult the wide-ranging survey of Andrew Granville [82].

It is important to stress that Zhang's work used multivariable exponential sum estimates based on the profound work of Deligne, solving the Riemann hypothesis for varieties over finite fields, and other techniques to derive his variation of Bombieri–Vinogradov's estimate. We don't cover most of these topics, but they should be treated fully elsewhere.

Under the leadership of Terence Tao, a Polymath question was formulated and group formed in 2013 (see some details in Section 1.4). Communicating over the Internet, this group and others began immediately to reduce the bound. A new concept, **densely divisible numbers** which generalize smooth integers, was formulated and extensive computations performed to optimize parameters, while remaining faithful to Zhang's approach. By the end of 2013, Polymath had reduced the bound to 4,680 (using a tuple of size 632), publishing this work in 2014 [173], with a more extensive write up on the archive [174], including extensive details of underlying computational methods. (There may be some doubt and confusion about the precise values of these bounds – at one stage, they were changing, both up and down, almost on a daily basis. For example, the author was able to reach a tuple size of 630 rather than 632 using Polymath's methods.)

Along the way, Polymath8a was able to reduce Zhang's bound significantly to 14,950 (using a tuple of size 1,783) without using Deligne, but with an estimate for exponential sums based on polynomials in a single variable, a so-called "Weil bound", which is much easier to derive and less reliant on very high-powered mathematical tools than Deligne's applications. This approach is included here in Chapter 8.

It is no longer news that this work was quickly superseded by James Maynard, who used a "mutivariable", or better to say a multidivisor, approach to Selberg's sieve, and some standard optimization over parameterized families of symmetric multinomials to reduce the bound to 600. Not only that, he completely sidestepped any use of an extended Bombieri–Vinogradov estimate or Deligne–Weil exponential sums. His work is the subject of Chapter 6 and was published in [138]. Of course, this discovery inspired another Polymath project under Terry Tao's leadership, and the bound was reduced to 246 with a tuple of size 50 by the end of 2014, with publication occurring [175] that year. The Polymath refinements of Maynard are described in Chapter 7.

The main part of the book concludes with Chapter 9, which gives a summary of some relatively recent (as of 2017) additional results proved using the methods of the principal authors using the Elliott–Halberstam conjecture, so these are conditional, or, for example, the best unconditional results on prime gaps, such as those of Maynard/Polymath. The appendices cover some of the more standard techniques which are employed by the authors, such as compact operators, Bessel functions, Stepanov's approach to the Weil bound for curves, some complex analysis needed for detailed estimates of the Riemann zeta function and an introduction to the dispersion method of Linnik. In addition, there is a minimanual for the suite of Mathematica programs, **PGpack**, which includes standard functions like DenselyDivisibleQ and VonMangoldt, but also Maynard and Polymath's algorithms, used in deriving their best results. There is also considerable support for Ignace Bogaert's Krylov method algorithm.

Finally, in this brief overview, it is good to keep in mind the dependencies of the work of the main authors on the work of many others. We represent some of these dependencies in the flow graph in Figure 1.1, adapted from that which was included in [150].

1.4 Polymath Projects and Members of Polymath8

A Polymath project played a significant role in lowering the proved gap between an infinite number of prime pairs. But what is a Polymath project and who were the named participants? In this section, an attempt will be made to answer these questions.

Polymath projects were initiated by Timothy Gowers in 2009. In the following I will summarize part of his thinking, which is extensively set out on Gowers' weblog "Is massively collaborative mathematics possible?", which enjoyed a very large amount of feedback. Within a few days, he had already formulated and

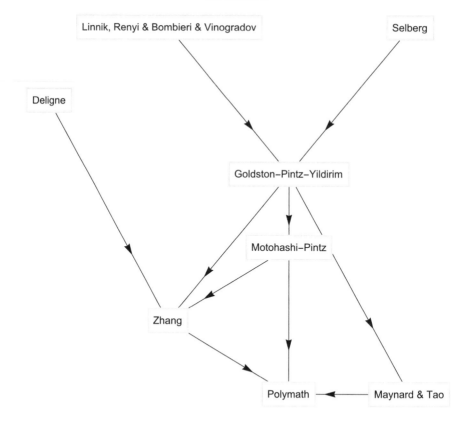

Figure 1.1 Some prime gaps developments dependencies.

revised a set of rules in "Questions of procedure", which described the process which was envisaged for Polymath projects. He had already listed several suitable topics and given some initial ideas, in "Background to a Polymath project", to start the process: the Hales–Jowett theorem, the Furstenberg–Katznelson theorem, Szemerédi's regularity lemma, the triangle removal lemma and so-called sparse regularity lemmas. By the time Tao had proposed Polymath8, seven other projects had been initiated.

Now Gowers, also known as Sir William Timothy Gowers, is a leading mathematician, having made many fundamental contributions to functional analysis, arithmetic combinatorics and graph theory. He was awarded the Fields Medal in 1998 for his research. Sitting on my bookshelf in pride of place is the *Princeton Companion to Mathematics* [80], a magnificent compendium for which he was the editor in chief.

Gowers' Concept of a Polymath Project
The idea of a Polymath project is anybody who had anything whatsoever to say about the project problem could contribute brief ideas, even if they were undeveloped and/or likely to be wrong. Some advantages:

• Sometimes luck is needed to have the idea that solves a problem. If lots of people think about a problem, then on probabilistic grounds there is more chance that one of them will have that bit of luck.

• Different people know different things, so the knowledge that a large group can bring to bear on a problem is significantly greater than the knowledge that one or two individuals will have. For example, consider a response, "The lemma you suggested trying to prove is known to be false." One can take weeks to discover this situation if one is on one's own.

• Different people have different characteristics when it comes to research. Some like to throw out ideas, others to criticize them, others to work out details, others to reexplain ideas in a different language, others to formulate different but related problems, others to step back from a big muddle of ideas and fashion some more coherent picture out of them and so on. Group members can specialize. In short, if a large group of mathematicians could connect their brains efficiently, they could perhaps solve suitable problems very efficiently.

• Why would anyone agree to share their ideas? Many work on problems in order to be able to publish solutions and get credit for them. And what if the Polymath collaboration resulted in a very good idea? Isn't there a danger that somebody would steal it from the Polymath to get the credit? Gowers rebutted these objections, but they are valid.

• Gowers doesn't believe that this approach is likely to be good for everything. Examples include the Riemann hypothesis in our current state of knowledge, or the solution of a very minor and specialized problem. He claimed there is a middle ground and gave examples. The first Polymath project, conceptualized and led by Gowers, was to find a new proof of the density Hales–Jewett theorem. (Hence the name D. H. J. Polymath as the author of some articles.) The statement is that for every $k > 0$ and $\epsilon > 0$, there is an $n \in \mathbb{N}$ such that for any subset A of words on $\{\{1, \ldots, k\} \cup \{x\}\}^n$ with $\#A > \epsilon k^n$, if $w(i)$ is the word obtained by replacing every occurrence of x by i, then A contains a set of the form, $\{w(i) : 1 \le i \le k\}$, a so-called "combinatorial line". The new proof came quickly [171] and showed that Gowers' concept was a positive contribution to mathematics research practice.

Polymath8 enjoyed a high level of participation by many mathematicians, some at the heart of the project and others more peripheral to the main developments, and they are not so easy to acknowledge. The following lists have been taken from the "Polymath8 grant acknowledgements" web page, accessible at the time of writing from `http://michaelneilsen.org/polymath1/index.php?title =Polymath8_grant_acknowledgements` (which should be entered into a search box as one expression with no spaces, returns or line-feeds):

Polymath8a authors or primary participants were Wouter Castryck, Etienne Fouvry, Gergely Harcos, Emmanuel Kowalski, Philippe Michel, Paul Nelson, Eytan Paldi, Janos Pintz, Andrew V. Sutherland, Terence Tao and Xiao-Feng Xie.

Polymath8b authors or primary participants were Ignace Bogaert, Aubrey de Grey, Gergely Harcos, Emmanuel Kowalski, Philippe Michel, James Maynard, Paul Nelson, Pace Nielsen, Eytan Paldi, Andrew V. Sutherland, Terence Tao and Xiao-Feng Xie.

Overviews of the work of Polymath8 have appeared in the *EMS Newsletter* in "The 'Bounded Gaps Between Primes' Polymath Project", by D. H. J. Polymath [172] and the *Notices of the AMS* in "Prime Numbers: A Much Needed Gap Is Finally Found", by J. Friedlander [59], and the cover description [25]. In addition, readers are strongly encouraged to first scan, and then consult from time to time as might be needed, the Polymath8 Home Page. This contains an impressive and expansive family of links and threads to the work of Polymath8 and others relating to prime gaps, including the latest results, write-ups, a timeline of bounds, code and data, errata, recent papers and notes, media reports and a bibliography. The home page is maintained by Michael Nielsen under the heading "Polymath1Wiki" with the title "Bounded Gaps Between Primes" and accessible at the time of writing from `http://michaelneilsen.org/polymath1/index.php?title=Main_Page`.

1.5 Timeline of Developments

This section is a brief but crucial element of this introductory chapter. The timeline of Table 1.1 shows how our knowledge of gaps between prime pairs has improved over almost 100 years, with improvements made by many leading mathematicians. It also shows how the work as published in regular journals splits into the period prior to 1988 and that between 2006 and 2015. This latter time range corresponds with the main contents of this book, but in Chapter 3 we do give some examples from the earlier period. The reader may wish to sometimes refer back to the table.

For all x sufficiently large, one can find primes in $[x, 2x]$ with $p_{n+1} - p_n$ less than the bound given in Table 1.1, and thus an infinite number of consecutive primes distant apart less than the given bound.

The dates attached to the bounds in Table 1.1 are not necessarily those of the first published form, which is often significantly earlier and often on the ArXiv preprint server, especially the dates for the work published during 2013 to 2015. The table is based on information in the online slides of Terence Tao, prepared for his lecture presented at the Latinos in the Mathematical Sciences Conference, UCLA, 2015.

Table 1.1 *Timeline of proved decreasing prime gaps.*

Year	Bound	People	Reference
1923	$(\frac{2}{3} + o(1)) \log x$	Hardy/Littlewood/GRH	[89]
1940	$(\frac{3}{5} + o(1)) \log x$	Rankin/GRH	[178, 179]
1940	$(1 - c + o(1)) \log x$	Erdős	[45]
1954	$(\frac{15}{16} + o(1)) \log x$	Ricci	[182]
1966	$(0.4665 + o(1)) \log x$	Bombieri/Davenport	[10]
1972	$(0.4571 + o(1)) \log x$	Pil'tjai	[164]
1973	$(0.4463 + o(1)) \log x$	Huxley	[109]
1975	$(0.4542 + o(1)) \log x$	Uchiyama	[209]
1977	$(0.4425 + o(1)) \log x$	Huxley	[110]
1984	$(0.4394 + o(1)) \log x$	Huxley	[111]
1988	$(0.2484 + o(1)) \log x$	Maier	[132]
2006	$o(\log x)$	Goldston/Motohashi/Pintz/Yildirim	[70]
2009	$C(\log x)^{\frac{1}{2}}(\log\log x)^2$	Goldston/Pintz/Yildirim	[72]
2013	7.0×10^7	Zhang	[215]
2013	$4,680$	Polymath8a	[26, 174]
2015	600	Maynard	[138]
2014	246	Polymath8b	[175]

1.6 Prime Patterns and the Hardy–Littlewood Conjecture

A study of the gaps between prime pairs is part of the more general search for patterns exhibited by finite sets of primes infinitely often. The ultimate strongly believed gap size is of course 2, and we begin this section with early approaches to the twin primes conjecture in a quantitative form. This is followed by the more general, and entirely relevant for the work of all contributors to the discoveries of this report, the so-called admissible k-tuples. Roughly speaking, these are sets of k distinct integers such that there are no modularity constraints which would prevent an infinite number of translates consisting of all primes. In a quantitative form, this gives rise to the Dickson–Hardy–Littlewood conjecture.

The **Hardy–Littlewood conjecture** of 1923 in particular gives the asymptotic relation that as $x \to \infty$

$$\pi_2(x) := \#\{p \le x : p + 2 \in \mathbb{P}\} \sim \eta C_2 \frac{x}{(\log x)^2} \text{ with } \eta = 2, \quad (1.1)$$

where the **twin primes constant** C_2 is the number

$$C_2 := \prod_{p \ge 3} \left(1 - \frac{1}{(p-1)^2}\right) = 0.6601618158468957\ldots.$$

Hardy and Littlewood derived their conjecture using Cramer's probability model, which is based on the assignment of a probability of $1/\log x$ to numbers in the range $[1, x]$ for x large, that they should be prime. If p is prime, the Cramer model gives a nonzero probability that $p + 1$ should also be prime, but this is impossible for odd p. If the model held, then the probability that p and $p + 2$ were prime would be $1/(\log x)^2$ leading to $x/(\log x)^2$ twin primes up to x. Hardy and Littlewood argued that the model could be used provided that it was corrected by a factor which is twice the twin primes constant.

Their argument was as follows: given a random integer n, then for both n and $n + 2$ to be prime both must be odd, which occurs with probability $1/2$ rather than $1/4$ for a pair of random integers. Neither must be divisible by 3, which means we must have $n \equiv 1 \bmod 3$, which has probability $1/3$ rather than $4/9$ if they were independent and obeyed that constraint. If the prime is 5, then we must have $n \not\equiv 0, 3 \bmod 5$, giving a probability of $1 - 2/5$ rather than $(1 - 1/5)^2$. In general, for a prime $p \geq 3$ the probability neither n or $n + 2$ is divisible by p would be $1 - 2/p$ rather than $(1 - 1/p)^2$ if they were independent. The factor 2 is a little different since both n and $n + 2$ are odd with probability $1 - 1/2$ rather than $(1 - 1/2)^2$. Thus, to get the adjusted asymptotic count, we must multiply $x/\log^2 x$ by

$$
\left(1 - \frac{1}{2}\right)\left(1 - \frac{1}{2}\right)^{-2} \prod_{p \geq 3}\left(1 - \frac{2}{p}\right)\left(1 - \frac{1}{p}\right)^{-2} = 2\prod_{p \geq 3}\left(1 - \frac{1}{(p-1)^2}\right) = 2C_2.
$$

Figure 1.2 is a plot of an exact count of the twin primes up to $n \in \mathbb{N}$, $T(n)$ times $(\log n)^2/n$, for $3 \leq n \leq 10^5$. Table 1.2 gives details of proved upper bounds for the leading coefficient η in (1.1), all for x sufficiently large.

Table 1.2 *Twin primes: upper bounds for the leading coefficient η.*

Year	Bound for η	People	Reference
1919	100	Brun	[24]
1966	$8 + \epsilon$	Bombieri and Davenport	[10]
1983	$\frac{68}{9} + \epsilon$	Fouvry and Iwaniec	[55]
1984	$\frac{128}{17}$	Fouvry	[53]
1986	$7 + \epsilon$	Bombieri, Friedlander and Iwaniec	[15]
1996	6.9075	Fouvry and Grupp	[56]
1990	6.8354	Wu	[214]
1999	6.8325	Haugland	[94]

Now we will consider more general patterns. Let $\mathcal{H} := \{h_1, \ldots, h_k\}$ be a finite set of distinct nonnegative integers with $k \geq 2$. For each prime p, let $v_p(\mathcal{H})$ be the number of distinct residue classes modulo p of the elements of $\mathcal{H}$, and define a product

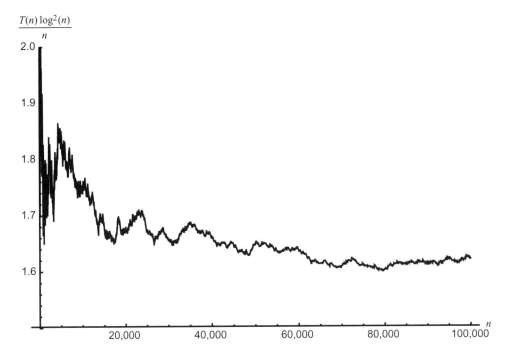

Figure 1.2 Twin primes constant.

$$\mathscr{G}(\mathscr{H}) := \prod_{p} \left(1 - \frac{v_p(\mathscr{H})}{p}\right)\left(1 - \frac{1}{p}\right)^{-k}.$$

This product converges since for all p sufficiently large we have $v_p(\mathscr{H}) = k$. If for some p we have $v_p(\mathscr{H}) = p$, then the 0 class occurs so for each $n \in \mathbb{N}$ one of the h_j satisfies $p \mid n + h_j$. Thus $n + h_j$ cannot be prime unless $p = n + h_j$, and this is so for at most k values of n. Thus the only sets $\mathscr{H}$ which are of interest, if we are seeking an infinite number of translates $n + \mathscr{H}$ consisting only of primes, are those with $v_p(\mathscr{H}) < p$ for all primes p. (We will later call such $\mathscr{H}$ **admissible**.) This is so if and only if $\mathscr{G}(\mathscr{H}) > 0$.

Generalizing the case $k = 2$ discussed previously, the **Hardy–Littlewood conjecture** is the asymptotic estimate

$$\pi(x, \mathscr{H}) := \#\{n \le x : (n + h_1, \ldots, n + h_k) \in \mathbb{P}^k\} \sim \mathscr{G}(\mathscr{H})\frac{x}{(\log x)^k}.$$

Consequences of assuming this conjecture are of course the twin primes conjecture, but also the patterns $(p, p+2, p+6)$, $(p, p+4, p+6)$, etc., consisting only of primes for an infinite number of primes p. Since $v_3(H) = 3$, $(p, p+2, p+4)$ has at most a finite number of such values p. Later we will see that for every k there are many k-tuples $\mathscr{H}$ which are admissible.

Indeed, a qualitative form of the Dickson–Hardy–Littlewood conjecture was first enunciated by Dickson [39] in the qualitative form that for every admissible k-tuple $\mathscr{H}$ there exists an infinite number of natural numbers n such that each member of the translate $n + \mathscr{H}$ is prime. Thus, as we have seen, it is more accurate and common to refer to the asymptotic form of the conjecture of this section as that of Dickson–Hardy–Littlewood (DHL).

There are a number of stronger forms of the Hardy–Littlewood conjecture which have been used in applications, for example that of Pat Gallagher [67, 68] to the distribution of primes in short intervals. The strongest is that, for fixed $k \geq 1$, as $x \to \infty$, we have

$$\pi(x, \mathscr{H}) = \mathscr{G}(\mathscr{H})\frac{x}{(\log x)^k} + O_k\left(x^{\frac{1}{2}+\epsilon}\right)$$

uniformly for all $\mathscr{H} \subset [1, x]$. Most stronger forms include some uniformity condition, weaker than the strongest form.

Some justification for at least the prime pair form of DHL could be obtained by considering Figure 1.3, which is a count of primes p up to 10^6 such that $p + 2m$ is also prime vs. m. Given positive integers m and n, then equal values at m and n, assuming DHL, should correspond to

$$\sum_{w(H)=2n} \mathscr{G}(H) = \sum_{w(H)=2m} \mathscr{G}(H),$$

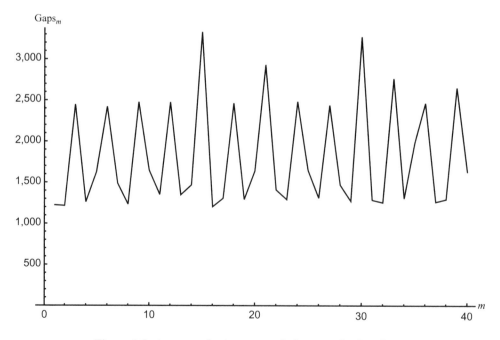

Figure 1.3 A count of primes p such that $p + 2m$ is prime.

where the sum is over admissible H with given diameter $w(H)$, which is $w(H) := h_k - h_1$ if $h_1 < h_2 < \cdots < h_k$. The interested reader is invited to check this for say $p + 6$ and $p + 12$ using the functions provided in **PGpack** – see Appendix I. The peak at 30, and its siblings, should be able to be explained, as should the apparently consistent lower bound.

A final additional note in this section: twin primes with more than 1,000 decimal digits are called "titanic". A simple Mathematica program found the titanic pair $p = 10^{1000} + 9705091$, $q = p + 2$ in a few hours.

1.7 Jumping Champions

In the search for infinite sets of prime pairs with a given gap, one might consider that the gap size which appears most frequently would be the easiest to find. In addition, it appears that 6 is a very frequently occurring gap. This is misleading. For $x_1 < x_2$ infinitely often the most frequently occurring gap up to x_2 can be (conjecturally?) larger than that up to x_1, with the most frequently occurring gap size tending to infinity with x. Here we discuss (and don't return to) studies of this topic since it relates to the distribution of gaps between primes, and our principal focus is only one aspect of that distribution. It relates to the question: is 6 the easiest prime gap to find, rather than 2 or 30, say?

In Figure 1.4, we consider primes up to 10^6 and for each even integer $2n$ count the number of primes p in that range for which the next prime is $p + 2n$. Easy observations are that 6 is the most common gap by far, that local gap maxima occur when the gap size is a multiple of 6 and some local minima when it is a power of 2. Some features could be proved subject to DHL; for example, $H_1 = \{0, 2\}$, $H_2 = \{0, 4\}$ has $\mathscr{G}(H_1) = \mathscr{G}(H_2)$ so, asymptotically, the main term for the number of gaps to the next prime of size 2 should be the same as for gap size 4.

Figure 1.4 seems to indicate 6 is the most common gap between consecutive primes. However, this is, apparently, far from the truth. In 1993, John Conway denoted the term "jumping champion" to mean the most frequently occurring gap between consecutive primes up to a given positive x. Already in 1980, Erdős and Strauss [47] had shown that the jumping champions tended to infinity, assuming the Hardy–Littlewood conjecture for prime pairs, i.e., $k = 2$.

In 1999, Odlyzko, Rubinstein and Wolf published the results of their extensive investigations into the nature of these numbers [157]. They had some evidence that 30 would take over from 6 as the jumping champion near 1.7427×10^{35} with 210 taking over around 10^{425}. Now 6 is a "primorial". By **primorial**, we mean a positive integer which has the form $2 \cdot 3 \cdot 5 \cdots p_j$, where p_j is the jth rational prime, i.e., the product of consecutive primes starting with $p_1 = 2$. They made two conjectures:

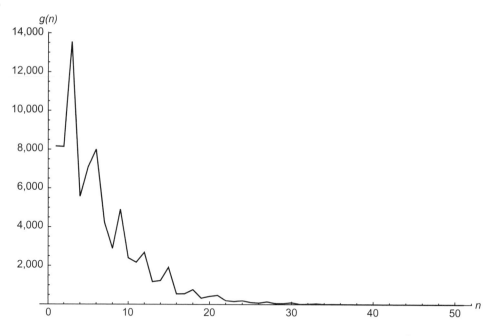

Figure 1.4 Next prime gap distribution for primes less than 10^6.

Conjecture A: (Jumping champions primorial) The jumping champions other than 4 are the primorials, 2,6,30,210,..., and

Conjecture B: The jumping champions tend to infinity. Any fixed prime p divides all jumping champions which are sufficiently large depending on p.

Of course, Conjecture B is a direct consequence of Conjecture A, and its first part had already been proved by Erdős and Strauss. In 2011, Goldston and Ledoan in [73] proved the second part of B, assuming the Hardy–Littlewood prime pair conjecture holds, for all sufficiently large x, uniformly for all integers d in the range $2 \le d \le (\log x)^2$, as described more precisely in what follows.

Conjecture C: (Hardy–Littlewood prime pairs) Let $d \ge 2$ be even. Then as $x \to \infty$, we have

$$\pi_2(x,d) := \sum_{\substack{p \le x \\ p - p' = d}} 1 = 2C_2 \prod_{\substack{p \mid d \\ p > 2}} \left(1 + \frac{1}{p-2}\right) \frac{x}{(\log x)^2} \left(1 + o\left(\frac{1}{(\log\log x)^2}\right)\right),$$

uniformly for $2 \le d \le (\log x)^2$.

Goldston and Ledoan used the property that the sequence $(\mathscr{G}(\{0,d\}))_{2\mid d}$ increases most rapidly when d is a primorial. In more detail [73, lemma 2.2], If

$p_k^{\#} := p_1, \ldots, p_k$ is the kth primorial, where $p_1 = 2$, then for all d with $2 \leq d < p_k^{\#}$, we have

$$\mathscr{G}(\{0,d\}) < \mathscr{G}(\{0, p_k^{\#}\}).$$

To see this first note from the form of $\mathscr{G}(\{0,d\})$, we can assume that d is square-free. If $d \mid p_k^{\#}$, since $d \neq p_k^{\#}$ there is a prime not greater than p_k which does not divide d, so the inequality follows. If however d has a prime factor greater than p_k, there must be at least one prime less than p_k which does not divide d. Replacing the larger prime factor in d by the smaller one gives a larger corresponding value for $\mathscr{G}(\{0,d\})$. We can do the same for each larger prime divisor of d which is larger than p_k until each prime factor of d is less than or equal to p_k, to which the first case applies.

Then in 2015, Goldston and Ledoan in [75] made another giant step forward towards a solution of Conjecture A. Again they assumed a form of the Hardy–Littlewood conjecture for prime pairs (Conjecture C) and for prime triples which we detail next. They showed that all sufficiently large jumping champions are primorials and that all sufficiently large primorials are jumping champions.

Conjecture D: (Hardy–Littlewood prime triples) Let $d, e \geq 2$ be even. Then as $x \to \infty$, uniformly for $2 \leq d < e \leq \log^2 x$, we have

$$\pi_3(x,d,e) := \sum_{\substack{p \leq x \\ p-p'=d \\ p-p''=e}} 1 = \mathscr{G}(\{0,d,e\}) \frac{x}{(\log x)^3}(1 + o(1)).$$

See also Section 1.10 on admissible tuples for some related observations.

1.8 The von Mangoldt Function

Each of the authors, GMPY, Zhang, Maynard and Polymath, use the von Mangoldt function to detect primes by detecting prime powers, at least some of the time. Here we develop a number of the properties of this function which have been found useful in the work on prime gaps.

(1) First we have the definition, which underlines the purpose, but not the structure, of the von Mangoldt function:

$$\Lambda(n) := \begin{cases} \log p, & n = p^m, \\ 0, & \text{otherwise}, \end{cases}$$

so $\Lambda(n) \geq 0$ and is zero when and only when n has more than one distinct prime factor.

Because

$$\sum_{\substack{p^m \le x \\ m \ge 2}} \Lambda(p^m) \ll \sqrt{x}\log(x),$$

the contribution to sums of the type $\sum_{n \le x} \Lambda(n)$ from primes to powers higher than 1 is "small", so $\Lambda(n)$ essentially detects primes, weighted by their logarithms.

(2) Next we see that the function is a divisor sum:

$$\Lambda(n) = \sum_{d|n} \mu(d) \log\left(\frac{n}{d}\right) = (\mu \star \log)(n).$$

To show this first note that, with a little thought, the fundamental theorem of arithmetic can be written

$$\log n = \sum_{d|n} \Lambda(d).$$

Applying Möbius inversion we obtain the given formula for $\Lambda(n)$.

(3) We generalize the function by using the kth power of the logarithm in property (2):

$$\Lambda_k(n) := \sum_{d|n} \mu(d) \left(\log\frac{n}{d}\right)^k = (\mu \star \log^k)(n) \iff (\log n)^k = \sum_{d|n} \Lambda_k(n).$$

(4) The function in (3) obeys a simple recurrence:

$$\Lambda_{k+1}(n) = (\log \cdot \Lambda_k)(n) + (\Lambda \star \Lambda_k)(n).$$

To see this, first differentiate the Dirichlet series for $\zeta(s)$ with $\Re s > 1$ to get

$$\zeta^{(k)}(s) = (-1)^k \sum_{n=1}^{\infty} \frac{\Lambda_k(n)}{n^s},$$

so

$$\frac{\zeta^{(k)}(s)}{\zeta(s)} = (-1)^k \sum_{n=1}^{\infty} \frac{(\log n)^k}{n^s} \sum_{n=1}^{\infty} \frac{\mu(n)}{n^s} = (-1)^k \sum_{n=1}^{\infty} \frac{\Lambda_k(n)}{n^s}.$$

Equating coefficients in the identity

$$\frac{d}{ds}\left(\frac{\zeta^{(k)}(s)}{\zeta(s)}\right) = \frac{\zeta^{(k+1)}(s)}{\zeta(s)} - \frac{\zeta^{(1)}(s)}{\zeta(s)} \cdot \frac{\zeta^{(k)}(s)}{\zeta(s)}$$

and cancelling $(-1)^{k+1}$ gives the recurrence.

(5) We have $\Lambda_k(n) = 0$ whenever n has more than $k \ge 1$ distinct prime factors: Since, in this case, using the definition $\Lambda_1(n) = \Lambda(n) = 0$, this property follows by induction from the recurrence Property (4). Similarly we have $\Lambda_k(n) \geq 0$ for all $n \in \mathbb{N}$.

(6) We want to detect primes in k-tuples of distinct integers, $\mathcal{H} = \{h_1, \ldots, h_k\}$, where $k \geq 2$. Define a polynomial based on $\mathcal{H}$ by

$$P(n) = P_{\mathcal{H}}(n) := (n + h_1) \cdots (n + h_k).$$

If $0 \leq h_1 < h_2 < \cdots < h_k$, n is such that $h_k^{k-1} < n$ and $\Lambda_k(P(n)) \neq 0$ then, we **claim** $P(n)$ has exactly k distinct prime factors. If so and we denote these factors by $p_1, \ldots, p_k$, we get in addition

$$\Lambda_k(P(n)) = k! \prod_{j=1}^{k} \log p_j.$$

To see this, by Property (5), since $\Lambda_k(P(n)) \neq 0$, $N := P(n)$ cannot have more than k distinct prime factors. Suppose, to get a contradiction, if it has $m \leq k-1$ such factors and we let $N = p_1^{\alpha_1} \cdots p_m^{\alpha_m}$. Following Granville [82], for $1 \leq i \leq m$ let

$$p_i^{j_i} \| n + h_j,$$

where j is chosen such that $1 \leq j_i \leq k$ is maximal. Since there are less than k of the j_i, there is a J with $1 \leq J \leq k$ such that for all i we have $J \neq j_i$. Then if $e_i \geq 0$ is such that

$$p_i^{e_i} \| n + h_J \implies p_i^{e_i} | (n + h_J) - (n + h_{j_i})| \prod_{\substack{1 \leq j \leq k \\ j \neq J}} (h_J - h_j).$$

Since $n + h_J = \prod_{i=1}^{m} p_i^{e_i}$, using the assumption $h_k^{k-1} < n$, we get

$$n \leq n + h_J | \prod_{\substack{1 \leq j \leq k \\ j \neq J}} |h_J - h_j| \leq h_k^{k-1} < n,$$

which is false. Thus $P(n)$ must have exactly k distinct prime factors.

To demonstrate the given identity, let $N = P(n)(n + h_{k+1})$ and suppose $\Lambda_{k+1}(N) \neq 0$. Assume, for a proof by induction, the given expression for $\Lambda_k(P(n))$ is true for some $k \geq 2$. By what we have shown, N has exactly $k+1$ distinct prime factors. Write

$$N = p_1^{\alpha_1} \cdots p_{k+1}^{\alpha_{k+1}}.$$

By the recurrence Property (4), and what we have shown here, we have

$$\Lambda_{k+1}(N) = (\Lambda \star \Lambda_k)(N) = \sum_{d|N} \Lambda(d) \Lambda_k\left(\frac{N}{d}\right).$$

All summands on the right are zero except for d, a prime power, and N/d, a product of k prime powers. Choosing in turn $d = p_j^{\alpha_j}$ and using the inductive assumption

$$\Lambda_{k+1}(N) = \sum_{1 \le j \le k+1} \log p_j(k!) \log p_1 \cdots \log p_{k+1}/\log p_j$$

$$= (k+1)! \log p_1 \cdots \log p_{k+1},$$

which completes the derivation.

This key result is strong motivation for using both Λ_k and $P = P_{\mathcal{H}}$ to detect primes in tuples.

1.9 The Bombieri–Vinogradov Theorem

All principal authors of these developments use the Bombieri–Vinogradov theorem, which we describe in this section. We need good estimates for a count of the primes in arithmetic progressions across a range of moduli. Dirichlet's theorem [4, chapter 7] can be written for $(q,a) = 1$ and $x \to \infty$

$$\theta(x,q,a) \sim \frac{x}{\varphi(q)},$$

where

$$\theta(x,q,a) = \sum_{\substack{p \le x \\ p \equiv a \bmod q}} \log p.$$

We need for applications this result to hold, not only for individual q, but uniformly for as wide a range of q values as possible.

Let $A > 0$ and $Q = x^{\vartheta}$ with $\vartheta > 0$. We seek values of ϑ such that as $x \to \infty$

$$\sum_{\substack{q \le Q \\ (a,q)=1}} \max_{a \bmod q} \left| \theta(x,q,a) - \frac{x}{\varphi(q)} \right| \ll_{A,\vartheta} \frac{x}{(\log x)^A}, \tag{1.2}$$

where $\ll_{A,\vartheta}$ means the implied constant depends at most on A and ϑ.

The early history of the Bombieri–Vinogradov theorem began with the seminal works of Linnik [126] and Rényi [181], but also those of Pan [163] and Barban [7]. Indeed, in 1948, Rényi, using the **large sieve** of Linnik, proved there was an absolute constant $\vartheta > 0$ such that the estimate (1.2) was true. In 1965, E. Bombieri [9], using the large sieve (see for example Section 2.8), and A. I. Vinogradov [211], using another of Linnik's concepts, the **dispersion method** [127] (see for example Appendix F), independently proved an estimate in a form equivalent to that given in the following equation, with a range of moduli up to $Q = \sqrt{x}/(\log x)^B$, and an implied constant depending at most on $A > 0$.

Since

$$0 \le \theta(x,q,a) \ll (x \log x)/q \implies -\frac{x}{\varphi(q)} \le \theta(x,q,a) - \frac{x}{\varphi(q)} \ll \frac{x \log x}{q},$$

we get an initial upper bound for a sum over q with $1 \le q \le Q \le x$ of $|\theta(x,q,a) - x/\varphi(q)|$ of $x(\log x)^2$. The Bombieri–Vinogradov theorem shows, that provided $Q \le \sqrt{x}/(\log x)^B$, where B is sufficiently large depending on $A > 0$, that as $x \to \infty$

$$\sum_{q \le Q} \max_{\substack{a \bmod q \\ (a,q)=1}} \left| \theta(x,q,a) - \frac{x}{\varphi(q)} \right| \ll_A \frac{x}{(\log x)^A},$$

where the implied constant depends at most on A. In this form of the theorem, we can choose $B = 2A + 5$. Note that Bombieri's form is stronger than that of Vinogradov in that the latter has $Q \le x^{\frac{1}{2}-\epsilon}$ for all $\epsilon > 0$. Often this distinction does not matter. In both cases, the power $1/2$ of x is referred to as "the $\sqrt{x}$ barrier", an obstacle to be "broken".

Goldston, Motohashi, Pintz and Yildirim (see Chapter 4) and Maynard (Chapter 6) use an equivalent form of the theorem to derive their results. Zhang, however (Chapter 5), breaks the "$\sqrt{x}$ barrier" by a small numerical quantity using smooth and squarefree moduli. A smooth positive integer has all prime divisors not greater than specified upper bound, the "smoothness". It is squarefree if each prime divisor has multiplicity one.

To state Zhang's theorem, which in some ways is potentially a more important breakthrough than its application to gaps between primes:

There are constants $\eta, \delta > 0$ such that if $Q = x^{\frac{1}{2}+\frac{\eta}{2}}$, we have

$$\sum_{\substack{q \le Q \\ 1 \le a < q : (q,a)=1 \\ q \ x^\delta-\text{smooth} \\ q \ \text{squarefree}}} \left| \theta(x,q,a) - \frac{x}{\varphi(q)} \right| \ll_A \frac{x}{(\log x)^A}.$$

This statement hides the meaning and evolution of the use of smooth moduli in Bombieri–Vinogradov-style estimates. There are several different uses of the term "smooth" in number theory. Sometimes a "smoothing technique" is used to reduce the error when an infinite series is truncated – see for example [117, page 73]. This is similar to the use "smooth function" which often is used for mappings $f : \Omega \to \mathbb{C}$, where $\Omega \subset \mathbb{R}^n$ is open. In 1980, Iwaniec [115] introduced a bilinear remainder form for a sieve due originally to Rosser, reducing its size with a smoothing technique. Then in 1983, Motohashi [147, sections 2.1 and 3.4] set out what he called "a smoothing device", based on a semigroup of subintervals to better control the error in the Selberg sieve. Once the work of GMPY and GPY had become available, Motohashi and Pintz [151] applied this smoothing device to the GPY method. Thus the stage was set for Zhang to use smooth moduli in [215]. But neither a smoothing technique or smoothing device can be identified easily with the use of "smooth integers".

A positive integer n is δ-**smooth** for some $\delta > 1$ if every prime divisor $p \mid n$ has $p \leq \delta$. It is really the set of all such integers, S_δ say, which should be called smooth because they have the property

$$\forall x \in S_\delta \text{ there exists } y \in S_\delta \text{ with } y \neq x \text{ such that } |y - x| \leq \delta x.$$

Thus the set of all such integers lacks relatively wide gaps. They might be compared with **rough** sets of integers R_δ, namely those with all prime factors $p \geq \delta$, so given

$$\forall x \in R_\delta \text{ and for all } y \in R_\delta \text{ with } y \neq x \text{ we have } |y - x| \geq \delta x,$$

so they are relatively distant from each other. See [144, chapter 7] for an analytic study of these sets.

It is the flexible multiplicative structure of the smooth integers which is useful in the application to prime gaps, i.e., if $Q_1 Q_2 \leq x^\vartheta$, then $q_1 q_2$ is a valid δ-smooth modulus whenever $q_1 \leq Q_1$ and $q_2 \leq Q_2$ and both are δ-smooth. This enables the dispersion method of Linnik (noted previously) to be applied in an inner sum over one of the variables, detecting more cancellation than otherwise.

Polymath8a clarified Zhang's variation, following Pintz, by using two parameters, ϵ and δ, and restating the variation of Bombieri–Vinogradov in the form $Q = x^{\frac{1}{2}+\epsilon}$ and

$$\sum_{\substack{q \leq Q \\ a \bmod q : (q,a)=1 \\ q \ x^\delta - \text{smooth} \\ q \ \text{squarefree}}} \left| \theta(x,q,a) - \frac{x}{\varphi(q)} \right| \ll_{A,\epsilon,\delta} \frac{x}{(\log x)^A},$$

with a combined bound for the sizes of ϵ and δ rather than explicit numerical values.

Polymath then went further than Zhang by introducing a new class of integers which they called **densely divisible**, a generalization of smooth integers. In that way, they were able to push the application of these ideas further, and reduce the known gap between consecutive primes at one stage in the progression of ideas and results. For details of these numbers and their application, see Chapter 8.

Some expect that the range of q values should be able to be extended to $x^{1-\epsilon}$ for any fixed ϵ with $0 < \epsilon < 1$, and this is the subject of the Elliott–Halberstam (EH) conjecture. It has been shown that extending the range all the way to x is impossible. Most of the authors reported here give conditional results, assuming EH is true, to show how far the application of their method could be extended should EH be proved, in whole or in part. The best result along these lines was derived by Polymath [175]: there are an infinite number of consecutive primes such that $p_{n+1} - p_n \leq 6$.

There have been a number of different proofs, improvements and generalizations of Bombieri–Vinogradov. These include those of Gallagher [65], Huxley–Iwaniec [112], Jutila [119] and Motohashi [146]. A version of Vaughan's proof [210] of the original Bombieri–Vinogradov theorem is given in [22, theorem 12.30], and in section 12.12 of that text we find a summary of other arithmetic applications, demonstrating its power and scope.

1.10 Admissible Tuples

We have see in Section 1.6 that for a k-tuple $\mathcal{H}$ to take all prime values for an infinite set of translates $n + \mathcal{H}$, the tuple must be what is called "admissible", i.e., $v_p(\mathcal{H}) < p$ for all primes p. This is an important notion and underlies the work in each of the main chapters in the book. In the first subsection, we give examples of admissible tuples with smallest width, denoted $H(k)$, for a given k. These are so called "narrow" k-tuples and are useful in studies of prime gaps since typically a method will show that at least two of the tuple elements for a given k are prime, so their distance apart will be not greater than the tuple width or diameter. Thus choosing a tuple which is as small as possible makes sense if we are to find the smallest possible gap size for a given method. The second subsection is devoted to obtaining bounds for $H(k)$, especially valuable since finding narrow tuples is currently only possible using exhaustive search, and the tighter the bounds, the smaller the search space. In the third subsection, we give the proof of the beautiful and surprising result that the Dickson–Hardy–Littlewood and, as defined in that section, the second Hardy–Littlewood conjectures cannot both be true. It is certainly a result which can be skipped at a first reading, given that it is not used elsewhere in the book.

1.10.1 Introduction

Recall the definition. Let $h \in \mathbb{N}$ and $\mathcal{H} = (h_1, h_2, \ldots, h_k)$, a k-tuple of distinct integers, with $0 \leq h_1 < h_2 < \cdots < h_k \leq h$ and for each prime p let $v_p(\mathcal{H})$ be the number of distinct residues $h_i \bmod p$. We say $\mathcal{H}$ is **admissible** if $v_p(\mathcal{H}) < p$ for all primes p. If $\mathcal{H} + n$ is to be prime for every $h_j + n$ for more than one integer n, it is necessary that $\mathcal{H}$ be admissible – see Section 1.6. Examples of admissible tuples of minimal width or "diameter" $(h_k - h_1)$ are

$$\{0, 2\}, \ \{0, 4, 6\}, \ \{0, 2, 6, 8\}, \ \{0, 2, 6, 8, 12\}, \ \{0, 4, 6, 10, 12, 16\}.$$

For example, $\mathscr{H} = \{0,2,4\}$ is *not* admissible: if $p = 3$ then modulo 3, we have the representatives $\mathscr{H} \bmod 3 = \{0,2,1\}$, so $v_3(\mathscr{H}) = 3$, and for every $n \in \mathbb{N}$, $n + \mathscr{H} \bmod 3 = \{0,1,2\}$, so at least one element is divisible by 3 and thus cannot be prime for $n \geq 4$.

Some further examples of admissible k-tuples of minimal width denoted $H(k)$ are given in Lemma 1.1, taken from the tables of Thomas Engelsma [43].

Lemma 1.1 (Smallest diameter k-tuples)

$$H(10) = 32, \; H(20) = 80, \; H(30) = 136, \; H(40) = 186, \; H(50) = 246,$$
$$H(54) = 270.$$

Proof The proof is by an exhaustive computation, considering all integer subsets of $[0, h_k]$ of size k, and checking each one's admissibility and width. Examples:

$\mathscr{H}_{10} = \{0,2,6,12,14,20,24,26,30,32\}$

$\mathscr{H}_{20} = \{0,2,8,12,14,18,24,30,32,38,42,44,50,54,60,68,72,74,78,80\}$

$\mathscr{H}_{30} = \{0,6,10,16,18,22,28,30,36,42,48,52,58,60,66,70,72,76,78,$
$\qquad 88,100,102,106,108,118,120,126,130,132,136\}$

$\mathscr{H}_{40} = \{0,4,10,18,24,28,30,34,40,46,48,54,58,66,70,76,84,88,90,$
$\qquad 96,100,108,114,118,124,130,136,138,144,150,154,156,100,166,$
$\qquad 168,174,178,180,184,186\}$

$\mathscr{H}_{50} = \{0,4,6,16,30,34,36,46,48,58,60,64,70,78,84,88,90,94,$
$\qquad 100,106,108,114,118,126,130,136,144,148,150,156,160,168,174,178,184,$
$\qquad 190,196,198,204,210,214,216,220,226,228,234,238,240,244,246\},$

$\mathscr{H}_{54} = \{0,4,10,18,24,28,30,40,54,58,60,70,72,82,84,88,94,102,108,$
$\qquad 112,114,118,124,130,132,138,142,150,154,160,168,172,174,$
$\qquad 180,184,192,198,202,208,214,220,222,228,234,238,$
$\qquad 240,244,250,252,258,262,264,268,270\}. \qquad \square$

By the term **narrow (admissible) tuple** is often meant a tuple of minimal diameter and given size. Sometimes the minimality of the diameter is conjectural. Here are examples of all narrowest tuples with $3 \leq k \leq 10$:

$\quad H[3] \;\; \{\{0,2,6\},\{0,4,6\}\}$

$\quad H[4] \;\; \{\{0,2,6,8\}\}$

$\quad H[5] \;\; \{\{0,2,6,8,12\},\{0,4,6,10,12\}\}$

$\quad H[6] \;\; \{\{0,4,6,10,12,16\}\}$

$H[7]$ $\{\{0, 2, 6, 8, 12, 18, 20\}, \{0, 2, 8, 12, 14, 18, 20\}\}$

$H[8]$ $\{\{0, 2, 6, 8, 12, 18, 20, 26\}, \{0, 2, 6, 12, 14, 20, 24, 26\},$
$\{0, 6, 8, 14, 18, 20, 24, 26\}\}$

$H[9]$ $\{\{0, 2, 6, 8, 12, 18, 20, 26, 30\}, \{0, 2, 6, 12, 14, 20, 24, 26, 30\},$
$\{0, 4, 6, 10, 16, 18, 24, 28, 30\}, \{0, 4, 10, 12, 18, 22, 24, 28, 30\}\}$

$H[10]$ $\{\{0, 2, 6, 8, 12, 18, 20, 26, 30, 32\}, \{0, 2, 6, 12, 14, 20, 24, 26, 30, 32\}\}$

Some observations related to the sets of narrow tuples H with the same k in the range $2 \leq k \leq 342$: for $p \leq k$, we have $v_p(H) = p - 1$, and for $p > k$, $v_p(H) = k$. All of the narrow tuples with the same value of k have the same gap distribution. The gap distributions are related by symmetries. Consecutive are apparently unequal. The PGpack functions which could be used to explore tuple properties are AdmissibleTupleQ, TupleGaps, TupleMinimalCompanions, NarrowestTuples, InductiveTuple and the lists HValues and InductiveTuplesOffsets1000, for example. Maybe the fine-scale combinatorics of these narrow tuples will play a role in further developments.

A short investigation, suggested by a lecture by Julia Wolf on randomness, led to the following exercise regarding successive equal gaps between primes giving rise to some "uniform" admissible tuples of preassigned length, and thus showing that the Dixon–Hardy–Littlewood conjecture implies the Green–Tao theorem on primes in arithmetic progression: For $n \geq 2$, let $L(n)$ be the maximum number of successive prime gaps (not necessarily consecutive) of size n. For example, $L(2) = L(4) = 1$. Then if $\#p_j$ is the j'th primorial with $\#p_1 = 2$, and $i_n :=$ $\max\{j : \#p_j \mid n\}$, then maybe $L(n) = p_{i_n+1} - 2$. Its easy to show that for $p \nmid \delta$ there are at most $p - 1$ gaps of size δ in any admissible tuple. So if primes occur somewhat at random, prime gaps definitely do not!

Finally in this section, for the possible amusement of the reader, we could not resist using Ulam's famous spiral to display a small number of prime gap patterns (see Figure 1.5). The small dots represent 1 and composite integers and the large the primes.

Note also that the tuples returned by the function InductiveTuple have the same width as those stated in Lemma 1.1 in the cases $\mathscr{H}_{10}$, $\mathscr{H}_{40}$, $\mathscr{H}_{50}$ and $\mathscr{H}_{54}$, and that the tuples returned by this function have the delightful property that the $(k+1)$th is found by adding one integer to the kth. Thus, if we call this integer h_{k+1} the sequence $(h_k)_{k\in\mathbb{N}}$ might be regarded as a "universal tuple" and have interesting properties.

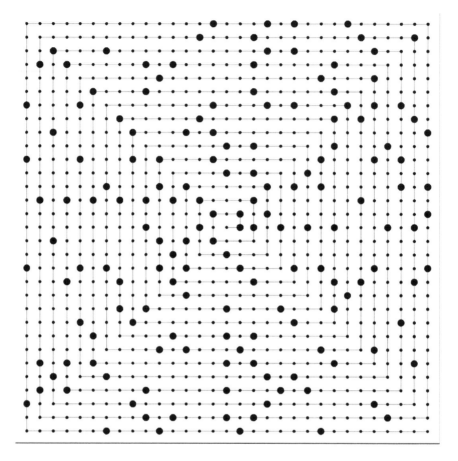

Figure 1.5 A plot of Ulam's spiral from 1 to 961.

1.10.2 Bounds for $H(k)$

In this subsection, we derive lower bound and asymptotic upper bounds for $H(k)$ and give a variation on a conjectured upper bound of Polymath. Recall the definition: $g(k) = o(f(k))$ means $\lim_{k \to \infty} g(k)/f(k) = 0$.

Lemma 1.2

$$H(k) \le k \log k + k \log\log k - k + o(k).$$

Proof   Let p_n be the nth prime with $p_1 = 2$. Then we have [174, equation (3.1)]

$$p_n = n \log n + n \log\log n - n + o(n).$$

We also have by the prime number theorem, as $k \to \infty$

$$\pi(k) = \frac{k}{\log k}(1 + o(1)).$$

It follows from this that if we set $n = \pi(k)$, we get

$$\log n = \log k - \log\log k + o(1),$$
$$\log\log n = \log\log k + o(1),$$
$$\log(n + k) = \log(k) + o(1),$$
$$\log\log(n + k) = \log\log(k) + o(1).$$

Consider the k-tuple

$$\mathcal{H} = \{p_{n+1}, \ldots, p_{n+k}\}.$$

Our choice $n = \pi(k)$ means that for any prime $p \le k$, the class 0 will be missed when $\mathcal{H}$ is taken modulo p, so this set is admissible. Hence $H(k) \le p_{n+k} - p_k$. We can bound this difference using the asymptotic expressions given earlier and recalling $n = \pi(k) = o(k)$:

$$
\begin{aligned}
p_{n+k} - p_n &= ((n + k)\log(n + k) + (n + k)\log\log(n + k) - n - k + o(n + k)) \\
&\quad - (n\log(n) + n\log\log(n) - n + o(n)) \\
&\le (n + k)(\log(k) + o(1)) + (n + k)(\log\log(k) + o(1)) - n\log(n) \\
&\quad - n\log\log(n) - k + o(k) \\
&\le k\log k + k\log\log k + n(\log k - \log n) + n(\log\log(k) \\
&\quad - \log\log(n)) - k + o(k) \\
&= k\log k + k\log\log k - k + o(k).
\end{aligned}
$$

This completes the proof.   □

We have the following adjusted supposition of Polymath, known to hold for $k \le$ 342: We have the bound $H(k) \le \lfloor k\log k + k + 14 \rfloor$.

Recall $H(k)$ is the width of a smallest admissible k-tuple. Now let y be a positive integer which is 3 or more. Define $K(y)$ to be the largest value of the integer $k \ge 2$ such that there exists an admissible k-tuple $\mathcal{H}$ with width less than y. For example, $K(4) = 2$ since $H(2) = 2$ and $H(3) = 6$.

Polymath8a give an argument, using $K(y)$, to show there is a lower bound $2H(k) \ge k\log k$ for all $k \ge 2$, which we have paraphrased in Lemma 1.3. Considering Table 1.3, one suspects that removing the factor 2 could be the goal of a further study.

Lemma 1.3   *([174, section 3.9]) For all integers $k \ge 2$, we have $2H(k) \ge k\log k$.*

Proof (**1**) First note that for all $k \geq 2$, since every subset of an admissible k-tuple of size less than k is admissible, we have $K(H(k)) = k - 1$, and it follows that for all $d \geq 3$ we have

$$K(d) \leq k - 1 \quad \Longleftrightarrow \quad H(k) \geq d.$$

(**2**) By [143, theorem 2], for $x > 0$ and $y > 1$ we have

$$\pi(x + y) - \pi(x) \leq \frac{2y}{\log y}.$$

(**3**) For x, y sufficiently large, we can choose an admissible k-tuple in $(x, x + y]$ consisting of only primes (say the first k primes greater than k). Therefore,

$$K(y) \leq \pi(x + y) - \pi(x) \leq \frac{2y}{\log y}.$$

Therefore, choosing $y = d := (k \log k)/2$, we get, for $k \geq 13$,

$$K(d) \leq \frac{k \log k}{\log((k \log k)/2)} < k - 1.$$

For $2 \leq k \leq 12$ we can check the inequality directly. See for example Table 1.3. Therefore, by Step (1) we have $H(k) \geq \frac{1}{2}k \log k$, which completes the proof. $\square$

See [174, section 3] for an overview of further work relating to bounds for $H(k)$.

In Table 1.3, we give the lower bound as computed from Lemma 1.3, the value of $H(k)$, a variation of the upper bound conjectured by Polymath, and the upper bound given by Lemma 1.2. These bounds are useful in reducing the computational burden to arrive at admissible tuple statistics and other related derivations.

1.10.3 The Second Hardy–Littlewood Conjecture

The second Hardy–Littlewood conjecture of 1923 is that for all real $x, y \geq 2$, we have $\pi(x + y) \leq \pi(x) + \pi(y)$. It is an interesting proved fact, which we show in Corollary 1.5, that the conjecture of Dickson–Hardy–Littlewood (DHL) and the second conjecture of Hardy–Littlewood cannot both be true. Montgomery and Vaughan have derived an alternative to the second conjecture, taking the form: $\pi(x + y) - \pi(y) \leq 2\pi(x)$ [143, corollary 2]. Many believe the conjecture DHL is true or close to true, whereas the second is false, even though one might expect $\pi(2x) \leq 2\pi(x)$ and $\pi(x + H(k)) \leq \pi(x) + \pi(H(k))$, for example.

Define for $x > 2$, $\rho_\star(x) := \max\{k : H(k) \leq x\}$ and for $x \in [1, 2)$ let $\rho_\star(x) := 1$. Note that $H : \mathbb{N} \cap [2, \infty) \to \mathbb{N}$ is strictly increasing, and $\rho_\star : [1, \infty) \to \mathbb{N}$

Table 1.3 *Some upper and lower bounds for H(k).*

k	$\left\lceil \frac{1}{2}k \log k \right\rceil$	$H(k)$	$\lfloor k \log k + k + 14 \rfloor$	$\lfloor k \log k + k \log\log k - k \rfloor$
2	1	2	17	−2
3	2	6	20	0
4	3	8	23	2
5	5	12	27	5
6	6	16	30	8
7	7	20	34	11
8	9	26	38	14
9	10	30	42	17
10	12	32	47	21
20	30	80	93	61
30	52	136	146	108
40	74	186	201	159
50	98	246	259	213
60	123	304	319	270
70	149	370	381	328
80	176	432	444	388
90	203	494	518	450
100	231	558	574	513

increasing with $\rho_\star(H(k)) = k$ for all k and $\rho_\star(x) = k$ for $H(k) \le x < H(k+1)$. A plot of $\rho_\star(x)$ on $[0, 12]$ is given as Figure 1.6. Then $\rho_\star(x)$ is the maximum size k of any admissible k-tuple which can lie in any translate of $[0, x]$.

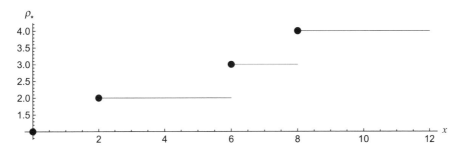

Figure 1.6 A plot of $\rho_\star(x)$ for $0 \le x \le 12$.

First we give the theorem of Hensley and Richards, which provides an explicit lower bound for $\rho_\star(x)$, and then deduce the incompatibility of the two conjectures as a corollary.

Theorem 1.4 **(Hensley–Richards' bound)** *[102, section 2]*

$$\rho_\star(x) - \pi(x) \ge (\log 2 - \epsilon) \frac{x}{\log^2 x}.$$

Proof **(1)** Let $N \geq 3$ be fixed and $x > 0$ be sufficiently large. Consider a set of integers

$$S_x := \left(-\frac{x}{2}, \frac{x}{2} \right] \cap \mathbb{Z} \setminus \{-1, 1\}.$$

Remove from S_x all positive and negative multiples of all primes in $(2, x/(N \log x)]$. The set of integers which remain following this sieving process is denoted T_x, and consists of all primes in $(x/(N \log x), x/2]$ together with their negatives.

Since we can write for some constant λ

$$\pi(x) = \frac{x}{\log x} + \frac{\lambda x}{\log^2 x} + O\left(\frac{x}{\log^3 x} \right),$$

we get

$$2\pi \left(\frac{x}{2} \right) - \pi(x) = \frac{x \log 2}{\log^2 x} + O\left(\frac{x}{\log^3 x} \right) \sim \frac{x \log 2}{\log^2 x}$$

and similarly

$$\pi \left(\frac{x}{N \log x} \right) = \frac{x}{N \log^2 x} + O\left(\frac{x \log\log x}{\log^3 x} \right).$$

Therefore,

$$\#T_x = 2\pi(x/2) - 2\pi \left(\frac{x}{N \log x} \right) + O\left(\frac{x \log\log x}{\log^3 x} \right)$$

$$= (2\pi(x/2) - \pi(x)) - 2\pi \left(\frac{x}{N \log x} \right) + O\left(\frac{x \log\log x}{\log^3 x} \right)$$

$$= \left(\log 2 - \frac{2}{N} \right) \frac{x}{\log^2 x} + \pi(x) + O\left(\frac{x \log\log x}{\log^3 x} \right),$$

which gives an asymptotic estimate for the size of T_x.

(2) Let $x > 0$ and $f(x)$ be the minimum value of y with $0 < y \leq x$ for which the set of primes $p \leq y$ can completely cover an interval of length x in the sense that there is an integer a_p such that the interval $[1, x]$ is contained in the union over $p \leq y$ of classes a_p mod p. Then we **claim** we have $f(x) = o(x)$.

To see this first note that the claim is implied by the estimate $\pi(f(x)) = o(\pi(x))$. Let $m > 1$ be sufficiently large and define

$$g(m) := \prod_{m < p \leq e^m} \left(1 - \frac{1}{p} \right).$$

By Mertens' theorem, we have

$$\prod_{p \le x} \left(1 - \frac{1}{p}\right) \sim \frac{e^{-\gamma}}{\log x} \implies g(m) \sim \frac{\log m}{m}.$$

Next we will prove the **subclaim** that for a sufficiently large fixed m, for x sufficiently large we can get

$$\pi(f(x)) < \pi\left(\frac{x}{m}\right) + g(m)\pi(x).$$

Assuming this, by using the fact that m can be chosen arbitrarily large, we get $\pi(f(x)) = o(\pi(x))$, which establishes the claim of Step (2).

To demonstrate the subclaim, take out from $[1, x]$ all multiples of all primes in the union

$$[1, m] \cup \left[e^m, \frac{x}{m}\right],$$

leaving in $[1, x]$ all primes greater than x/m together with all integers with all prime factors from (m, e^m). When x is large (recall m is fixed), the first set is large in comparison with the second, so we can say, asymptotically, both sets have at most $\pi(x)$ elements.

Now sieve out elements using primes in (m, e^m), reducing the number of elements remaining in $[1, x]$ by a factor less than or equal to $g(m)$.

Finally remove the remaining at most $g(m)\pi(x)$ points in (m, e^m) using at most $g(m)\pi(x)$ primes. Since we have completely covered $[1, x]$ and used at most $\pi(x/m) + g(m)\pi(x)$ primes, this shows that

$$\pi(f(x)) \le \pi(x/m) + g(m)\pi(x),$$

completing the proof of Step (2).

(3) Now we **claim** that given a positive integer b there is an integer a such that for each positive integer n sufficiently large, each term in the arithmetic progression

$$a + b, \ a + 2b, \ a + 3b, \dots, a + nb$$

is divisible by at least one prime $p \le f(n)$.

To see this first, using the definition established in Step (2), for each prime $p \le f(n)$ select an integer a_p such that $[1, n]$ is covered by the union of the classes a_p mod p. By the Chinese remainder theorem there is a simultaneous solution c to the congruences

$$c \equiv -a_p \text{ mod } p, \ 1 \le p \le f(n).$$

This implies $p \mid c + a_p \mid bc + ba_p$ for all $p \in [1, f(n)]$. Let $a = bc$. Since the classes cover $[1, n]$ for each $j \in [1, n]$, there is a p such that $j \equiv a_p$ mod p, and therefore $p \mid a + jb$, completing the proof of the claim.

(4) Let $N > 0$ be a large fixed integer. We now **claim** that there is an integer x_N such that for every integer $x \geq x_N$, and every arbitrarily large integer y and all integers $b > 0$ there is an arithmetic progression

$$a + b, \; a + 2b, \; a + 3b, \ldots, a + Mb$$

such that (a) $M \geq N \log x$, (b) $a + b \in (y, y + x]$, and (c) for all $1 \leq j \leq M$ there exists a $p \leq (\log x)/N$ such that $p \mid a + jb$.

To see this, first use Step (3) to construct the arithmetic progression of length n with each term divisible by some prime $p \leq f(n)$. By Step (2), we have $f(n) = o(n)$. Hence, because N is fixed, when x and M are sufficiently large (choose $M/N^2 \leq \log x \leq M/N$, for example), we have $M \geq N \log x$ and $p \leq (\log x)/N$, which gives (a) and (c).

Finally in this step we show that (b) is also true. Note that by Step (3) we can add any multiple of

$$\prod_{p \leq (\log x)/N} p$$

to the first term $a + b$ of the progression. By the prime number theorem, for given $\epsilon > 0$ and u sufficiently large

$$e^{u-\epsilon} \leq \prod_{p \leq u} p \leq e^{u+\epsilon} \leq x^{1/N} < x.$$

Thus we can find a multiple such that the first term lies in $(y, y + x]$, completing the proof of (b).

(5) Recall from Step (1) that we sieved out of the set $(-x/2, x/2] \cap \mathbb{Z} \setminus \{\pm 1\}$ all multiples of all primes $p \leq x/(N \log x)$. We must show that the remaining elements of T_x form an admissible set. To see this, first note that in T_x, for all primes $p \leq x/(N \log x)$, the zero class is missing. We thus need to show that for each prime $p > x/(N \log x)$, there is at least one empty class.

First dilate the original interval by a factor of 3 and replace the N of Step (4) by $3N$ and b with the prime $p > x/(N \log x)$. Let $M = \lceil 3N \log 3x \rceil$, and the first term of the associated arithmetic progression

$$a + p, \; a + 2p, \; a + 3p, \ldots, a + Mp$$

satisfy $-3x/2 < a + p \le 3x/2$. The last term $a + Mp$ satisfies

$$a + Mp > a + \lceil 3N \log(3x) \rceil \cdot \frac{3x}{3N \log(3x)} \ge a + 3x > \frac{3x}{2}.$$

Taking the intersection of T_x with the congruence class $a \bmod p$ with $(x/2, x/2]$, since each member of the progression is divisible by some prime $p \le (\log 3x)/3N < (\log x)/N$, and multiples of primes up to $3x/(3N \log(3x))$ have been removed, including multiples of primes up to $(\log x)/N$, we see that the intersection is empty. Thus T_x, the set which remains after sieving, is admissible and, by Step (1), given $\epsilon > 0$ and choosing N sufficiently large (and fixed) as $x \to \infty$ we get

$$\#T_x - \pi(x) \sim (\log 2 - \epsilon) \frac{x}{\log^2 x}.$$

This completes the proof. $\square$

Corollary 1.5 **(Hensley–Richards)** *The Dickson–Hardy–Littlewood and second Hardy–Littlewood conjectures cannot both be true.*

Proof Assume DHL is true. Assume also that there exists a real number x such that $k = \rho_\star(x)$, and suppose that $y \le h_1 < \cdots < h_k < y + x$ is such that $\{h_1, \ldots, h_k\}$ is admissible. Then by DHL there is a positive integer n (which can be sufficiently large) such that each integer $n + h_1, \ldots, n + h_k$ is prime. By Theorem 1.4, we have $\rho_*(x) - \pi(x) > 0$, so

$$\pi(x) < \rho_*(x) = k,$$

and since there may be additional primes in the interval $(y + n, y + n + x]$ other than the $n + h_j$, we get

$$\pi(x) < k \le \pi(n + x + y) - \pi(y + n) \implies \pi(y + n) + \pi(x) < \pi(x + y + n),$$

so the second Hardy–Littlewood conjecture is false. $\square$

1.11 A Brief Guide to the Literature

(1) To begin this very brief guide, we note there is a movie which has an extended interview with Yitang Zhang, but also some of the other main players appear: Zhala films *Counting from Infinity*, a movie by George Csicsery [34].

(2) An elementary book but also containing interesting information about how some of the main players came to their results and interacted: *Closing the Gap: The Quest to Understand Prime Numbers*, by Vicky Neal, [154].

(3) An excellent short preliminary reading, touching on the sociology of mathematics reviewing applied especially to Zhang's *Annals of Mathematics* article: John Friedlander *Prime Numbers: A Much Needed Gap Is Finally Found* [59].

(4) An article by Kannen Soundararajan which is very expository and celebrates GPY's achievement: "Small Gaps Between Prime Numbers: The Work of Goldston–Pintz–Yildirim" [198].

(5) An extended article by Andrew Granville, offering deep insights, variations and applications: "Primes in Intervals of Bounded Length" [82].

Figure 1.7 Kannan Soundararajan and Andrew Granville. This photograph was taken by the author.

Figure 1.7 is an image of two of these principal commentators, and also contributors to these developments, taken by the author during the Gauss–Dirichlet conference in Göttingen in 2005. Note that the argument used for GPY's theorem 4.19 is attributed to Granville and Soundararajan. Granville, by careful reading, also made a significant contribution.

(6) Motohashi's excellent expository article setting the developments in an historical context and explaining where the ideas came from (but modestly downplaying his own role as part of GMPY in [70]) "The Twin Primes Conjecture" [150].

(7) GPY's main presentation of their result $\liminf_{n\to\infty}(p_{n+1} - p_n)/\log n = 0$: Goldston, Pintz and Yildirim, "Primes in Tuples I" [72].

(8) Zhang's article giving the proof of his variation of Bombieri–Vinogradov and the $p_{n+1} - p_n \leq 7 \times 10^7$ for infinitely many $n \in \mathbb{N}$: "Bounded Gaps Between Primes" [215].

(9) Polymath8a's write-up of their work improving Zhang's result and introducing the "densely divisible integer" concept: "New Equidistribution Estimates of Zhang Type" [173, 174]. Note that version 2 of the preprint on ArXiv contains considerable additional information than the journal article presentation.

(10) The absolutely essential paper wherein, with his "multidimensional sieve" approach, James Maynard reduces the gap to 600 and proves quite a few other outstanding results: "Small Gaps Between Primes" [138].

(11) The article where Polymath8b improve Maynard's gap size by perturbing his method, but also demonstrate some limits to their approach to the problem and deriving the bound 246: "Variants of the Selberg Sieve, and Bounded Intervals Containing Many Primes" [175].

1.12 End Notes

This work is focused on small prime gaps, where the ultimate goal is a solution to the twin primes conjecture. However, its context is the distribution of gaps between consecutive primes, where recent progress has been made also in understanding the higher end of the distribution. A brief summary comment is in order before we return to small gaps. By the prime number theorem, asymptotically the average gap between the primes in $[2, x]$ is $\log x$. One might expect there to exist consecutive primes distant apart significantly more than the average, using the average as a scale. Indeed, Westzynthius [213] showed in 1931 that for any real number $\rho > 1$, if $G(x)$ represents the size of the largest gap up to x, then $G(x) > \rho \log x$ for an infinite number of $x \to \infty$. Thus began a succession of improvements described in Table 1.4 where

$$g(x) := \frac{\log x \, \log_2 x \, \log_4 x}{(\log_3 x)^2}$$

with $\log_1 x := \log x$ and $\log_{n+1} x := \log(\log_n(x))$ for all $n \in \mathbb{N}$, and where

$$g_M(x) := f(x)g(x), \text{ where } f(x) \to \infty \text{ as } x \to \infty, \text{ and}$$
$$g_F(x) := \frac{\log x \, \log_2 x \, \log_4 x}{\log_3 x}.$$

The following is a list of PGpack functions related to the work of this chapter. See Appendix I for the manual entries.

AdmissibleTupleQ, BrunsConstant, ContractTuple, DenseTuple, EnlargeAdmissibleTuple, CountResidueClasses, EratosthenesTuple, GreedyTuple, HensleyRichardsTuple, HValues, InductiveTuple, InductiveTuplesOffset1000,

Table 1.4 *Large prime gaps.*

Year	Bound ρ	People	Reference
1931	$\forall\, \rho > 1\ G(x) \gg \rho \log x$	Westzynthius	[213]
1935	$G(x) \gg \log x \log_2 x / (\log_3 x)^2$	Erdős	[44]
1938	$G(x) \geq (\frac{1}{3} + o(1))g(x)$	Rankin	[178]
1963	$G(x) \geq (e^\gamma/2 + o(1))g(x)$	Schönhage	[187]
1962/63	$G(x) \geq (e^\gamma + o(1))g(x)$	Rankin	[180]
1990	$G(x) \geq (1.31256e^\gamma + o(1))g(x)$	Maier/Pomerance	[133]
1997	$G(x) \geq (2e^\gamma + o(1))g(x)$	Pintz	[165]
2015	$\forall\, \rho > 1\ G(x) \geq \rho g(x)$	Ford/Green/Konyagin/Tao	[52]
2016	$G(x) \gg g_M(x)$	Maynard	[140]
2018	$G(x) \gg g_F(x)$	Ford/Green/Konyagin/ Maynard/Tao	[51, 52, 140]

NarrowestTuples, NarrowH, NarrowTuples, NextPrimeGapDist, PlotHValues, PlotTwins, PrimeTuple, RhoStar, SchinzelTuple, ShiftedPrimeTuple, TupleDiameter, TupleGaps, TupleMinimalCompanions, TwinPrimesConstant, VonMangoldt.

2

The Sieves of Brun and Selberg

2.1 Introduction

Chapter 1 introduced some of the main concepts needed for recent studies of prime gaps, and this provides an introduction to sieve theory, which provides a most important technique. We begin with the two modern founders of the subject, Brun and Selberg. Section 2.2 summarizes the material given in the body of the chapter. It can be skipped by readers with a background in sieve theory.

Viggo Brun (1885–1978) lived and died in his native Norway, apart from several years spent in Göttingen during the time of Hilbert, Klein and Landau. He worked on some outstanding problems in number theory, including the twin primes and Goldbach conjectures, proving that there were an infinite number of natural numbers n such that n and $n + 2$ each had at most nine prime factors. To do this he introduced "Brun's pure sieve" and "Brun's sieve" [23, 24], with the latter sometimes called the "Brun's combinatorial sieve", which still find applications up until the present day. He was appointed to a chair at the Technical University of Tronheim and later at the University of Oslo. He taught and wrote extensively on the history of mathematics, and supported many social causes, such as political parties and the peace movement. In this chapter, we give his proof that the sum of the reciprocals of twin primes converges, [24].

Brun and Selberg could be regarded as the founders of the theory of sieves, which flourishes in this century and is at the foundation of the breakthroughs recorded in this book. Both Norwegian, both experiencing longevity, both were members of large families, and both made outstanding contributions to mathematics, especially number theory. However, their life experiences were very different. Whereas Brun's parents died when he was very young, Selberg's father was a mathematics teacher who obtained a doctorate in number theory at the University of Oslo, and Selberg learnt a great deal reading his father's books.

Selberg (1917–2007) was a mathematical prodigy and published an article while still at high school. He undertook doctoral studies at the University of Uppsala and defended his thesis "On the Zeros of the Riemann Zeta Function" in 1943. The Second World War raged during this period, and Norway was occupied. Selberg was to spend some time in prison, and then returned to his family to work for a number of years on the Riemann hypothesis. He married and moved to the USA in 1947, with a temporary membership of the Institute for Advanced Studies (IAS) Princeton and work at other US institutions, before becoming a permanent IAS member in 1949, where he remained for the rest of his life.

He was awarded a fields medal in 1950 for his work on sieve theory and the zeros of the Riemann zeta function, where he had shown that a positive proportion of the non-trivial zeros were on the critical line. He made many outstanding contributions, including a proof of the prime number theorem which did not use complex analysis (unfortunately giving rise to a dispute with Paul Erdős over attribution), but also to the Selberg sieve, Selberg trace formula for $SL_2(\mathbb{R})$, Selberg's zeta function, Dirichlet series and many other topics [193, 194].

This chapter gives classical results and methods to set the scene for what is to follow. References are given in Section 2.9, which takes the form of an introductory readers' guide to sieve theory.

2.2 Summary of This Chapter

Section 2.3, on Brun's pure sieve, gives a little of the history of its discovery and several of the main results which demonstrate its strength.

In Theorem 2.3, the inclusion-exclusion identity is used to show how the sieve can be used to derive both lower and upper bounds by truncating sums. In applications, the sets A_j, which appear in the statement, can normally only be counted approximately and the errors accumulate, reducing the effectiveness of the sieve. However, it still finds good applications.

> **Theorem 2.3   (Brun's pure sieve)** *Let X be a set with $\infty > |X| = N \geq 1$ and let $A_1, \ldots A_r$ be subsets of X. Let*
>
> $$N_0 = |X \setminus \cup_{j=1}^r A_j|$$
>
> *be the number of elements of X that are in none of the subsets. Then for any given* **even** $m \geq 0$, *we have*
>
> $$N + \sum_{j=1}^{m+1}(-1)^j \sum_{|S|=j} |\cap_{i \in S} A_i| \leq N_0 \leq N + \sum_{j=1}^{m}(-1)^j \sum_{|S|=j} |\cap_{i \in S} A_i|.$$

Then the sieve is used to derive the first upper bound we give for the number of prime twins up to $x > 0$:

Theorem 2.6 (Brun's twin prime bound) *As $x \to \infty$ we have for an absolute implied constant*

$$\pi_2(x) = \#\{p \le x : p + 2 \in \mathbb{P}\} \ll \frac{x (\log\log x)^2}{(\log x)^2}.$$

This bound is sufficient to show that the sum of the reciprocals of prime twins converges in $\mathbb{R}$,

Theorem 2.7 (Brun) *The sum of the reciprocals of twin primes converges.*

The sum of twin prime reciprocals is called "Brun's constant" (Section 2.4). The sum has a number of definitions and a small industry which computes it. If the constant was found to be irrational, then the number of twin primes would necessarily be infinite.

Section 2.5 introduces a form of the Selberg sieve giving something of its history and the definition of essential quantities. Note that it is variations of the form of the sieve that have led to prime gap breakthroughs, so this section is an introduction to its spirit rather than definitions which should not be taken literally elsewhere. Theorem 2.15 gives an application to counting primes p up to $x > 0$ such that $p + b$ is also prime, denoted $\pi_b(x)$, improving on the result obtained using Brun's pure sieve and giving what is believed to be the correct order as a function of x of the upper bound. It is important to remark that a more flexible result may be obtained using Brun's sieve described in the text of Halberstam and Richert [88, corollary 2.4.1] (see Section 2.8 for the difference between Brun's pure sieve and Brun's sieve).

Theorem 2.15 (Selberg sieve prime fixed gap bound) *Let b be an even positive integer. For an absolute implied constant, as $x \to \infty$ we have*

$$\pi_b(x) \ll \prod_{p|b} \left(1 + \frac{1}{p}\right) \frac{x}{(\log x)^2}.$$

In Section 2.6, the Selberg sieve is used in a rather hidden manner to give an explicit value, as a function of $2n = b$, for the leading terms which appear in Theorem 2.15. This is the theorem of Bombieri and Davenport where

$$K(n) := \prod_{p>2} \left(1 - \frac{1}{(p-1)^2}\right) \prod_{p|n, \ p>2} \left(1 + \frac{1}{p-2}\right),$$

so, with a bit of work using the notation of Section 1.6, $2K(n) = \mathscr{G}(\{0, 2n\})$.

Theorem 2.24 (**Bombieri–Davenport's prime fixed gap constant**) *Given $\epsilon > 0$ for all N sufficiently large, we have*

$$\pi_{2n}(N) < (8 + \epsilon)K(n)\frac{N}{(\log N)^2}.$$

In Section 2.7, we give an application of the Selberg sieve to derive a weak form of the Brun–Titchmarsh inequality. This is used much later in this book, in fact in Appendix D, to derive an inequality for multiplicative functions. In the application given here, $\pi(x,q,a)$ is a count of the number of primes up to x which are congruent to a modulo q:

Corollary 2.26 (**A weak Brun–Titchmarsh inequality**) *For all $z \geq 2$, $q \in \mathbb{N}$, $(a,q) = 1$ and $0 < x \leq y$, we have*

$$\pi(x,q,a) - \pi(x - y,q,a) \leq \frac{y}{\varphi(q)\log z} + z^2 + \pi(z).$$

The final sections of the chapter are more discursive, with an attempt to give the reader an overview of sieve theory. The first, Section 2.8, gives a description of ten sieve types with some key applications: included are the sieve of Eratosthenes, Legendre's sieve, Brun's (combinatorial) sieve, Selberg's sieve, the large sieve and Bombieri's asymptotic sieve. This is followed by a brief reader's guide to some of the main texts on this subject. In Section 2.10, on twin almost primes and the sieve parity problem, there is a discussion of the limits of sieve theory when it comes to capturing primes, namely the famous "parity problem" of the theory.

Finally we have a short overview of Section 2.9. Whole books, and extensive parts of books, have been written on the topic of this chapter, and the reader might want at some stage when necessary to dig and delve into some of these. At a more introductory level, we have the texts by Nathanson [152] and Cojocaru and Ram Murty [33].

The classic text is that of Halberstam and Richert [88]. There is also the authoritative and elegant more recent work of Friedlander and Iwaniec [64], and the specialist texts of Harman [91] and Ramaré [177]. The list and descriptions in Section 2.9 are more complete.

2.3 Brun's Pure Sieve

It appears that Viggo Brun began work on his sieve while at Göttingen in the 1910s. Even though Hilbert, Klein and Landau were active there at that time, Brun's ideas were his own, and might have been regarded as being too simple to give good results. He attacked both the twin primes and Golbach's conjectures using his sieve, and by the end of the decade had shown and published significant new results, including the convergence of the reciprocals of twin primes and that there are an infinite number of integers n for which both n and $n + 2$ have at most nine prime factors.

This sieve is an application of the set theoretical inclusion–exclusion principle, and is limited in its application because of the difficulty in controlling errors arising from approximating the integer part function. In Section 2.8, we discuss what has become known as simply "Brun's sieve", in which these errors are reduced by limiting the number of terms arising in approximating sums using some combinatorics. We begin this section with the inclusion–exclusion principle and then give the double inequality which is at the heart of his sieve.

Lemma 2.1   *For integers l and m with $l \geq 1$ and $0 \leq m \leq l$, we have*

$$\sum_{j=0}^{m}(-1)^j \binom{l}{j} = (-1)^m \binom{l-1}{m}.$$

Proof   If $m = 0$, both sides of the equation are 1, and if $m = 1$ both sides are $1 - l$, so we can assume $m \geq 2$ and the equation is true for $m - 1$. Then using induction on m and the Pascal's triangle binomial identity, we get

$$\sum_{j=0}^{m}(-1)^j \binom{l}{j} = \sum_{j=0}^{m-1}(-1)^j \binom{l}{j} + (-1)^m \binom{l}{m}$$

$$= (-1)^m \left(\binom{l}{m} - \binom{l-1}{m-1}\right) = (-1)^m \binom{l-1}{m}.$$

Hence the given equation is true for all m and l satisfying the given constraints.   □

The inclusion–exclusion principle is set theoretic and used throughout mathematics.

Lemma 2.2   **(inclusion–exclusion principle)** *Given a finite family of finite sets $A_1, \ldots A_r$, which are all subsets of a given set, then*

$$\left| \cup_{j=1}^r A_j \right| = \sum_{j=1}^{r} |A_j| - \sum_{1 \leq i < j \leq r} |A_i \cap A_j| + \cdots + (-1)^{r-1} |A_1 \cap A_2 \cap \cdots \cap A_r|$$

$$= \sum_{\emptyset \neq J \subset \{1,2,\ldots,r\}} (-1)^{|J|-1} \left| \cap_{j \in J} A_j \right|.$$

Proof   The left-hand side counts the number of elements in the union of the r subsets. The right-hand side sums the cardinalities of each of the sets, then subtracts off 1 for each element in two or more sets, then adds back in elements which are in three sets. This process stops since no element can be in more than r sets.   □

Inclusion–exclusion gives an expression for the size of the union of r sets, but the size on the right in Lemma 2.2 has $2^r - 1$ terms, i.e., far too many generally. In Brun's pure sieve, the size of the sum is reduced by simply truncating the sum in terms of the number of terms in the set intersections, sacrificing the equality for upper and lower bounds. Varying the manner in which such truncations are performed goes to the heart of distinctions between different sieve techniques.

> **Theorem 2.3** (**Brun's pure sieve**) *Let X be a set with $\infty > |X| = N \geq 1$ and let $A_1, \ldots A_r$ be subsets of X. Let*
>
> $$N_0 = |X \setminus \cup_{j=1}^r A_j|$$
>
> *be the number of elements of X that are in none of the subsets. Then for any given* **even** $m \geq 0$, *we have*
>
> $$N + \sum_{j=1}^{m+1} (-1)^j \sum_{|S|=j} |\cap_{i \in S} A_i| \leq N_0 \leq N + \sum_{j=1}^{m} (-1)^j \sum_{|S|=j} |\cap_{i \in S} A_i|,$$
>
> *where the inner sums are over all subsets $S \subset \{1, 2, \ldots r\}$ with precisely j elements.*

Proof Let x be an element of X which is in precisely l of the subsets A_j, and let B_l be the subset of X of all such elements.

(**1**) If $l = 0$, then x counts 1 for N_0 and once for N and 0 for all the other terms in the identity of the lemma, so the lemma holds for all $x \in B_0$.

(**2**) If $l \geq 1$, then x does not contribute to N_0. Assume, renumbering if necessary, that $x \in A_1 \cap \cdots \cap A_l$. If $i > l$ and $|S| = i$, then x is not counted by any term

$$|\cap_{i \in S} A_i|.$$

If $S \subset \{1, 2, \ldots, l\}$, then $x \in S$ contributes 1 to each corresponding term. For each j with $0 \leq j \leq l$, there are precisely $\binom{l}{j}$ subsets with $|S| = j$.
(**2.1**) If $m \geq l$, x contributes

$$\sum_{j=0}^{l} (-1)^j \binom{l}{j} = 0$$

to the left-hand side and right-hand side of the inequalities.
(**2.2**) If $m < l$, then the contribution of x to the left-hand side and right-hand side of the inequalities is

$$\sum_{j=0}^{m} (-1)^j \binom{l}{j},$$

which by Lemma 2.1 is positive if m is even and negative if m is odd.

(3) Since $X = \cup_{l \geq 0} B_l$, where the union is disjoint, we have

$$|X| = \sum_{l \geq 0} |B_l|,$$

so, since the inequalities are true for each $l \geq 0$, they are true for X. This completes the proof. □

The sets A_j and their intersections which appear in Brun's pure sieve are often associated with congruence classes modulo primes or squarefree integers. We need to count the number of elements and this normally involves an approximation represented by the parameter θ, which can take different values depending on the context.

Lemma 2.4 *Let $m \geq 1$ be an integer and $x \geq 1$ real. Let a be any integer. Then in the congruence class of integers congruent to a modulo m, the number of integers*

$$|\{1 \leq n \leq x : n \equiv a \bmod m\}| = \frac{x}{m} + \theta,$$

where $|\theta| < 1$ and $\theta = 0$ if $m \mid x$.

Proof (1) When $x/m = q$ is an integer, then $qm \leq x$ and $\{1, 2, \ldots, qm\}$ contains exactly q elements in the congruence class $a \bmod m$, and we can choose $\theta = 0$. If x/m is not an integer, let $\lfloor x \rfloor = qm + r$ with $0 \leq r < m$. Thus if $\theta = \{x\}$

$$qm < x = qm + r + \{x\} \leq qm + (m - 1) + \theta < (q + 1)m.$$

Therefore, $q < x/m < q + 1$.

(2) Next partition the integers up to x setting $A := \{1, \ldots, m\}$ so

$$\{1, \ldots, \lfloor x \rfloor\} = A \cup (A + m) \cup (A + 2m) \cup \cdots \cup (A + (q - 1)m) \cup B,$$

where B has r elements. Each translate of A contains one integer congruent to a, and B at most one. This shows there are either q or $q + 1$ elements congruent to a, and since

$$\max\left(\frac{x}{m} - q, q + 1 - \frac{x}{m}\right) < 1,$$

this completes the proof. □

Next we use Lemma 2.4 with squarefree odd moduli to count sets which we will show are well adapted to the twin primes problem.

Lemma 2.5 *Let $x \geq 1$ and $p_1, \ldots, p_k$ distinct odd primes. Then for some θ with $|\theta| < 1$, with $\theta = 0$ if $p_1 \cdots p_k \mid x$, we have*

$$N := |\{n \leq x : n(n + 2) \equiv 0 \bmod p_1 \cdots p_k\}| = \frac{2^k x}{p_1 \cdots p_k} + 2^k \theta.$$

Proof For each j, if $n(n+2) \equiv 0 \bmod p_j$, then either $n \equiv 0 \bmod p_j$ or $n+2 \equiv 0 \bmod p_j$. In addition, since p_j is odd, $0 \not\equiv -2 \bmod p_j$. Thus, if $n(n+2) \equiv 0 \bmod p_1 \cdots p_k$, then there exist unique integers $u_1, \ldots, u_k$ in $\{0, -2\}$ with $n \equiv u_j \bmod p_j$, $1 \le j \le k$ (*).

By the Chinese Remainder Theorem, for each of the 2^k choices of the u_j, there is a unique $a \bmod p_1 \cdots p_k$ such that n is a solution of the congruences (*) when and only when

$$n \equiv a \bmod p_1 \cdots p_k.$$

By Lemma 2.4, for fixed a there are

$$\frac{x}{p_1 \cdots p_k} + \theta_a$$

solutions to this congruence with $1 \le n \le x$, where $\theta_a = 0$ if $p_1 \cdots p_k \mid x$. Thus, summing over the 2^k choices for the u_j, for some θ with $|\theta| < 1$ we have, with $\theta = 0$ if $p_1 \cdots p_k \mid x$,

$$N = \frac{2^k x}{p_1 \cdots p_k} + 2^k \theta.$$

This completes the proof.   □

Next, for $x \ge 2$ and positive even b, let $\pi_b(x)$ be the number of primes $p \le x$ such that $p + b$ is also prime. The main theorem of this section is Theorem 2.6, wherein $b = 2$. It is proved in seven steps using Brun's pure sieve, and depends critically on the upper bound from Lemma 2.1 and the counting expression from Lemma 2.5. Recall that $\mathbb{P}$ represents the set of prime rational integers. The reader might compare this result with the prime gap upper bound obtained later in the chapter using the Selberg sieve in Theorem 2.15 and its corollary, as well as Bombieri–Davenport's estimate Theorem 2.24, where the implied constant is explicit.

> **Theorem 2.6** **(Brun's twin prime bound)** *As $x \to \infty$ we have for an absolute implied constant*
>
> $$\pi_2(x) \ll \frac{x(\log\log x)^2}{(\log x)^2}.$$

Proof (1) Let $2 < y < x$ and $r := \pi(y) - 1$ the number of odd primes less than or equal to y, which we denote $3 = p_1 < p_2 < \cdots < p_r \le y$. (We note here the potential value of choosing y to be the rth odd primorial.) Let

$$\pi_2(y, x) := \{p \in \mathbb{P} : y < p \le x, \, p + 2 \in \mathbb{P}\}.$$

For n such that $y < n \le x$ with $n, n+2 \in \mathbb{P}$ and $n > p_r$ then $p_i \nmid n(n+2)$ for $1 \le i \le r$.

Therefore, if $N_0(y, x)$ is the number of positive integers n such that $n \le x$ with

$$n(n + 2) \not\equiv 0 \bmod p_i, \ 1 \le i \le r,$$

it follows, since each $p_i \le y$, that

$$\pi_2(x) \le y + \pi_2(y, x) \le y + N_0(y, x).$$

In Steps (2) through (5), we will derive an upper bound for $\pi_2(x)$ by finding an upper bound for $N_0(y, x)$ using Lemma 2.5.

(2) Let $X := \{n \in \mathbb{N} : n \le x\}$ and let $A_i := \{n \in X : p_i \mid n(n + 2)\}$. For any $k \ge 1$ and subset $S = \{i_1, \dots, i_k\} \subset \{1, \dots, r\} =: R$, let

$$N(S) := |\cap_{i \in S} A_i|.$$

By Lemma 2.5, we can write

$$N(S) = \frac{2^k x}{p_{i_1} \cdots p_{i_k}} + 2^k \theta_S \text{ and } N = |X| = \lfloor x \rfloor = N(\emptyset).$$

Using this estimate, if m is an even integer such that $1 \le m \le r$ then by Theorem 2.3 we can write

$$N_0(y, x) \le x + \sum_{j=1}^{m} (-1)^j \sum_{|S|=j} N(S)$$

$$\le \sum_{j=0}^{m} (-1)^j \sum_{\{i_1, \dots, i_j\} \subset R} \left(\frac{2^j x}{p_{i_1} \cdots p_{i_j}} + O(2^j) \right)$$

$$\le x \sum_{j=0}^{m} \sum_{\{i_1, \dots, i_j\} \subset R} \frac{(-2)^j}{p_{i_1} \cdots p_{i_j}} + \sum_{j=0}^{m} (-1)^j \binom{r}{j} O(2^j)$$

$$\le x \sum_{j=0}^{r} \sum_{\{i_1, \dots, i_j\} \subset R} \frac{(-2)^j}{p_{i_1} \cdots p_{i_j}} + x \sum_{j=r+1}^{m} \sum_{\{i_1, \dots, i_j\} \subset R} \frac{(-2)^j}{p_{i_1} \cdots p_{i_j}}$$

$$+ O\left(\sum_{j=0}^{m} \binom{r}{j} 2^j \right).$$

Each of the terms in the expression on the right will be bounded separately in Steps (3) through (5).

(3) Using the well-known estimate of Merten's to get the last line, we can bound the first term:

$$x \sum_{j=0}^{r} \sum_{\{i_1,\dots,i_j\}\subset R} \frac{(-2)^j}{p_{i_1}\cdots p_{i_k}} = x \prod_{2<p\leq y} \left(1 - \frac{2}{p}\right)$$

$$< x \prod_{2<p\leq y} \left(1 - \frac{1}{p}\right)^2$$

$$\ll \frac{x}{\log^2 y}.$$

(4) Bounding the second term requires a more elaborate construction using symmetric polynomials. For $j \geq 0$, let $s_j(x_1, \dots, x_r)$ be the jth elementary symmetric polynomial in r real variables, which are restricted to being nonnegative, so $s_0(x_1, \dots, x_r) = 1$ and for $j \geq 1$, where as before $R = \{1, 2, \dots, r\}$,

$$s_j(x_1, \dots, x_r) := \sum_{\{i_1,\dots,i_j\}\subset R} x_{i_1} \cdots x_{i_j}.$$

Then, because for all $j \geq 0$ we have $(j/e)^j < j!$, we get

$$s_j(x_1, \dots, x_r) \leq \frac{(x_1 + \cdots + x_r)^j}{j!}$$

$$= \frac{s_1(x_1, \dots, x_r)^j}{j!}$$

$$< \frac{e^j}{j^j}(x_1 + \cdots + x_r)^j.$$

Using this bound for the fourth line in the following equation, we derive, for some absolute constant $c_1 > 0$,

$$x \left| \sum_{j=r+1}^{m} \sum_{\{i_1,\dots,i_j\}\subset R} \frac{(-2)^j}{p_{i_1}\cdots p_{i_j}} \right| \leq x \sum_{j=r+1}^{m} \sum_{\{i_1,\dots,i_j\}\subset R} \frac{2^j}{p_{i_1}\cdots p_{i_j}}$$

$$\leq x \sum_{j=r+1}^{m} \sum_{\{i_1,\dots,i_j\}\subset R} \frac{2}{p_{i_1}} \cdots \frac{2}{p_{i_j}}$$

$$= x \sum_{j=m+1}^{r} s_j\left(\frac{2}{p_1}, \dots, \frac{2}{p_r}\right)$$

$$< x \sum_{j=m+1}^{r} \frac{e^j}{j^j} s_1\left(\frac{2}{p_1}, \dots, \frac{2}{p_r}\right)^j$$

$$\leq x \sum_{j=m+1}^{r} \left(\frac{2e}{m}\right)^j \left(\sum_{p\leq y} \frac{1}{p}\right)^j$$

$$< x \sum_{j=m+1}^{r} \left(\frac{c_1 \log\log y}{m}\right)^j.$$

Finally in this step we choose m to be an even integer satisfying $m > 2c_1 \log\log y$, so the right-hand side of the second term of the bound for $N_0(y,x)$ derived in Step (2) is then bounded above by

$$x \sum_{j=m+1}^{r} \frac{1}{2^j} < \frac{x}{2^m}.$$

(5) Since $r \leq \pi(y) \leq y/2$ we get a bound for the third term:

$$\sum_{j=0}^{m} \binom{r}{j} 2^j < \sum_{j=0}^{m} (2r)^j \ll (2r)^m \ll y^m,$$

where the implied constant is absolute.

(6) Combining these three estimates with the result of Step (2) gives for $5 \leq y < x$, $m \geq 2$ even with $m > 2c_1 \log\log y$, for an absolute implied constant

$$\pi_2(x) \ll y + \frac{x}{(\log y)^2} + \frac{x}{2^m} + y^m \ll \frac{x}{(\log y)^2} + \frac{x}{2^m} + y^m. \qquad (2.1)$$

(7) We now choose y as a function of x to derive the bound of the theorem. First let $c_2 := \max(2c_1, 1/\log 2)$, set $m := 2 \lfloor c_2 \log\log x \rfloor$ and define

$$y := \exp\left(\frac{\log x}{3c_2 \log\log x}\right).$$

Then for x sufficiently large, we have $5 \leq y < x$ and $m > 2c_1 \log\log y$.
   Because $\log y = \log x/(3c_2 \log\log x)$, we have

$$\frac{x}{(\log y)^2} \ll \frac{x(\log\log x)^2}{(\log x)^2}.$$

Because $c_2 \geq 1/\log 2$ and $m > 2c_2 \log\log x - 2$, we get for all x sufficiently large

$$\frac{x}{2^m} < \frac{4x}{2^{2c_2 \log\log x}} = \frac{4x}{(\log x)^{2c_2 \log 2}} \leq \frac{4x}{(\log x)^2} \ll \frac{x(\log\log x)^2}{(\log x)^2}.$$

And because $m < 2c_2 \log\log x$

$$y^m \leq y^{2c_2 \log\log x} = x^{2/3} \ll \frac{x(\log\log x)^2}{(\log x)^2}.$$

This completes the proof.   □

   Only a weak form of Brun's estimate from Theorem 2.6 is used to prove easily the famous result of Theorem 2.7.

Theorem 2.7 (Brun)   *The sum of the reciprocals of twin primes converges in* $\mathbb{R}$.

Proof　Let $(p_n)_{n \in \mathbb{N}}$ be the sequence of prime numbers where each $p_n + 2$ is also prime. By Theorem 2.6, for all $x \geq 2$ we have for $0 < \epsilon < 1$

$$\pi_2(x) \ll \frac{x}{(\log x)^{1+\epsilon}}.$$

Thus

$$n = \pi_2(p_n) \ll \frac{p_n}{(\log p_n)^{1+\epsilon}} \leq \frac{p_n}{(\log n)^{1+\epsilon}}.$$

Therefore, $n(\log n)^{1+\epsilon} \ll p_n$ for $n \geq 2$ so, using the integral test for convergence,

$$\sum_{n=1}^{\infty} \frac{1}{p_n} \leq \frac{1}{3} + \sum_{n=1}^{\infty} \frac{1}{n(\log n)^{1+\epsilon}} < \infty,$$

which completes the proof.　　　　　　　　　　　　　　　　　　□

2.4 Brun's Pure Sieve Addendum

With a simple lemma, we can also derive a lower estimate for the first term in the sum on the right from Step (2) of the proof of Theorem 2.6. Unfortunately, it is too large by a factor $(\log\log x)^2$, as estimates given later using Selberg's sieve will show.

Lemma 2.8　*As $x \to \infty$, we have for some absolute implied constant*

$$\prod_{3 \leq p \leq x} \left(1 - \frac{2}{p}\right) \gg \frac{1}{(\log x)^2}.$$

Proof　We get for $x \geq 3$,

$$\prod_{3 \leq p \leq x} \frac{\left(1 - \frac{1}{p}\right)^2}{1 - \frac{2}{p}} = \prod_{3 \leq p \leq x} \frac{1 - \frac{2}{p} + \frac{1}{p^2}}{1 - \frac{2}{p}} = \prod_{3 \leq p \leq x} \left(1 + \frac{1}{p^2(1 - 2/p)}\right)$$

$$= \prod_{3 \leq p \leq x} \left(1 + \frac{1}{p(p-2)}\right) \leq \prod_{3 \leq p \leq x} \left(1 + \frac{1}{(p-2)^2}\right)$$

$$< \prod_{n \in \mathbb{N}} \left(1 + \frac{1}{n^2}\right) = \lambda,$$

where $\lambda > 0$. Therefore, using Merten's theorem,

$$\prod_{3 \leq p \leq x} \left(1 - \frac{2}{p}\right) \gg (1/\lambda) \prod_{2 \leq p \leq x} \left(1 - \frac{1}{p}\right)^2 \gg \frac{1}{(\log x)^2}.$$

This completes the proof.　　　　　　　　　　　　　　　　　　□

Then replace x by y in Step (3) of the theorem and use the definition of y from Step (7) to obtain the lower estimate for the first term of the same order as the upper estimate.

Brun's constant is defined to be the sum of the reciprocals of prime twins, which was shown to converge by Theorem 2.7. More precisely set as has become customary

$$B := \left(\frac{1}{3} + \frac{1}{5}\right) + \left(\frac{1}{5} + \frac{1}{7}\right) + \left(\frac{1}{11} + \frac{1}{13}\right) + \left(\frac{1}{p_n} + \frac{1}{p_n + 2}\right) + \cdots,$$

where p_n is the smaller member of the nth pair of twin primes. We call B Brun's constant. It has a value

$$B = 1.902160577783278\ldots.$$

2.5 The Selberg Sieve

The Selberg sieve has many applications and often gives superior results to those of Brun's pure or combinatorial sieves, although in its simplest forms it gives only upper bounds but not lower bounds. In this section, we will first define the sieve in a particular context and derive its principal inequality, Estimate (2.2). Then we will give a classical application to bounding the number of primes p such that $p + b$ is also prime, where b is a fixed even positive integer, improving and generalizing Theorem 2.6. This presentation is based on that of Nathanson [153].

The history of the Selberg sieve is somewhat obscure, but apparently it arose as part of Selberg's work on the Riemann hypothesis in the early 1940s. In this work, he showed for example that a positive proportion of nontrivial zeros of $\zeta(s)$ was on the critical line, by introducing a so-called "mollifier" or convergence factor. This is very well described in Edwards [40, section 11.3], where an approximation to the Dirichlet series of $1/\sqrt{\zeta(s)}$ taking the form of a modified series with a finite number of nonzero terms is used.

Before he introduced the sieve, Selberg used the underlying idea in several applications, which don't appear at all to use such a device. For example (see Heath-Brown [100, section 2.1]), he studied the mean value of the integral

$$I_1(T) := \int_0^T \left|\zeta\left(\tfrac{1}{2} + it\right)\right|^2 \, dt.$$

To make this problem more tractable, he used a mollifier M_X to dampen the oscillations in zeta, considering instead the integral

$$I_2(T) := \int_0^T \left|\zeta\left(\tfrac{1}{2} + it\right) M_X\left(\tfrac{1}{2} + it\right)\right|^2 \, dt \text{ where } M_X(s) := \sum_{n \leq X} \frac{\alpha(n)}{n^s},$$

and where $\alpha(1) = 1$ and the real coefficients $\alpha(n)$ are chosen to minimize $I_2(T)$ regarded as a quadratic form in the $\alpha(n)$. The resulting values of the $\alpha(n)$ are close to

$$\mu(n)\left(1 - \frac{\log(n)}{\log(X)}\right),$$

the "damped" coefficients of $\zeta(s)^{-1}$.

The first explicit definition of the sieve apparently is Selberg's article [189]. The reader might well consult this and his other extensive writing on this subject including [189, 190, 191, 192] or/and the excellent article of Bombieri [17] which uses current notation. However note that both Bombieri and Selberg refer to the sieve "density" rather than "dimension".

Here we give the essential definitions and setup for one version of the Selberg sieve. Not all applications in later chapters use this exact same set-up, and as we will see in Section 3.4, some applications are rather implicit. Following the definitions we give a summary of the contents of the remainder of this chapter.

To begin, let $\mathscr{P}$ be an infinite set of primes and $A = (a_n)$ a nonempty finite sequence of distinct natural numbers, which may depend on variables such as a real $x > 0$. Later we will choose $\mathscr{P} = \mathbb{P}$, the set of all rational primes, but this choice is not at all essential. Let, for $z \geq 2$,

$$\mathscr{P}(z) := \prod_{p<z,\; p\in\mathscr{P}} p,$$

the product of all primes in $\mathscr{P}$ which are less than z. Define the fundamental counting function

$$S(A, \mathscr{P}, z) := \#\{n : \text{for all } p \in \mathscr{P} \text{ with } p < z,\; p \nmid a_n\} = \#\{a \in A : (a, \mathscr{P}(z))=1\},$$

which is the number of elements of A not divisible by any prime in $\mathscr{P}$ less than z. The second form follows since the elements of A are distinct. When $\mathscr{P} = \mathbb{P}$, we set $S(A, z) := S(A, \mathbb{P}, z)$.

For example, if $A = \mathbb{N} \cap [2, x]$ and $\sqrt{x} \notin \mathbb{P}$, then

$$S(A, \sqrt{x}) = \pi(x) - \pi(\sqrt{x})$$

is the sieve of Eratosthenes.

Let $q \in \mathbb{N}$ be squarefree (we will in this chapter generally restrict q and sometimes d to stand for a squarefree positive integer including 1) and

$$A_q = \{a_n \in A : q \mid a_n\}.$$

For example, if $A = \mathbb{N} \cap [1, x]$, then $|A_q| = \lfloor x/q \rfloor$. Next we define a most critical ingredient of Selberg's sieve. Let $g(n)$ be a completely multiplicative function with $0 < g(p) < 1$ for $p \in \mathscr{P}$ and $g(p) = 0$ for $p \in \mathbb{P} \setminus \mathscr{P}$. This function is fixed depending on a given application of the sieve. Define for squarefree q

$$r(q) := |A_q| - g(q)|A| \text{ and } G(z) := \sum_{\substack{m < z \\ p|m \implies p \in \mathscr{P}}} g(m),$$

so $G(z)$ is a sum of values of g over all of the positive integers less than z with all prime factors in $\mathscr{P}$. The particular definition of $g(p)$ depends on the application and is a key component of the sieve structure. It is chosen to minimize the remainder $r(p)$ and represents the long-run proportion of the elements of A which are divisible by p. For example, if $A = \mathbb{N} \cap [1, x]$, then $|A_p| = \lfloor x/p \rfloor$, so $|A_p| = x/p + r(p)$, with $|r(p)| < 1$.

The next definition sets the Selberg sieve apart from other sieve methods. Let $q \mid \mathscr{P}(z)$, so q is squarefree, and define $\lambda_1 := 1$ and $\lambda_q = 0$ for $q \in \mathbb{N}$ and $q \geq z$ or $q \nmid \mathscr{P}(z)$, so for λ_q to be nonzero we must have q squarefree. Otherwise, each λ_q is a real number.

The key to understanding the Selberg sieve is the simple inequality

$$S(A, \mathscr{P}, z) = \sum_{\substack{a \in A \\ (a, \mathscr{P}(z))=1}} 1 = \sum_{a \in A} \left(\sum_{q|(a, \mathscr{P}(z))=1} \lambda_q \right)^2$$

$$\leq \sum_{a \in A} \left(\sum_{q|(a, \mathscr{P}(z))=1} \lambda_q \right)^2 + \sum_{a \in A} \left(\sum_{q|(a, \mathscr{P}(z))\neq 1} \lambda_q \right)^2$$

$$\leq \sum_{a \in A} \left(\sum_{q|(a, \mathscr{P}(z))} \lambda_q \right)^2.$$

In Lemma 2.9, we will show, using the λ_q, how we can upper bound $S(A, \mathscr{P}, z)$ in the form $|A|Q + R$, where Q is a quadratic form. In applications of the Selberg sieve, the actual values of the λ_q are then chosen to minimize this quadratic form.

The reader might note that we can also write using these notations

$$S(A, \mathscr{P}, z) = \sum_{a \in A} \sum_{q|(a, \mathscr{P}(z))} \mu(q) = \sum_{q|\mathscr{P}(z)} \mu(q) \sum_{\substack{a \in A \\ q|a}} 1 = \sum_{q|\mathscr{P}(z)} \mu(q)|A_q|,$$

which is a representation of Brun's pure sieve.

Lemma 2.9 **(Selberg)** *With the notation defined earlier, we have*

$$S(A, \mathscr{P}, z) \leq |A|Q + R, \text{ where}$$

$$Q = \sum_{[q_1,q_2]|\mathscr{P}(z)} \frac{1}{g((q_1,q_2))} g(q_1)\lambda_{q_1} g(q_2)\lambda_{q_2}, \text{ and}$$

$$R = \sum_{[q_1,q_2]|\mathscr{P}(z)} \lambda_{q_1}\lambda_{q_2} r([q_1,q_2]).$$

Proof   We use the properties of the λ_q given before in this section, expanding the square, changing the order of summation, using the definitions of $r(q)$ and A_q, and using the complete multiplicativity of $g(n)$, to derive an upper bound:

$$S(A, \mathscr{P}, z) \leq \sum_{a \in A} \left(\sum_{q | (a, \mathscr{P}(z))} \lambda_q \right)^2$$

$$= \sum_{a \in A} \sum_{\substack{q_1 | a \\ q_1 | \mathscr{P}(z)}} \sum_{\substack{q_2 | a \\ q_2 | \mathscr{P}(z)}} \lambda_{q_1} \lambda_{q_2}$$

$$= \sum_{[q_1, q_2] | \mathscr{P}(z)} \lambda_{q_1} \lambda_{q_2} \sum_{\substack{a \in A \\ [q_1, q_2] | a}} 1$$

$$= \sum_{[q_1, q_2] | \mathscr{P}(z)} \lambda_{q_1} \lambda_{q_2} |A_{[q_1, q_2]}|$$

$$= \sum_{[q_1, q_2] | \mathscr{P}(z)} \lambda_{q_1} \lambda_{q_2} \left(g([q_1, q_2]) |A| + r([q_1, q_2]) \right)$$

$$= |A| \sum_{[q_1, q_2] | \mathscr{P}(z)} \lambda_{q_1} \lambda_{q_2} (g([q_1, q_2]) + \sum_{[q_1, q_2] | \mathscr{P}(z)} \lambda_{q_1} \lambda_{q_2} r([q_1, q_2])$$

$$= |A| Q + R,$$

where Q and R are defined in the statement of the lemma. This completes the proof. $\square$

Next we give a standard lemma for the optimization of a quadratic form subject to a single linear constraint. We need this to minimize the right-hand side of the upper bound we have derived for $S(A, \mathscr{P}, z)$. This minimization will often give explicit expressions for the free parameters λ_q.

Lemma 2.10   *Let a_i, $1 \leq i \leq n$ be strictly positive real numbers and b_i, $1 \leq i \leq n$, real numbers such that at least one is nonzero. Then the minimum value of the quadratic form*

$$Q(x_1, \ldots, x_n) = \sum_{i=1}^{n} a_i x_i^2 \quad \text{subject to} \quad \sum_{i=1}^{n} b_i x_i = 1$$

is

$$\theta = \left(\sum_{i=1}^{n} \frac{b_i^2}{a_i} \right)^{-1}$$

and is attained at the unique point in $(0, \infty)^n$ with coordinates

$$x_i = \frac{\theta b_i}{a_i}, \quad 1 \leq i \leq n.$$

Proof We use the method of Lagrange multipliers, with one multiplier λ, to optimize

$$f(x, \lambda) = \sum_{i=1}^{n} a_i x_i^2 + \lambda \left(\sum_{i=1}^{n} b_i x_i - 1 \right).$$

At an extremum $\partial f(x)/\partial x_i = 0$ for all i and $\partial f(x)/\partial \lambda = 0$. Solving the resulting equations leads directly to the solution given in the statement of the lemma, which is thus unique. $\square$

Because the LCM appears in the definitions of Q and R, we often need, for $n \in \mathbb{N}$ squarefree, to count pairs (d, e) with $[d, e] = n$.

Lemma 2.11 *If $n \in \mathbb{N}$ is squarefree, there are precisely $3^{\omega(n)}$ ordered pairs (d, e) of positive integers such that the least common multiple of d and e is n, that is to say $[d, e] = n$.*

Proof If $[d, e] = n$, then d, e are squarefree and $d \mid n$. Identify n with its set of prime divisors $T = \{p_1, \ldots, p_m\}$ say, so identifying $d \mid n$ with subsets $S \subset T$ and e with subsets of S and counting pairs we get

$$\#\{(d, e) : [d, e] = n\} = \sum_{S \subset T} 2^{|S|} = \sum_{j=0}^{m} \sum_{\substack{S \subset T \\ |S| = j}} 2^j = \sum_{j=0}^{m} \binom{m}{j} 2^j = (1 + 2)^m = 3^{\omega(n)},$$

which completes the proof. $\square$

We are now in a position to derive the main inequality of this form of the Selberg sieve. The proof is in five steps. First recall the definitions

$$S(A, \mathscr{P}, z) := \#\{a \in A : (a, \mathscr{P}(z)) = 1\} \text{ and}$$

$$G(z) := \sum_{\substack{m < z \\ p \mid m \implies p \in \mathscr{P}}} g(m).$$

Theorem 2.12 **(Selberg's sieve)** *With the notation introduced earlier, we have*

$$S(A, \mathscr{P}, z) \leq \frac{|A|}{G(z)} + \sum_{\substack{q < z^2 \\ q \mid \mathscr{P}(z)}} 3^{\omega(q)} |r(q)|. \tag{2.2}$$

Proof **(1)** In the first step, we show that $Q = 1/G(z)$ and can be expressed as a quadratic form. Let

$$\mathscr{D} := \{k \mid \mathscr{P}(z) : 1 \leq k < z\},$$

be divisors of $\mathscr{P}(z)$, so, is a set of squarefree positive integers. We have $0 < g(k) < 1$ for all $k \in \mathscr{D}$ other than 1. Define a function $f : \mathscr{D} \to \mathbb{Q}$ by

$$f(k) := \sum_{q|k} \frac{\mu(q)}{g(k/q)} = \frac{1}{g(k)} \sum_{q|k} \mu(q)g(q) = \frac{1}{g(k)} \prod_{p|k}(1 - g(p)) > 0, \quad (2.3)$$

and being a divisor sum of multiplicative functions is multiplicative on $\mathscr{D}$. We get $f(p) = 1/g(p) - 1$ so, using the multiplicativity of f and g, we get

$$\frac{1}{g(k)} = \prod_{p|k}(1 + f(p)) = \sum_{q|k} f(q).$$

This enables us to rewrite Q as follows:

$$Q = \sum_{q_1,q_2 \in \mathscr{D}} \frac{1}{g((q_1,q_2))} g(q_1)\lambda_{q_1} g(q_2)\lambda_{q_2}$$

$$= \sum_{[q_1,q_2]\in\mathscr{D}} \sum_{\substack{k|q_1 \\ k|q_2}} f(k)g(q_1)\lambda_{q_1} g(q_2)\lambda_{q_2}$$

$$= \sum_{k\in\mathscr{D}} f(k) \left(\sum_{\substack{q\in\mathscr{D} \\ k|q}} g(q)\lambda_q \right)^2.$$

Therefore, if we define $x_k := \sum_{\substack{q\in\mathscr{D} \\ k|q}} g(q)\lambda_q$, Q has been expressed as a diagonal quadratic form in the variables x_k, $k \in \mathscr{D}$, i.e.,

$$Q = \sum_{k\in\mathscr{D}} f(k)x_k^2.$$

(2) Now we express each x_k in terms of the function $f(k)$ by minimizing the quadratic form. First let $h(q) = g(q)\lambda_q$, and note that $x_k = \sum_{q\in\mathscr{D}, \, k|q} h(q)$ implies since we get $q = q'$

$$\sum_{\substack{k\in\mathscr{D} \\ q'\in\mathscr{D} \\ q|k|q'}} \mu\left(\frac{k}{q}\right) h(q') = \sum_{q'\in\mathscr{D}} h(q') \left(\sum_{k/q|q'/q} \mu\left(\frac{k}{q}\right) \right) = h(q),$$

which is a form of Möbius inversion, we can write for $q \in \mathscr{D}$

$$g(q)\lambda_q = \sum_{\substack{k\in\mathscr{D} \\ q|k}} \mu(k/q)x_k = \mu(q) \sum_{\substack{k\in\mathscr{D} \\ q|k}} \mu(k)x_k, \qquad (2.4)$$

which gives an expression for the λ_q in terms of the values x_k. In particular, if $q = 1$, we get the constraint

$$\sum_{k \in \mathscr{D}} \mu(k) x_k = 1. \tag{2.5}$$

Let $F(z) := \sum_{k \in \mathscr{D}} 1/f(k)$. Then, using Lemma 2.10, the minimum value of Q subject to (2.5) is

$$\theta = \left(\sum_{k \in \mathscr{D}} \frac{\mu(k)^2}{f(k)} \right)^{-1} = \left(\sum_{k \in \mathscr{D}} \frac{1}{f(k)} \right)^{-1} = \frac{1}{F(z)},$$

and is attained when $x_k = \mu(k)/(f(k)F(z))$.

(3) Let

$$F_q(z) := \sum_{\substack{e < z/q \\ qe | \mathscr{P}(z)}} \frac{1}{f(e)}.$$

We now use (2.4) evaluating the λ_q at the minimum derived in Step (2), and using the fact that if $qe \mid \mathscr{P}(z)$, then $(q, e) = 1$, so by (2.4):

$$\lambda_q = \frac{\mu(q)}{g(q)} \sum_{\substack{k \in \mathscr{D} \\ q|k}} \mu(k) x_k = \frac{\mu(q)}{g(q)} \sum_{\substack{qe < z \\ qe | \mathscr{P}(z)}} \mu(qe) x_{qe}$$

$$= \frac{\mu(q)}{g(q)} \sum_{\substack{e < z/q \\ qe | \mathscr{P}(z)}} \mu(qe) \frac{\mu(qe)}{f(qe)F(z)} = \frac{\mu(q)}{f(q)g(q)F(z)} \sum_{\substack{e < z/q \\ qe | \mathscr{P}(z)}} \frac{1}{f(e)}$$

$$= \frac{\mu(q) F_q(z)}{f(q) g(q) F(z)}.$$

(4) We **claim** that with this assignment we have each $|\lambda_q| \le 1$, which is very useful when it comes to bounding the error term R. This follows from the expression derived in Step (3) once we have shown $F(z) \ge F_q(z)/(f(q)g(q))$ for each $q \in \mathscr{D}$, which we now do. To obtain the inequality, we simply replace z/e by z/q to obtain a uniform generally shorter range for the inner summation to get the fourth line in the following equation. We also exploit the fact that divisors of $\mathscr{P}(z)$ are all squarefree. Let $q \mid \mathscr{P}(z)$ be fixed. Then, using Step (1) in the derivation of the final line,

$$F(z) = \sum_{k \in \mathcal{D}} \frac{1}{f(k)} \geq \sum_{e|q} \sum_{\substack{k \in \mathcal{D} \\ (k,q)=e}} \frac{1}{f(k)}$$

$$= \sum_{e|q} \sum_{\substack{em<z \\ em|\mathcal{P}(z) \\ (em,q)=e}} \frac{1}{f(em)} = \sum_{e|q} \frac{1}{f(e)} \sum_{\substack{m<z/e \\ em|\mathcal{P}(z) \\ (m,q/e)=1}} \frac{1}{f(m)}$$

$$= \sum_{e|q} \frac{1}{f(e)} \sum_{\substack{m<z/e \\ m|\mathcal{P}(z) \\ (m,q)=1}} \frac{1}{f(m)} \geq \sum_{e|q} \frac{1}{f(e)} \sum_{\substack{m<z/e \\ qm|\mathcal{P}(z)}} \frac{1}{f(m)}$$

$$\geq \sum_{e|q} \frac{1}{f(e)} \sum_{\substack{m<z/q \\ qm|\mathcal{P}(z)}} \frac{1}{f(m)} = F_q(z) \sum_{e|q} \frac{1}{f(e)}$$

$$= \frac{F_q(z)}{f(q)} \sum_{e|q} f(q/e) = \frac{F_q(z)}{f(q)g(q)},$$

and therefore

$$|\lambda_q| = \frac{F_q(z)}{f(q)g(q)F(z)} \leq 1.$$

(5) By Lemma 2.11 for squarefree q there are $3^{\omega(q)}$ ordered pairs (q_1, q_2) of positive integers such that the least common multiple $[q_1, q_2] = q$. Let $q_1, q_2 < z$ and $q_1, q_2 \mid \mathcal{P}(z)$. In particular, $[q_1, q_2] \mid \mathcal{P}(z)$ and is squarefree. Thus, using the inequality $|\lambda_q| \leq 1$ derived in Step (4), and recalling $\lambda_{qq} = 0$ when $q \geq z$ and the definition of R given in the statement of Lemma 2.9, we get

$$|R| = \left| \sum_{[q_1, q_2]|\mathcal{P}(z)} \lambda_{q_1} \lambda_{q_2} r([q_1, q_2]) \right| \leq \sum_{\substack{q_1, q_2 < z \\ q_1, q_2|\mathcal{P}(z)}} |r([q_1, q_2])| \leq \sum_{\substack{q<z^2 \\ q|\mathcal{P}(z)}} 3^{\omega(q)} |r(q)|.$$

Therefore, by Step (2) and Lemma 2.9

$$S(A, \mathcal{P}, z) \leq Q|A| + R \leq \frac{|A|}{F(z)} + \sum_{\substack{q<z^2 \\ q|\mathcal{P}(z)}} 3^{\omega(q)} |r(q)|.$$

(6) Finally we **claim** $F(z) \geq G(z) > 0$. Recall from (2.3) that with k squarefree

$$f(k) = \frac{1}{g(k)} \prod_{p|k} (1 - g(p)).$$

Then recalling $0 \le g(p) < 1$ and in the final sum in the following equation, make the choice k as the largest squarefree divisor of m to get:

$$F(z) = \sum_{k \in \mathscr{D}} \frac{1}{f(k)} = \sum_{k \in \mathscr{D}} g(k) \prod_{p|k} (1 - g(p))^{-1}$$

$$= \sum_{k \in \mathscr{D}} g(k) \prod_{p|k} \sum_{j=0}^{\infty} g(p^j) \ge \sum_{k \in \mathscr{D}} g(k) \sum_{\substack{e=1 \\ p|e \implies p|k}}^{\infty} g(e)$$

$$= \sum_{k \in \mathscr{D}} \sum_{\substack{e=1 \\ p|e \implies p|k}}^{\infty} g(ke) = \sum_{k \in \mathscr{D}} \sum_{\substack{m=1 \\ k|m \\ p|m/k \implies p|k}}^{\infty} g(m)$$

$$\ge \sum_{\substack{m<z \\ p|m \implies p \in \mathscr{P}}} g(m) \sum_{\substack{k|m \\ k \in \mathscr{D}}} 1 \ge \sum_{\substack{m<z \\ p|m \implies p \in \mathscr{P}}} g(m) = G(z).$$

Feeding this inequality back into the result of Step (5) completes the proof.   □

Next we simplify the upper bound in case we have a bound for the error $|r(q)|$.

Corollary 2.13 *Suppose that in the notation of Theorem 2.12 we have $|r(q)| \le w(q)$ for some $w(p)$ supported on squarefree integers. If we define*

$$W(z) := \prod_{\substack{p \le z \\ p \in \mathscr{P}}} \left(1 - \frac{w(p)}{p}\right) \quad \text{then} \quad \sum_{\substack{q \le z^2 \\ q | \mathscr{P}(z)}} 3^{\omega(q)} |r(q)| \le \frac{z^2}{W(z)^3}.$$

Proof This is a straightforward derivation. We begin by replacing $|r(d)|$ by $w(d)$ in the expression for the remainder term:

$$\sum_{\substack{q \le z^2 \\ q | \mathscr{P}(z)}} 3^{\omega(q)} |r(q)| \le z^2 \sum_{\substack{q | \mathscr{P}(z) \\ q \le z^2}} \frac{3^{\omega(q)} w(d)}{q}$$

$$\le z^2 \prod_{\substack{p \le z \\ p \in \mathscr{P}}} \left(1 + \frac{3w(p)}{p}\right)$$

$$\le z^2 \prod_{\substack{p \le z \\ p \in \mathscr{P}}} \left(1 + \frac{w(p)}{p}\right)^3$$

$$\le z^2 \prod_{\substack{p \le z \\ p \in \mathscr{P}}} \left(1 - \frac{w(p)}{p}\right)^{-3} = \frac{z^2}{W(z)^3}.$$ □

Lemma 2.14   *For an absolute implied constant, as $x \to \infty$ we have*

$$\sum_{n \leq x} \frac{\tau(n)}{n} = \tfrac{1}{2}(\log x)^2 + O(\log x).$$

Proof   Use Dirichlet's divisor sum [4, theorem 3.3]

$$\sum_{n \leq x} \tau(n) = x \log x + x(2\gamma - 1) + O(\sqrt{x})$$

with Abel's theorem [4, theorem 4.2].   □

Next we illustrate Selberg's sieve in the context of prime gaps by deriving an upper bound for the number of primes p up to x such that p plus a specified even natural number b is also prime, i.e., for

$$\pi_b(x) := \#\{p \leq x : p \in \mathbb{P}, p + b \in \mathbb{P}\}.$$

Note, as we have remarked before, that this bound might also be derived using Brun's (combinatorial) sieve.

Theorem 2.15   (**Selberg sieve prime fixed gap bound**) *Let b be an even positive integer. For an absolute implied constant, as $x \to \infty$ we have*

$$\pi_b(x) \ll \prod_{p \mid b} \left(1 + \frac{1}{p}\right) \frac{x}{\log^2 x}.$$

Proof   **(1)** Let $x \geq 1$ and let $A = \{a_n := n(n + b) : 1 \leq n \leq x\}$ be a finite sequence of integers. Let $2 < z \leq \sqrt{x}$ and

$$S(A, z) := |\{a_n : p \mid a_n \implies p \geq z\}.$$

If $n > \sqrt{x}$ and $p \mid a_n$ for some $p < z$, so a_n is not counted by $S(A, z)$, then either n or $n + b$ must be composite, so a_n is not counted by $\pi_b(x)$. Thus

$$\pi_b(x) \leq \sqrt{x} + S(A, z). \tag{2.6}$$

(2) We now employ the Selberg sieve with $\mathscr{P} = \mathbb{P}$ to derive an upper bound for $S(A, z)$. Let q be squarefree and write

$$q = p_1 \cdots p_k q_1 \cdots q_l,$$

where the p_i are primes dividing b and q_j primes not dividing b. Let as usual $|A_q|$ be the number of terms in the sequence A divisible by q. We **claim**

$$|A_q| = |A|g(q) + r(q),$$

where $g(n)$ is the completely multiplicative function defined on primes by $g(p) = 1/p$ if $p \mid b$ and $2/p$ if $p \nmid b$. To see this, note that elements $n(n + b)$ of A are divisible by p with $p \nmid b$, since $(n, n+b) = (n, b)$ are in one-to-one correspondence with the multiples of p in $\mathbb{N}$ together with the distinct multiples of p in $\mathbb{N} + b$, whereas when $p \mid b$ an element $n(n + b)$ is a multiple of p when and only when n is a multiple of p. Thus $g(q) = 2^l/q$. In addition, by a similar derivation to that of Lemma 2.5,

$$|r(q)| \le 2^l \le 2^{\omega(q)}. \tag{2.7}$$

By Theorem 2.12, since we have $\mathscr{P} = \mathbb{P}$ we have for

$$G(z) := \sum_{m < z} g(m),$$

the bound

$$S(A, z) \le \frac{|A|}{G(z)} + \sum_{\substack{q < z^2 \\ q \mid \mathbb{P}(z)}} 3^{\omega(q)} |r(q)|. \tag{2.8}$$

(3) Let m be an integer with factorization analogous to that of q

$$m = \prod_{i=1}^{k} p_i^{\alpha_i} \prod_{j=1}^{l} q_j^{\beta_j}$$

such that $p_i \mid b$ and $(q_j, b) = 1$. Then by complete multiplicativity

$$g(m) = \prod_{i=1}^{k} \left(\frac{1}{p_i}\right)^{\alpha_i} \prod_{j=1}^{l} \left(\frac{2}{q_j}\right)^{\beta_j} = \frac{2^{\beta_1 + \cdots + \beta_l}}{m}.$$

Let

$$\tau_b(m) := \#\{d \mid m : (d, b) = 1\}.$$

Then

$$\tau_b(m) = \prod_{j=1}^{l} (\beta_j + 1) \le \prod_{j=1}^{l} 2^{\beta_j} = 2^{\beta_1 + \cdots + \beta_l}.$$

It follows that $g(m) \ge \tau_b(m)/m$ and therefore

$$G(z) = \sum_{m < z} g(m) \ge \sum_{m < z} \frac{\tau_b(m)}{m}.$$

(4) Next we use the identity

$$\prod_{p \mid b} \left(1 - \frac{1}{p}\right)^{-1} = \sum_{\substack{n=1 \\ p \mid n \implies p \mid b}}^{\infty} \frac{1}{n}$$

and the result of Step (3), where $m < z$ means $1 \le m < z$. Interchanging the order of summation to get the fourth line in the following equation, we can derive

$$\prod_{p|b}\left(1-\frac{1}{p}\right)^{-1} G(z) \ge \sum_{m<z} \frac{\tau_b(m)}{m} \sum_{\substack{n=1 \\ p|n \implies p|b}}^{\infty} \frac{1}{n}$$

$$= \sum_{m<z} \tau_b(m) \sum_{\substack{n=1 \\ p|n \implies p|b}}^{\infty} \frac{1}{mn}$$

$$= \sum_{m<z} \tau_b(m) \sum_{\substack{u=1,\ m|u \\ p|u/m \implies p|b}}^{\infty} \frac{1}{u}$$

$$= \sum_{u=1}^{\infty} \frac{1}{u} \sum_{\substack{m<z,\ m|u \\ p|u/m \implies p|b}} \tau_b(m) \quad \text{so}$$

$$\prod_{p|b}\left(1-\frac{1}{p}\right)^{-1} G(z) \ge \sum_{u<z} \frac{1}{u} \sum_{\substack{m|u \\ p|u/m \implies p|b}} \tau_b(m).$$

(5) Now, with the same meaning for the p_i and q_j as before, let

$$u = \prod_{i=1}^{k} p_i^{\eta_i} \prod_{j=1}^{l} q_j^{\rho_j} \quad \text{and } m = \prod_{i=1}^{k} p_i^{\alpha_i} \prod_{j=1}^{l} q_j^{\beta_j}.$$

Then $m \mid u$ implies $\alpha_i \le \eta_i$ and $\beta_j \le \rho_j$ for all i, j, and we can write

$$\frac{u}{m} = \prod_{i=1}^{k} p_i^{\eta_i - \alpha_i} \prod_{j=1}^{l} q_j^{\rho_j - \beta_j}.$$

Because we require every prime $p \mid u/m$ to also divide b, it follows that none of the q_j can divide u/m, so

$$m = \prod_{i=1}^{k} p_i^{\alpha_i} \prod_{j=1}^{l} q_j^{\rho_j} \implies \tau_b(m) = \prod_{j=1}^{l} (\rho_j + 1).$$

In addition, for every fixed integer u the number of such divisors m of u is $\prod_{i=1}^{k}(\eta_i + 1)$. Thus for each u we can write

$$\sum_{\substack{m|u \\ p|u/m \implies p|b}} \tau_b(m) = \sum_{\substack{m|u \\ p|u/m \implies p|b}} \prod_{j=1}^{l}(\rho_j + 1) = \prod_{i=1}^{k}(\eta_i + 1)\prod_{j=1}^{l}(\rho_j + 1) = \tau(u).$$

(6) Next we make the choice $z = x^{1/8}$. By Lemma 2.14 and the inequality of Step (4) and equality of Step (5), we can write

$$\prod_{p|b}\left(1-\frac{1}{p}\right)^{-1} G(z) \geq \sum_{u<z} \frac{\tau(u)}{u} \gg \log^2 z \gg (\log x)^2.$$

Therefore, since $|A| = \lfloor x \rfloor$,

$$\frac{|A|}{G(z)} \ll \frac{x}{(\log x)^2} \prod_{p|b}\left(1-\frac{1}{p}\right)^{-1}$$

$$\ll \frac{x}{(\log x)^2} \prod_{p|b}\left(1-\frac{1}{p^2}\right)^{-1} \prod_{p|b}\left(1+\frac{1}{p}\right)$$

$$\ll \frac{x}{(\log x)^2} \prod_{p|b}\left(1+\frac{1}{p}\right).$$

(7) In this final step, we derive an upper bound for the remainder terms. Firstly, by Estimate (2.7) we have $|r(q)| \leq 2^{\omega(q)}$. Therefore,

$$R := \sum_{\substack{q<z^2 \\ q|\mathbb{P}(z)}} 2^{\omega(q)}3^{\omega(q)} \leq \sum_{q<z^2} 6^{\omega(q)}.$$

But $2^{\omega(q)} \leq q$ so, because $q < z^2$, we get $6^{\omega(q)} < z^{2\log 6/\log 2}$. Therefore, since we have chosen $z = x^{1/8}$, we get

$$R \leq \sum_{q<z^2} z^{2\log 6/\log 2} < z^{2+2\log 6/\log 2} < z^{7.2} = x^{9/10}.$$

Thus we get by the result of Step (6),

$$S(A,z) \ll \frac{x}{(\log x)^2} \prod_{p|b}\left(1+\frac{1}{p}\right) + x^{9/10} \ll \frac{x}{(\log x)^2} \prod_{p|b}\left(1+\frac{1}{p}\right).$$

Therefore, by (2.6)

$$\pi_b(x) \leq \sqrt{x} + S(A,z) \ll \frac{x}{(\log x)^2} \prod_{p|b}\left(1+\frac{1}{p}\right),$$

which completes the proof. □

Corollary 2.16 *As $x \to \infty$ there is an absolute implied constant such that*

$$\pi_2(x) \ll \frac{x}{(\log x)^2}.$$

2.6 Making the Constant Explicit

Theorem 2.15 gives an upper bound for π_{2n} which reveals some dependence on $n \in \mathbb{N}$ but leaves unstated an implicit constant. It is the purpose of this section to give a value to this constant, at least for N sufficiently large. This is done in the theorem of Bombieri and Davenport [10], theorem 2.24, where it is shown the constant can be made arbitrarily close to $4\mathcal{G}(\{0, 2n\}) = 8K(n)$, where K is defined in the following equation and $\mathcal{G}$ in Section 1.6. The use of a Selberg sieve is somewhat hidden – see the remark before Lemma 2.18.

We begin with some new definitions. Recall that p denotes a prime number. Let $z > 0$ be real to be assigned in terms of $N \in \mathbb{N}$ later. Then we define

$$K(n) := \prod_{p>2}\left(1 - \frac{1}{(p-1)^2}\right) \prod_{p|n,\ p>2}\left(1 + \frac{1}{p-2}\right),$$

$$\pi_{2n}(N) := \sum_{\substack{p,\,p'\leq N \\ p'-p=2n}} 1,$$

$$\eta(m) := \begin{cases} 1, & \text{when } p \mid m \implies p > z, \\ 0, & \text{there is a } p \mid m \text{ with } p \leq z, \end{cases}$$

$$\pi'_{2n}(N) := \sum_{\substack{m\leq N,\ p\leq N \\ m-p=2n}} \eta(m),$$

$$\varphi_1(m) := \sum_{d|m}\mu(d)\varphi(m/d) \implies \varphi(m) = \sum_{d|m}\varphi_1(d),$$

$$T(z,n) := \sum_{\substack{m\leq z \\ (m,2n)=1}} \frac{\mu(m)^2}{\varphi_1(m)},$$

$$\rho(m) := \begin{cases} \frac{\mu(m)}{T(z,n)\varphi_1(m)}, & \text{if } m \leq z \text{ and } (m,2n)=1, \\ 0, & \text{otherwise,} \end{cases}$$

$$\lambda(m) := \varphi(m)\sum_{v=1}^{\infty}\mu(v)\rho(mv) \implies \rho(m) = \sum_{v=1}^{\infty}\frac{\lambda(mv)}{\varphi(mv)},$$

$$\lambda_1(d) = \sum_{\substack{d_1,d_2\in\mathbb{N} \\ d=[d_1,d_2]}} |\lambda(d_1)\lambda(d_2)|.$$

The definition of π_{2n} is not identical to that of π_b used before, but is chosen to match π'_{2n}. Note also that if $p', p \leq N$ and $p' - p = 2n$, then (p', p) is counted by $\pi_{2n}(N)$. If it is not counted by $\pi'_{2n}(N)$, then we must have $\eta(p') = 0$ so $p' < z$. Therefore, $\pi_{2n}(N) \leq \pi'_{2n}(N) + z$. We will find an upper bound for π_{2n} by first upper bounding π'_{2n}.

By manipulating the Euler product representations, we have $2K(n) = \mathcal{G}(\{0, 2n\})$ for all $n \in \mathbb{N}$, a not completely trivial exercise.

We also need a form of the Bombieri–Vinogradov theorem – see Section 1.9 but also Lemmas 3.7 and 4.6.

Lemma 2.17 **(Bombieri–Vinogradov)** *Let $x \in \mathbb{R}, q, a \in \mathbb{N}$ have $(a, q) = 1$ and define*

$$E(x, q, a) := \left(\sum_{\substack{p \leq x \\ p \equiv a \bmod q}} 1 \right) - \frac{\mathrm{li}(x)}{\varphi(q)} = \pi(x, q, a) - \frac{\mathrm{li}(x)}{\varphi(q)},$$

$$E(x, q) := \max_{\substack{1 \leq a \leq q \\ (a, q) = 1}} |E(x, q, a)|.$$

Then for all $A > 0$ there is a $B > 0$, dependent only on A, such that if $Q \leq x^{\frac{1}{2}}/(\log x)^B$, we have for each $y \leq x$,

$$\sum_{q \leq Q} E(y, q) \ll_A \frac{x}{(\log x)^A}.$$

Recall we have defined

$$\eta(m) := \begin{cases} 1, & \text{when } p \mid m \implies p > z, \\ 0, & \text{there is a } p \mid m \text{ with } p \leq z. \end{cases}$$

The following lemma shows the method used in this section is a variation of the Selberg sieve. However, there is no optimization step – the definition of $\lambda(d)$ is explicit and has been given already at the start of this section. We see from the definition that it is supported on squarefree integers since that is true for $\rho(m)$.

Lemma 2.18 *For all $m \in \mathbb{N}$, we have*

$$\eta(m) \leq \left(\sum_{d \mid m} \lambda(d) \right)^2.$$

Proof First $\eta(1) = 1$. Using the given definitions of λ, ρ and T, we get $\lambda(1) = 1$ so the inequality is true. If $m > 1$, from the definition of $\eta(m)$ we need only consider m with all prime factors $p > z$. In that case, if $d \mid m$ and $d > 1$ we have $\lambda(d) = 0$, so the inequality holds in this case also. This completes the proof. □

Next we derive a bound for $\pi'_{2n}(N)$. This comes in two parts which we treat separately.

Lemma 2.19 *Let* $d = [d_1, d_2]$ *be the LCM and*

$$\mathscr{B} := \sum_{\substack{d_1, d_2 \in \mathbb{N} \\ d=[d_1,d_2]}} |\lambda(d_1)\lambda(d_2)| E(N - 2n, d),$$

where $E(x,q)$ *is defined in Lemma 2.17. Then*

$$\pi'_{2n}(N) \leq \mathrm{li}(N - 2n) \sum_{\substack{d_1, d_2 \in \mathbb{N} \\ d=[d_1,d_2]}} \frac{\lambda(d_1)\lambda(d_2)}{\varphi(d)} + \mathscr{B}.$$

Proof By the definition of π'_{2n} and the inequality of Lemma 2.18, we can write

$$\pi'_{2n}(N) = \sum_{\substack{m, p \leq N \\ m-p=2n}} \eta(m) \leq \sum_{\substack{m, p \leq N \\ m-p=2n}} \left(\sum_{d|m} \lambda(d) \right)^2 = \sum_{\substack{m, p \leq N \\ m-p=2n}} \sum_{d_1,d_2|m} \lambda(d_1)\lambda(d_2).$$

If d_1, d_2 are fixed with $(d_i, 2n) = 1$, setting $d = [d_1, d_2]$, then the conditions on p are equivalent to

$$p \leq N - 2n \text{ and } p \equiv -2n \bmod d.$$

In addition, p determines m because $2n$ is fixed. Thus by Lemma 2.17, we get

$$\pi'_{2n}(N) \leq \sum_{d_1,d_2 \in \mathbb{N}} \lambda(d_1)\lambda(d_2) \left(\frac{\mathrm{li}(N - 2n)}{\varphi(d)} + E(N - 2n, d, -2n) \right),$$

and the result of the lemma follows from the definition of $\mathscr{B}$. □

Next we see that the coefficient of the first term in the upper bound derived in Lemma 2.19 is simply the inverse of $T(z,n)$, which we estimate in Lemma 2.22.

Lemma 2.20 *We have the identity*

$$\frac{1}{T(z,n)} = \sum_{\substack{d_1, d_2 \in \mathbb{N} \\ d=[d_1,d_2]}} \frac{\lambda(d_1)\lambda(d_2)}{\varphi(d)}.$$

Proof Let $e = (d_1, d_2)$ be the GCD. Then using the definition of $\varphi(m)$, we have, as before with $d = [d_1, d_2]$

$$\varphi(d_1)\varphi(d_2) = \varphi(d)\varphi(e) \implies \sum_{\substack{d_1, d_2 \in \mathbb{N} \\ d=[d_1,d_2]}} \frac{\lambda(d_1)\lambda(d_2)}{\varphi(d)} = \sum_{d,d_2 \in \mathbb{N}} \frac{\lambda(d_1)\lambda(d_2)}{\varphi(d_1)\varphi(d_2)}\varphi(e).$$

Since $\varphi(m) = \sum_{d|m} \varphi_1(m)$, this double sum can be written

$$\sum_{d_1,d_2 \in \mathbb{N}} \frac{\lambda(d_1)\lambda(d_2)}{\varphi(d_1)\varphi(d_2)} \sum_{m|(d_1,d_2)} \varphi_1(m) = \sum_{m \in \mathbb{N}} \varphi_1(m) \left(\sum_{l \in \mathbb{N}} \frac{\lambda(ml)}{\varphi(ml)} \right)^2$$

$$= \sum_{l \in \mathbb{N}} \varphi_1(l) \rho(l)^2$$

$$= \frac{1}{T(z,n)^2} \sum_{\substack{l \leq z \\ (l,2n)=1}} \frac{\mu(l)^2}{\varphi_1(l)}$$

$$= \frac{1}{T(z,n)},$$

which completes the proof. $\qquad\Box$

We need for Step (2) of the asymptotic evaluation of $T(z,n)$ Lemma 2.22, a standard estimate for the sum of primes p dividing a natural number weighted by $(\log p)/p$.

Lemma 2.21 *For all $n \geq 3$, we have*

$$\sum_{p|n} \frac{\log p}{p} \ll \log\log n.$$

Proof Let $2 = p_1 < p_2 < p_3 < \cdots$ be the ordered sequence of primes and suppose that

$$n = p_{n_1}^{\alpha_1} p_{n_2}^{\alpha_2} \cdots p_{n_m}^{\alpha_m}$$

is the standard factorization into primes with $n_i < n_{i+1}$ for $1 \leq i < m$ so $m \ll \log n$. Then, since $p_n \sim n \log n$, there is a constant $c_1 > 0$ such that for all j we have

$$\frac{\log p_{n_j}}{p_{n_j}} \leq \frac{c_1}{n_j}.$$

Then

$$\sum_{p|n} \frac{\log p}{p} = \sum_{j=1}^{m} \frac{\log p_{n_j}}{p_{n_j}} \leq c_1 \sum_{j=1}^{m} \frac{1}{n_j} \leq c_1 \sum_{j=1}^{m} \frac{1}{j} \ll \log m \ll \log\log n.$$

This completes the proof. $\qquad\Box$

The key step of Bombieri–Davenport's proof is the derivation of the expression for $T(z,n)$. It uses the Riemann zeta function and some interesting manipulations of Dirichlet series and associated Euler products, taking four steps. This is where the coefficient $K(n)$ comes into the picture.

Lemma 2.22   **(Bombieri–Davenport)** *Uniformly in n for $n \le z^5$ and $z \le N^{\frac{1}{4}} (\log N)^{-B}$, we have the equation*

$$T(z,n) = \frac{1}{2} \frac{\log z}{K(n)} + O((\log\log n)^2).$$

Proof **(1)** Let $q \in \mathbb{N}$ be squarefree. Then using the definition of $\varphi_1(m)$, the multiplicativity of φ and $d \mid q$ implies $(d, q/d) = 1$, we get

$$\varphi_1(q) = \sum_{d \mid q} \frac{\mu(d)\varphi(q)}{\varphi(d)}$$

$$= \varphi(q) \sum_{d \mid q} \frac{\mu(d)}{\varphi(d)}$$

$$= \varphi(q) \prod_{p \mid q} \left(1 - \frac{1}{\varphi(p)} \right)$$

$$= q \prod_{p \mid q} \left(1 - \frac{2}{p} \right).$$

Therefore, since only squarefree values of m are nonzero in the sum, for $\Re s > 0$

$$\sum_{\substack{m=1 \\ (m,2n)=1}}^{\infty} \frac{\mu(m)^2}{\varphi_1(m)} \frac{1}{m^s} = \prod_{p \nmid 2n} \left(1 + \frac{1}{(p-2)p^s} \right).$$

If we define for $\Re s > 0$

$$\alpha_n(s) := \prod_{p \mid 2n} \left(1 - \frac{1}{p^{s+1}} \right), \quad \text{and}$$

$$\beta_n(s) := \prod_{p \nmid 2n} \left(\left(1 + \frac{1}{(p-2)p^s} \right) \left(1 - \frac{1}{p^{s+1}} \right) \right),$$

we get

$$\prod_{p \nmid 2n} \left(1 + \frac{1}{(p-2)p^s} \right) = \zeta(s+1)\alpha_n(s)\beta_n(s).$$

(2) Next expand α_n, β_n as Dirichlet series so we can write for some real coefficients a_n, b_n

$$\alpha_n(s) = \sum_{m=1}^{\infty} \frac{a_n(m)}{m^s} \quad \text{and} \quad \beta_n(s) = \sum_{m=1}^{\infty} \frac{b_n(m)}{m^s}.$$

This enables us to write, using the definition of $T(z,n)$ and $\sum_{m \leq x} 1/m = \log x + O(1)$,

$$T(z,n) = \sum_{\substack{m \leq z \\ (m,2n)=1}} \frac{\mu(m)^2}{\varphi_1(m)}$$

$$= \lim_{\epsilon \to 0+} \sum_{mkl \leq z} \frac{a_n(k)b_n(l)}{m^{1+\epsilon}}$$

$$= \sum_{kl \leq z} a_n(k)b_n(l) \sum_{m \leq z/(kl)} \frac{1}{m}$$

$$= (\log z) \sum_{kl \leq z} a_n(k)b_n(l) - \sum_{kl \leq z} a_n(k)b_n(l) \log(kl) + O\left(\sum_{kl \leq z} |a_n(k)b_n(l)|\right).$$

$$(2.9)$$

Expanding the product definition of $\beta_n(s)$ gives

$$\beta_n(s) = \prod_{p \nmid 2n} \left(1 + \frac{2}{p(p-2)p^s} - \frac{1}{p(p-2)p^{2s}}\right).$$

If we define a Dirichlet series $\varphi_f(s)$, and associated arithmetic function $f(m)$, by

$$\varphi_f(s) := \sum_{m \in \mathbb{N}} \frac{f(m)}{m^s} = \prod_{p>2} \left(1 + \frac{2}{p(p-2)p^s} - \frac{1}{p(p-2)p^{2s}}\right)$$

we get for all $m,n \in \mathbb{N}$, $|b_n(m)| \leq f(m)$ and the Dirichlet series and product converge absolutely for $\Re s > -\frac{1}{2}$. Therefore, since $f(m)$ does not depend on n, we have, using (2.9),

$$T(z,n) = (\log z) \sum_{k,m \in \mathbb{N}} a_n(k)b_n(m) + O\left((\log z) \sum_{km>z} |a_n(k)| f(m)\right)$$

$$+ O\left(\sum_{k,m \in \mathbb{N}} |a_n(k)| f(m) \log(km)\right).$$

(3) From the convergence of the Dirichlet series $\varphi_f(s)$, we get

$$\sum_{m=1}^{\infty} f(m) = O(1), \quad \sum_{m=1}^{\infty} f(m) \log m = O(1).$$

In addition, we have

$$\sum_{k=1}^{\infty} \frac{|a_n(k)|}{k^s} = \prod_{p|2n}\left(1 + \frac{1}{p^{s+1}}\right) \implies \sum_{k=1}^{\infty} |a_n(k)| = \prod_{p|2n}\left(1 + \frac{1}{p}\right) = O(\log\log n).$$

Logarithmic differentiation of this Dirichlet series, evaluated at $s = 0$, together with Lemma 2.21, which gives

$$\sum_{p|n} \frac{\log p}{p} \ll \log\log n,$$

implies

$$\sum_{k=1}^{\infty} |a_n(k)| \log k = -\prod_{p|2n}\left(1 + \frac{1}{p}\right)\sum_{p|2n} \frac{\log p}{p+1} = O((\log\log n)^2).$$

Using the definitions of $\alpha_n(s)$ and $\beta_n(s)$ given in Step (1), we have by the definition of $K(n)$ and some rather miraculous Euler product manipulation

$$\sum_{k,m=1}^{\infty} a_n(k)b_n(m) = \alpha_n(0)\beta_n(0)$$

$$= \prod_{p|2n}\left(1 - \frac{1}{p}\right)\prod_{p\nmid 2n}\left(\left(1 + \frac{1}{p-2}\right)\left(1 - \frac{1}{p}\right)\right)$$

$$= \frac{1}{2K(n)}.$$

Substituting these evaluations in the result of Step (2) gives

$$T(z,n) = \frac{\log z}{2K(n)} + O((\log\log n)^2) + O\left((\log z)\sum_{kl>z} |a_n(k)| f(l)\right).$$

(4) Finally we estimate the last term in the expression for $T(z,n)$ derived in Step (3). Since the Dirichlet series $\varphi_f(s)$ converges absolutely for $\Re s > -\frac{1}{2}$, we have

$$\sum_{l=1}^{\infty} f(l) l^{1/3} = O(1),$$

which, noting the Euler product for $\sum_k |a_n(k)|/k^s$ is finite, implies

$$\sum_{kl>z} |a_n(k)| f(l) = \sum_{k=1}^{\infty} |a_n(k)| \sum_{l>z/k} f(l)$$

$$\leq \sum_{k=1}^{\infty} |a_n(k)| \frac{k^{1/3}}{z^{1/3}} \sum_{l>z/k} f(l) l^{1/3}$$

$$\ll \frac{1}{z^{1/3}} \sum_{k=1}^{\infty} |a_n(k)| k^{1/3}$$

$$\ll \frac{1}{z^{1/3}} \prod_{p|2n} \left(1 + \frac{1}{p^{2/3}} \right)$$

$$\ll \frac{\tau(2n)}{z^{1/3}}.$$

Hence, since we have $\log z \ll \log N$ and $n \leq z^5$

$$(\log z) \sum_{kl>z} |a_n(k)| f(l) \ll \frac{(\log z)\tau(2n)}{z^{1/3}} \ll 1.$$

This completes the proof of the lemma. $\qquad\square$

Next we show that we can get a good upper bound for $\mathscr{B}$ provided z is chosen sufficiently small. First, recall the definitions

$$E(x,q) := \max_{(a,q)=1} |E(x,q,a)|,$$

$$\mathscr{B} := \sum_{\substack{d_1,d_2\in\mathbb{N} \\ d=[d_1,d_2]}} |\lambda(d_1)\lambda(d_2)| E(N-2n,d),$$

$$\lambda_1(d) := \sum_{\substack{d_1,d_2\in\mathbb{N} \\ [d_1,d_2]=d}} |\lambda(d_1)\lambda(d_2)|.$$

Lemma 2.23 *We have for $N \to \infty$,*

$$z \leq \frac{N^{\frac{1}{4}}}{(\log N)^B} \implies \mathscr{B} \ll \frac{N}{(\log N)^{A/2-5/2}}.$$

Proof **(1)** The definitions of $\mathscr{B}$ and $\lambda_1(d)$ imply

$$\mathscr{B} = \sum_{d\in\mathbb{N}} \lambda_1(d) E(N-2n,d).$$

Note, from the definition of $\rho(m)$ at the start of this section, that $\lambda_1(d) = 0$ when $d > z^2$ or if $(d, 2n) \neq 1$. Using the Cauchy–Schwarz inequality, we have

$$\mathcal{B} \leq \left(\sum_{d \leq z^2} \frac{\lambda_1(d)^2}{\varphi(d)} \right)^{\frac{1}{2}} \left(\sum_{d \leq z^2} \varphi(d) E(N - 2n, d)^2 \right)^{\frac{1}{2}}.$$

Next, counting all integers coprime to d rather than just primes, we have $\varphi(d) E(N - 2n, d) \ll N$, so using Bombieri–Vinogradov's Lemma 2.17 we get, since we are assuming $z^2 \leq N^{\frac{1}{2}}/(\log N)^{2B}$,

$$\sum_{d \leq z^2} \varphi(d) E(N - 2n, d)^2 \ll \frac{N^2}{\log^A N}.$$

(2) In this step, we estimate $\sum_{d \in \mathbb{N}} \lambda_1(d)^2 / \varphi(d)$. Using the definitions of ρ and λ, we get

$$\lambda(m) = \varphi(m) \frac{1}{T(z, n)} \sum_{\substack{l \leq z/m \\ (lm, 2n)=1}} \mu(l) \frac{\mu(lm)}{\varphi_1(lm)}$$

$$\implies |\lambda(m)| \leq \varphi(m) \frac{1}{T(z, n)} \sum_{\substack{l \leq z/m \\ (lm, 2n)=1}} \frac{\mu(lm)^2}{\varphi_1(lm)}.$$

If $\mu(lm)^2 = 1$, then $(l, m) = 1$ so, since φ_1 is multiplicative, $\varphi_1(lm) = \varphi_1(l)\varphi_1(m)$. Therefore,

$$|\lambda(m)| \leq \varphi(m) \frac{1}{T(z, n)} \frac{\varphi(m)\mu(m)^2}{\varphi_1(m)} \sum_{\substack{l \leq z \\ (l, 2n)=1}} \frac{\mu(l)^2}{\varphi_1(l)} = \frac{\varphi(m)\mu(m)^2}{\varphi_1(m)},$$

and so, using the first part of Step (1) of Lemma 2.22,

$$|\lambda(m)| \leq \prod_{\substack{p \mid m \\ p > 2}} \frac{p - 1}{p - 2} \ll \log\log m.$$

Using the definition

$$\lambda_1(m) := \sum_{\substack{d_1, d_2 \in \mathbb{N} \\ [d_1, d_2]=m}} |\lambda(d_1)\lambda(d_2)|,$$

we get

$$\lambda_1(m) \leq \left(\sum_{d \mid m} |\lambda(d)| \right)^2 \ll \tau(m)^2 (\log\log m)^2.$$

Therefore, by the definition of $T(z, n)$, since by Merten's estimate we have $m/\varphi(m) \ll \log z$,

$$\sum_{m \in \mathbb{N}} \frac{\lambda_1(m)^2}{\varphi(m)} \ll \sum_{m \leq z^2} \frac{\tau(m)^4 (\log\log m)^4}{\varphi(m)} \ll (\log z)^5.$$

(3) Combining the result of Step (1) with that of Step (2) and the given upper bound for z gives

$$\mathcal{B} \ll \left(\log^5 z\right)^{\frac{1}{2}} \left(N^2 (\log N)^{-A}\right)^{\frac{1}{2}} \ll \frac{N}{(\log N)^{A/2 - 5/2}}.$$

This completes the proof of the lemma. $\qquad\qquad\qquad\qquad\qquad\qquad\square$

Recall the definitions

$$\pi_{2n}(N) := \sum_{\substack{p, p' \leq N \\ p' - p = 2n}} 1,$$

$$\pi'_{2n}(N) := \sum_{\substack{m \leq N, \ p \leq N \\ m - p = 2n}} \eta(m),$$

$$K(n) := \prod_{p > 2} \left(1 - \frac{1}{(p-1)^2}\right) \prod_{p \mid n, \ p > 2} \left(1 + \frac{1}{p-2}\right).$$

We are now able to give the final derivation of Bombieri and Davenport's estimate based on lemmas proved using Selberg sieve style techniques. Note that this is superior to the bounds which can be obtained using either Brun's pure sieve or Brun's (combinatorial) sieve with their implicit constants, or the bound we have derived using Selberg's sieve Theorem 2.15.

Theorem 2.24 **(Bombieri–Davenport's prime fixed gap constant)** *Given $\epsilon > 0$ for all N sufficiently large, we have*

$$\pi_{2n}(N) < (8 + \epsilon) K(n) \frac{N}{(\log N)^2}.$$

Proof If $2n \geq N$, then $\pi'_{2n}(N) = 0$, so we will assume $2n < N$. Let

$$z = N^{\frac{1}{4}}/(\log N)^B \implies n \leq z^5,$$

and use Lemmas 2.19, 2.20, 2.22 and 2.23 to write for a constant $c_1 > 0$, choosing $A = 11$, so that for N sufficiently large

$$\pi'_{2n}(N) \leq \frac{N}{\log N} \frac{(2+\epsilon) K(n)}{\log z} + \frac{c_1 N}{(\log N)^{A/2 - 5/2}} \leq (8 + 5\epsilon) K(n) \frac{N}{\log^2 N}.$$

Because $\pi_{2n}(N) \leq \pi'_{2n}(N) + z$, this gives for N sufficiently large

$$\pi_{2n}(N) < (8 + 6\epsilon)K(n)\frac{N}{(\log N)^2},$$

which completes the proof.   □

2.7 An Application to a Brun–Titchmarsh Inequality

In Appendix D, we give the proof of a type of Brun–Titchmarsh inequality due to
Shiu, which we need in Chapter 8. This relies on a Selberg sieve estimate which we
prove in this section. We use the Selberg sieve form set out in Section 2.5.

Theorem 2.25   *Let l and a be fixed integers with $l \geq 1$ and $(a, l) = 1$. Let the
symbol q, with or without a subscript, represent squarefree integers, and, using the
notation of Lemma 2.9, let*

$$\mathscr{P} := \{p \in \mathbb{P} : (p, l) = 1\},$$
$$A = \{n \geq 0 : a + nl \in (x - y, x]\},$$
$$\mathscr{P}(z) := \prod_{\substack{p < z \\ p \in \mathscr{P}}} p,$$
$$S(A, \mathscr{P}, z) := \sum_{\substack{a \in A \\ (a, \mathscr{P}(z)) = 1}} 1,$$
$$|A_p| = g(p)|A| + r(p), \text{ with } 0 < g(p) < 1,$$
$$R = \sum_{q_1, q_2 | \mathscr{P}(z)} \lambda_{q_1} \lambda_{q_2} r([q_1, q_2]),$$
$$G(z) := \sum_{\substack{q < z \\ p | q \implies p \in \mathscr{P}}} g(q),$$

*where $g(p)$ is defined in Step (1) of the following proof. Then for all $z \geq 2$ and
$0 < y \leq x$, we have*

$$S(A, \mathscr{P}, z) \leq \frac{y}{\varphi(l) \log z} + z^2.$$

Note that A is the arithmetic progression $\{a, a + l, a + 2l, \ldots\} \cap (x - y, x]$ and
$\mathscr{P}$ the set of primes which do not divide the modulus l.

Proof **(1)** In this application of Selberg's sieve (Theorem 2.12), we define the two
basic multiplicative functions defined on primes in $\mathscr{P}$ and supported on squarefree
integers:

$w(p) := 1$ for $p \nmid l$ else 0 for $p \mid l$,

$$g(p) := \frac{w(p)}{p - w(p)} \text{ for } p \in \mathscr{P} \text{ so } g(q) = 1/\varphi(q) \text{ for } (q,l) = 1, \text{ else } 0 \text{ for } (q,l) \neq 1,$$

which is as expected the asymptotic average proportion of elements of A divisible by q. Then, by Lemma 2.4, we can write for $(q,l) = 1$, $|A_q| = (w(q)/q)x + r(q)$ and in this situation $|r(q)| \leq 1$.

(2) Using these assignments, we have

$$G(z) = G_l(z) := \sum_{\substack{q < z \\ (q,l)=1}} \frac{\mu(q)^2}{\varphi(q)}.$$

Then, recalling q is squarefree and $G_l(z)$ is nondecreasing in z, we get

$$G_1(z) \leq \sum_{m|l} \sum_{\substack{q < z \\ (q,l)=m}} \frac{\mu(q)^2}{\varphi(q)}$$

$$= \sum_{m|l} \frac{\mu(m)^2}{\varphi(m)} \sum_{\substack{n \leq z/m \\ (n,l)=1}} \frac{\mu(n)^2}{\varphi(n)}$$

$$= \sum_{m|l} \frac{\mu(m)^2}{\varphi(m)} G_l(z/m)$$

$$\leq G_l(z) \sum_{m|l} \frac{\mu(m)^2}{\varphi(m)}.$$

Therefore,

$$G(z) = G_l(z) \geq \frac{G_1(z)}{\sum_{m|l} \frac{\mu(m)^2}{\varphi(m)}} = \frac{G_1(z)}{\prod_{p|l} \left(1 + \frac{1}{p-1}\right)} = G_1(z) \prod_{p|l} \left(1 - \frac{1}{p}\right).$$

(3) We next get a lower bound for $G_1(z)$. Since we can write any positive integer $n < z$ in the form $n = qe$ where q is squarefree, $q < z$ and e a product of primes each of which divides q, we can write

$$G_1(z) = \sum_{q \leq z} \frac{\mu(q)^2}{\varphi(q)} = \sum_{q \leq z} \frac{\mu(q)^2}{q} \prod_{p|q} \left(1 + \frac{1}{p} + \frac{1}{p^2} + \cdots\right) \geq \sum_{n \leq z} \frac{1}{n} \geq \log z.$$

Replacing $G_1(z)$ in the lower bound we derived in Step (2) gives

$$G(z) \geq \frac{\varphi(l)}{l} \log z.$$

(4) Finally, by Theorem 2.12 Step (4) $|\lambda_q| \leq 1$, by Step (1) $|r(q)| \leq 1$, the bound of Step (3), Lemma 2.9, and in addition $|A_l| \leq y$, we get

$$S(A, \mathcal{P}, z) \leq \frac{|A|}{G(z)} + \sum_{\substack{[q_1,q_2]|\mathcal{P}(z) \\ q_1,q_2 \leq z}} 1 \leq \frac{y}{\varphi(l) \log z} + z^2,$$

which completes the proof.   □

If a prime p is in $(x - y, x]$ and is not counted by $S(A, \mathcal{P}, z)$, then we must have $p < z$. Therefore:

Corollary 2.26 **(A weak Brun–Titchmarsh estimate)** *For all $z \geq 2$, $l \in \mathbb{N}$, $(a, l) = 1$ and $0 < y \leq x$, we have*

$$\pi(x, l, a) - \pi(x - y, l, a) \leq \frac{y}{\varphi(l) \log z} + z^2 + \pi(z).$$

Montgomery and Vaughan [143] improved the weak estimate to get

$$\pi(x, l, a) \leq \frac{2x}{\varphi(l) \log(x/l)},$$

which like the weak estimate has the advantage of being valid for all $l < x$. Motohashi [147] discovered a bilinear structure in the Selberg sieve error to derive an improved form of the inequality valid for a restricted range of moduli.

2.8 Brun's, Selberg's and Other Sieves

We have given just two examples, Brun's pure sieve and Selberg's sieve. It would be misleading not to point out that sieve theory has flourished and sieves have become many and various in recent times, mostly tailored to solve a particular range of problems in number theory. This is true in the case of the prime gaps breakthroughs. Future history might tell that this is the early days. That sieve theory, analogous to the beginnings of subjects like topology, has an abstract formulation which has yet to be written down and agreed on. Mostly one has the feeling that it should be a general mathematical tool rather than restricted to number theory. But what we now have is a range of important examples and applications, with technicalities and assumed contexts which could be bewildering, especially for the beginner.

In this section, we give a very brief overview of some of the better-known sieves, leading up to that of Maynard and Tao. There are a small number of key references, but the reader is alerted to the following section, which is a readers' guide, listing some texts which could be easier to access.

(1) The **sieve of Eratosthenes**, dates back to ancient times. Take the set of primes up to a given $z > 2$ and cast out each multiple of each prime in the set from the interval $[z, z^2) \cap \mathbb{N}$, leaving just the subset of primes in that interval. This sieve remains useful, for example for generating large tables of primes inductively.

(2) Legendre's sieve [125] systemizes the inclusion–exclusion underlying the sieve of Eratosthenes. If we let A be the set of all integers in $[1, x]$, $P(z) = \prod_{p<z} p$ and denote by $S(\mathcal{A}, z)$ the number of integers left after all multiples of all primes less than z have be cast out, i.e.,

$$S(A, z) := |\{n \in A : 1 \le n \le x, (n, P(z)) = 1\}|,$$

then we can write as in Section 2.5

$$S(A, z) := \sum_{d | P(z)} \mu(d) \left\lfloor \frac{x}{d} \right\rfloor.$$

This gives an exact formula for the number of natural numbers up to x not divisible by any prime less than or equal to z. Taking $z = \sqrt{x}$ and assuming it is not prime, the sieve of Eratosthenes then has the form

$$\#\mathcal{P} \cap (\sqrt{x}, x] = \pi(x) - \pi(\sqrt{x}) = \sum_{1 \le n \le x} \delta((n, P(\sqrt{x}))),$$

where $\delta(1) = 1$ and $\delta(n) = 0$ for $n > 1$ where (a, b) is the GCD.

Because each approximation of the integer part function has an error r with $0 \le r < 1 = O(1)$, the error in estimations of $S(A, z)$ is exponential in z, and the exact formula is useful only for z up to $\log x$, i.e., in a very small range.

(3) Brun's pure sieve [23] was the first major attempt to extend the range of z, replacing the exact formula with truncated exclusion–inclusion-based sums of an at most polynomial number of terms. See Sections 2.3 and 2.4. Many of the ideas inherent in Brun's approach have found their way into modern sieve theory. For example, since $\mu \star \mathbf{1} = \delta$, we can describe Brun's sieve as the introduction of two arithmetic functions μ_1 and μ_2 which vanish sufficiently often to significantly reduce the number of terms which are nonzero, and such that

$$\mu_1 \star \mathbf{1} \le \delta \le \mu_2 \star \mathbf{1}.$$

These inequalities can be used to obtain upper and lower bounds for the quantity being sieved. For example, if $\mathcal{A}$ is a finite set of integers and $\mathcal{P}$ a set of primes (not necessarily all primes), and we let

$$A_q := \{a \in \mathcal{A} : q \mid a\},$$

$$\mathcal{P}(z) := \prod_{\substack{p \in \mathcal{P} \\ p < z}} p,$$

$$S(A, \mathcal{P}, z) := |\{a \in A : (a, \mathcal{P}(z)) = 1\}|,$$

then for each $m \geq 0$, we have

$$\sum_{\substack{q \mid \mathcal{P}(z) \\ \omega(q) \leq 2m+1}} \mu(q)|A_q| \leq S(\mathcal{A}, \mathcal{P}, z) \leq \sum_{\substack{q \mid \mathcal{P}(z) \\ \omega(q) \leq 2m}} \mu(d)|A_q|.$$

To make the errors in the sums on the left and right tractable (because each term will normally need to be approximated with an error), we must limit the number of terms, but now we can limit m as well as z, making the method more flexible.

As we have seen in Theorem 2.6, one application of the sieve was to find an upper estimate for the number of prime pairs $(p, p+2)$ for fixed $n \in \mathbb{N}$, up to x, namely

$$\pi_2(x) := \#\{p \leq x : p+2 \in \mathbb{P}\} \ll \frac{x(\log\log x)^2}{(\log x)^2}.$$

(4) Brun's sieve [24], sometimes referred to as **Brun's combinatorial sieve**, included a more complex approach to truncating sums than Brun's pure sieve. This was improved and further developed by Rosser and Iwaniec, but we will describe only the original form here, with a brief description of Rosser's variation at the end of this part. For complete details, see [85, section 3.4]. As for Selberg's sieve, there are quite a few parameters to be described.

For each integer $r \in \mathbb{N}$, a positive constant B_r is defined so that the logarithms are in geometric progression. In particular, if the "level" of the sieve is D and "range" $z > 2$ satisfies $z^s = D$, then Brun set

$$B_n := z^{\exp(1-n/\beta)} \text{ with } 2 \leq \beta \leq s/e.$$

(In the important case of the linear sieve with dimension or density $\kappa = 1$, the parameter $\beta = 2$.) Then the functions which appear in approximations to $\mu(n)$, when $q = p_r \cdots p_1$ with $p_1 > \cdots > p_r$, are defined by

$$\chi^-(q) = 1, \quad p_{2i} < B_{2i}, \ 2 \leq 2i \leq r, \text{ and}$$
$$\chi^+(q) = 1, \quad p_{2i+1} < B_{2i+1}, \ 1 \leq 2i+1 \leq r.$$

If the constraints are not satisfied for $\chi^\pm$, then the function is set to zero. Next, as is common for this type of sieve, we take $\mathcal{P} \subset \mathbb{P}$ an infinite set of primes, and $\mathcal{P}(z)$ the product of all of the primes from $\mathcal{P}$ up to *and including* the sieve range z. It can then be shown that if $q \mid \mathcal{P}(z)$ and $\chi^\pm(q) \neq 0$, then $q \leq D$. Then $0 \leq \chi^\pm(q) \leq 1$ and we define "sieve lambdas" by $\lambda^\pm(q) := \mu(q) \cdot \chi^\pm(q)$. In this case in the notation of Step (3), we choose $\mu_1 = \lambda^-$ and $\mu_2 = \lambda^+$, so

$$\sum_{q \mid n} \lambda^-(q) \leq \sum_{q \mid n} \mu(q) \leq \sum_{d \mid n} \lambda^+(q).$$

Nonzero values of the arithmetic functions occur only when the argument q is squarefree.

Recall some further standard notations for this sieve. Let $X = |A|$,

$$|A_q| = X \frac{\rho(q)}{q} + r_A(q) \text{ where } A_q = \{a \in A : q \mid a\},$$

with $\rho(q)$ multiplicative such that $0 \leq \rho(p) < p$, defining the error r_A. Then as we have seen,

$$S(A, \mathscr{P}, z) = \sum_{q \mid \mathscr{P}(z)} \mu(q) |A_q|.$$

Furthermore, for $2 \leq w < z$,

$$V(\mathscr{P}(z)) := \prod_{p \mid \mathscr{P}(z)} \left(1 - \frac{\rho(p)}{p}\right) = \sum_{d \mid \mathscr{P}(z)} \frac{\mu(d)\rho(d)}{d} \text{ so,}$$

$$\frac{V(\mathscr{P}(w))}{V(\mathscr{P}(z))} = \prod_{p \mid \mathscr{P}(z)/\mathscr{P}(w)} \left(1 - \frac{\rho(p)}{p}\right)^{-1} \leq \left(\frac{\log z}{\log w}\right)^{\kappa} \left(1 + \frac{L}{\log w}\right) \leq K \left(\frac{\log z}{\log w}\right)^{\kappa},$$

for constants L and K, in this manner defining the sieve dimension κ. If for a particular ρ the constants L, K, κ don't all exist, then the method does not apply to the particular problem under consideration.

We also need to define

$$V^{\pm}(\mathscr{P}(z)) := \sum_{q \mid \mathscr{P}(z)} \frac{\mu(q) \chi^{\pm}(q) \rho(q)}{q}, \text{ and}$$

$$R^{\pm}(\mathscr{P}(z)) := \sum_{q \mid \mathscr{P}(z)} \mu(d) \chi^{\pm}(q) r_A(q).$$

In this setting, one then derives a "squeezing" for the main sum

$$X V^{-}(\mathscr{P}(z)) + R^{-}(\mathscr{P}(z)) \leq S(A, \mathscr{P}(z)) \leq X V^{+}(\mathscr{P}(z)) + R^{+}(\mathscr{P}(z)),$$

and then, for $s \geq e\beta$ and $z = D^{1/s}$, the estimate

$$S(A, \mathscr{P}, z) = X V(\mathscr{P}(z)) \left(1 + \theta_1 \frac{a^{\beta}}{1 - a} \frac{K^{1-b}}{e^{\kappa+\beta+1}}\right) + \theta_2 \sum_{q \leq D} |r_A(q)|,$$

where $|\theta_1|, |\theta_2| \leq 1$, the parameters a, b satisfy $\beta = b\kappa$, $a = e^{1+1/b}/b$, and $b > c$, where c is the solution to $c = e^{1+1/c}$ and necessarily $0 < a < 1$.

Brun's applications of the sieve justified its complexity. They included showing that every interval $(n, n + \sqrt{n})$, with n sufficiently large, contained a P_{11} almost prime; that every sufficiently large even integer could be expressed as the sum of two P_9 almost primes; and that there were an infinite number of $n \in \mathbb{N}$ such that both n and $n + 2$ were P_9's.

Brun, as we have described earlier, made a particular choice of the parameters B_r upper bounding the primes p_r. This was investigated by Rosser, who chose for $\beta \geq 1$ the alternative

$$p_r \leq \left(\frac{D}{p_1 \cdots p_r} \right)^{1/\beta} =: B_r, \; r \in \mathbb{N}.$$

This choice gave an improved estimate for $S(A, \mathscr{P}, z)$. For example, the sieve could be used to show that there are an infinite set of n's such that n and $n + 2$ are P_8's. See [85, section 3.3] for details. Experts are needed to shed light on the deeper meaning of these choices for the B_r, and one is left feeling that the last word has yet to be voiced. Special cases, such as $\kappa = 1$, have been especially influential for theoretical reasons – see Step (5).

(4) Selberg's upper bound sieve [190, 191, 192], or the Λ^2-sieve, also is intrinsically derived from inclusion–exclusion. It replaces the Möbius function values $\mu(n)$ with weights $\lambda(n)$, chosen to relate to a given problem under study and then optimized to provide a valid upper bound for the number of elements in the set being sifted. A typical setup and example application are given in Section 2.5. In both Brun's and Selberg's methods, only a small number of residue classes are removed from the sifted set. Thus they are sometimes called **small sieves**.

Selberg's sieve has a host of applications. Here we give three:

The Brun–Titchmarsh theorem, as proved by Montgomery and Vaughan, [143], provides an explicit upper bound for primes in arithmetic progression without any additional hypothesis and for a wide range of parameters. For all positive integers $l < x$ with $(l, a) = 1$, we have

$$\pi(x, l, a) \leq \frac{2x}{\varphi(l) \log(x/l)}.$$

The upper bound was improved for a restricted range $l \leq x^{9/20}$ by Motohashi [147], who obtained

$$\pi(x, l, a) \leq \frac{(2 + o(1))x}{\varphi(l) \log(x/l^{3/8})}.$$

For the twin primes conjecture, Selberg's sieve enables us to derive an upper bound which is of the expected order of magnitude:

$$\#\{p \leq x : p + 2 \in \mathbb{P}\} \ll \frac{x}{(\log x)^2}.$$

Regarding Goldbach's conjecture, that every even integer can be expressed as the sum of two primes, we get

$$\#\{(p,q) \in \mathbb{P}^2 : 2n = p + q\} \ll \frac{n}{(\log n)^2} \prod_{p|n}\left(1 + \frac{2}{p}\right).$$

(5) The **linear sieve** of Jurkat and Richert [118] will take a bit to explain. We will give an overview of the form described in detail by Nathanson [153, chapter 9], which was based on notes of Rutgers lectures of Iwaniec. It is a special type of β-sieve, namely that with dimension $\kappa = 1$ and $\beta = 2$. The dimension will be defined later. Suffice it to say the β-sieve provides a way of reducing the sizes of the sums used in the bounds for $S(A, P, z)$ using a tuning parameter $\beta > 1$ based on the parity and size of prime products which appear in the divisor sums. See Step (4). We need to define a few terms:

Let $A = (a_n)_{n \in \mathbb{N}}$ be a sequence of distinct nonnegative real numbers (weights) based on the problem of interest and such that

$$|A| := \sum_{n=1}^{\infty} a_n < \infty.$$

Normally one would have elements/values of A nonzero on a finite set of integers in an interval such as $[1, x]$, where $x > 1$ will tend to infinity. The other main parameters, such as z or D, would normally be (fractional) powers of x, so the whole problem is x dependent. In the statement of the main estimates for the linear sieve, this dependence is often suppressed, but in applications it becomes of vital importance.

Let $\mathcal{P}$ be a subset of primes, D the **sieving range**, $z \geq 2$ the **sieving limit**,

$$\mathcal{P}(z) := \prod_{\substack{p \in \mathcal{P} \\ p < z}} p.$$

Note that the product here does not include z, but this is a minor matter.

Let the **sieving function** S be defined by

$$S(A, \mathcal{P}, z) := \sum_{\substack{n \in \mathbb{N} \\ (n, \mathcal{P}(z))=1}} a_n,$$

which is a weighted sum of values of the sequence indexed up to x such that the **index** is not divisible by any prime in $\mathcal{P}$, which is less than z. The theorem of Jurkat and Richert provides upper and lower bounds for the sieving function. We need to define the terms which are used in the theorem statement. Let, for $q \in \mathbb{N}$,

$$\mathscr{A}_q = \{a_n : n \leq x, q \mid n\} \text{ and } |A_q| = \sum_{\substack{n \leq x \\ q|n}} a_n.$$

Let $g: \mathbb{N} \to \mathbb{R}$ be a multiplicative function, supported on squarefree products of primes in $\mathscr{P}$, such that $0 \leq g(p) < 1$ for all $p \in \mathscr{P}$. Let $\mathscr{Q} \subset \mathscr{P}$ be a finite subset and denote by Q the product of all primes in $\mathscr{Q}$. The function g is required to satisfy the following **linear sieve inequality**:

$$\prod_{\substack{p \in \mathscr{P} \backslash \mathscr{Q} \\ w \leq p < z}} \frac{1}{1 - g(p)} < (1 + \epsilon) \frac{\log z}{\log w}$$

for some fixed $\epsilon > 0$ in $1 < \epsilon < 1/200$ for all $1 < w < z$. A density or dimension parameter κ is used to introduce flexibility into this inequality replacing the right-hand-side ratio by $(\log z / \log w)^\kappa$.

Define the remainder term r_q by

$$|A_q| = \left(\sum_{n \in \mathbb{N}} a_n \right) g(q) + r_q \text{ and } R := \sum_{\substack{q | \mathscr{P}(z) \\ q < DQ}} |r_q|.$$

Also let

$$X := |\mathscr{A}| \prod_{p | \mathscr{P}(z)} (1 - g(p)).$$

We now define two continuous and nonnegative functions which play a central role in the development. Let $s = \log D / \log z$, and let $f(s)$ and $F(s)$ be defined by convergent series

$$f(s) := 1 - \sum_{n \in \mathbb{N} \text{ odd}} f_n(s),$$

$$F(s) := 1 + \sum_{n \in \mathbb{N} \text{ even}} f_n(s)$$

respectively, where the functions $f_n(s)$ are defined by

$$s f_n(s) := \int_{R_n(s)} \frac{dx_1 \cdots dx_n}{t_1 \cdots t_{n-1} t_n^2},$$

where $R_n(s)$ is an open convex subset of $\mathbb{R}^n$ consisting of points $(x_1, \ldots, x_n)$, which satisfy the inequalities

$$0 < x_n < \cdots < x_1 < \frac{1}{s},$$

$$x_1 + \cdots + x_n + 2x_n > 1,$$

$$x_1 + \cdots + x_m + 2x_m < 1, \ m < n \text{ and } m \equiv n \bmod 2.$$

(The parameter β of (4) introduces more flexibility replacing the $2x_n$ by βx_n and $2x_m$ by βx_m.) Then the theorem of Jurkat and Richert [152, theorem 9.7] gives for $D \geq z$ (i.e., $s \geq 1$) an upper bound for the sieving function

$$S(A, \mathcal{P}, z) < (F(s) + \epsilon e^{14-s})X + R,$$

and for $D \geq z^2$ (i.e., $s \geq 2$) the lower bound

$$(f(s) - \epsilon e^{14-s})X - R < S(A, \mathcal{P}, z),$$

where the definitions of X and R for this sieve are given earlier. We give just two applications of the linear sieve:

Chen announced his famous theorem $2n = p + P_2$, in 1966, and the details of a proof are given in [29, chapter 20, theorem I]. Chen included a proof that $p + 2 = P_2$ infinitely often using his implementation of the linear sieve described earlier. Let $h \in \mathbb{N}$ be even and $x_h(1, 2)$ represent the number of primes $p \leq x$ such that there exist primes p_1, p_2 such that $p + h = p_1$ or $p + h = p_1 p_2$. Then [29, chapter 20, theorem II]

$$x_h(1, 2) \geq 0.67 C_h \frac{x}{(\log x)^2} \text{ where } C_h := \prod_{\substack{p|h \\ p>2}} \frac{p-1}{p-2} \prod_{p>2} \left(1 - \frac{1}{(p-1)^2}\right).$$

Heath-Brown [98] in 1986 used the linear sieve and a modified form of the Bombieri–Vinogradov theorem, as well as ideas of Murty and Ram Murty and others, to show that with the possible exception of just two primes p there are infinitely many primes q depending on p such that p is a primitive root modulo q.

(6) The **large sieve** of Linnik [126], but also contributed to by Roth, Davenport, Montgomery, Gallagher, Bombieri and others, looks completely different. In contrast to small sieves, this enables the removal of many residue classes, up to modulo $p \leq \sqrt{N}$ from intervals of length N. It takes the form of a so-called "large sieve inequality", or better to say inequalities, since they come in a variety of forms. For example, if we let $(a_n)_{n \in \mathbb{N}}$ be a sequence of complex numbers and for all integers $M, N \geq 0$ and $\alpha \in \mathbb{R}$ we define the weighted exponential sum of N terms

$$S(\alpha) = \sum_{M < n \leq M+N} a_n e^{2\pi i \alpha n},$$

then for all r-tuples of real numbers $(\alpha_j)_{1 \leq j \leq r}$ which are δ-**well-spaced**, in the sense that for a given $\delta > 0$ we have

$$\min_{1 \leq i < j \leq r} \|\alpha_i - \alpha_j\| \geq \delta$$

with $\|x\|$ being the distance of real x to the nearest integer, we have

$$\sum_{1\leq i\leq r} \|S(\alpha_i)\|^2 \leq \left(N + \frac{1}{\delta} - 1\right) \sum_{M<n\leq M+N} |a_n|^2.$$

We are able to derive from this that if $N \in \mathbb{N}$ and for each prime $p \leq N$ we remove $r(p)$ residue classes from $\mathbb{N}$ with $0 \leq r(p) < p$, in any interval $(M, M + N] \cap \mathbb{N}$, then the set of integers S which remains satisfies

$$\#S \leq \frac{(1+\pi)N}{g(N)} \text{ where } g(N) := \sum_{p\leq\sqrt{N}} \frac{r(p)}{p - r(p)}.$$

Gallagher's version of the large sieve also finds its realization in the form of inequalities. For example [22, theorem 12.19] or [65]: if a_n are complex numbers, $Q > 1$, $M, N \in \mathbb{N}$, and we define $S(\alpha)$ as before, then

$$\sum_{q\leq Q} \sum_{\substack{1\leq a\leq q \\ (a,q)=1}} \left|S\left(\frac{a}{q}\right)\right| \leq (Q^2 + \pi N) \sum_{M<n\leq M+N} |a_n|^2.$$

Gallagher gives an expressive set of examples showing how large sieve inequalities can be used to derive sieve-type estimates. For example [22, section 12.10], the number of integers in $[1, N]$ which are primitive roots for no primes less than or equal to $\sqrt{N}$ is $O(\sqrt{N} \log N)$. This implies almost every $n \in \mathbb{N}$ which is not a square is a primitive root for some prime.

Using the large sieve, and quite a lot else besides, Bombieri [9] and Vinogradov [211] quite independently found in 1965 their well-known estimate, which is regarded as an unconditional Riemann hypothesis for primes in arithmetic progressions on average. See Section 1.9 for the statement and references to proofs. It has many applications, for example:

The Titchmarsh divisor problem [206] is to estimate the average number of divisors of primes shifted by a given integer $a \neq 0$. Using Bombieri–Vinogradov's estimate, we get, for a constant $c_a > 0$,

$$\sum_{p\leq x} \tau(p + a) = c_a x + O\left(\frac{x \log\log x}{\log x}\right).$$

(7) The **larger sieve** of Gallagher [66] enables an arbitrary number of classes modulo prime powers, in addition to primes, to be deleted. For each prime power q in a finite set of prime powers Q, let $0 \leq r(q) < q$ and set $g(q) = q - r(q)$. If

all but $g(q)$ residue classes modulo q are removed from a set of natural numbers of length N, then the set of integers R which remains satisfies

$$\#R \le \frac{\sum_{q \in Q} \Lambda(q) - \log N}{\sum_{q \in Q} \frac{\Lambda(q)}{g(q)} - \log N}$$

whenever the denominator is positive. In particular, he shows that if for a fixed $G > 0$ all but at most G classes are removed modulo q for each $q \in Q$, then the size of the remaining subset of any interval of length N is at most G if $\sum_{q \in P} \Lambda(q) > G^2 \log N$ and at most $2G - 1$ if $\sum_{q \in P} \Lambda(q) > 2G \log N$. Another example application shows that uniformly for $0 \le \theta \le 1$, the number of integers $n \le N$ for which the order $\exp_p(n) \le N^\theta$ for all primes $p \le N^{\theta + \epsilon}$ is $O(N^\theta)$.

(8) The **asymptotic sieve of Bombieri** has an important role in studies of potential solutions to the twin primes conjecture. During the 1970s, Bombieri [12, 13] described this sieve, which we describe briefly later in this section. It provides asymptotic formulas for the number of primes p which satisfy $p + 2 = P_k$ for fixed $k \ge 2$, giving the expected order of magnitude assuming the Elliott–Halberstam conjecture, but just failing when $k = 1$, i.e., in the case of twin primes. Bombieri's unconditional theorem gives for x sufficiently large and all $k > 1$ the estimate

$$\sum_{n \le x} \Lambda(n + 2)\Lambda_k(n) = 2C_2 \left(k + O \left(\frac{k^{4/3}}{2^{k/3}} \right) \right) x (\log x)^{k-1}. \qquad (2.10)$$

Assuming the Elliott–Halberstam conjecture, i.e., for all $\epsilon > 0$ and $A > 0$, we have

$$\sum_{q < x^{1-\epsilon}} \max_{(a,q)=1} \max_{y \le x} \left| \psi(y, q, a) - \frac{y}{\varphi(q)} \right| \ll_{\epsilon, A} \frac{x}{(\log x)^A},$$

then for $k \ge 2$, Bombieri showed that

$$\sum_{n \le x} \Lambda(n + 2)\Lambda_k(n) \sim 2C_2 kx (\log x)^{k-1},$$

where the von Mongoldt functions Λ and Λ_k are described in Section 1.8. To obtain these results, he modified the sum on the left of (2.10) and replaced it with

$$\sum_{n \le x} \Lambda(n + 2)\Lambda_k(n) \left(\sum_{\substack{q | n \\ q < z}} \lambda_q \right)^2,$$

where $z < x^{\frac{1}{20k}}$ and the weights λ_q obey the standard rules for a Selberg sieve, i.e., $\lambda_1 = 1$, $\lambda_q = 0$ for $q \ge z$ and $\lambda_q = 0$ if q is not squarefree. Bombieri uses standard properties of the $\Lambda_k(n)$, such as $\Lambda_k(n) = 0$ if n has more than k distinct prime factors, and for $(a, b) = 1$ we have

$$\Lambda_k(ab) = \sum_{n=1}^{k} \binom{k}{b} \Lambda_n(a) \Lambda_{k-n}(b).$$

Note that Chen's result takes the form

$$\sum_{n=1}^{\infty} \Lambda(n+2)\Lambda_2(n) = \infty.$$

Given $k \geq 2$, a sequence (a_n) of complex numbers and constant $H := \prod_{p \in \mathbb{P}}(1 - f(p))/(1 - 1/p)$, with $f(p)$ a given multiplicative function, one way of writing the asymptotic sieve is

$$\sum_{n \leq x} a_n \Lambda_k(x) \sim H \left(\sum_{n \in \mathbb{N}} a_n \right) (\log x)^{k-1}.$$

Friedlander and Iwaniec used this sieve, and a lot else, to show that there are infinitely many primes of the form $x^2 + y^4$. See [62].

(9) The evolution of the **GPY sieve**, and its application to detecting primes in admissible tuples, is outlined in the introduction to Chapter 4.

(10) The **multidimensional sieve of Maynard and Tao**, while suggested by Selberg, was developed in a more arithmetic form by Maynard, and an analytic form based in the Fourier transform by Tao. These are described in Chapters 6 and 7 respectively. This might also be called a "multidivisor sieve" to avoid confusion with beta sieves of integral dimension higher than 1.

2.9 A Brief Reader's Guide to Sieve Theory

This brief guide is intended to give an overview of books on the subject, and is necessarily quite incomplete. Fortunately, most of the authors in the prime gaps developments define the particular sieve they are using and derive many of the properties that are needed.

One place to start would be Melvin Nathanson's *Additive Number Theory* [152]. This is written in an expository manner and places the sieve theory which is developed in the context of applications. It covers Brun's pure sieve and twin primes (Section 6.4), Selberg's upper bound sieve with several applications including an upper bound estimate for twin primes (Sections 7.2 and 7.3), and the linear sieve (Chapter 9) applied to Chen's famous theorem ($2n = p + P_2$ in Nathanson's chapter 10).

Gérald Tenenbaum's *Introduction to Analytic and Probabilistic Number Theory* [204, chapter I.4], covers in just a few pages Brun's pure sieve, Legendre's sieve and two forms of the large sieve, together with applications.

The reader might follow these texts with at least parts of George Greaves' *Sieves in Number Theory* [85]. This lays out in the introduction the history of the evolution of sieve theory up until about the year 2000, including Brun's pure sieve, Brun's sieve, Rosser's sieve, weighted sieves and the linear sieve. (Note that Greaves calls the important parameter κ the "sifting density", whereas others denote this same quantity the "sieve dimension" – see Section 2.8(4).) There is also an excellent account of the evolution of sieve theory in Motohashi's *An Overview of the Sieve Method and Its History* [149] and his *Lectures on Sieve Methods and Prime Number Theory* [147]. There are many applications given in the first chapter, but these tail off. The presentation is clear and not too complex. It covers Brun's sieve, the Rosser–Iwaniec sieve (the beta sieve and its special case the linear sieve), sieves with weights, sieves with a bilinear form for the error and Selberg's lower bound method.

The book *Sieve Methods* by Halberstam and Richert [88] was for many years the standard text. Its distinguishing feature is the number and breadth of applications. This writer found the use of special symbols very dense, making this valuable book more of a reference than something to read cover to cover.

The comprehensive Iwaniec and Kowalowski's *Analytic Number Theory* [117] devotes two chapters of its vast coverage to sieve theory. Iwaniec would be regarded by many as a master of this subject. There is a discussion of the combinatorial sieve, Selberg's sieve and the large sieve.

Heath-Brown in his MathSciNet review rates Iwaniec and Friedlander's *Oper de Cribro* [64] as the best overview of what sieves are all about. It is enriched with applications and includes all of the main tools and methods. Brun's sieve, Bombieri's asymptotic sieve (but see also Terry Tao's excellent account "Notes on Bombieri's Asymptotic Sieve" on his Wordpress blogsite), Selbergs Λ^2-sieve, sieves with weights, upper and lower bound sieves including an applications to GPY's gaps between primes work, the large sieve, the Rosser–Iwaniec beta sieve, the linear sieve and the sieve parity problem (see Section 2.10). It is a reference and a book to read following some less comprehensive texts. As always, one needs to be vigilant regarding the contexts of results and the use of implied variables.

2.10 End Note: Twin Almost Primes and the Sieve Parity Problem

The parity of a natural number n is odd if $\Omega(n)$ is odd and even otherwise. The sieves of Brun and Selberg both have an inherent deficiency, in that they cannot detect parity; in other words, they have a parity problem. To paraphrase Tao [200, section 3.10.2], if r is in a given fixed even (or odd) set of integers and A a set all of whose elements n have $\Omega(n) = r$, then without additional information regarding the elements of A, sieve theory is unable to provide nontrivial lower bounds on $|A|$, and any upper bounds must differ from the actual size of $|A|$ by at least a factor or 2.

Iwaniec and Friedlander [64, chapter 16, p. 337] describe the phenomenon as the **Parity Barricade**:

For a general sequence (a_n) one can establish using only the sieve axioms (the basic ones given in [64] in terms of congruence sums) the existence neither of integers with an even number of prime factors nor of integers with an odd number of prime factors.

We say a natural number is P_r, or a P_r almost prime, if it is the product of at most r prime factors, which need not be distinct. For example, n is a P_2 **almost prime** if it is a prime or the product of two primes. Thus we are not able to lower bound nontrivially or accurately asymptotically upper bound using sieves an infinite number of primes since the parity is 1. However, we bound nontrivially an infinite number of P_2 almost primes with parity 1 or 2.

This is beautifully illustrated by Chen's famous theorem, proved using Selberg's sieve: there is an infinite number of primes p such that $p+2 \in P_2$, [88, chapter 11].

Semiprimes are products of exactly two primes, so we cannot use sieves to show that the corresponding set A contains an infinite number of semiprimes with their even parity. This phenomenon was first described by Atle Selberg in the late 1940s, who called it the "principle of parity". One explanation, which has become quite standard, comes from a study of Liouville's arithmetic function $\lambda(n)$ with value $+1$ if n has an even number of prime factors and -1 otherwise.

We need a property of the **Liouville function**, for all $\epsilon > 0$ and integers a, q with $(a, q) = 1$ and fixed $q > 0$, as $N \to \infty$, we have

$$\sum_{n \leq N} \chi_{a \bmod q}(n) \lambda(n) = \sum_{\substack{n \leq N \\ n \equiv a \bmod q}} \lambda(n) = o(N) = O_\epsilon(N^{\frac{1}{2}+\epsilon}),$$

where the third equality holds under the generalized Riemann hypothesis (GRH).

We say here a set of integers is **smooth** if its description has low complexity. This notion is informal, but roughly speaking, following Tao, a set is smooth if it can be defined using only the most significant digits for some base of a set of integers, and the most significant (i.e., small) p-adic digits for small primes p. This implies smooth sets are sets with low complexity. For example,

$$\{n \in \mathbb{N} : N < n \leq 2N : n \equiv a \bmod q\}$$

is smooth in the given sense. Iwaniec and Friedlander define the sets under consideration as those which can be defined using congruence relations. Current sieve theories approximate nonsmooth sets, such as the primes in $(N, 2N]$, which are difficult to count, with smooth sets A_q, with $A = (N, 2N]$, for example, which we can count with good approximations because of their low complexity. Sets such as the primes are called **rough**. One way of thinking about the Liouville function estimate is that $\lambda(n)$ is approximately orthogonal to all characteristic functions of smooth sets. Since the function embodies parity, this property is at the heart of the parity problem.

To show this more concretely, we give an overview of an example due essentially to Selberg, who exhibited optimal sequences based on $\lambda(n)$ for the linear sieve, which was outlined in Section 2.8. Let $A \subset (N, 2N]$ be a set on which $\lambda(n)$ is the constant -1, and we have $R > 0$ with

$$\sum_{\substack{d \mid n \\ d < R}} c_d \leq \mathbf{1}_A(n). \tag{2.11}$$

The bound R is the so-called *sieve level* and c_d the sieve weights. We assume $R < N^{\frac{1}{2}-2\epsilon}$ and $|c_d| \leq 1$, but are otherwise arbitrary. Summing inequality (2.11) over $n \in (N, 2N]$ gives

$$\sum_{N<n\leq 2N} \sum_{\substack{d<R \\ d\mid n}} c_d = \sum_{d<R} c_d \sum_{\substack{N<n\leq 2N \\ d\mid n}} 1 = \sum_{d<R} c_d \frac{N}{d} + error \leq |A|. \tag{2.12}$$

In order that we might derive a lower bound for $|A|$ which is strictly positive, we need the leading term to be strictly positive and the order of the error strictly smaller. Multiplying the inequality (2.11) by $1 + \lambda(n)$, which recall is 0 on A, gives

$$\sum_{N<n\leq N} \sum_{\substack{d\mid n \\ d<R}} c_d(1 + \lambda(n)) \leq 0.$$

Using the GRH-based sum given above gives

$$\sum_{N<n\leq 2N} \sum_{\substack{d\mid n \\ d<R}} c_d \lambda(n) \leq \sum_{d<R} |c_d| \sum_{\substack{N<n\leq 2N \\ n\equiv 0 \bmod d}} \lambda(n) \ll RN^{\frac{1}{2}+\epsilon} \leq N^{1-\epsilon} = o(N).$$

This is the approximate orthogonality of $\lambda(n)$ to divisor sums. Thus we get

$$\sum_{N<n\leq 2N} (1 + \lambda(n)) \sum_{\substack{d<R \\ d\mid n}} c_d = \sum_{d<R} c_d \frac{N}{d} + error \leq 0,$$

so the lower bound of (2.12) cannot do better than $0 \leq |A|$, no matter what choices are made for the c_d. Similarly, when $\lambda(n) = 1$ on A, using $1 - \lambda(n)$, upper bounds must be at least as large as $2|A|$.

The reader will note that we can estimate primes and semiprimes,

$$\pi(x) \sim \frac{x}{\log x} \text{ and } \#\{n \le x : n = pq \text{ with } p,q \in \mathbb{P}\} \sim \frac{x \log\log x}{\log x}.$$

The parity problem in this case says in particular we cannot hope to show that either set is even infinite using current sieve theories!

Bombieri [12, 13] shed light on the parity issue in the context of his asymptotic sieve (which assumes the Elliott–Halberstam conjecture is true – see Section 2.8 (8)). The sieve gives an asymptotic formula for sums

$$\sum_{n \le x} a_n f(n)$$

whenever the given function $f(n)$ gives equal weight to integers with an even number of prime factors and those with an odd number of prime factors. More precisely, the sufficient condition on f is that it is supported on squarefree integers and has the form

$$f(p_1 \cdots p_r) = g_r \left(\frac{\log p_1}{\log n}, \ldots, \frac{\log p_1}{\log n} \right)$$

for $1 \le r \le r_0$ and some continuous and symmetric functions $g_r : \Delta_r \to \mathbb{C}$ vanishing on the boundary of the standard simplex $\Delta_r \subset [0,\infty)^r$, and such that

$$\sum_{r \text{ odd}} \int_{\Delta_r} g_r \, d\mu = \sum_{r \text{ even}} \int_{\Delta_r} g_r \, d\mu.$$

There are situations in which the parity problem can be overcome. Some of these are set out in Terence Tao's Blog Notes [200, section 3.10.2], from which I have taken parts of the following list.

(1) If a central problem is to show explicit polynomials in one or several variables and integer coefficients have an infinite number of prime values, then for affine forms $a + nq = f(n)$ of course Dirichlet's theorem gives a complete answer. In the case of $f(x, y) = x^2 + y^2$, the norm of the ring of integers of the field $\mathbb{Q}(i)$, rational primes have a representation $p = f(x, y)$ if and only if $p \equiv 1 \mod 4$. This, together with Dirichlet's theorem, also gives a complete answer.

It is a classical result of Dirichlet from 1840, completed by Weber in 1882, that any quadratic form with integer coefficients

$$f(x, y) = ax^2 + bxy + cy^2,$$

with discriminant $D := b^2 - 4ac < 0$ and coefficients having no common prime divisor, represents an infinite number of rational primes as x, y vary over $\mathbb{Z}$.

The general case of

$$f(x, y) = ax^2 + bxy + cy^2 + \alpha x + \beta y + \gamma$$

with suitable restrictions on $f(x, y)$ such as $(a, b, c, \alpha, \beta, \gamma) = 1$, $f(x, y)$ irreducible over $\mathbb{Q}$, the forms $\partial f / \partial x, \partial f / \partial y$ linearly independent over $\mathbb{Q}$ and representing arbitrarily large positive values, is quoted in [64, section 21.2], with a very brief sketch of the proof, and which uses sieve theory results from [114]:

Theorem 2.27 *(Friedlander–Iwaniec) Define*

$$\Delta := b^2 - 4ac \text{ and } D := a\beta^2 - b\alpha\beta + c\alpha^2 - \Delta\gamma.$$

If $f(x, y) \in \mathbb{Z}[x, y]$ satisfies the conditions set out in the previous paragraph, and in addition the discriminants $D \neq 0$ and $\Delta \neq \square$, then as $x \to \infty$ we have

$$\#\{p \leq x : p = f(m, n)\} \asymp \frac{x}{(\log x)^{3/2}}.$$

For polynomials in one variable, such as $f(x) = x^2 + 1$, an infinite number of prime values are expected but a proof remains elusive.

(2) Again exploiting some algebraic structure and using sophisticated new techniques, Friedlander and Iwaniec showed [62] there were an infinite number of primes of the form $p = x^2 + y^4$. Their method was taken up by others, the first being Heath-Brown with a similar result for $p = x^3 + 2y^3$, [99].

(3) This item on the sieve parity problem requires background on Dirichlet quadratic characters and Dirichlet L-functions, exceptional characters and exceptional (Siegel) zeros. The interested reader is directed to the literature, especially [144], [35] and [22, section 12.5], for definitions and examples. Suffice it to say here that a Siegel zero β is real and close to $s = 1$ with $\frac{1}{2} < \beta < 1$. A character χ with such a zero for $L(s, \chi)$ has at most one, although there could exist an infinite number of such characters (with different conductors). None have ever been found, or look as though they ever will be. We do not expect Siegel zeros for Dirichlet L-functions to exist, and of course they fail to do so under GRH, which implies there are no zeros off the line $\Re s = \frac{1}{2}$, let alone real zeros. However, we can sometimes break the parity barrier by assuming there is an exceptional quadratic character, say $\chi(n)$ with an exceptional associated zero. Heath-Brown showed the way forward with this idea by showing that an infinite number of Siegel zeros would imply an infinite number of twin primes [96]. To see how the barrier might be broken, consider an exceptional character $\chi(n)$ with Siegel zero $\frac{1}{2} < \beta < 1$ so

$$L(\beta, \chi) = \sum_{n=1}^{\infty} \frac{\chi(n)}{n^\beta} = 0.$$

For this to happen, there must be an infinite number of prime values p such the $\chi(p) = (p|q) = -1$, where q is the conductor of χ. Since on a semiprime we would have $\chi(n) = 1$, this enables the dense set of primes to be recovered from the P_2 almost primes.

(4) Finally we give Tao's (hypothetical) combinatorial example. Let

$$A = \{N < n \le 2N : n \in P_2, \, n + 2 \in P_2, \, n + 6 \in P_2\},$$
$$A_1 = \{n \in A : n \in \mathbb{P}\},$$
$$A_2 = \{n \in A : n + 2 \in \mathbb{P}\},$$
$$A_3 = \{n \in A : n + 6 \in \mathbb{P}\}.$$

Then $A_1 \cup A_2 \cup A_3 \subset A$. Suppose we have sufficiently accurate lower bounds for the $|A_j|$ and an upper bound for $|A|$ to be able to say

$$|A| < |A_1| + |A_2| + |A_3|.$$

If for all n in A we had at most one of the three $n, n + 2, n + 6$ prime, then the A_j would be pairwise disjoint, so $|A| \ge |A_1| + |A_2| + |A_3|$, which is impossible. Thus there is an n such that at least two of $n, n + 2, n + 6$ are prime, so there are at least two primes in $(N, 2N]$ which are at most six apart.

For Friedlander and Iwaniec's fascinating overview see [63] in which in particular a link is sketched between the "parity phenomenon" and Siegel zeros.

We conclude this chapter with an example of Tao's thinking on this subject [200]:

It is probably premature, with our current understanding, to try to find a systematic way to get around the parity problem in general. It seems likely that we will be able to find some further ways in special cases, and perhaps when we have assembled enough of these special cases, it will become clearer what to do in general.

3

Early Work

3.1 Introduction

This chapter gives some information on the progress made and published on prime gaps between the years 1923 and 1988. See the timeline Table 1.1 in Chapter 1. We have made a selection from this work – the discovery of Erdős which initiated the progress, the substantial breakthrough of Bombieri and Davenport, and the clever trick of Maier with his "matrix method". This is included to give some context to the difficulty of the problem and the achievement of GMPY/Zhang/Maynard/Tao and Polymath which was to follow in the twenty-first century. We have included this account to give context to the later chapters, to show the issue of prime gaps became a problem worthy of the serious attention of some of the best number theorists, and to put a little flesh on the bones of the early years of the timeline of Table 1.1. It could be skipped on a first reading.

Paul Erdős was the most prolific and connected mathematician of all time, in terms of the number of articles he published and the number of joint authors. He travelled the world discussing problems with his joint authors and mathematical audiences, mostly in set theory, graph theory and combinatorial, probabilistic and analytic number theory. He never held a regular position.

Enrico Bombieri's (1940–) supervisor at the University of Milan was Giovanni Ricci, and he followed this with time at Cambridge studying with Harold Davenport. We have already seen some of the fruits of their interaction in Chapter 2, Theorem 2.24. His work has covered a wide set of fields, including algebraic geometry, number theory and analysis. He has received numerous awards, including the Fields Medal in 1974. Following a position as professor in Pisa, he accepted a position at the Princeton Institute for Advanced Study, where he is now a professor emeritus.

Harold Davenport (1907–1969) graduated in mathematics first at Manchester University and then Cambridge, where he undertook studies and research for the

Ph.D. under Littlewood. He wrote a thesis on the distribution of quadratic residues. He was strongly influenced by Mordell and worked with him, Hasse and Heilbronn, and focused on the geometry of numbers and diophantine approximation, as well as analytic number theory. He had a succession of chairs: Bangor, University College London and finally Cambridge. Generous to a fault with his time, he gave detailed assistance to many students and young mathematicians throughout his career. One of his better-known sayings is, "Great mathematics is achieved by solving difficult problems, not by fabricating elaborate theories in search of a problem", which of course is easy to refute.

Helmut Maier was born 17 October 1953 in Germany. He is currently a professor at the University of Ulm. His fields are analytic number theory and mathematical analysis.

3.2 Chapter Summary

The average gap between primes in $(x, 2x]$, by the prime number theorem is asymptotically $\log x$. Thus it is no surprise to have an infinite number of consecutive primes strictly less than this average. This was first proved unconditionally by Paul Erdős [45] in 1940, and his proof is given in Section 3.3.

> **Theorem 3.1** **(Erdős)** *If $x \geq 2$ and $p_1, p_2, \ldots, p_n$ are all of the primes in $[x/2, x]$, then there is an $i < n$ and constant c with $0 < c < 1$ with*
>
> $$p_{i+1} - p_i < (1 - c + o(1)) \log x.$$

Even though his constant c is absolute, he did not assign an explicit numerical value. His proof relies on the prime number theorem, the upper bound for the number of solutions to $a = p - q$ for primes up to x, and little else.

Section 3.4 sets out the method of Bombieri and Davenport, which uses quite a few tools and concepts. These include the method of van der Corput for exponential sums, Ramanujan sums, the sums up to x of $(\log p)^2$, $\mu(n)^2/\varphi(n)$ and $\tau(n)/\varphi(n)^2$, Fejérs kernel from Fourier theory and the Bombieri–Vinogradov theorem. Their fundamental lemma has ten steps. Even though this might seem to the uninitiated quite a mouthful, it is an excellent prelude to what lies ahead in some of the chapters, especially Chapters 4, 5 and 8. A remark following the proof gives an improvement.

> **Theorem 3.10** **(Bombieri–Davenport)** *We have the upper bound*
>
> $$\liminf_{n \to \infty} \frac{p_{n+1} - p_n}{\log p_n} \leq E = \frac{2 + \sqrt{3}}{8} = 0.46650\ldots$$

Section 3.5 gives Andrew Granville and Kannan Soundararajan's rendition of the work of Helmut Maier, which has applications to primes in intervals as well as prime gaps. Theorem 3.12 shows that the simple probabilistic model for primes is false in general. Maier showed that for primes in intervals $(x, x + (\log x)^A]$, the model infinitely often both significantly overestimates and infinitely often underestimates their number.

> **Theorem 3.12 (Maier)** *For all fixed $A > 2$, there is a constant $\delta_A > 0$ and two real positive sequences $x_n \to \infty$ and $y_n \to \infty$ such that for all $n \in \mathbb{N}$ we have*
>
> $$\pi(x_n + (\log x_n)^A) - \pi(x_n) > (1 + \delta_A)(\log x_n)^{A-1} \text{ and}$$
> $$\pi(y_n + (\log y_n)^A) - \pi(y_n) < (1 - \delta_A)(\log y_n)^{A-1}.$$

> **Theorem 3.13 (Granville)** *For all x sufficiently large, there are two distinct primes in $[x, 2x)$ which differ by less than $e^{-\gamma} \log x$, where γ is Euler's constant.*

The work of Maier has been given a modern treatment and greatly enlarged scope by Granville and Soundararajan in their article [84].

3.3 Erdős and the First Unconditional Step

> **Theorem 3.1 (Erdős)** *If $x \geq 2$ and $p_1, p_2, \ldots, p_n$ are all of the primes in $[x/2, x]$, then there is an $i < n$ and constant c with $0 < c < 1$ with*
>
> $$p_{i+1} - p_i < (1 - c + o(1)) \log x.$$

Proof **(1)** Let $\epsilon > 0$ be given. By the prime number theorem, for x sufficiently large, we have

$$\left(\tfrac{1}{2} - \epsilon\right) \frac{x}{\log x} < n.$$

(2) Next set

$$b_1 = p_2 - p_1, \ b_2 = p_3 - p_2, \ldots, b_{n-1} = p_n - p_{n-1},$$

so in particular $b_1 + \cdots + b_{n-1} \leq x/2$, the length of the interval.

(3) It follows from Theorem 2.15 in Chapter 2 that if we let a be a nonzero integer, the number of solutions N_a of $a = p - q$ with primes $p, q \leq x$ satisfies for some $c_5 > 0$

$$N_a \leq c_5 \prod_{p|a} \left(1 + \frac{1}{p}\right) \frac{x}{(\log x)^2}.$$

If $0 < c < 1$, define the interval

$$J_c := [(1 - c) \log x, (1 + c) \log x].$$

Then we also have, using the bound

$$\prod_{p|a}\left(1+\frac{1}{p}\right) \le \sum_{d|a}\frac{1}{d}$$

and interchanging the order of summation, for sufficiently small c_4 with $0 < c_4 < 1$, we get

$$\sum_{a\in J_{c_4}}\prod_{p|a}\left(1+\frac{1}{p}\right) \le \sum_{\substack{d\le(1+c_4)\log x}}\frac{1}{d}\sum_{\substack{a\in J_{c_4}\\ d|a}}1$$

$$\le \sum_{d\le(1+c_4)\log x}\frac{1}{d}\left(\frac{2c_4\log x}{d}+1\right)$$

$$< c_6\log x + \sum_{d\le(1+c_4)\log x}\frac{1}{d}$$

$$< \frac{1}{6c_5}\log x.$$

Therefore, using this estimate and the bound for N_a from Step (3), we can choose c_4 sufficiently small so the number of i with b_i in the interval J_{c_4} is bounded above by

$$c_5\frac{x}{(\log x)^2}\sum_{a/\log x\in[1-c_4,1+c_4]}\prod_{p|a}\left(1+\frac{1}{p}\right) < \frac{x}{6\log x}.$$

(4) If for every i we had $b_i = p_{i+1} - p_i \ge (1 - c_4)\log x$, then by Step (1) we would have at least

$$n - \frac{x}{6\log x} > \left(\frac{1}{2}-\epsilon\right)\frac{x}{\log x} - \frac{x}{6\log x} = \left(\frac{1}{3}-\epsilon\right)\frac{x}{\log x}$$

values $b_i > (1+c_4)\log x$. Therefore, in this situation we would have

$$b_1 + \cdots + b_{n-1} \ge \frac{x}{6\log x}(1-c_4)\log x + \left(\frac{1}{3}-\epsilon\right)\frac{x}{\log x}(1+c_4)\log x$$

$$= x\left(\frac{1}{2}+\frac{c_4}{6}-\epsilon(1+c_4)\right) > \frac{1}{2}x,$$

which by Step (2) is (only just!) false. Therefore, for at least one i we have $p_{i+1} - p_i < (1 - c_4)\log x$, which completes the proof. $\square$

3.4 The Beautiful Method of Bombieri and Davenport

That the prime gaps problem would require some serious mathematics to solve was certainly illustrated by the estimate of Bombieri and Davenport reported in this section. They used a considerable number of mathematical tools: Ramanujan sums, an old lemma of Ward for a sum involving $\mu(n)$ and $\varphi(n)$ and other arithmetic sums, the Bombieri–Vinogradov theorem and Fourier theory. For the early work on prime gaps, this is not the simplest, but it is certainly the most intricate.

Lemma 3.2 *As $x \to \infty$, we have*

$$\sum_{n \le x} (\log p)^2 = x \log x + O(x).$$

Proof This follows directly from Abel's theorem [4, theorem 4.2] setting $f(t) = (\log t)^2$ and $a_n = 1$ if $n \in \mathbb{P}$ and 0 otherwise. □

Lemma 3.3 *Let*

$$I(\beta) := \sum_{n \le x} e^{2\pi i n \beta}.$$

As $\beta \to 0$, uniformly in $x \ge 1$, we have the estimate

$$|I(\beta)| \ll \frac{1}{|\beta|}.$$

Proof Sum the geometric series for $I(\beta)$ and use $e^{2\pi i \beta} - 1 \asymp \beta$ in the denominator of the sum. □

Define the **Ramanujan sum** $c(n,q)$ for $n \in \mathbb{Z}$ and $q \in \mathbb{N}$ by

$$c(n,q) := \sum_{\substack{a=1 \\ (a,q)=1}}^{q} e_q(an),$$

where $e_q(n) := \exp\left(2\pi i \frac{n}{q}\right)$.

Lemma 3.4 *The function $c(n,q)$ is periodic in n with period q, and for fixed n is a multiplicative function of q. We can write*

$$c(n,q) = \sum_{d \mid (n,q)} d\mu\left(\frac{q}{d}\right).$$

Proof The periodicity and multiplicativity of the sum is a simple derivation. To show the alternative expression holds, let

$$\theta(n,q) := \sum_{a=1}^{q} e_q(an) = \begin{cases} q \text{ if } q \mid n, \\ 0 \text{ if } q \nmid n. \end{cases}$$

Then we can write

$$c(n,q) = \sum_{a=1}^{q} \mathbf{e}_q (an) \sum_{d|(a,q)} \mu(d) = \sum_{d|q} \mu(d) \sum_{\substack{a=1 \\ d|a}}^{q} \mathbf{e}_q (an)$$

$$= \sum_{d|q} \mu(d) \sum_{m=1}^{q/d} \mathbf{e}_q (mnd) = \sum_{d|q} \mu(d)\theta(n,q/d)$$

$$= \sum_{d|q} \mu(q/d)\theta(n,d) = \sum_{d|(q,n)} \mu(q/d)d.$$

This completes the proof. $\qquad\qquad\qquad\qquad\qquad\qquad\qquad\qquad\qquad\qquad\square$

Lemma 3.5   (Ward's sum) *[212, Section 2] As $x \to \infty$, we have*

$$\sum_{n \leq x} \frac{\mu(n)^2}{\varphi(n)} = \log x + O(1).$$

Proof **(1)** We begin with a Dirichlet series and Euler product where u is any complex number:

$$\sum_{v=1}^{\infty} \frac{\mu(v)^2 v^u}{\varphi(v)^u v^s} = \frac{\zeta(s)}{\zeta(2s)} \prod_p \left(1 + \frac{(1 - (1 - \frac{1}{p})^u}{(p^s + 1)(1 - \frac{1}{p})^u} \right) = \frac{\zeta(s)}{\zeta(2s)} \sum_{v=1}^{\infty} \frac{a_u(v)}{v^s}. \quad (3.1)$$

The first factor converges for $\sigma > 1$, and the Euler product and Dirichlet series with coefficients $a_u(v)$ for $\sigma > 0$, which is fundamental to the proof.

(2) Using $6x/\pi^2 + o(\sqrt{x})$ for the number of squarefree integers up to $x > 0$, we can derive, using Abel's theorem [4, theorem 4.2], for some constant $R \in \mathbb{R}$,

$$\sum_{v \leq x} \frac{\mu(v)^2}{v} = \frac{6 \log x}{\pi^2} + R + O\left(\frac{1}{\sqrt{x}} \right).$$

(3) Next we write

$$\frac{\zeta(s)}{\zeta(2s)} = \sum_{v=1}^{\infty} \frac{\mu^2(v)}{v^s},$$

so therefore

$$\sum_{v=1}^{n} \frac{\mu(v)^2 v^u}{\varphi(v)^u v} = \sum_{v=1}^{n} \frac{a_u(v)}{v} \sum_{m=1}^{n/v} \frac{\mu(m)^2}{m}.$$

Using Step (2) we get

$$\left| \sum_{v=1}^{n} \frac{\mu(v)^2 v^u}{\varphi(v)^u v} - \left(\frac{6\log n}{\pi^2} + R \right) \sum_{v=1}^{\infty} \frac{a_u(v)}{v} + \frac{6}{\pi^2} \sum_{v=1}^{\infty} \frac{a_u(v)\log v}{v} \right|$$

$$\leq \left| \sum_{v=1}^{n} \frac{a_u(v)}{v} \left(\sum_{m=1}^{n/v} \frac{\mu(m)^2}{m} - \frac{6}{\pi^2} \log \frac{n}{v} - R \right) \right|$$

$$+ \left(\frac{6\log n}{\pi^2} + R \right) \sum_{v=n+1}^{\infty} \left| \frac{a_u(v)}{v} \right| + \frac{6}{\pi^2} \sum_{v=n+1}^{\infty} \left| \frac{a_u(v)\log v}{v} \right|$$

$$= S_1 + S_2 + S_3, \text{ say.}$$

Then for any $\epsilon > 0$, using the absolute convergence and Euler product from Step (1), we have $S_2 = O(1/n^{1-\epsilon})$ and $S_3 = O(1/n^{1-\epsilon})$. Estimating S_1 is more difficult and will be undertaken in Step (4).

(4) The sum derived in Step (2) implies, for all $\eta > 0$, n sufficiently large and $v \leq \sqrt{n}$, that

$$\left| \sum_{m=1}^{n/v} \frac{\mu(m)^2}{m} - \frac{6}{\pi^2} \log \frac{n}{v} - R \right| \leq \eta \sqrt{\frac{v}{n}}.$$

For $\sqrt{n} < v \leq n$, there is an absolute constant $K > 0$ such that

$$\left| \sum_{m=1}^{n/v} \frac{\mu(m)^2}{m} - \frac{6}{\pi^2} \log \frac{n}{v} - R \right| \leq K \sqrt{\frac{v}{n}}.$$

It follows, splitting the sum, that

$$S_1 \leq \sum_{v=1}^{\sqrt{n}} \left| \frac{a_u(v)}{v} \right| \left| \sum_{m=1}^{n/v} \frac{\mu(m)^2}{m} - \frac{6}{\pi^2} \log \frac{n}{v} - R \right|$$

$$+ \sum_{v=\lfloor \sqrt{n} \rfloor + 1}^{n} \left| \frac{a_u(v)}{v} \right| \left| \sum_{m=1}^{n/v} \frac{\mu(m)^2}{m} - \frac{6}{\pi^2} \log \frac{n}{v} - R \right|$$

$$\leq \frac{\eta}{\sqrt{n}} \sum_{v=1}^{\infty} \left| \frac{a_u(v)}{\sqrt{v}} \right| + \frac{K}{\sqrt{n}} \sum_{v=\lfloor \sqrt{n} \rfloor}^{\infty} \left| \frac{a_u(v)}{\sqrt{v}} \right|$$

$$= O(n^{-\frac{1}{2}}).$$

(5) Using the estimates from Steps (3) and (4), we can therefore write

$$\sum_{v=1}^{n} \frac{\mu(v)^2 v^u}{\varphi(v)^u v} = \left(\frac{6\log n}{\pi^2} + R \right) \sum_{v=1}^{\infty} \frac{a_u(v)}{v} - \frac{6}{\pi^2} \sum_{v=1}^{\infty} \frac{a_u(v)\log v}{v} + O(1/\sqrt{n}).$$

If we define

$$A(u) := \prod_p \left(1 + \frac{(1 - (1 - \frac{1}{p})^u}{(p+1)(1 - \frac{1}{p})^u} \right),$$

then we have the simpler expression, for a constant $C(u)$ depending only on u,

$$\sum_{v \leq x} \frac{\mu(v)^2 v^u}{\varphi(v)^u v} = A(u-1)(\log x + \gamma) + C(u) + O(1/\sqrt{x}).$$

Setting $u = 1$ and noting that $A(0) = 1$ completes the proof of the lemma. □

Lemma 3.6   *As $x \to \infty$, we have*

$$\sum_{n > x} \frac{\tau(n)}{\varphi(n)^2} \ll \frac{(\log x)^3}{x}.$$

Proof   By Lemma 2.14, we have

$$\sum_{n \leq x} \frac{\tau(n)}{n} = \frac{1}{2}(\log x)^2 + O(\log x).$$

Using this estimate with Abel's theorem [4, theorem 4.2], we get

$$\sum_{n > x} \frac{\tau(n)}{n^{3/2}} \ll \frac{(\log x)^2}{\sqrt{x}}.$$

Using $n/\log\log n \ll \varphi(n)$ (e.g. [21, theorem 5.25]), for x sufficiently large we then get

$$\sum_{x < n} \frac{\tau(n)}{\varphi^2(n)} \ll \sum_{x < n} \frac{\tau(n)}{n^2}(\log\log n)^2 = \sum_{x < n} \frac{\tau(n)}{n^{3/2}} \frac{(\log\log n)^2}{n^{\frac{1}{2}}}$$

$$\ll \frac{(\log\log x)^2 (\log x)^2}{\sqrt{x}} \ll \frac{(\log x)^3}{x},$$

which completes the proof of the lemma. □

Next we give a form of the Bombieri–Vinogradov theorem. Note that the definitions of $E(x,q,a)$ and $E(q,a)$ are different from those used in Lemma 2.17, which would require a larger value of B, but still dependent only on A.

Lemma 3.7   **(Bombieri–Vinogradov)** *Let $a \in \mathbb{Z}$ and $q \in \mathbb{N}$ have $(a,q) = 1$ and define $E(x,q,a)$ by*

$$\sum_{\substack{p \leq x \\ p \equiv a \bmod q}} \log p = \frac{x}{\varphi(q)} + E(x,q,a).$$

Set

$$E(x,q) := \max_{\substack{a:1\leq a\leq q \\ (a,q)=1}} |E(x,q,a)|.$$

Let $A > 0$, and if $B := 24A + 46$, let $Q := \sqrt{x}/(\log x)^B$. Then there is a constant $C > 0$ such that

$$\sum_{q\leq Q} max_{y\leq x}|E(y,q)| \leq \frac{Cx}{(\log x)^A}.$$

In particular, the average value

$$\frac{\sum_{q\leq Q} E(x,q)}{Q} \leq \frac{C\sqrt{x}}{(\log x)^{B-A}}.$$

Proof For a discussion or proof of the Bombieri–Vinogradov theorem, see Section 1.9 or [22, chapter 12]. □

We have some definitions which are used throughout the remainder of this section. Let $e(\theta) := e^{2\pi i\theta}$ and (a_n) be a sequence defined on $\mathbb{Z}$ which has the symmetry $a_{-n} = a_n$ for all $n \in \mathbb{Z}$. Define for $\alpha \in \mathbb{R}$,

$$T(\alpha) := \sum_{n=-k}^{k} a_n e(2n\alpha) = a_0 + 2\sum_{n=1}^{k} a_n \cos(4\pi n\alpha),$$

and suppose the a_n are such that this function is nonnegative for all $\alpha \in \mathbb{R}$. Let

$$S(\alpha) := \sum_{p\leq x} e(p\alpha) \log p,$$

where the sum is over primes, and

$$Z(x,2n) := \sum_{\substack{p,\, p'\leq x \\ p'-p=2n}} (\log p)(\log p').$$

Finally let

$$K(n) := \prod_{p>2} \left(1 - \frac{1}{(p-1)^2}\right) \prod_{2<p|n} \left(\frac{p-1}{p-2}\right), \tag{3.2}$$

and note as before in Chapter 2, manipulating the respective Euler products, that for $n \in \mathbb{N}$ we have

$$\mathscr{G}(\{0,2n\}) = 2K(n).$$

We now give the fundamental lemma of Bombieri and Davenport published in 1966 [10]. It takes ten steps, but uses, in its application to Theorem 3.10, the simple assignment $a_n := k - |n|$ for $-k \leq n \leq k$ and $a_n = 0$ otherwise.

Lemma 3.8 **(Bombieri–Davenport)** *[10] Using the notation and definitions given previously, if for some $C > 0$ we have $k < (\log x)^C$, then for any fixed $\epsilon > 0$ and $x \to \infty$, we get*

$$\sum_{n=1}^{k} a_n Z(x, 2n) > 2x \sum_{n=1}^{k} a_n K(n) - \left(\tfrac{1}{4} + \epsilon\right) a_0 x \log x. \tag{3.3}$$

Proof **(1)** First since $T(\alpha) \geq 0$, by Fourier inversion for $-k \leq n \leq k$

$$|a_n| = \left| \int_0^1 T(\alpha) e(-2n\alpha) \, d\alpha \right| \leq \int_0^1 T(\alpha) \, d\alpha = a_0. \tag{3.4}$$

It follows from this using the definition of $T(\alpha)$

$$T(\alpha) \leq (2k+1)a_0. \tag{3.5}$$

Then using the Mean Value Theorem, we can write for $|\beta| \leq 1$

$$|T(\alpha + \beta) - T(\alpha)| = |\beta| |T'(\xi)| \leq 4\pi k(k+1)a_0. \tag{3.6}$$

(2) Next we derive an expression for

$$\int_0^1 |S(\alpha)|^2 T(\alpha) \, d\alpha$$

in terms of $Z(x, 2n)$. We evaluate this integral using

$$m = 0 \implies \int_0^1 e(m\alpha) \, d\alpha = 1 \text{ and } m \neq 0 \implies \int_0^1 e(m\alpha) \, d\alpha = 0.$$

It follows that

$$\int_0^1 |S(\alpha)|^2 T(\alpha) \, d\alpha = \sum_{n=-k}^{k} a_n \sum_{\substack{p, p' \leq x \\ p' - p = 2n}} (\log p)(\log p')$$

$$= a_0 \sum_{p \leq x} (\log p)^2 + 2 \sum_{n=1}^{k} a_n Z(x, 2n).$$

Then using Lemma 3.2, we get

$$\int_0^1 |S(\alpha)|^2 T(\alpha) \, d\alpha = a_0(x \log x + O(x)) + 2 \sum_{n=1}^{k} a_n Z(x, 2n). \tag{3.7}$$

(3) Now let $1 \leq Q \leq x$ and $0 < \eta < 1/(2Q^2)$. For coprime integers a and q with $1 \leq q \leq Q$, define the open interval

$$M_{a,q} := \left(\frac{a}{q} - \eta, \frac{a}{q} + \eta \right).$$

If the pairs $(a,q) \neq (a',q')$, because

$$2\eta < \frac{1}{Q^2} \leq \frac{1}{qq'} \leq \left| \frac{a}{q} - \frac{a'}{q'} \right|,$$

we have $M_{a,q} \cap M_{a',q'} = \emptyset$. Thus, since we have specified $T(\alpha) \geq 0$ on $\mathbb{R}$,

$$\int_0^1 |S(\alpha)|^2 T(\alpha)\, d\alpha \geq \sum_{q \leq Q} \sum_{\substack{a=1 \\ (a,q)=1}}^q \int_{M_{a,q}} |S(\alpha)|^2 T(\alpha)\, d\alpha.$$

Next we change variables in the integrals over $M_{a,q}$ by setting $\alpha = \beta + a/q$ and replace $T(\alpha)$ by its central value on $M_{a,q}$, namely $T(a/q)$. The error in making these substitutions, using the result of Step (1) and Lemma 3.2, is

$$O\left(\sum_{q \leq Q} \sum_{\substack{a=1 \\ (a,q)=1}}^q \eta k^2 a_0 \int_{M_{a,q}} |S(\alpha)|^2\, d\alpha \right) \ll \eta k^2 a_0 \int_0^1 |S(\alpha)|^2\, d\alpha \ll \eta k^2 a_0 x \log x.$$

Therefore,

$$\int_0^1 |S(\alpha)|^2 T(\alpha)\, d\alpha \geq \sum_{q \leq Q} \sum_{\substack{a=1 \\ (a,q)=1}}^q T\left(\frac{a}{q} \right) \int_{-\eta}^{\eta} |S(\beta + a/q)|^2\, d\beta + O(\eta k^2 a_0 x \log x).$$

$$(3.8)$$

(4) Recall $(a,q) = 1$ and, as before set $\alpha = \beta + a/q$. If $(m,q) = 1$ and n is prime, say $n = p$, and $n \equiv m \bmod q$, let

$$\rho(n,q,m) := \log p - \frac{1}{\varphi(q)}.$$

If n is not prime or $n \not\equiv m \bmod q$, set $\rho(n,q,m) := -1/\varphi(q)$.
Recall the definition

$$S(\alpha) := \sum_{p \leq x} e(p\alpha) \log p = \sum_{p \leq x} e(p(\beta + a/q)) \log p = S(\beta + a/q),$$

and define

$$S_1(\beta,q,a) := \sum_{\substack{m=1 \\ (m,q)=1}}^q e\left(\frac{ma}{q} \right) \sum_{n \leq x} \rho(n,q,m) e(n\beta).$$

Split the sum in $S(\beta+a/q)$ into arithmetic progressions m modulo q with $(m,q) = 1$ and $p \nmid q$, and define as in Lemma 3.3 $I(\beta) := \sum_{n \le x} e(n\beta)$. Since $(a,q) = 1$, we can use (see [4, exercise 14(b) chapter 2])

$$\mu(q) = \sum_{\substack{m=1 \\ (m,q)=1}}^{q} e\left(\frac{am}{q}\right)$$

to derive

$$S\left(\beta + \frac{a}{q}\right) = \sum_{\substack{m=1 \\ (m,q)=1}}^{q} e\left(\frac{ma}{q}\right) \sum_{\substack{p \le x \\ p \equiv m \bmod q}} (\log p)e(p\beta) + O\left(\sum_{p|q} \log p\right)$$

$$= \sum_{\substack{m=1 \\ (m,q)=1}}^{q} e(ma/q) \sum_{n \le x}\left(\rho(n,q,m) + \frac{1}{\varphi(q)}\right)e(n\beta) + O(\log q)$$

$$= S_1(\beta,q,a) + \frac{\mu(q)I(\beta)}{\varphi(q)} + O(\log q).$$

Thus

$$S\left(\beta + \frac{a}{q}\right) = \frac{\mu(q)I(\beta)}{\varphi(q)} + S_1(\beta,q,a) + O(\log q). \tag{3.9}$$

(5) Using the equation derived in Step (4), the trivial bounds $|S(\alpha)| \ll x$ by the Prime Number Theorem, $\log q \le \log Q \le \log x$, assuming x is sufficiently large so $k \ll Q$ and dropping the term $|S_1|^2 \ge 0$, we can write

$$\left|S\left(\beta + \frac{a}{q}\right)\right|^2 = \left|\frac{\mu(q)I(\beta)}{\varphi(q)} + S_1(\beta,q,a)\right|^2 + O(x \log x)$$

$$\ge \frac{\mu(q)^2}{\varphi(q)^2}|I(\beta)|^2 + 2\Re\frac{\mu(q)}{\varphi(q)}\overline{I(\beta)}S_1(\beta,q,a) + O(x \log x). \tag{3.10}$$

Next, substitute this lower bound into the integral on the right-hand side of estimate derived in Step (3). The first term of the lower bound gives

$$\sum_{q \le Q} \frac{\mu(q)^2}{\varphi(q)^2} \sum_{\substack{a=1 \\ (a,q)=1}}^{q} T\left(\frac{a}{q}\right) \int_{-\eta}^{\eta} |I(\beta)|^2 \, d\beta,$$

and since we have defined $I(\beta) := \sum_{n \le x} e(n\beta)$ and we have, by Lemma 3.3, $|I(\beta)| \ll 1/|\beta|$, we also have

$$\int_{-\eta}^{\eta} |I(\beta)|^2 \, d\beta = \int_{-\frac{1}{2}}^{\frac{1}{2}} |I(\beta)|^2 \, d\beta + O\left(\int_{\eta}^{\frac{1}{2}} \frac{1}{\beta^2} \, d\beta\right) = x + O(1/\eta).$$

Now let

$$\mathscr{A} := \sum_{q \leq Q} \frac{\mu(q)}{\varphi(q)} \sum_{\substack{a=1 \\ (a,q)=1}}^{q} T\left(\frac{a}{q}\right) \int_{-\eta}^{\eta} \overline{I(\beta)} S_1(\beta, q, a) \, d\beta, \qquad (3.11)$$

so the contribution of the second terms of the lower bound of (3.10) to the integral

$$\int_0^1 |S(\alpha)|^2 T(\alpha) \, d\alpha$$

is $2\Re\mathscr{A} \geq -2|\mathscr{A}|$.

Next using $|a_n| \leq a_0$ from Step (1) in the third line, the assumption $k \ll Q$, changing the order in the double summation to get the second line, the contribution to the integral of the $O(x \log x)$ term in (3.10) is bounded above by a positive quantity:

$$x \log x \sum_{q \leq Q} \sum_{\substack{a=1 \\ (a,q)=1}}^{q} T\left(\frac{a}{q}\right) \eta \ll \eta x \log x \sum_{q \leq Q} \sum_{\substack{a=1 \\ (a,q)=1}}^{q} \sum_{n=-k}^{k} a_n e\left(\frac{2na}{q}\right)$$

$$\leq \eta x \log x \sum_{q \leq Q} q \sum_{\substack{n=-k \\ 2n \equiv 0 \bmod q}}^{k} a_n$$

$$\ll \eta x \log x \sum_{q \leq Q} a_0 q (1 + k/q)$$

$$\ll \eta x (\log x) a_0 Q^2.$$

Finally, we combine these estimates with the result of Step (3) to get

$$\int_0^1 |S(\alpha)|^2 T(\alpha) \, d\alpha \geq (x + O(1/\eta)) \sum_{q \leq Q} \frac{\mu(q)^2}{\varphi(q)^2} \sum_{\substack{a=1 \\ (a,q)=1}}^{q} T\left(\frac{a}{q}\right) - 2|\mathscr{A}|$$

$$+ O(\eta Q^2 a_0 x \log x). \quad (3.12)$$

(6) Next we will derive an upper bound for $|\mathscr{A}|$ as defined in Step (5). First, by the definition of $I(\beta)$ and $S_1(\beta, q, a)$, we can write, with $N = \lfloor x \rfloor$,

$$j = n - n', \ n_1 = n_1(j) := \max(1, j+1) \text{ and } n_2 = n_2(j) := \min(N, N+j),$$

$$\int_{-\eta}^{\eta} \overline{I(\beta)} S_1(\beta, q, a) \, d\beta$$

$$= \sum_{n \leq x} \sum_{n' \leq x} \sum_{\substack{m=1 \\ (m,q)=1}}^{q} e\left(\frac{ma}{q}\right) \sum_{n=n_1}^{n_2} \rho(n, q, m) \int_{-\eta}^{\eta} e((n - n')\beta) \, d\beta$$

$$= \sum_{j=-N+1}^{N-1} \int_{-\eta}^{\eta} e(j\beta) \, d\beta \sum_{\substack{m=1 \\ (m,q)=1}}^{q} e\left(\frac{ma}{q}\right) \sum_{n=n_1(j)}^{n_2(j)} \rho(n, q, m).$$

Also, by the definition of ρ and $E(x,q,a) := \left(\sum_{\substack{p \leq x \\ p \equiv a \bmod q}} \log p \right) - x/\varphi(q)$, we have

$$\sum_{n=1}^{n_1} \rho(n,q,m) = E(n_1,q,m).$$

Substituting these expressions in the definition of $\mathscr{A}$, and defining

$$B(m,q) := \sum_{j=-N+1}^{N-1} \int_{-\eta}^{\eta} e(\beta j) \, d\beta \, (E(n_2(j),q,m) - E(n_1(j)+1,q,m)),$$

gives

$$\mathscr{A} := \sum_{q \leq Q} \frac{\mu(q)}{\varphi(q)} \sum_{\substack{a=1 \\ (a,q)=1}}^{q} T\left(\frac{a}{q}\right) \int_{-\eta}^{\eta} \overline{I(\beta)} S_1(\beta,q,a) \, d\beta$$

$$= \sum_{q \leq Q} \frac{\mu(q)}{\varphi(q)} \sum_{\substack{a=1 \\ (a,q)=1}}^{q} T\left(\frac{a}{q}\right) \sum_{\substack{m=1 \\ (m,q)=1}}^{q} e\left(\frac{ma}{q}\right) B(m,q).$$

Therefore, if we define

$$\mathscr{A}_1(q) := \sum_{\substack{m=1 \\ (m,q)=1}}^{q} \left| \sum_{\substack{a=1 \\ (a,q)=1}}^{q} T\left(\frac{a}{q}\right) e\left(\frac{ma}{q}\right) \right|$$

it follows that

$$|\mathscr{A}| \leq \sum_{q \leq Q} \frac{\mu(q)^2}{\varphi(q)^2} \mathscr{A}_1(q) \max_{1 \leq m \leq q} |B(m,q)|. \tag{3.13}$$

In addition, we have

$$|B(m,q)| \ll \eta \sum_{j=1}^{N} |E(j,q,m)|.$$

Recalling the definition and bound for $E(n,q) := \max_{1 \leq m \leq q} |E(n,q,m)|$ from Lemma 3.7 and defining

$$E^*(n,q) := \max_{1 \leq m \leq n} E(m,q)$$

and setting $\mathscr{A}_2(q) := x + x E^*(x,q)$ the upper bound for $|B(m,q)|$ which we have derived gives $|B(m,q)| \ll \eta \mathscr{A}_2(q)$. Therefore, by the estimate (3.13), we have

$$|\mathscr{A}| \le \eta \sum_{q \le Q} \frac{\mu(q)^2}{\varphi(q)^2} \mathscr{A}_1(q) \mathscr{A}_2(q). \tag{3.14}$$

(**7**) In this step, we will derive an upper bound for $\mathscr{A}_1(q)$. Recall from Lemma 3.4 the definition and property of a Ramanujan sum

$$c(m,q) := \sum_{\substack{a=1 \\ (a,q)=1}}^{q} e(am/q) = \sum_{d|(m,q)} d\mu(q/d).$$

We can write

$$\sum_{\substack{a=1 \\ (a,q)=1}}^{q} T\left(\frac{a}{q}\right) e\left(\frac{ma}{q}\right) = \sum_{n=-k}^{k} a_n c(2n+m,q).$$

Hence, from the definition of $\mathscr{A}_1$ in Steps (1) and (6), using the periodicity in q of $c(m,q)$ as a function of m, noting that

$$\{2n+1, 2n+2, \dots, 2n+q\} \bmod q \equiv \{1, 2, \dots, q\}$$

to get the third line, using $|a_n| \le a_0$ and Lemma 3.4, we get

$$\mathscr{A}_1(q) := \sum_{\substack{m=1 \\ (m,q)=1}}^{q} \left| \sum_{\substack{a=1 \\ (a,q)=1}}^{q} T(a/q) e(ma/q) \right|$$

$$\le \sum_{\substack{m=1 \\ (m,q)=1}}^{q} \sum_{n=-k}^{k} |a_n c(2n+m,q)|$$

$$\le (2k+1) a_0 \sum_{m=1}^{q} |c(m,q)|$$

$$\ll k a_0 \left(\sum_{d|q} d \sum_{\substack{m=1 \\ d|m}}^{q} 1 \right)$$

$$= k a_0 \sum_{d|q} d\frac{q}{d} = k a_0 q \tau(q).$$

Therefore,

$$\mathscr{A}_1(q) \ll k a_0 q \tau(q). \tag{3.15}$$

(8) We are now able to derive a more explicit bound for $|\mathscr{A}|$. Assume

$$Q = \frac{\sqrt{x}}{(\log x)^B},$$

where B occurs in Lemma 3.7. Now

$$\sum_{q \leq Q} \frac{q\tau(q)}{\varphi(q)} \ll \log\log Q \sum_{q \leq Q} \tau(q)$$

$$\ll (\log\log Q) Q \log Q \ll \sqrt{x}.$$

Hence, using the Cauchy–Schwarz inequality, and the divisor bound

$$\sum_{n \leq x} \frac{\tau(n)^2}{n} \ll (\log x)^4,$$

which we can derive from Ramanujan's sum [153, theorem 7.8]

$$\sum_{n \leq x} \tau(n)^2 \sim \frac{1}{\pi^2} x (\log x)^3$$

using Abel's theorem [4, theorem 4.1], we get

$$\sum_{q \leq Q} \frac{q\tau(q)}{\varphi(q)} \mathscr{A}_2(q) = \sum_{q \leq Q} \frac{q\tau(q)}{\varphi(q)} (x + x E^*(N, q))$$

$$\leq \left(\sum_{q \leq Q} \frac{q^2 \tau(q)^2}{\varphi(q)^3} \right)^{\frac{1}{2}} \left(\sum_{q \leq Q} \varphi(q) E^*(x, q)^2 \right)^{\frac{1}{2}} + O(x^{3/2})$$

$$\ll \left((\log\log Q)^3 (\log Q)^4 \right)^{\frac{1}{2}} \left(\sum_{q \leq Q} \varphi(q) E^*(x, q)^2 \right)^{\frac{1}{2}} + O(x^{3/2}).$$

We also have from the definitions of E and E^* trivially and using Bombieri–Vinogradov, we get

$$E^*(x, q) \ll \frac{x \log x}{\varphi(q)} \implies \varphi(q) E^*(x, q)^2 \ll \frac{x^2}{(\log x)^{(A-1)}}.$$

Thus

$$\sum_{q \leq Q} \frac{q\tau(q)}{\varphi(q)} \mathscr{A}_2(q) \ll (\log x)^{5/2} (x^2 (\log x)^{1-A})^{\frac{1}{2}} \ll x (\log x)^{3-A/2},$$

and it follows from this, using the estimates (3.14) and (3.15) from Steps (6) and (4) respectively, that

$$|\mathscr{A}| \ll \eta k a_0 x^2 (\log x)^{3-A/2}.$$

(9) Referring back again to the lower bound derived in Step (5), we now proceed to find a good estimate for the sum

$$\sum_{q\le Q} \frac{\mu(q)^2}{\varphi(q)^2} \sum_{\substack{a=1\\(a,q)=1}}^{q} T\left(\frac{a}{q}\right). \tag{3.16}$$

From the definition of $T(\alpha)$, we have

$$\sum_{\substack{a=1\\(a,q)=1}}^{q} T\left(\frac{a}{q}\right) = \sum_{n=-k}^{k} a_n \sum_{\substack{a=1\\(a,q)=1}}^{q} e\left(\frac{2na}{q}\right) = a_0\varphi(q) + \sum_{\substack{n=-k\\n\ne0}}^{k} a_n c(2n,q).$$

Substituting this in the sum we are estimating, and using Lemma 3.5, which gave

$$\sum_{n\le x} \frac{\mu(n)^2}{\varphi(n)} = \log x + O(1),$$

we get

$$\sum_{q\le Q} \frac{\mu(q)^2}{\varphi(q)^2} \sum_{\substack{a=1\\(a,q)=1}}^{q} T\left(\frac{a}{q}\right) = a_0(\log Q + O(1)) + \sum_{\substack{n=-k\\n\ne0}}^{k} a_n \sum_{q\le Q} \frac{\mu(q)^2}{\varphi(q)^2} c(2n,q).$$

Using the multiplicativity in q of the $c(2n,q)$, the definition of $K(n)$ in (3.2), manipulating the Euler products, and using Lemma 3.4 to evaluate $c(2n,p)$, the sum over q with Q infinite in this expression can be written

$$\prod_p \left(1 + \frac{c(2n,p)}{\varphi(p)^2}\right) = \prod_{p|2n}\left(1 + \frac{1}{p-1}\right) \prod_{p\nmid 2n}\left(1 - \frac{1}{(p-1)^2}\right) =: 2K(n).$$

Thus if the sum over q were infinite it would contribute to Equation (3.16)

$$\sum_{\substack{n=-k\\n\ne0}}^{k} 2K(n)a_n = 4\sum_{n=1}^{k} K(n)a_n.$$

The remainder for the infinite sum, by Lemmas 3.4 and 3.6, is bounded above by

$$\sum_{q>Q} \frac{|c(2n,q)|}{\varphi(q)^2} \ll \sum_{q>Q} \frac{1}{\varphi(q)^2} \sum_{d|(2n,q)} d \ll n \sum_{q>Q} \frac{\tau(q)}{\varphi(q)^2} \ll n\frac{(\log Q)^3}{Q},$$

so the contribution of the tail to the error in (3.17), when we assume $k \le \sqrt{Q}/(\log Q)^{3/2}$, is bounded by

$$\sum_{n=-k}^{k} a_n n\frac{(\log Q)^3}{Q} \ll k^2 a_0 \frac{(\log Q)^3}{Q} \ll a_0.$$

Therefore, if $k \leq \sqrt{Q}/(\log Q)^{3/2}$ we get

$$\sum_{q \leq Q} \frac{\mu(q)^2}{\varphi(q)^2} \sum_{\substack{a=1 \\ (a,q)=1}}^{q} T\left(\frac{a}{q}\right) = a_0(\log Q + O(1)) + 4\sum_{n=1}^{k} a_n K(n). \qquad (3.17)$$

(10) In this final step, we derive the lower bound of the lemma. First let as before

$$Q = \frac{\sqrt{x}}{(\log x)^B}$$

so we have satisfied the condition for Step (8). Recall we have specified $k < (\log x)^C$, so certainly we have $k \leq \sqrt{Q}/(\log Q)^{3/2}$, needed for Step (9). Using the result of Step (2) for the first line, Steps (5) and (9) for the second, Step (8) for the error and using the value of B from Lemma 3.7 so the stated error dominates the error from Step (5), we get the lower bound

$$\int_0^1 |S(\alpha)|^2 T(\alpha)\, d\alpha = a_0(x \log x + O(x)) + 2\sum_{n=1}^{k} a_n Z(x, 2n)$$

$$\geq (x + O(1/\eta))\left(a_0(\log Q + O(1)) + 4\sum_{n=1}^{k} a_n K(n)\right)$$

$$+ O(\eta k a_0 x^2 (\log x)^{3-A/2}).$$

$$(3.18)$$

Next take $\eta := (\log x)^{C+1}/x$ so $0 < \eta < 1/(2Q^2)$, which we needed for Step (3), assuming B is large compared with C, which we can achieve by making A sufficiently large. Then, using Step (1) and Lemma 3.9 and noting that $K(n) \ll \log\log n$ to get the second estimate, we have

$$\log Q = \frac{1}{2}\log x + O(\log\log x),$$

$$\frac{1}{\eta}\sum_{n=1}^{k} |a_n| K(n) \ll \frac{x}{(\log x)^{C+1}} k a_0 \ll x a_0,$$

$$\eta k a_0 x^2 (\log x)^{3-A/2} \ll x(\log x)^{4+2C-A/2} a_0 \ll x a_0.$$

Hence for any fixed $\epsilon > 0$, provided we choose x sufficiently large, the right-hand side of the inequality (3.18) is bounded below by

$$\left(\tfrac{1}{2} - \epsilon\right) a_0 x \log x + 4x \sum_{n=1}^{k} a_n K(n).$$

Inserting this lower bound and simplifying gives

$$\sum_{n=1}^{k} a_n Z(x, 2n) > 2x \sum_{n=1}^{k} a_n K(n) - \left(\tfrac{1}{4} + \epsilon\right) a_0 x \log x,$$

which completes the proof of the lemma. $\square$

We need one final lemma to derive Bombieri and Davenport's main result. It relies on the multiplicative nature of the Ramanujan sums $c(n, q)$ and the expression for $K(n)$ given in the proof of Step (9) of Lemma 3.8. The reader might wish to compare this with Theorem 4.14.

Lemma 3.9 **(Bombieri–Davenport)** *Define as before*

$$K(n) := \prod_{p>2} \left(1 - \frac{1}{(p-1)^2}\right) \prod_{2<p|n} \frac{p-1}{p-2} = \frac{1}{2} \prod_{p|2n} \left(1 + \frac{1}{p-1}\right) \prod_{p\nmid 2n} \left(1 - \frac{1}{(p-1)^2}\right).$$

Then for $k \to \infty$, we have

$$\sum_{n=1}^{k} K(n) = k + O\left((\log k)^4\right) = k(1 + o(1)).$$

Proof From the proof of Step (9) of Lemma 3.8, we can write

$$2K(n) = \prod_{p} \left(1 + \frac{c(2n, p)}{\varphi(p)^2}\right) = \sum_{q=1}^{\infty} \frac{\mu(q)^2 c(2n, q)}{\varphi(q)^2}. \qquad (3.19)$$

In Step (9), we showed that, when $q > X := k^2$ and $n \leq k$, this sum is $O(n(\log X)^3/X) = O(\log^3 k)$.

Let $\|x\|$ be the distance of real x to the nearest integer. Then, by Lemma 3.3, for $q \geq 3$ we have

$$\left|\sum_{n=1}^{k} c(2n, q)\right| = \left|\sum_{\substack{a=1 \\ (a,q)=1}}^{q} \sum_{n=1}^{k} e\left(\frac{2na}{q}\right)\right| \ll \sum_{\substack{a=1 \\ (a,q)=1}}^{q} \frac{1}{\|2a/q\|} \ll q \log q.$$

(Compare the derivation of Polya's inequality [4, theorem 8.21].) Thus, using the bound $1/\varphi(q) \ll (\log\log q)/q$, we have

$$\sum_{q=3}^{k^2} \frac{\mu(q)^2}{\varphi(q)^2} \sum_{n=1}^{k} c(2n, q) \ll \sum_{q=3}^{k^2} \frac{(\log\log q)^2}{q^2} \times q \log q \ll (\log k)^4.$$

The terms of the sum $\sum_{n=1}^{k} 2K(n)$ with $q = 1$ and $q = 2$, by the formula in Step (7) of Lemma 3.8 and (3.19), give a total contribution $2k$. Therefore,

$$\sum_{n=1}^{k} K(n) = k + O\left((\log k)^4\right),$$

which completes the proof.   □

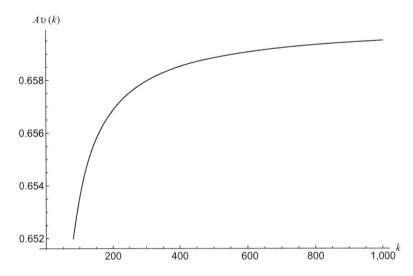

Figure 3.1 The average value of $K(k)$ for $1 \leq n \leq 1,000$.

Figure 3.1 is a plot of the average value of $K(k)$, i.e., of

$$Av(k) := \frac{\sum_{n=1}^{k} K(n)}{k} \quad \text{for } 2 \leq k \leq 10^3.$$

Theorem 3.10 (Bombieri–Davenport) [10] *For all x sufficiently large, there is a prime p_n with $x/\log^2 x < p_n$ and*

$$p_{n+1} - p_n < \left(\tfrac{1}{2} + o(1)\right) \log p_n.$$

Hence

$$\liminf_{n \to \infty} \frac{p_{n+1} - p_n}{\log p_n} \leq \frac{1}{2}.$$

Proof In the result of Lemma 3.8, take $a_n := k - |n|$. With this choice, $T(\alpha)$ is an instance of Fejér's kernel used in Fourier theory (see for example [22, appendix E, lemma E.6]):

$$T(\alpha) = \left(\frac{\sin(2\pi k\alpha)}{\sin(2\pi \alpha)}\right)^2 = \sum_{n=-k}^{k} (k - |n|)e(2n\alpha).$$

Lemma 3.8, when $k < (\log x)^C$, then gives

$$\sum_{n=1}^{k}(k-n)Z(x,2n) > 2x\sum_{n=1}^{k}(k-n)K(n) - \left(\tfrac{1}{4}+\epsilon\right)kx\log x. \tag{3.20}$$

By Lemma 3.9, the average value of $K(n)$ is 1, so for fixed ϵ with $0 < \epsilon < \tfrac{1}{4}$ and sufficiently large k we have

$$\sum_{n=1}^{k}(k-n)K(n) = \sum_{j=1}^{k-1}\sum_{n=1}^{j}K(n) > \frac{1}{2}(1-\epsilon)k^2.$$

Substitute this in the inequality (3.20) to get

$$\sum_{n=1}^{k}(k-n)Z(x,2n) > (1-\epsilon)k^2x - \left(\tfrac{1}{4}+\epsilon\right)kx\log x. \tag{3.21}$$

If we set $k := \left\lceil (\tfrac{1}{4}+3\epsilon)\log x\right\rceil$, we **claim** the right-hand side of this equation is larger than $\epsilon kx\log x$. To see this, note that $k \geq (\tfrac{1}{4}+3\epsilon)\log x$, so using the assumption $\epsilon < \tfrac{1}{4}$, the right-hand side is bounded below by

$$kx\log x\left((1-\epsilon)\left(\frac{1}{4}+3\epsilon\right)-\left(\frac{1}{4}+\epsilon\right)\right)=kx\log x\left(\frac{7}{4}\epsilon - 3\epsilon\times\epsilon\right) > \epsilon kx\log x.$$

Thus there is some $n \leq k$ with $Z(x,2n) > 2\epsilon x$, or, in other words, for that n

$$\sum_{\substack{p,p'\leq x \\ p'-p=2n}}(\log p)(\log p') > 2\epsilon x.$$

This inequality shows that we are not able to have all of the primes in the given sum bounded above by $x/(\log x)^2$, because in that case the sum would be bounded above by

$$\sum_{p\leq x/(\log x)^2}(\log x)^2 \ll \frac{x}{\log x} \implies 2\epsilon x \ll \frac{x}{\log x},$$

which is false for x sufficiently large. Therefore, for x sufficiently large there is at least one pair of primes p, p' with

$$\frac{x}{(\log x)^2} < p \leq x, \text{ and } p' - p = 2n \leq 2k.$$

If p is the jth prime $p = p_j$, then because $x/(\log x)^2 \leq p_j \leq x$ we must have $\log p_j \sim \log x$ so

$$p_{j+1} - p_j \leq p' - p \leq 2k < \left(\tfrac{1}{2}+7\epsilon\right)\log x \leq \left(\tfrac{1}{2}+8\epsilon\right)\log p_j.$$

This implies

$$\liminf_{n \to \infty} \frac{p_{n+1} - p_n}{\log p_n} \leq \frac{1}{2}$$

and completes the proof of the theorem. ☐

Recall the definition for sums of prime pairs (p', p)

$$Z(N, 2n) := \sum_{\substack{p', p \leq N \\ p' - p = 2n}} (\log p')(\log p).$$

In Chapter 2, Theorem 2.24, we gave the proof of Bombieri and Davenport that for all $\epsilon > 0$ we have as $N \to \infty$

$$\pi_{2n}(N) < (8 + \epsilon)K(n)\frac{N}{\log^2 N}, \text{ which implies } Z(N, 2n) < (8 + \epsilon)K(n)N.$$

Bombieri and Davenport used this bound to finesse the method of Theorem 3.10, to derive in the mid-1960s an improved prime gap estimate using "variation of parameters", a method used in Chapter 6 to show that a number of Maynard's choices are in a sense optimal:

Theorem 3.11 (**Bombieri–Davenport**) *We have the upper bound*

$$\liminf_{n \to \infty} \frac{p_{n+1} - p_n}{\log p_n} \leq E = \frac{2 + \sqrt{3}}{8} = 0.46650\dots$$

Proof (**1**) In the proof of Theorem 3.10, (3.21), we showed that for $k < (\log N)^C$, all $\epsilon > 0$ and fixed and all N sufficiently large, we have the inequality

$$\sum_{n=1}^{k}(k - n)Z(N, 2n) > (1 - \epsilon)k^2 N - \left(\frac{1}{4} + \epsilon\right)kN \log N. \tag{3.22}$$

Assume, in order that we might obtain a contradiction in Step (2), that for all m with $1 < m < k$, for all $1 \leq n \leq m$, we have

$$Z(N, 2n) \leq 2\epsilon N.$$

Using the inequality derived in Theorem 2.24, namely

$$Z(N, 2n) \leq (8 + \epsilon)K(n)\frac{N}{(\log N)^2} \leq (8 + \epsilon)NK(n),$$

and the estimate (3.22), we get

$$\sum_{n=1}^{k}(k - n)Z(N, 2n) < 2\epsilon N \sum_{n=1}^{m}(k - n) + (8 + \epsilon)N \sum_{n=m+1}^{k}(k - n)K(n).$$

Using the result of Lemma 3.9, namely

$$\sum_{n=1}^{k} K(n) = k + O((\log k)^4),$$

enables us to write for m sufficiently large

$$\sum_{n=m+1}^{k} (k-n)K(n) = \sum_{v=m-1}^{k-1} \sum_{n=m+1}^{v} K(n)$$

$$< \sum_{v=m+1}^{k-1} (v - m + \epsilon v)$$

$$< \frac{1}{2}(k-m)^2 + \epsilon k^2.$$

Therefore,

$$\sum_{n=1}^{k}(k-n)Z(N,2n) < N(4(k-m)^2 + 11\epsilon k^2).$$

Substituting the right hand side for the upper bound in (3.22), we get

$$4(k-m)^2 > (1 - 12\epsilon)k^2 - \left(\tfrac{1}{4} + \epsilon\right)k \log N.$$

(2) Now let $\kappa > \frac{1}{4}$ be a constant to be determined and choose $k = \lfloor \kappa \log N \rfloor$. Then if λ is the smallest positive solution to $4(\kappa - \lambda)^2 = \kappa^2 - \kappa/4$ and where ϵ_0, ϵ_1 are small and go to zero with ϵ, by the result of Step (1), we have

$$\kappa = \frac{k}{\log N} + \epsilon_0 \implies \frac{m}{\log N} < \lambda + \epsilon_1.$$

Let $\kappa := (3 + \sqrt{12})/24$ to obtain $\lambda = (2 + \sqrt{3})/16$. Choosing ϵ_2 sufficiently small and $m = \lfloor (\lambda + \epsilon_2) \log N \rfloor$ with $\epsilon_2 > \epsilon_1$, we obtain a contradiction when N is sufficiently large.

(3) Therefore, for some $n \le m$, we must have $Z(N, 2n) > 2\epsilon N$. It follows, using the same approach as in the proof of Theorem 3.10, that there are primes p', p with

$$\frac{N}{\log^2 N} < p \le N \text{ and } p' - p = 2n \le 2m,$$

and consequently

$$\liminf_{j \to \infty} \frac{p_{j+1} - p_j}{\log p_j} \le 2\lambda = \frac{2 + \sqrt{3}}{8}.$$

This completes the proof. □

Remark: Time has moved on and these results have now been hugely superseded. However one feels that at the time the method could have been exploited to obtain better results.

Indeed, the explicit solution they used for κ, λ in 1966 was not optimal. We have

$$\kappa \to \infty \ implies \ \lambda = \kappa - \sqrt{\kappa(\kappa - 1/4)} \to 1/8.$$

Thus, tweaking the method of Bombieri and Davenport and all things being equal we should get

$$\liminf_{n\to\infty} \frac{p_{n+1} - p_n}{\log p_n} \le 1/4,$$

a result not improved until 1988.

3.5 Maier's Matrix Method

A direct application of Cramer's rule, assigning a probability of $1/\log x$ to an integer in the neighbourhood of x that it is prime, would give the asymptotic relation

$$\pi(x + (\log x)^A) - \pi(x) \sim (\log x)^{A-1}.$$

However, Maier showed in 1985 [132] that this relation is false in that there are intervals near x of logarithmic length containing more than the expected number of prime numbers, and intervals containing fewer than the expected number.

Theorem 3.12   (Maier) *For all fixed $A > 2$, there is a constant $\delta_A > 0$ and two real positive sequences $x_n \to \infty$ and $y_n \to \infty$ such that for all $n \in \mathbb{N}$, we have*

$$\pi(x_n + (\log x_n)^A) - \pi(x_n) > (1 + \delta_A)(\log x_n)^{A-1} \ and$$
$$\pi(y_n + (\log y_n)^A) - \pi(y_n) < (1 - \delta_A)(\log y_n)^{A-1}.$$

We don't include the proof of Maier's interesting result since it would take us too far away from our prime gaps topic. However, we do provide a proof of a nice application of his method, which is in essence based on counting primes in an interval in two different ways. Even though it was published relatively recently and is not part of the timeline Table 1.1, a variant of the so-called "Maier's matrix method" [132], due to Andrew Granville, readily gives a pair of primes closer than $e^{-\gamma} \le 0.565$ times the average separation:

Theorem 3.13   (Granville) *There are two distinct primes in $(x, 2x]$ which differ by less than $e^{-\gamma} \log x$.*

Proof Let $N < n \leq 2N$ with $N = m^2$ and $m = \prod_{p \leq y} p$ a product of all primes up to y. Set $L = y \log y$, and $x = mN = m^3$. By the prime number theorem, we have $y \sim \log m = (\log x)/3$. We count the primes in the intervals $(mn+1, mn+L)$, for $N < n \leq 2N$ in two ways. Since $m \leq x^{1/3}$, we get

$$\sum_{n=N+1}^{2N} (\pi(mn + L) - \pi(mn + 1))$$

$$= \sum_{j=1}^{L} (\pi(2x, m, j) - \pi(x, m, j)) \sim \sum_{\substack{1 \leq j \leq L \\ (j,m)=1}} \frac{x}{\varphi(m) \log x}.$$

Taking the average of the m^2 terms in the sum on the left, using the prime number theorem and Merten's theorem, and noting that

$$\#\{1 \leq j \leq L : (j,m) = 1\} \geq \pi(y \log y) - \pi(y) = y(1 + o(1)),$$

and replacing x by m^3 on the right and cancelling, we get

$$\max_{n \in (N, 2N]} (\pi(mn + L) - \pi(mn + 1))$$

$$\gg \frac{L}{\log x} \frac{\#\{1 \leq j \leq L : (j,m) = 1\}}{(\varphi(m)L/m)} \geq \frac{L}{e^{-\gamma} \log x}.$$

Thus there is an interval of length L in $(x, 2x]$ which has at least $L/(e^{-\gamma} \log x)$ primes, so there must be two of these which differ by at most $e^{-\gamma} \log x$. This completes the proof. $\square$

3.6 End Notes

This completes our sample of early work on prime gaps. Table 1.1 in Chapter 1 shows a long period between 1988 and 2006 during which little progress was made. Dan Goldston, as a graduate student in the 1970s, became interested in the prime gaps problem. This was to eventuate in the breakthrough of the famous gang of four, GMPY, and its subset GPY, reported on in the next chapter, Chapter 4.

The PGpack function BombieriDavenportK relates to the work of this chapter. See Appendix I for the manual entry.

4

The Breakthrough of Goldston, Motohashi, Pintz and Yildirim

4.1 Introduction

In this chapter, our description of modern (twenty-first century) developments begins in earnest. However, we have already seen some of the intense early work and implicit use of the Selberg sieve in Chapter 3. This, plus the Dickson–Hardy–Littlewood conjecture and discussion of admissible tuples of Chapter 1, provides some context and preparation for what is to follow.

We begin with brief biographical details of the four mathematicians whose discoveries initiated modern developments on gaps between primes. Dan Goldston (Figure 4.1) was born in 1954. He completed a Ph.D. at the University of California at Berkeley in 1981, supervised by R. Sherman Lehman, with a thesis entitled "Large Differences Between Consecutive Primes". He has been a faculty member at San Jose State University since 1983.

The Japanese number theorist Yoichi Motohashi (Figure 4.2) obtained his Ph.D. from the University of Kyoto in 1974 with a thesis entitled "On Some Improvements to the Brun–Titchmarsh Theorem" [145]. He is a foreign member of the Finnischen Academy of Science and Letters and an honorary doctor of Turku University. He currently is a professor of Nihon University in Tokyo (retired).

János Pintz was born in 1950 in Budapest, Hungary. He completed his Ph.D. at Eőtvős Loránd University (more precisely from the Hungarian Academy of Sciences) in 1975 with a thesis entitled "Investigations in the Theory of Dirichlet's L-Functions", with Paul Turán as his supervisor. Since 1977, he has been a research fellow at the Alfréd Renyi Mathematical Institute of the Hungarian Academy of Sciences.

Cem Yildirim was born in 1961 in Indiana, but grew up and now works in Ankara, Turkey. He graduated Ph.D. from Toronto University in 1990 with a thesis entitled "Pair Correlation and Value Distribution", supervised by John Friedlander. He is currently at Boğazici University, Istanbul.

114

Figure 4.1 Pintz, Yildirim and Goldston (clockwise from the left). Printed with the permission of Ryoko Goldston.

Figure 4.2 Yoichi Motohashi (1944–). Courtesy of Haruko Motohashi.

4.2 Outline of the GPY Method

Goldston, Pintz and Yildirim (GPY) commence with an admissible k-tuple $\mathcal{H}$, i.e., $k \geq 2$ is fixed, $\mathcal{H} = \{h_1, h_2, \ldots, h_k\}$ is a set of distinct integers such that for each prime p the integers represent less than p congruence classes (see Section 1.10). The goal is to show that there are an infinite number of integers n such that each $n + \mathcal{H}$ contains at least two primes. This will be so if for all $N \in \mathbb{N}$ sufficiently large, we can find at least two primes in $n + \mathcal{H}$ for some n with $N < n \leq 2N$.

GPY chose to detect primes using the von Mangoldt function $\Lambda(n)$ (see Figure 4.3). This function, the reader will recall, has the value $\log p$ at prime powers p^m, and is zero otherwise. It satisfies ([4, theorem 2.11]) for $n \in \mathbb{N}$,

$$\Lambda(n) = \sum_{d|n} \mu(d) \log \left(\frac{n}{d}\right).$$

If the power $m > 1$, the prime power evaluations of $\Lambda(n)$ up to $2N$ make little difference, being in number for each prime $O(\log N)$ and with only $O(\pi(\sqrt{N}))$ primes to consider. The von Mangoldt function is not multiplicative, but, along with its modifications mentioned later, is sufficient for GPY's and GMPY's purposes, where GMPY = GPY + Motohashi.

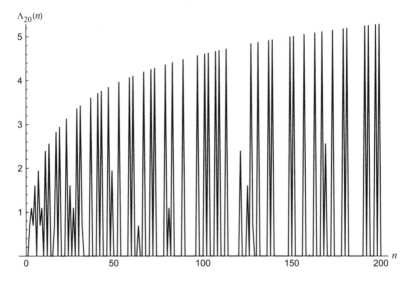

Figure 4.3 The von Mangoldt function for $0 \leq n \leq 200$.

Their key idea, and used in essence by others throughout these developments, is that if for some n with $2N \geq n > N > h_k$ and $r \geq 1$, we had

$$\left(\sum_{j=1}^{k} \theta(n + h_i) \right) - r \log(3N) > 0,$$

where $\theta(p) = \log p$ and $\theta(n) = 0$ when $n \notin \mathbb{P}$, then since each $\theta(n + h_j) < \log(3N)$ we must have at least $r + 1$ of the $\theta(n + h_j) > 0$. Therefore, we must have at least $r + 1$ primes in $n + \mathcal{H}$.

Introducing a nonnegative weight function $w(n)$ enables us to define

$$S_1(N) := \sum_{n=N+1}^{2N} \left(\sum_{j=1}^{k} \theta(n + h_j) - r \log(3N) \right) w(n)$$

so, if for all N sufficiently large, $S_1(N) > 0$, we must have, for at least one n with $N < n \le 2N$, both $w(n) > 0$ and at least $r + 1$ primes in $n + \mathcal{H}$. (A note added in passing: we don't need all N sufficiently large, but only a sequence $N_j \to \infty$.)

Finding a viable definition for the weights $w(n)$ is difficult. Using squares makes the nonnegativity criteria easy to attain, but evaluating the two sums $\sum_{n \le N} \theta(n + h_j) w(n)$ and $\sum_{n \le N} w(n)$ goes to the heart of the challenge this problem provides.

GMPY succeeded using a number of strategies. They "went global" making the weights depend on all k-tuples with $h_k \le h$, h being a fixed but arbitrary parameter. They also "went local" by truncating the von Mangoldt function, reducing the number of terms in the divisor sum representation, and introducing a truncation level $R \ll N^{\frac{1}{4k} - \epsilon}$, which is very small relative to N when k is large (see Figure 7.4). They use

$$\Lambda_R(n) := \sum_{\substack{d|n \\ d \le R}} \mu(d) \log\left(\frac{R}{d}\right)$$

and

$$\Lambda_R(n, \mathcal{H}) := \Lambda_R(n + h_1) \cdots \Lambda_R(n + h_k),$$

and then define

$$w(n) := \sum_{\substack{(h_1, \ldots, h_k) \in [1, h]^k \\ h_j \text{ distinct}}} \Lambda_R(n, \{h_1, \ldots, h_k\})^2.$$

With these weights, we have

$$S_2(N) := \sum_{n=N+1}^{2N} \left(\sum_{1 \le h_0 \le h} \theta(n + h_0) - r \log(3N) \right) w(n).$$

According to GPY, if we define

$$\Delta := \liminf_{n \to \infty} \frac{p_{n+1} - p_n}{\log p_n}$$

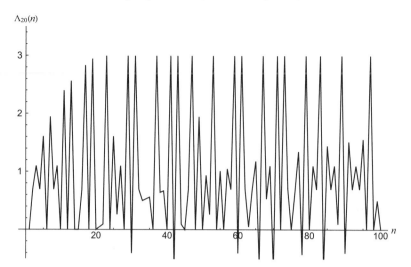

Figure 4.4 The truncated von Mangoldt function on $[0, 100]$ with $R = 20$.

in the preceding approach described, GPY show that to get $S_2(N) > 0$ with $r = 1$ requires $h \geq \frac{3}{4} \log N$. Thus, giving h its minimum value

$$\Delta = \liminf_{n \to \infty} \frac{p_{n+1} - p_n}{\log p_n} \leq \liminf_{N \to \infty} \frac{h}{\log N} = \frac{3}{4} \implies \Delta \leq \frac{3}{4},$$

a result already superseded in 1966 – see Table 1.1. Using linear combinations of tuples of lengths between 2 and k and optimizing enabled GPY to reduce this to $\Delta \leq \frac{1}{4}$, ie., about the level attained by 1988 – see Table 1.1.

A distinct disadvantage of this approach is that to form the weights, because of the squares many sums over divisors are multiplied together, forcing them to be made very short, i.e., choosing R very small relative to N. GPY overcame this by first, as is the case in many sieve theory applications, defining a polynomial

$$P(n) = (n + h_1) \cdots (n + h_k),$$

and then using the generalized von Mangoldt function $\Lambda_k(n)$, which is zero if n has more than k distinct prime factors (see Lemma 4.1). It can be defined by

$$\Lambda_k(n) := \sum_{d \mid n} \mu(d) \left(\log \frac{n}{d} \right)^k,$$

so $\Lambda_k(P(n)) \neq 0$ whenever each of the k translates $n + h_j$ is a prime, and we detect prime tuples and some other tuples. Truncating at level R leads to the definition

$$\Lambda_R(n, \mathcal{H}) := \frac{1}{k!} \sum_{\substack{d \mid P(n) \\ d \leq R}} \mu(d) \left(\log \frac{R}{d} \right)^k,$$

where in this definition of Λ_R we have reassigned the role of the subscript to be that of the truncation level R with k being picked up from the tuple $\mathcal{H}$. This allows us to detect a prime tuple with a single divisor sum. However, when applied it resulted in $\Delta \leq 0.1339$. To quote from Dan Goldston's paper [69]:

This was disappointing to me and Yildirim in 2004, but we should have taken to heart the advice, "Never give up!". It turns out there was only one more idea needed to break through the barrier, and this was discovered by Pintz.

The new idea was to introduce an additional parameter l, as small as possible, with $1 \leq l \leq k - 2$ and consider k-tuples $\mathcal{H}$, detecting prime tuples with the revised truncated divisor sum

$$\Lambda_R(n, \mathcal{H}, l) := \frac{1}{(k+l)!} \sum_{\substack{d|P(n) \\ d \leq R}} \mu(d) \left(\log \frac{R}{d} \right)^{k+l}.$$

If $\Lambda_R(n, \mathcal{H}, l) \neq 0$, then $\mathcal{H}$ should have sufficiently often $k - l \geq 2$ prime terms. This plausible deduction relies on being able to estimate the sums which are required, being able to control the errors arising from the von Mangoldt function detecting distinct prime powers rather than primes, having a value of R in terms of N which enables the error arising from the von Mangoldt function being truncated to be controlled and of course there being the expected proportion of k-tuples with $k - l$ prime values.

The expression to check for positivity then becomes

$$S_3(N) := \sum_{n=N+1}^{2N} \left(\sum_{1 \leq h_0 \leq h} \theta(n + h_0) - r \log(3N) \right) \sum_{\substack{(h_1, \ldots, h_k) \in [1,h]^k \\ h_j \text{ distinct}}} \Lambda_R(n, \mathcal{H}, l)^2.$$

If this expression is positive for some N, arbitrarily large, then the first factor must be positive for at least one $n \in [N, 2N]$ and thus $\theta(n + h_j) > 0$ for at least $r + 1$ values of j, so at least $r + 1$ of the $n + h_j$ must be prime. Therefore this enabled $\Delta = 0$ to be deduced, using a much greater value of $R \ll N^{\frac{1}{4}-\epsilon}$.

One is led to remark that the achievement here of GMPY, and it was rather wonderful, was to derive asymptotic expressions for the sums that were required to obtain a tractable value for $S_3(N)$. They used a mixture of combinatorics, complex analysis and sheer determination. Their intuition was that this particular $w(n)$ would make the completion of this goal possible.

This determination and intuition has been recorded in a succession of articles in the magazine *Science* by Barry Cipra and Dana Mackenzie [30, 31, 130, 131] published between April 2003 and May 2005. Of particular interest is the discovery of an error by Andrew Granville and Kannan Soundararajan in an early draft, but the correct ideas were most helpful to Ben Green and Terence Tao in proving their great theorem on the existence of primes in arithmetic progressions of arbitrary length [86].

Following this introduction, we now give the technical details of the GMPY method, guided by GPY's work set out in [72]. The main theorem is proved in Section 4.7.

Then in Section 4.8, we give the details of a simpler and shorter proof of the main result. This was due in part to a significant contribution of Yoichi Motohashi, who, among others, had been sent a form of the draft manuscript to check. The simplified form of the proof has been very influential, and the busy reader might with benefit skip ahead to the section.

The final main Section 4.9 exploits the use of double tuples and an optimization step, under the Elliott–Halberstam (EH) conjecture on the distribution of primes (defined in that section), to give the proof that there are an infinite number of $n \in \mathbb{N}$ with $p_{n+1} - p_n \leq 16$. This result relies on lemmas given for the longer proof and is conditional. The longer proof has been included, at least in part, because the use of EH has been very influential and, in this regard, GMPY's lead has been followed by each of the other authors.

4.3 Definitions and Summary

In this section, we assemble most of the special definitions which are used in the proof of the main Theorem 4.19 for ease of reference. (An individual definition is often repeated near the place where it is used.) This is followed by a summary of the principal lemmas used in the proof of the main result.

The symbol p is reserved for rational primes, and $k \in \mathbb{N}$ is a fixed natural number with $k \geq 2$. The arithmetic function $\omega(n)$ is the number of distinct prime divisors of $n \in \mathbb{N}$. The symbol N will be a natural number, which can be arbitrarily large. The expression $\lceil y \rceil$ is the smallest integer greater than or equal to real y and $\lfloor y \rfloor$ the integer part of y.

Let $h \in \mathbb{N}$, $h \geq 2$, and $\mathcal{H} = (h_1, h_2, \ldots, h_k)$, a k-tuple of distinct integers, with $0 \leq h_1 < h_2 < \cdots < h_k \leq h$ and $v_p(\mathcal{H})$ the number of distinct residues $h_i \bmod p$. Recall the definition: we say $\mathcal{H}$ is **admissible** if $v_p(\mathcal{H}) < p$ for all primes p. If $\mathcal{H} + n$ is to be prime for every $h_j + n$ for more than one integer n, it is necessary that $\mathcal{H}$ be admissible. For background on admissible tuples, see Section 1.10.

Following GPY, we consider two tuples $\mathcal{H}_1$ and $\mathcal{H}_2$ simultaneously. Let $l_1, l_2 \geq 0$ and

$$\mathcal{H} = \mathcal{H}_1 \cup \mathcal{H}_2, \ k_1 := |\mathcal{H}_1|, \ k_2 := |\mathcal{H}_2|, \ r := |\mathcal{H}_1 \cap \mathcal{H}_2|, k := k_1 + k_2,$$

$$M := k + l_1 + l_2. \quad (4.1)$$

It follows that $|\mathcal{H}| = k - r$.

Although employing two tuples makes the arguments more intricate and taxing for the reader, and can be avoided (see the simpler proof in Section 4.8), it is a useful strategy with an application in Section 4.9 which may have been the initial spark for the minimization techniques used by Maynard and Polymath8b described in Chapters 6 and 7 respectively. There may be further applications, yet to be discovered, using multiple tuples and a mixture of GMPY/Zhang/Tao/Maynard's methods, although this would take considerable courage to undertake.

In addition to these specifications, we have h_0 satisfying $1 \leq h_0 \leq h$ and set $\mathcal{H}^0 = \mathcal{H} \cup \{h_0\}$.

Let

$$\Delta := \prod_{1 \leq i < j \leq k} |h_j - h_i|. \tag{4.2}$$

In addition, if we define, for a fixed constant $C \geq \frac{1}{2}$, (it may need to be larger)

$$U := Ck^2 \log(2h) \implies \log \Delta \leq U. \tag{4.3}$$

Define as in Section 1.6 an infinite product which converges for all $k \geq 2$:

$$\mathcal{G}(\mathcal{H}) := \prod_p \left(1 - \frac{1}{p}\right)^{-k} \left(1 - \frac{v_p(\mathcal{H})}{p}\right). \tag{4.4}$$

Then $\mathcal{H}$ is admissible if and only if $\mathcal{G}(\mathcal{H}) \neq 0$.

Next we consider the genesis of the key approximation used by Goldston, Pintz and Yildirim. Define the von Mangoldt function

$$\Lambda(n) := \begin{cases} \log p, & n = p^m, \\ 0, & \text{otherwise,} \end{cases}$$

so [4, theorem 2.11],

$$\Lambda(n) = \sum_{d|n} \mu(d) \log\left(\frac{n}{d}\right) = (\mu \star \log)(n).$$

We can approximate this function by restricting the sum to small divisors, defining for $R \in \mathbb{N}$ (we will later adopt the restriction $R \leq N^{\frac{1}{4}-\epsilon}$),

$$\Lambda_R(n) := \sum_{\substack{d|n \\ d \leq R}} \mu(d) \log\left(\frac{R}{d}\right). \tag{4.5}$$

Define

$$\Lambda(n, \mathcal{H}) := \Lambda(n + h_1) \cdots \Lambda(n + h_k), \tag{4.6}$$

so this function is nonzero only if the tuple $n + \mathcal{H}$ is a prime power in every component $n + h_i$. We can approximate this function in turn by defining

$$\Lambda_R(n, \mathcal{H}) := \Lambda_R(n + h_1) \cdots \Lambda_R(n + h_k). \tag{4.7}$$

Goldston, Pintz and Yildirim succeed in proving their theorem 4.19 by defining a better adapted approximation than this. First, let

$$P(n) := (n + h_1) \cdots (n + h_k) \tag{4.8}$$

be a polynomial function in n with positive integer coefficients, and define the so-called kth generalized von Mangoldt function:

$$\Lambda_k(n) = \sum_{d|n} \mu(d) \left(\log \frac{n}{d}\right)^k = (\mu \star \log^k)(n). \tag{4.9}$$

For example [153, theorem 9.1], $\Lambda_2(n) = \Lambda(n) \log n + (\Lambda \star \Lambda)(n)$.

Then by Lemma 4.1, $\Lambda_k(n) = 0$ if and only if n has more than k distinct prime factors. We set

$$\Lambda_k(n, \mathcal{H}) := \frac{1}{k!} \Lambda_k((n + h_1) \cdots (n + h_k)) = \frac{1}{k!} \Lambda_k(P(n)), \tag{4.10}$$

and we truncate this function for a given $R \in \mathbb{N}$ by

$$\Lambda_{k,R}(n, \mathcal{H}) := \frac{1}{k!} \sum_{\substack{d|P(n) \\ d \leq R}} \mu(d) \left(\log \frac{R}{d}\right)^k = \frac{1}{k!} \sum_{\substack{d|P(n) \\ d \leq R}} \mu(d) L_{R/d}(k), \tag{4.11}$$

where it is often convenient to use the notation $L_X(Y) := (\log X)^Y$, and to drop the subscript k where this can be clearly deduced from the context.

These definitions are designed to detect prime powers in every component of a k-tuple $\mathcal{H}$, which is too much to ask with the methods we have at hand. Indeed, we seek to detect prime (powers) in some components, so introduce an additional parameter l with $0 \leq l \leq k$ enabling the k-tuple to have components which are not prime powers:

$$\Lambda_{k,R}(n, \mathcal{H}, l) := \frac{1}{(k + l)!} \sum_{\substack{d|P(n) \\ d \leq R}} \mu(d) L_{R/d}(k + l). \tag{4.12}$$

If $\mathcal{H} = \emptyset$, set $\Lambda_R(n, \emptyset, 0) := 1$.

Next let

$$\mathscr{S}_R(N, \mathcal{H}_1, \mathcal{H}_2, l_1, l_2) := \sum_{n=1}^{N} \Lambda_R(n, \mathcal{H}_1, l_1) \Lambda_{k,R}(n, \mathcal{H}_2, l_2). \tag{4.13}$$

As usual, let $\theta(n) := \log n$ when n is prime and 0 otherwise, and then define

$$\widetilde{\mathscr{F}}_R(N, \mathscr{H}_1, \mathscr{H}_2, l_1, l_2, h_0) := \sum_{n=1}^{N} \Lambda_R(n, \mathscr{H}_1, l_1) \Lambda_R(n, \mathscr{H}_2, l_2) \theta(n + h_0). \quad (4.14)$$

We also have

$$\mathcal{T}_R(\mathscr{H}) := \frac{1}{k!} \sum_{d \leq R} \frac{\mu(d) v_d(\mathscr{H})}{d} L_{R/d}(k),$$

where $v_d(\mathscr{H}) := \prod_{p|d} v_p(\mathscr{H})$.

Generalized divisor functions $\tau_k(n)$ for $n, k \in \mathbb{N}$ are normally defined as the number of ways of expressing n as k distinct, other than the factor 1, factors, with order counting. For example, $6 = 1 \cdot 6 = 2 \cdot 3$, so $\tau_2(6) = 4 = 2^2$, and

$$\tau_3(5) = \#\{5 \cdot 1 \cdot 1, \ 1 \cdot 5 \cdot 1, \ 1 \cdot 1 \cdot 5\} = 3^1.$$

This is the so-called Piltz function, and in general, if $m = p^{e_1} \cdots p_n^{e_n}$, then

$$\tau_k(m) = \prod_{j=1}^{n} \binom{k + e_j - 1}{k - 1}.$$

Hence for squarefree q, we have $\tau_k(q) = k^{\omega(q)}$.

GPY extend this definition for squarefree q and real m by setting

$$\tau_m(q) := m^{\omega(q)}. \quad (4.15)$$

It follows directly from the extended definition that for fixed $q \in \mathbb{N}$ and real α, β, η, we have

$$\tau_\alpha(q) \tau_\beta(q) = \tau_{\alpha\beta}(q) \text{ and } \tau_\alpha(q)^\eta = \tau_{\alpha^\eta}(q). \quad (4.16)$$

In terms of these divisor functions, define

$$D(x, m) := \sum_{\substack{q \leq x \\ \text{squarefree}}} \tau_m(q) \text{ and } D'(x, m) := \sum_{\substack{q \leq x \\ \text{squarefree}}} \frac{\tau_m(q)}{q}. \quad (4.17)$$

Chapter Summary:

We now give a summary of the contents of this chapter. A network flow of dependencies between the lemmas and theorems is given in Figure 4.5. The reader is advised to skim this part and refer back to it as progress is made through the detail which follows.

Section 4.1 contains brief biographical information about Goldston, Motohashi, Pintz and Yildirim, denoted "GMPY". Section 4.2 summarizes Dan Goldston's account of how GPY arrived at their method, giving some of the blocked, and then unblocked, paths to the proof. This Section 4.3 gives an overview of the essential definitions which are used. Section 4.4 gives preliminary lemmas and Section 4.5 fundamental lemmas for the proof.

The first main lemma gives an asymptotic value for the sum of the Λ_R with fixed $\mathcal{H}$. Note that the cutoff R is chosen to be a positive power less than 1 of N. This lemma is used many times in the derivation of the main result of GPY, namely Theorem 4.19,

$$\liminf_{n \to \infty} \frac{p_{n+1} - p_n}{\log p_n} = 0.$$

Lemma 4.7 *[72, proposition 3] Let $k \geq 1$, $l = 0$, $h \geq 2$, $\eta_0 > 0$ be fixed and arbitrarily small, $C > 1$ a fixed real number and*

$$k \ll_{\eta_0} (\log R)^{\frac{1}{2} - \eta_0} \text{ and } h_k \leq h \leq R^C.$$

Then as $N, R \to \infty$

$$\sum_{n=1}^{N} \Lambda_R(n, \mathcal{H}) = \mathcal{G}(\mathcal{H})N + O\left(Ne^{-c\sqrt{\log R}}\right) + O\left(R(2 \log R)^{2k}\right).$$

Lemma 4.8 is also an essential prelude to Theorem 4.19. It is a derivation of the main term of an integral which depends on a very wide range of possible functions $G(s_1, s_2)$. For the precise conditions, see the statement of the lemma.

Lemma 4.8 *[72, lemma 3] Let M be a large constant and a, b, d, u, v real parameters such that*

$$0 \leq a, b, d, u, v \leq M, \ a + u \geq 1, \ b + v \geq 1, \ d \leq \min(a, b).$$

Let $C > 0$ be a fixed positive constant and $h \ll R^C$. Then as $R \to \infty$, we have

$$T_R^*(a, b, d, u, v, h) := \frac{1}{(2\pi i)^2} \int_{(1)} \int_{(1)} \frac{D(s_1, s_2) R^{s_1 + s_2}}{s_1^{u+1} s_2^{v+1} (s_1 + s_2)^d} \, ds_1 \, ds_2,$$

$$= \binom{u+v}{u} \frac{L_R(u + v + d)}{(u + v + d)!} G(0, 0)$$

$$+ \sum_{j=1}^{u+v+d} \mathcal{D}_j(a, b, d, u, v, h) L_R(u + v + d - j) + O_M\left(\exp(-c\sqrt{\log R})\right),$$

where $W(s) := s\zeta(s + 1)$ with $D(s_1, s_2) := \frac{G(s_1, s_2) W^d(s_1 + s_2)}{W^a(s_1) W^b(s_2)}$, and where G and $\mathcal{D}_j$ are functions which satisfy some specified regularity and boundedness conditions.

As before, let

$$\mathcal{H} = \mathcal{H}_1 \cup \mathcal{H}_2, \ k_1 = |\mathcal{H}_1|, \ k_2 = |\mathcal{H}_2|, \ r = |\mathcal{H}_1 \cap \mathcal{H}_2|, k = k_1 + k_2,$$

$$M = k + l_1 + l_2.$$

Lemma 4.9 *[72, proposition 4] Let $C > 0$ be a fixed constant and $h \ll R^C$. Then when $R, N \to \infty$, we have*

$$\sum_{n=1}^{N} \Lambda_R(n, \mathcal{H}_1, l_1) \Lambda_R(n, \mathcal{H}_2, l_2) = \frac{N}{(r + l_1 + l_2)!} \binom{l_1 + l_2}{l_1} \mathcal{G}(\mathcal{H}) L_R(r + l_1 + l_2)$$

$$+ N \sum_{j=1}^{r+l_1+l_2} \mathcal{D}_j(l_1, l_2, \mathcal{H}_1, \mathcal{H}_2) L_R(r + l_1 + l_2 - j)$$

$$+ O_M \left(N \exp\left(-c\sqrt{\log R}\right)\right) + O(R^2 (3 \log R)^{3k+M}).$$

We also need sums of products of the Λ_R with possibly different tuple and l values with values of $\theta(n)$. The reader might wish to refer back to see the structure of the sums $S_j(N)$ in Section 4.2. Note our continued use of the notation $L_R(k) = (\log R)^k$, which emphasizes that the "action" is normally with the power of the logarithm. The corresponding sum which is derived has an explicit dependency on primes through θ, as well as Λ_R, and a similar form to that found in Lemma 4.9. However, note the relatively large size of the error term.

Lemma 4.10 *[72, proposition 5] Let $\mathcal{H}_1 \neq \emptyset$, $\mathcal{H}_2 \neq \emptyset$, $h \ll R$, $h_0 \notin \mathcal{H}$ and $\mathcal{H}^0 := \mathcal{H} \cup \{h_0\}$. For all $A > 0$, there exists a B dependent at most on A and M such that if R and N are sufficiently large and $R \ll_{M,A} N^{\frac{1}{4}} / L_N(B)$, then*

$$\sum_{n=1}^{N} \Lambda_R(n, \mathcal{H}_1, l_1) \Lambda_R(n, \mathcal{H}_2, l_2) \theta(n + h_0)$$

$$= \frac{N}{(r + l_1 + l_2)!} \binom{l_1 + l_2}{l_1} \mathcal{G}(\mathcal{H}^0) L_R(r + l_1 + l_2)$$

$$+ N \sum_{j=1}^{r} \mathcal{D}_j(l_1, l_2, \mathcal{H}_1, \mathcal{H}_2, h_0) L_R(r + l_1 + l_2 - j) + O_{M,A}\left(\frac{N}{L_N(A)}\right).$$

Lemma 4.11 is just a simplification of Lemma 4.9.

Lemma 4.11 *[72, proposition 1] If $R \ll N^{\frac{1}{2}} / L_N(4M)$, and $h \leq R^C$ for any fixed $C > 0$. Then as $R, N \to \infty$, we have*

$$\sum_{n=1}^{N} \Lambda_R(n, \mathcal{H}_1, l_1) \Lambda_R(n, \mathcal{H}_2, l_2) = N \binom{l_1 + l_2}{l_1} \frac{L_R(r + l_1 + l_2)}{(r + l_1 + l_2)!} (\mathcal{G}(\mathcal{H}) + o_M(1)).$$

Lemma 4.12 is a derivation from Lemma 4.10 which incorporates estimates for the $\mathcal{D}_j$. It provides essential information for GPY's main result by showing that the main term of the prime sum depends on whether h_0 is in both $\mathcal{H}_1$ and $\mathcal{H}_2$ or in neither. We don't actually need the third possibility.

Lemma 4.12 *[72, proposition 2] Let $B(M) > 0$ be sufficiently large, $R \ll_M N^{\frac{1}{4}}/L_N(B(M))$ with $B(M)$ sufficiently large and positive and $h \le R$. Then as $R, N \to \infty$, we have in case (a) $h_0 \in \mathcal{H}_1 \cap \mathcal{H}_2$, so $\mathcal{H}^0 = \mathcal{H} := \mathcal{H}_1 \cup \mathcal{H}_2$, or case (b) $h_0 \notin \mathcal{H}$, we have respectively for the "prime" sum*

$$\sum_{n=1}^{N} \Lambda_R(n, \mathcal{H}_1, l_1) \Lambda_R(n, \mathcal{H}_2, l_2) \theta(n + h_0),$$

the estimate

$$(a) \ N \binom{l_1 + l_2 + 2}{l_1 + 1} \frac{L_R(r + l_1 + l_2 + 1)}{(r + l_1 + l_2 + 1)!} \left(\mathcal{G}(\mathcal{H}^0) + o_M(1) \right),$$

$$(b) \ N \binom{l_1 + l_2}{l_1} \frac{L_R(r + l_1 + l_2)}{(r + l_1 + l_2)!} \left(\mathcal{G}(\mathcal{H}^0) + o_M(1) \right).$$

Section 4.6 gives Gallagher's tuples theorem, with two alternative proofs, one by Ford and the other by Pintz, underlining its broad applicability in work with k-tuples. It is an essential part of GPY's and GMPY's arguments, but is not used explicitly in the other works reported here. (But consider Lemma 3.9.) Note that $\mathcal{H}$ is regarded as a set with a fixed number of elements, so each $\mathcal{H}$ gives one term on the left of the theorem identity, i.e., order does not count here.

Theorem 4.14   (Gallagher) *[67, 68] Let $k \in \mathbb{N}$. Then as $h \to \infty$, we have*

$$\sum_{\substack{\mathcal{H} \subset [1, h] \\ |\mathcal{H}| = k}} \mathcal{G}(\mathcal{H}) = (1 + o(1)) \frac{h^k}{k!}.$$

Section 4.7 includes the proof of the main GMPY theorem:

Theorem 4.19   (Goldston–Motohashi–Pintz–Yildirim's prime gap estimate) *Let p_n be the nth prime. Then*

$$\Delta_1 := \liminf_{n \to \infty} \frac{p_{n+1} - p_n}{\log p_n} = 0.$$

The chapter also includes, in Section 4.8, GMPY's simplified version of the main lemmas, i.e., the case where $\mathcal{H} := \mathcal{H}_1 = \mathcal{H}_2$ and $l := l_1 = l_2$. This was published before the *Annals of Mathematics* paper – see Section 4.8 for the reference. The only situation in which distinct l_1 and l_2 are essential is Theorem 4.19, where the Elliott–Halberstam conjecture is used. The simplified version is embodied in Lemmas 4.20 and 4.21.

Lemma 4.20 *If $h \ll \log N$ with $R \le \sqrt{N}/L_N(C)$ for $C > 0$ sufficiently large, and k, l arbitrary and fixed integers with $k \ge 2$ and $l > 0$, then*

$$\sum_{N < n \le 2N} \Lambda_R(n, \mathcal{H}, k+l)^2$$

$$= \frac{\mathcal{G}(\mathcal{H})}{(k+2l)!} \binom{2l}{l} N L_R(k+2l) + O\left(N L_N(k+2l-1)(\log\log N)^c\right).$$

Lemma 4.21 *Assume the primes have a level of distribution ϑ for some fixed ϑ with $0 < \vartheta < 1$. Assume also $h \ll \log N$ and k, l are arbitrary fixed integers with $k \ge 2$ and $l > 0$, and that $R \le N^{\vartheta/2}$. Define the "prime sum"*

$$\mathcal{S} := \sum_{N < n \le 2N} \theta(n+h) \Lambda_R(n, \mathcal{H}, k+l)^2.$$

If $h \notin \mathcal{H}$, we have, for some $c > 0$,

$$\mathcal{S} = \frac{\mathcal{G}(\mathcal{H} \cup \{h\})}{(k+2l)!} \binom{2l}{l} N L_R(k+2l) + O\left(N L_N(k+2l-1)(\log\log N)^c\right).$$

If however we have $h \in \mathcal{H}$, we get

$$\mathcal{S} = \frac{\mathcal{G}(\mathcal{H})}{(k+2l+1)!} \binom{2l+2}{l+1} N L_R(k+2l+1) + O\left(N L_N(k+2l)(\log\log N)^c\right).$$

Then in the final Section 4.9, reverting back to using l_1 and l_2, we give the proof of GPY's application using the Elliott–Halberstam conjecture [72, theorem 1], which could have provided the initial spark for further developments (see Chapter 9) and even Maynard's method (see Chapter 6). This has been included because others, following GMPY's lead, often use the conjecture to derive the best possible conjectural application of their method.

Theorem 4.23 **(GPY's bounded gaps assuming EH)** *Assume the prime numbers have a level of distribution $\vartheta > \frac{1}{2}$. Then there is an effective constant c_ϑ such that any admissible k-tuple with $k \ge c_\vartheta$ contains at least two primes infinitely often. In particular, if $\vartheta \ge 0.971$, then we can choose $c_\vartheta = 6$. Using the admissible $\mathcal{H} = \{0, 4, 6, 10, 12, 16\}$. This shows assuming EH there are an infinite number of integers n such that $p_{n+1} - p_n \le 16$.*

The flow graph Figure 4.5 shows most of the dependencies between the results in this chapter.

4.4 General Preliminary Results

By Property (5) in Section 1.8, we have

Lemma 4.1 *For all $k \ge 1$, the generalized von Mangoldt function $\Lambda_k(n)$ is zero if n has more than k distinct prime factors.*

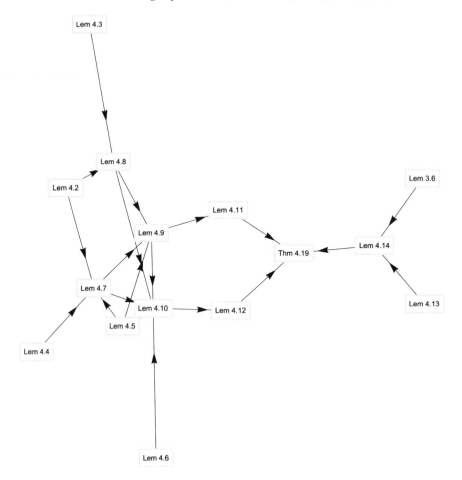

Figure 4.5 Some Chapter 4 dependencies.

We also need some estimates for the zeta function and its logarithmic derivative.

Lemma 4.2 *([144, theorem 6.6, 6.7] and [4, section 13.4]) There is a positive constant $\eta \geq 1/20$ such that the Riemann zeta function $\zeta(s)$ is never zero in the region*

$$\Omega := \left\{ s = \sigma + it \in \mathbb{C} : \sigma \geq 1 - \frac{\eta}{\log(|t| + 3)} \right\}.$$

In Ω, we have the following estimates:

$$(a) \quad \zeta(\sigma + it) - \frac{1}{\sigma - 1 + it} \ll \log(|t| + 3),$$

$$(b) \quad \frac{1}{\zeta(\sigma + it)} \ll \log(|t| + 3),$$

$$(c) \quad \frac{\zeta'(\sigma + it)}{\zeta(\sigma + it)} + \frac{1}{\sigma - 1 + it} \ll \log(|t| + 3).$$

Proof **(1)** The zero free region Ω follows for example from [21, theorem 2.4]. For the estimate (a), using Euler–Maclaurin summation we get for $\sigma > 0$ ([4, theorem 12.21]),

$$\zeta(s) = \sum_{n=1}^{N} \frac{1}{n^s} - s \int_{N}^{\infty} \frac{x - \lfloor x \rfloor}{x^{s+1}} \, dx + \frac{N^{1-s}}{s-1}. \tag{4.18}$$

If $\sigma \geq 2$, then $|\zeta(s)| \leq \zeta(2)$, so $\zeta(s) \ll \log(|t|)$, and we can assume $\sigma < 2$. We have for $t \geq e$,

$$|s| \leq \sigma + t \leq 2|t| \text{ and } |s - 1| \geq |t| \implies \frac{1}{|s - 1|} \leq \frac{1}{|t|}.$$

Therefore,

$$|\zeta(s)| \leq \sum_{n=1}^{N} \frac{1}{n^\sigma} + 2|t| \int_{N}^{\infty} \frac{1}{x^{\sigma+1}} \, dx + \frac{N^{1-\sigma}}{|t|}.$$

Next, let $N = \lfloor |t| \rfloor$ so $N \leq |t| < N + 1$ and for $n \leq N$ we have $\log n \leq \log |t|$. Thus, using

$$\sigma \geq 1 - \frac{\eta}{\log(|t|)},$$

we get for some fixed $A > 0$,

$$\frac{1}{n^\sigma} = \frac{1}{n} \exp((1 - \sigma) \log n) < \frac{\exp(A \log n / \log |t|)}{n} = O\left(\frac{1}{n}\right).$$

Therefore, since in Ω we have σ bounded away from 0,

$$\frac{2|t|}{\sigma N^\sigma} \ll \frac{N + 1}{N} \ll 1 \text{ and } \frac{N^{1-\sigma}}{|t|} \ll \frac{1}{N} \ll 1.$$

It follows that

$$\left| \zeta(s) - \frac{1}{s - 1} \right| \ll \left(\sum_{n=1}^{N} \frac{1}{n} \right) + O\left(\frac{1}{|t|}\right) + O(1) \ll \log(|t| + 3).$$

On $[0, 2] \times [-e, e]$, $\zeta(s) - 1/(s - 1)$ is entire, and therefore bounded, so (a) holds on all of Ω.

(2) By (4.18), for $\sigma > 0$ setting $N = 1$ we get

$$\zeta(s) = 1 + \frac{1}{s - 1} - s \int_{1}^{\infty} \frac{\{u\}}{u^{s+1}} \, du. \tag{4.19}$$

Since for all $u \in \mathbb{R}$ we have $0 \leq \{u\} < 1$, we have for $\sigma > 0$,

$$0 \leq \int_1^\infty \frac{\{u\}}{u^{\sigma+1}} \, du < \int_1^\infty \frac{1}{u^{\sigma+1}} \, du = \frac{1}{\sigma}. \tag{4.20}$$

(3) For $\sigma > 1$, logarithmic differentiating the absolutely convergent Euler product of $\zeta(s)$ gives

$$-\frac{\zeta'(s)}{\zeta(s)} = \sum_{n=1}^\infty \frac{\Lambda(n)}{n^s}.$$

Therefore, since $-\zeta'(s)/\zeta(s)$ and $1/(s-1)$ both have simple poles at $s = 1$ with residues 1, their difference is holomorphic at that point, so as $\sigma \to 1+$ we get

$$\left| \frac{\zeta'(s)}{\zeta(s)} \right| \leq \sum_{n=1}^\infty \frac{\Lambda(n)}{n^\sigma} = -\frac{\zeta'(\sigma)}{\zeta(\sigma)} \ll \frac{1}{\sigma - 1}.$$

This estimate is true for $\sigma \geq 1 + 1/\log \tau$, where $\tau = |t| + 4$, and substituting $z = 1 + 1/\log \tau + it$, we get

$$\frac{\zeta'(z)}{\zeta(z)} \ll \log \tau.$$

(4) Next we apply the Borel–Carathéodory Theorem F.3 to $f(z) := \zeta(z+3/2+it)$ setting $r = 2/3$ and $R = 5/6$. Firstly, because the Euler product for $\zeta(s)$ converges absolutely in $\sigma > 1$ to a nonzero value, we get $|f(0)| \gg 1$. From (4.19), we also have $f(z) \ll \tau$ for $|z| \leq 1$. Therefore, for $|t| \geq 7/8$ and $5/6 \leq \sigma \leq 2$, and using Lemma F.5, we get

$$\frac{\zeta'(s)}{\zeta(s)} = \sum_{\rho \in D} \frac{1}{s - \rho} + O(\log \tau),$$

where D is the disk $B(3/2 + it, 5/6]$ and the sum is over the zeros of $\zeta(s)$ in that disk.

Combining this with the result of Step (3) gives

$$\sum_{\rho \in D} \Re \frac{1}{z - \rho} \ll \log \tau.$$

(5) Now if

$$1 - \frac{\eta}{2 \log \tau} \leq \sigma \leq 1 + \frac{1}{\log \tau},$$

then using Step (4) we can write

$$\frac{\zeta'(s)}{\zeta(s)} - \frac{\zeta'(z)}{\zeta(z)} = \sum_{\rho \in D} \left(\frac{1}{s - \rho} - \frac{1}{z - \rho} \right) + O(\log \tau).$$

But there are absolute constants $\alpha, \beta > 0$ such that for all zeta zeros $\rho \in D$ we have $\alpha|s - \rho| \leq |z - \rho| \leq \beta|s - \rho|$. This implies we can write

$$\frac{1}{s - \rho} - \frac{1}{z - \rho} \ll \frac{1}{|z - \rho|^2 \log \tau} \ll \mathfrak{R}\frac{1}{z - \rho}.$$

Combining these estimates then gives $\zeta'(s)/\zeta(s) \ll \log \tau$.

(6) Again from the Euler product, we get for $\sigma > 1$,

$$\log \zeta(s) = \sum_{p} \sum_{n=1}^{\infty} \log\left(1 - \frac{1}{p^s}\right) = \sum_{p} \sum_{n=1}^{\infty} \frac{1}{np^{ns}}.$$

Therefore,

$$|\log \zeta(s)| \leq \sum_{n=2}^{\infty} \frac{\Lambda(n)}{\log(n)n^\sigma} = \log \zeta(\sigma).$$

Inserting the bound from (4.20) in (4.19), we get

$$\zeta(\sigma) < 1 + \frac{1}{\sigma - 1}.$$

Therefore, if $\sigma \geq 1 + 1/\log \tau$ we have $|\log \zeta(s)| \leq \log\log \tau + O(1)$ and the same estimate then applies at $z = 1 + 1/\log \tau + it$. If $\sigma < 1 + 1/\log \tau$ and $s \in \Omega$, then consider

$$\log \zeta(s) - \log \zeta(z) = \int_z^s \frac{\zeta'(w)}{\zeta(w)} \, dw$$

taking the path of integration as the line segment $[z, s]$. Using the bound from Step (5) then gives for all $s \in \Omega$, $|\log \zeta(s)| \leq \log\log \tau + O(1)$.

(7) Since $\log 1/|\zeta(s)| = -\mathfrak{R} \log \zeta(s)$, using the result of Step (6) we get

$$\frac{1}{\zeta(s)} \ll \log \tau \ll \log(|t| + 3).$$

This completes the proof of (b).

(8) Finally, consider $g(s) := \log(s-1)\zeta(s)$ on Ω. Note that $\mathfrak{R} \log \zeta(s) \leq |\log \zeta(s)|$, so using the result of Step (6),

$$\mathfrak{R}g(s) = \mathfrak{R} \log((s - 1)\zeta(s)) = \log|s - 1| + \log|\zeta(s)| = \log|s - 1| + \mathfrak{R} \log \zeta(s)$$
$$\leq \log|s - 1| + |\log \zeta(s)| \ll \log \tau.$$

Therefore, by Theorem F.3 we have derived

$$\frac{\zeta'(s)}{\zeta(s)} + \frac{1}{s - 1} \ll \log \tau.$$

This completes the proof of (c). $\qquad\square$

The GPY method depends on the truth of a particularly nasty combinatorial binomial identity. A proof was found using symbolic computation (Mathematica), which readers may wish to verify, or provide their own derivation. None of the other methods described in this book use this result, at least directly. We employ Krummer's generalized hypergeometric series $_1F_1(a,b,x)$, which is defined for $a,b,z \in \mathbb{C}$ in terms of so-called rising factorials:

$$a^{(0)} := 1,$$
$$a^{(n)} := a(a+1)\cdots(a+n-1), \ n > 0,$$
$$_1F_1(a,b,x) := \sum_{n=0}^{\infty} \frac{a^{(n)}z^n}{b^{(n)}n!}.$$

See for example [161, chapter 5].

Lemma 4.3 *For all $u,v,d \geq 0$, we have the identity*

$$\binom{u+v}{u}\frac{1}{(u+v+d)!} = \frac{1}{u!}\sum_{j=0}^{u}\binom{u}{j}(-1)^j\frac{d(d+1)\cdots(d+j-1)}{(v+d+j)!}. \qquad (4.21)$$

Proof Multiply both sides by x^u and expand in a power series for $0 \leq u < \infty$. Then on the right interchange the summations over u and j. We obtain, with the help of Mathematica, on the left

$$\frac{_1F_1(1+v,1+d+v,x)}{(d+v)!}.$$

On the right, we obtain

$$e^x\frac{_1F_1(d,1+d+v,-x)}{(d+v)!}.$$

Expanding each of these as Taylor series gives the same expression:

$$\frac{1}{(d+v)!} + \frac{(v+1)x}{(d+v+1)(d+v)!} + \frac{(v+1)(v+2)x^2}{2(d+v+1)(d+v+2)(d+v)!}$$
$$+ \frac{(v+1)(v+2)(v+3)x^3}{6(d+v+1)(d+v+2)(d+v+3)(d+v)!}$$
$$+ \frac{(v+1)(v+2)(v+3)(v+4)x^4}{24(d+v+1)(d+v+2)(d+v+3)(d+v+4)(d+v)!} + O\left(x^5\right),$$

which shows, using the hypergeometric differential equation, all coefficients are equal. This completes the proof. ☐

The following lemma provides an estimate for a contour integral, where the contour is the shifted boundary of the zero free region from Lemma 4.2.

Lemma 4.4 *Let $C > 0$ be a constant, $R \geq C$, $k \geq 2$, $B \leq Ck$. Let $\mathscr{L}$ be the complex contour*

$$\mathscr{L} := \left\{ s : s = -\frac{\eta}{\log(|t|+3)} + it \right\},$$

which is the boundary of the region Ω defined in Lemma 4.2 shifted by 1 to the left. Then there exist constants c_1, c_2 dependent only on C such that

$$\int_{\mathscr{L}} (\log(|s|+3))^B \left| \frac{R^s}{s^k} \right| ds \ll_C c_1^k R^{-c_2} + \exp(-\sqrt{\eta \log R}/2). \tag{4.22}$$

If we also have $k \leq c_3 \log R$, with c_3 dependent only on C and taken sufficiently small, then we have the simpler estimate

$$\int_{\mathscr{L}} (\log(|s|+3))^B \left| \frac{R^s}{s^k} \right| ds \ll_C \exp(-\sqrt{\eta \log R}/2). \tag{4.23}$$

Proof Let c_4 be sufficiently large and dependent only on C, with ω a real parameter to be assigned later, and let $\sigma(t) := -\eta/\log(|t|+3)$, to get

$$\int_{\mathscr{L}} (\log(|s|+3))^B \left| \frac{R^s}{s^k} \right| ds \ll \int_0^\infty R^{\sigma(t)} \frac{(\log|t|+4)^B}{(|t|+\eta/2)^k} dt$$

$$\ll \int_0^{c_4} c_1^k R^{-c_2} dt + \int_{c_4}^{\omega-3} \frac{R^{-\frac{\eta}{\log(|t|+3)}}}{t^{3/2}} dt$$

$$+ \int_{\omega-3}^\infty \frac{1}{t^{3/2}} dt$$

$$\ll c_1^k R^{-c_2} + e^{-\frac{\eta \log R}{\log(\omega)}} + \frac{1}{\sqrt{\omega}}.$$

Finally, choose ω such that $\log \omega = \sqrt{\eta \log R}$ to complete the proof of the first part. The second part follows directly from this. $\qquad\square$

Recall we have defined for real $m > 0$ and squarefree q, the generalized divisor function $\tau_m(q) := m^{\omega(q)}$. Next we derive an estimate for a sum which depends on this function.

Lemma 4.5 *For all integer $m > 0$ and all $x \geq 1$, we have the bound*

$$\sum_{\substack{q \leq x \\ \text{squarefree}}} \frac{\tau_m(q)}{q} \leq (m + \log x)^m. \tag{4.24}$$

Proof Equation (4.24) is true for $m = 1$ since $\tau_1(q) = 1$ for all squarefree q. Assume it is true for $m - 1 \in \mathbb{N}$ with $m \geq 2$, and write $q = jq'$, representing a factorization of q into m factors with the smallest being $j \leq x^{1/m}$, where q' is a factorization of q/j into $m - 1$ factors. Then, because in an ordered factorization of q', j can sit in at most m places in the order, we have $\tau_m(jq') \leq m\tau_{m-1}(q')$ so using the inductive assumption

$$\sum_{\substack{q \leq x \\ \text{squarefree}}} \frac{\tau_m(q)}{q} \leq m \sum_{\substack{j=1 \\ j \text{ squarefree}}}^{x^{1/m}} \frac{1}{j} \sum_{\substack{q' \leq x/j \\ q' \text{squarefree}}} \frac{\tau_{m-1}(q')}{q'}$$

$$\leq m(1 + \log(x^{1/m}))(m - 1 + \log x)^{m-1}$$
$$\leq (m + \log x)(m + \log x)^{m-1} = (m + \log x)^m.$$

The result then follows by induction. $\qquad\square$

It follows directly from Lemma 4.5, with the given definitions of $D'(x,m)$ and $D(x,m)$, that for real $m > 0$ we have

$$D'(x,m) := \sum_{\substack{q \leq x \\ \text{squarefree}}} \frac{\tau_m(q)}{q} \leq (\lceil m \rceil + \log x)^{\lceil m \rceil} \leq (m + 1 + \log x)^{m+1}, \quad (4.25)$$

since $m \leq \lceil m \rceil < m + 1$ and since $D(x,m) \leq xD'(x,m)$, we also have

$$D(x,m) := \sum_{\substack{q \leq x \\ \text{squarefree}}} \tau_m(q) \leq x(\lceil m \rceil + \log x)^{\lceil m \rceil} \leq x(m + 1 + \log x)^{m+1}. \quad (4.26)$$

We use these estimates later.

The following lemma is a form of the well-known Bombieri–Vinogradov theorem; see Section 1.9 for an overview. For a proof and applications, see for example [22, chapter 12].

Lemma 4.6 (Bombieri–Vinogradov) *Let*

$$E(x,q,a) := \pi(x,q,a) - \frac{\text{li}(x)}{\varphi(q)} \text{ and } E(x,q) := \max_{(a,q)=1} |E(x,q,a)|.$$

Then for a given $A > 0$, there exists a constant $B(A) > 0$ and $C > 0$ such that if $Q := \sqrt{x}/(\log x)^B$ then

$$\sum_{q \leq Q} \max_{y \leq x} |E(y,q)| \leq C \frac{x}{(\log x)^A}.$$

Any $B \geq 24A + 46$ will work.

GPY define the **level of distribution of rational primes** to be

$$\vartheta = \sup_{\substack{\epsilon > 0 \\ 0 < \theta < 1}} \{\theta : Q = x^{\theta - \epsilon}\},$$

where Q is the upper bound for the sum over q in the Bombieri–Vinogradov theorem. By Lemma 4.6, we can choose $0 < \vartheta \leq \frac{1}{2}$. There are various, normally equivalent, definitions for this concept, which we call, along with Polymath, EH[ϑ]. For their unconditional results, GPY use EH[$\frac{1}{2}$].

4.5 Special Preliminary Results

Recall the definitions:

$$\mathscr{G}(\mathscr{H}) := \prod_{p} \left(1 - \frac{1}{p}\right)^{-k} \left(1 - \frac{v_p(\mathscr{H})}{p}\right),$$

$$P(n) := \prod_{j=1}^{k} (n + h_j),$$

$$\Lambda_R(n, \mathscr{H}, l) := \frac{1}{(k+l)!} \sum_{\substack{d|P(n) \\ d \le R}} \mu(d) \left(\log \frac{R}{d}\right)^{k+l},$$

$$\mathscr{S}_R(N, \mathscr{H}) := \sum_{n=1}^{N} \Lambda_R(n, \mathscr{H}, l)$$

$$= \sum_{n=1}^{N} \frac{1}{(k+l)!} \sum_{\substack{d|P(n) \\ d \le R}} \mu(d) \left(\log \frac{R}{d}\right)^{k+l},$$

$$\Lambda_R(n, \mathscr{H}) := \Lambda_R(n, \mathscr{H}, 0) = \frac{1}{k!} \sum_{\substack{d|P(n) \\ d \le R}} \mu(d) \left(\log \frac{R}{d}\right)^{k}.$$

Also recall

$$\mathscr{H} = \mathscr{H}_1 \cup \mathscr{H}_2, \; k_1 = |\mathscr{H}_1|, \; k_2 = |\mathscr{H}_2|, \; r = |\mathscr{H}_1 \cap \mathscr{H}_2|, k = k_1 + k_2,$$
$$M = k + l_1 + l_2.$$

Lemma 4.7 is a major part of the GPY method. It has eight main steps and makes extensive use of Euler product manipulations and zeta estimates, as well as some simple inequalities. Its main focus is the estimate of a function $G_{\mathscr{H}}(s)$, which extends $\mathscr{G}(\mathscr{H}) = G_{\mathscr{H}}(0)$. The quantity R should be regarded as a normally small, fixed power of N.

Lemma 4.7 *[72, proposition 3] Let $k \ge 1$, $l = 0$, $h \ge 2$, $\eta_0 > 0$ be fixed and arbitrarily small, $C > 0$ a fixed sufficiently large real number,*

$$k \ll_{\eta_0} (\log R)^{\frac{1}{2} - \eta_0} \text{ and } h_k \le h \le R^C. \tag{4.27}$$

Then as $N, R \to \infty$

$$\mathscr{S}_R(N, \mathscr{H}) = \sum_{n=1}^{N} \Lambda_R(n, \mathscr{H}) = \mathscr{G}(\mathscr{H})N + O\left(Ne^{-c\sqrt{\log R}}\right) + O\left(R(\log R)^{2k}\right).$$

(4.28)

Proof **(1)** Expanding the definition, we have

$$\mathscr{S}_R(N, \mathscr{H}) = \frac{1}{k!} \sum_{d \le R} \mu(d) L_{R/d}(k) \sum_{\substack{1 \le n \le N \\ d \mid P(n)}} 1.$$

(4.29)

If $p \mid P(n)$, the equations $n \equiv -h_i \bmod p$ for $1 \le i \le k$ will have a total of $v_p(\mathscr{H})$ distinct solutions for n modulo p, and using the Chinese Remainder Theorem and definition of $v_q(\mathscr{H}) = \prod_{p \mid q} v_p(\mathscr{H})$, for squarefree q, $v_q(\mathscr{H})$ distinct solutions for $n \bmod q$ which satisfy $q \mid P(n)$. Thus

$$\sum_{\substack{1 \le n \le N \\ q \mid P(n)}} 1 = v_q(\mathscr{H}) \left(\frac{N}{q} + O(1) \right).$$

(4.30)

In addition, if q is squarefree, we have $v_q(\mathscr{H}) \le k^{\omega(q)} = \tau_k(q)$. Therefore, inserting this into (4.29), using Lemma 4.5 to simplify the error, and defining

$$\mathcal{T}_R(\mathscr{H}) := \frac{1}{k!} \sum_{q \le R} \frac{\mu(q) v_q(\mathscr{H})}{q} L_{R/q}(k),$$

we get

$$\mathscr{S}_R(N, \mathscr{H}) = N \left(\frac{1}{k!} \sum_{q \le R} \frac{\mu(q) v_q(\mathscr{H})}{q} L_{R/q}(k) \right) + O\left(\frac{L_R(k)}{k!} \sum_{\substack{q \le R \\ \text{squarefree}}} v_q(\mathscr{H}) \right)$$

$$= N\mathcal{T}_R(\mathscr{H}) + O\left(\frac{1}{k!} R(k + \log R)^{2k} \right),$$

(4.31)

$$= N\mathcal{T}_R(\mathscr{H}) + O\left(RL_R(2k) \right).$$

(2) Next, we let an integral lower limit of the form "(a)" represent the vertical line contour $\{s \in \mathbb{C} : s = a + it, \ t \in \mathbb{R}\}$ and use the integral valid for all $c > 0$ (cf. [4, section 11.12])

$$\frac{1}{2\pi i} \int_{(c)} \frac{x^s}{s^{k+1}} \, ds = \begin{cases} \frac{(\log x)^k}{k!}, & x \ge 1, \\ 0, & 0 < x < 1. \end{cases}$$

(4.32)

To see this for $x > 1$, integrate x^s/s^{k+1} up part of the line $s = c + it$ and around part of a circular contour centre 0 and radius $R > 2c$ to make a closed contour, and let $R \to \infty$. The circular part of the contour integral has limit zero. To evaluate the residue of the pole at $s = 0$, one can use the Laurent series of the integrand:

$$\frac{1}{s^{k+1}} + \frac{\log x}{s^k} + \frac{(\log x)^2}{s^{k-1}} + \cdots + \frac{(\log x)^k}{k!\, s} + g(s),$$

where $g(s)$ is entire. For $x = 1$, the formula is also true, since both sides are zero when $k \geq 1$.

Now define for $\sigma = \Re s > 0$,

$$F(s) := \sum_{\substack{q=1 \\ \text{squarefree}}}^{\infty} \frac{\mu(q)v_q(\mathscr{H})}{q^{1+s}} = \prod_p \left(1 - \frac{v_p(\mathscr{H})}{p^{1+s}}\right), \qquad (4.33)$$

we have

$$T_R(\mathscr{H}) = \frac{1}{2\pi i} \int_{(1)} \frac{F(s)R^s}{s^{k+1}}\, ds. \qquad (4.34)$$

Since for all $p > h \geq h_k$, we have $v_p(\mathscr{H}) = k$ and can define (compare with the definition of $\mathscr{G}(\mathscr{H})$, (4.4))

$$G_{\mathscr{H}}(s) := \prod_p \left(1 - \frac{1}{p^{1+s}}\right)^{-k} \left(1 - \frac{v_p(\mathscr{H})}{p^{1+s}}\right)$$

$$= \zeta(1+s)^k \prod_p \left(1 - \frac{v_p(\mathscr{H})}{p^{1+s}}\right). \qquad (4.35)$$

Then

$$F(s) = \frac{G_{\mathscr{H}}(s)}{\zeta(1+s)^k}, \qquad (4.36)$$

and expanding the definition of $\mathscr{G}_{\mathscr{H}}(s)$, we can write

$$G_{\mathscr{H}}(s) = \prod_p \left(1 + \frac{k - v_p(\mathscr{H})}{p^{1+s}} + O_h\left(\frac{k^2}{p^{2+2\sigma}}\right)\right).$$

The function $G_{\mathscr{H}}(s)$ is homomorphic and uniformly bounded in any subset of $\mathbb{C}$ with $\sigma > -\frac{1}{2} + \delta$ for any fixed $\delta > 0$. In addition, (4.4) implies $G_{\mathscr{H}}(0) = \mathscr{G}(\mathscr{H})$.

By the estimate of Lemma 4.2 and (4.36), we get on and to the right of the contour $\mathscr{L}$ the bound

$$F(s) \ll |G_{\mathscr{H}}(s)|(C \log(|t| + 3))^k. \qquad (4.37)$$

Note that in this region $G_{\mathscr{H}}(s)$ is holomorphic and bounded and depends in particular on h and k.

(3) Next recall we have defined

$$\Delta := \prod_{1 \leq i < j \leq k} |h_j - h_i|, \qquad (4.38)$$

and note that if $p > h$ or $p \nmid \Delta$, then $v_p(\mathscr{H}) = k$ and that $\Delta \leq h^{k(k-1)/2} \leq h^{k^2}$.
   Recall also the definition of the key parameter U:

$$U := Ck^2 \log(2h) \implies \log \Delta \leq U \text{ when } C \geq \frac{1}{2}. \tag{4.39}$$

Recall $\sigma = \Re s$ and let $-\frac{1}{4} < \sigma \leq 1$ and $\delta = \max(-\sigma, 0)$, so $0 \leq \delta < \frac{1}{4}$. We will show in Steps (4) through (6) that in this vertical strip

$$|G_{\mathscr{H}}(s)| \ll \exp(5kU^\delta \log\log U). \tag{4.40}$$

(4) Consider the primes p such that $p \leq U$. Using the inequality $\log(1 + x) \leq x$, $x \geq 0$, we get

$$\left| \prod_{p \leq U} \left(1 - \frac{v_p(\mathscr{H})}{p^{1+s}} \right) \right| \leq \prod_{p \leq U} \left(1 + \frac{k}{p^{1-\delta}} \right)$$

$$\leq \prod_{p \leq U} \left(1 + \frac{v_p(\mathscr{H})}{|p^{1+s}|} \right)$$

$$= \exp\left(\sum_{p \leq U} \log\left(1 + \frac{k}{p^{1-\delta}} \right) \right)$$

$$\leq \exp\left(\sum_{p \leq U} \frac{k}{p^{1-\delta}} \right)$$

$$\leq \exp\left(kU^\delta \sum_{p \leq U} \frac{1}{p} \right)$$

$$\ll \exp\left(kU^\delta \log\log U \right).$$

Then using

$$\frac{1}{p^{1-\delta}} \leq \frac{1}{2^{3/4}} \leq \frac{2}{3} \text{ and } \frac{1}{1-x} \leq 1 + 3x, \ 0 \leq x \leq 2/3,$$

we get

$$\left| \prod_{p \leq U} \left(1 - \frac{1}{p^{1+s}} \right)^{-k} \right| \leq \left(\prod_{p \leq U} \left(1 + \frac{3}{p^{1-\delta}} \right) \right)^k$$

$$\ll \exp\left(3kU^\delta \log\log U \right).$$

Therefore, using this inequality and the result of Step (4), the product of all Euler factors in $G_{\mathscr{H}}(s)$ having $p \leq U$ is $O(\exp(4kU^\delta \log\log U))$.

(5) Next consider primes such that $p > U$ and $p \mid \Delta$. Then

$$\left| \prod_{\substack{p>U \\ p|\Delta}} \left(1 - \frac{v_p(\mathcal{H})}{p^{1+s}}\right)\left(1 - \frac{1}{p^{1+s}}\right)^{-k} \right| \leq \prod_{\substack{p>U \\ p|\Delta}} \left(1 + \frac{k}{p^{1-\delta}}\right)\left(1 + \frac{3}{p^{1-\delta}}\right)^k$$

$$\leq \exp\left(\sum_{\substack{p>U \\ p|\Delta}} \frac{4k}{p^{1-\delta}}\right).$$

There are less than $(\log \Delta) < U$ primes such that $p \mid \Delta$ when $C \geq 3/4$. Hence the argument of the exponential in the preceding expression is increased if we replace the primes in the sum by all the integers in $(U, 2U]$. Thus

$$\exp\left(\sum_{\substack{p>U \\ p|\Delta}} \frac{4k}{p^{1-\delta}}\right) \leq \exp\left(4k \sum_{U<n\leq 2U} \frac{1}{n^{1-\delta}}\right)$$

$$\leq \exp\left(4k(2U)^\delta \sum_{U<n\leq 2U} \frac{1}{n}\right) \ll \exp(4k2^\delta U^\delta).$$

(6) Finally, for the primes with $p > U$ and $p \nmid \Delta$, we have

$$\left|\frac{k}{p^{1+s}}\right| \leq \frac{k}{U^{1-\delta}} \leq \frac{1}{2}$$

for k sufficiently large, which we use in the following expression to get the final inquality. Expanding the logarithms to get the second line

$$\left| \prod_{\substack{p>U \\ p\nmid\Delta}} \left(1 - \frac{v_p(\mathcal{H})}{p^{1+s}}\right)\left(1 - \frac{1}{p^{1+s}}\right)^{-k} \right| = \left| \prod_{\substack{p>U \\ p\nmid\Delta}} \left(1 - \frac{k}{p^{1+s}}\right)\left(1 - \frac{1}{p^{1+s}}\right)^{-k} \right|$$

$$= \left| \exp\left(\sum_{\substack{p>U \\ p\nmid\Delta}} \left(-\sum_{n=1}^{\infty} \frac{1}{n}\left(\frac{k}{p^{1+s}}\right)^n + k\sum_{n=1}^{\infty} \frac{1}{n}\left(\frac{1}{p^{1+s}}\right)^n\right)\right) \right|$$

$$\leq \exp\left(\sum_{p>U}\sum_{n=2}^{\infty} \frac{2}{n}\left(\frac{k}{p^{1-\delta}}\right)^n\right) \leq \exp\left(\sum_{p>U}\sum_{n=2}^{\infty} \left(\frac{k}{p^{1-\delta}}\right)^n\right)$$

$$\leq \exp\left(2k^2 \sum_{n>U} \frac{1}{n^{2-2\delta}}\right) \leq \exp\left(\frac{4k^2U^\delta}{U^{1-\delta}}\right) \leq \exp(2kU^\delta).$$

Combining this estimate with that from Step (5), it follows that the terms with $p > U$ give

$$\left| \prod_{p > U} \left(1 - \frac{v_p(\mathcal{H})}{p^{1+s}} \right) \left(1 - \frac{1}{p^{1+s}} \right)^{-k} \right| \le \exp(6kU^\delta),$$

from which the estimate (4.40) claimed in Step (3) then follows.

(7) Thus, by the estimate (4.37), we get for any fixed but arbitrarily large $C > 0$ and then for $h \ll R^C$, and for s on or to the right of the contour $\mathscr{L}$ as defined in Lemma 4.4, the estimate

$$F(s) \ll \exp(5kU^\delta \log\log U)(C \log(|t| + 3))^k. \tag{4.41}$$

(8) We now use this estimate in the integral we derived in Step (2), namely

$$T_R(\mathcal{H}) = \frac{1}{2\pi i} \int_{(1)} \frac{F(s) R^s}{s^{k+1}} \, ds. \tag{4.42}$$

For s in the strip $-\frac{1}{4} < \sigma \le 1$, the integrand tends to zero as $t \to \pm\infty$. If we assume $\mathcal{H}$ is admissible, when the contour (1) is shifted to $\mathscr{L}$, we pass over a simple pole at $s = 0$ with residue $\mathscr{G}_\mathcal{H}(0)$. If however $\mathscr{G}(\mathcal{H}) = 0$, the integrand is holomorphic in a neighbourhood of $s = 0$. In each of these possible cases, using $G_\mathcal{H}(0) = \mathscr{G}(\mathcal{H})$,

$$\zeta(s) = \frac{1}{s-1} + \gamma + O(|s-1|), \quad s \to 1,$$

Lemma 4.4 estimate (4.23) and the estimates (4.39) and (4.41), for all k with

$$k \ll_{\eta_0} L_R(\tfrac{1}{2} - \eta_0),$$

we get

$$T_R(\mathcal{H}) = G_\mathcal{H}(0) + \frac{1}{2\pi i} \int_{\mathscr{L}} \frac{F(s) R^s}{s^{k+1}} \, ds = \mathscr{G}(\mathcal{H}) + O\left(\exp(-c\sqrt{\log R}) \right). \tag{4.43}$$

Therefore, by (4.31)

$$\sum_{n=1}^{N} \Lambda_R(n, \mathcal{H}) = N T_R(\mathcal{H}) + O(R(\log R)^{2k})$$

$$= N\mathscr{G}(\mathcal{H}) + O\left(N e^{-c\sqrt{\log R}} \right) + O\left(R(\log R)^{2k} \right),$$

which completes the proof.   □

The next result, Lemma 4.8, is "heavy artillery" indeed. It takes ten main steps to prove. Note the flexible nature of the function $G(s_1, s_2)$ and the simplicity of the bound which the lemma gives for the $\mathscr{D}_j$. The reader may wish to follow Artin's general advice and skip a complete analysis of the proof at a first reading. It uses no fewer than eight (closely related) integration contours and the nasty combinatorial identity of Lemma 4.3! It is an essential step in the derivation of estimates for both nonprime and prime sums.

Recall the definition of $\mathscr{L}$: it is the complex contour

$$\mathscr{L} := \left\{ s : s = -\frac{\eta}{\log(|t| + 3)} + it \right\},$$

which is the boundary of the Riemann zeta zero free region Ω defined in Lemma 4.2 shifted by one to the left.

Also recall

$$\mathscr{H} = \mathscr{H}_1 \cup \mathscr{H}_2, \ k_1 = |\mathscr{H}_1|, \ k_2 = |\mathscr{H}_2|, \ r = |\mathscr{H}_1 \cap \mathscr{H}_2|, k = k_1 + k_2,$$
$$M = k + l_1 + l_2,$$

and note that all we assume from this is that M is a large constant.

Lemma 4.8 [72, lemma 3] *Let M be a large constant and a, b, d, u, v integer parameters such that*

$$0 \leq a, b, d, u, v \leq M, \ a + u \geq 1, \ b + v \geq 1, \ d \leq \min(a, b). \tag{4.44}$$

Let $C > 0$ be a sufficiently large constant and $h \ll R^C$. Then as $R \to \infty$, we have

$$T_R^*(a, b, d, u, v, h) := \frac{1}{(2\pi i)^2} \int_{(1)} \int_{(1)} \frac{D(s_1, s_2) R^{s_1 + s_2}}{s_1^{u+1} s_2^{v+1} (s_1 + s_2)^d} \, ds_1 \, ds_2,$$

$$= \binom{u + v}{u} \frac{L_R(u + v + d)}{(u + v + d)!} G(0, 0)$$

$$+ \sum_{j=1}^{u+v+d} \mathscr{D}_j(a, b, d, u, v, h) L_R(u + v + d - j) + O_M \left(\exp(-c\sqrt{\log R}) \right),$$

$$\tag{4.45}$$

where $W(s) := s \zeta(s + 1)$ and

$$D(s_1, s_2) := \frac{G(s_1, s_2) W^d(s_1 + s_2)}{W^a(s_1) W^b(s_2)}, \tag{4.46}$$

for any $G(s_1, s_2)$ which is regular to the right and on $\mathscr{L}$ and satisfies the bound

$$G(s_1, s_2) \ll_M g(U) := \exp(CMU^{\delta_1 + \delta_2} \log\log(U)), \tag{4.47}$$

and where

$$\delta_i = \max(-\sigma_i, 0) \ \textit{with} \ \sigma_i > -\frac{1}{4} + \delta, \ \delta > 0 \ \textit{and} \ U := CM^2 \log(2h), \ C \geq \frac{1}{2},$$

and where the $\mathscr{D}_j$ *are functions not depending on* R *and which satisfy for some constants* C_j, C'_j *depending at most on* M,

$$\mathscr{D}_j(a,b,d,u,v,h) \ll_M (\log U)^{C_j} \ll_M (\log\log(10h))^{C'_j}. \tag{4.48}$$

Proof **(1)** By the hypotheses, the estimates from Lemma 4.2 and the estimates (4.47), we can write, for $s_1, s_2, s_1 + s_2$ on or to the right of $\mathscr{L}$,

$$\frac{D(s_1, s_2)}{s_1^{u+1} s_2^{v+1} (s_1 + s_2)^d}$$

$$\ll_M \frac{\exp\left(CMU^{\delta_1 + \delta_2} \log\log U\right) \log^{2M}(|t_1| + 3) \log^{2M}(|t_2| + 3) \max(1, |s_1 + s_2|^{-d})}{|s_1|^{a+u+1} |s_2|^{b+v+1}}.$$

$$\tag{4.49}$$

(2) Next, set $V := \exp\left(\sqrt{\log R}\right)$ and define eight contours:

$$L'_1 = \left\{\frac{c}{4\log V} + it : t \in \mathbb{R}\right\}, \qquad L'_2 = \left\{\frac{c}{16\log V} + it : t \in \mathbb{R}\right\},$$

$$L_1 = \left\{\frac{c}{4\log V} + it : |t| \leq V/4\right\}, \qquad L_2 = \left\{\frac{c}{16\log V} + it : |t| \leq V/16\right\},$$

$$\mathscr{L}_1 = \left\{-\frac{c}{4\log V} + it : |t| \leq V/4\right\}, \qquad \mathscr{L}_2 = \left\{-\frac{c}{16\log V} + it : |t| \leq V/16\right\},$$

$$H_1 = \left\{\sigma_1 \pm iV/4 : |\sigma_1| \leq \frac{c}{4\log V}\right\}, \qquad H_2 = \left\{\sigma_2 \pm iV/16 : |\sigma_2| \leq \frac{c}{16\log V}\right\}.$$

$$\tag{4.50}$$

Figure 4.6 is a representation of these contours.

If s_1, s_2 are to the right of L'_2 then the estimate (4.49) implies the integrand for (4.45) tends to zero when $|t_1|$ or $|t_2| \to \infty$. Shift the contours $(1 + it : t \in \mathbb{R})$ for the integrals over s_1, s_2 to L'_1, L'_2 and then replace them with the bounded contours L_1, L_2 respectively. This replacement introduces errors $E_{11}, E_{12}, E_{21}, E_{22}$, which may be estimated by (4.49). Calculating the error in the double integral coming from L'_1 and the truncated piece from L'_2 we get from the definitions that $U \ll \log h \ll \log R = (\log V)^2$ and $\log V = \sqrt{\log R}$ so $U \ll \log R = (\log V)^2$. Thus

$$g(U) \ll \exp(CMU^{\frac{1}{2}-\epsilon} \log\log U) \ll \exp(CM(\log V)^{1-\epsilon} \log\log V) \ll (\log V)^{CM \log V},$$

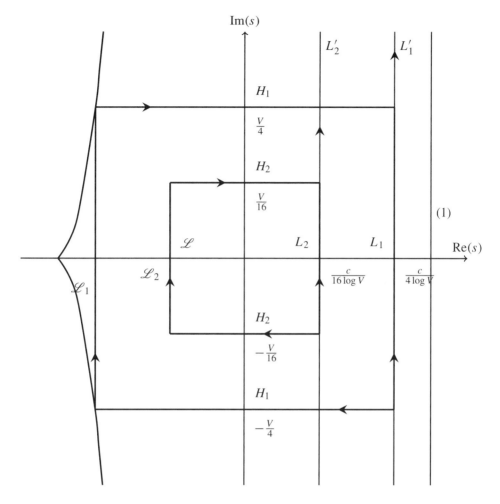

Figure 4.6 The eight contours of Lemma 4.8.

and therefore,

$$E_{12} \ll_M (\log V)^{CM \log V}$$

$$\times \left(\int_{-\infty}^{\infty} \frac{(\log |t| + 3)^{2M}}{|c/(4 \log V) + it|^{a+u+1}} \, dt \right) \left(\int_{V/16}^{\infty} \frac{\log^{2M} t}{t^2} \, dt \right)$$

$$\ll_M \frac{(\log V)^{CM \log V}}{V} \ll_M \exp\left(-c\sqrt{\log R}\right).$$

Therefore, with $\mathcal{T}_R^* = \mathcal{T}_R^*(a,b,d,u,v,h)$, we get

$$\mathcal{T}_R^* = -\frac{1}{4\pi^2} \int_{L_2} \int_{L_1} \frac{D(s_1,s_2) R^{s_1+s_2}}{s_1^{u+1} s_2^{v+1} (s_1 + s_2)^d} \, ds_1 \, ds_2 + O_M\left(e^{-c\sqrt{\log R}}\right). \quad (4.51)$$

(3) Now we replace the contour L_1 for the integral of s_1 by $\mathcal{L}_1$. The rectangle formed by L_1, H_1 and $\mathcal{L}_1$ contains poles for the integrand at $s_1 = 0$ and for given s_2 at $s_1 = -s_2$. Hence, orienting the $\mathcal{L}_1$, H_1 contours clockwise,

$$
T_R^* = \frac{1}{2\pi i} \int_{L_2} \operatorname*{Res}_{s_1=0} \left(\frac{D(s_1,s_2) R^{s_1+s_2}}{s_1^{u+1} s_2^{v+1} (s_1+s_2)^d} \right) ds_2
$$

$$
+ \frac{1}{2\pi i} \int_{L_2} \operatorname*{Res}_{s_1=-s_2} \left(\frac{D(s_1,s_2) R^{s_1+s_2}}{s_1^{u+1} s_2^{v+1} (s_1+s_2)^d} \right) ds_2
$$

$$
+ \frac{1}{(2\pi i)^2} \int_{L_2} \int_{\mathcal{L}_1 \cup H_1} \left(\frac{D(s_1,s_2) R^{s_1+s_2}}{s_1^{u+1} s_2^{v+1} (s_1+s_2)^d} \right) ds_1\, ds_2 + O_M(\exp(-c\sqrt{\log R})).
$$

$$(4.52)$$

Then in the first and third of these integrals, we move the contour over L_2 to $\mathcal{L}_2$, passing over the pole at $s_2 = 0$ only. Naming each successive term I_j, $0 \le i \le 4$, we get

$$
T_R^* = \operatorname*{Res}_{s_2=0} \operatorname*{Res}_{s_1=0} \left(\frac{D(s_1,s_2) R^{s_1+s_2}}{s_1^{u+1} s_2^{v+1} (s_1+s_2)^d} \right)
$$

$$
+ \frac{1}{2\pi i} \int_{\mathcal{L}_2 \cup H_2} \operatorname*{Res}_{s_1=0} \left(\frac{D(s_1,s_2) R^{s_1+s_2}}{s_1^{u+1} s_2^{v+1} (s_1+s_2)^d} \right) ds_2
$$

$$
+ \frac{1}{2\pi i} \int_{\mathcal{L}_1 \cup H_1} \operatorname*{Res}_{s_2=0} \left(\frac{D(s_1,s_2) R^{s_1+s_2}}{s_1^{u+1} s_2^{v+1} (s_1+s_2)^d} \right) ds_1
$$

$$
+ \frac{1}{2\pi i} \int_{L_2} \operatorname*{Res}_{s_1=-s_2} \left(\frac{D(s_1,s_2) R^{s_1+s_2}}{s_1^{u+1} s_2^{v+1} (s_1+s_2)^d} \right) ds_2
$$

$$
+ \frac{1}{(2\pi i)^2} \int_{\mathcal{L}_2 \cup H_2} \int_{\mathcal{L}_1 \cup H_1} \left(\frac{D(s_1,s_2) R^{s_1+s_2}}{s_1^{u+1} s_2^{v+1} (s_1+s_2)^d} \right) ds_1\, ds_2 + O_M\left(\exp\left(-c\sqrt{\log R} \right) \right)
$$

$$
= I_0 + I_1 + I_2 + I_3 + I_4 + O_M \left(\exp\left(-c\sqrt{\log R} \right) \right).
$$

$$(4.53)$$

In this expression, we will show, by considering each term separately, in Steps (4) and (5) that I_0 is the main term plus some terms of lower order, in Step (9) that I_3 is a lower order term and that the remaining terms I_2, I_4 can be absorbed by the error.
(4) In this step, we evaluate the residue I_0. Considering the pole of order $u + 1$ at $s_1 = 0$, we have using Leibniz's rule for the multiple derivative of a product

$$
\operatorname*{Res}_{s_1=0} \left(\frac{D(s_1,s_2) R^{s_1}}{s_1^{u+1} (s_1+s_2)^d} \right) = \frac{1}{u!} \sum_{j=0}^{u} \binom{u}{j} L_R(u-j) \frac{\partial^j}{\partial s_1^j} \left(\frac{D(s_1,s_2)}{(s_1+s_2)^d} \right) \Bigg|_{s_1=0},
$$

and by Leibniz again, assigning 1 as usual as the value of any empty product,

$$\frac{\partial^j}{\partial s_1^j}\left(\frac{D(s_1,s_2)}{(s_1+s_2)^d}\right)\bigg|_{s_1=0} = (-1)^j\frac{D(0,s_2)d(d+1)\cdots(d+j-1)}{s_2^{d+j}}$$

$$+\sum_{i=1}^{j}\binom{j}{i}\frac{\partial^i}{\partial s_1^i}D(s_1,s_2)\bigg|_{s_1=0}(-1)^{j-i}\frac{d(d+1)\cdots(d+j-i-1)}{s_2^{d+j-i}}.$$

Hence, for functions $a(i,j)$, which can be computed explicitly,

$$\operatorname*{Res}_{s_1=0}\left(\frac{D(s_1,s_2)R^{s_1}}{s_1^{u+1}(s_1+s_2)^d}\right) = \sum_{i=0}^{u}\sum_{j=0}^{i}\frac{a(i,j)L_R(u-i)}{s_2^{d+i-j}}$$

$$\times\left(\frac{\partial^j}{\partial s_1^j}D(s_1,s_2)\right)\bigg|_{s_1=0}. \quad (4.54)$$

Thus the (i,j)th term contributes to I_0 a pole at $s_2 = 0$ of order at most $v+1+d+i-j$. Thus

$$\operatorname*{Res}_{s_2=0}\frac{R^{s_2}}{s_2^{v+1+d+i-j}}\frac{\partial^j}{\partial s_1^j}D(s_1,s_2)\bigg|_{s_1=0} = \frac{1}{(v+d+i-j)!}$$

$$\times\sum_{m=0}^{v+d+i-j}\binom{v+d+i-j}{m}L_R(v+d+i-j-m)\frac{\partial^m}{\partial s_2^m}\frac{\partial^j}{\partial s_1^j}D(s_1,s_2)\bigg|_{\substack{s_1=0\\s_2=0}}.$$

Finally in this step, if we define

$$b(i,j,m) := (-1)^{i-j}\binom{u}{i}\binom{i}{j}\binom{v+d+i-j}{m}\frac{d(d+1)\cdots(d+i-j-1)}{u!\,(v+d+i-j)!}$$

$$(4.55)$$

and use the reductions of I_0 we have derived, we get

$$I_0 = \sum_{i=0}^{u}\sum_{j=0}^{i}\sum_{m=0}^{v+d+i-j}b(i,j,m)\left(\frac{\partial^m}{\partial s_2^m}\frac{\partial^j}{\partial s_1^j}D(s_1,s_2)\bigg|_{\substack{s_1=0\\s_2=0}}\right)L_R(u+v+d-j-m).$$

$$(4.56)$$

(5) We see the main term, where $j = m = 0$, is of order $(\log R)^{u+v+d}$. Using Lemma 4.3, this is

$$G(0,0)L_R(u+v+d)\left(\frac{1}{u!}\sum_{j=0}^{u}\binom{u}{j}(-1)^j\frac{d(d+1)\cdots(d+j-1)}{(v+d+j)!}\right)$$

$$= G(0,0)L_R(u+v+d)\binom{u+v}{u}\frac{1}{(u+v+d)!}.$$

$$(4.57)$$

(6) We now use Cauchy's derivative bound, valid for functions holomorphic on the disk $\{z : |z - z_0| \le \eta\}$, namely

$$|f^{(j)}(z_0)| \le \max\left\{\frac{|f(z)|j!}{\eta^j} : |z - z_0| = \eta\right\},$$

to bound the partial derivatives of $D(s_1, s_2)$ at $(0,0)$. Here we let z_0 be on or to the right of the contour $\mathcal{L}$ and set

$$T := |s_1| + |s_2| + 3 \text{ with } \eta := \frac{1}{C\log U \log T}.$$

With this choice for η, the region

$$\sigma \ge 1 - \frac{4\eta}{\log(|t| + 3)}$$

contains the circle $|z - z_0 - 1| = \eta$ and the estimates for $\zeta(z)$ given by Lemma 4.2 all hold inside and on this circle. Therefore, for s_1, s_2 on or to the right of $\mathcal{L}$, with $j \le M, m \le 2M$ and $\max(|s_1|, |s_2|) \le C$, and recalling

$$\delta_i = \max(-\sigma_i, 0) \text{ with } \sigma_i > -\frac{1}{4} + \delta, \ \delta > 0,$$

we have

$$\frac{\partial^m}{\partial s_2^m}\frac{\partial^j}{\partial s_1^j}D(s_1, s_2) \le j!\, m!\, (C\log U \log T)^{j+m}$$

$$\times \max\{|D(w_1, w_2)| : |w_i - s_i| \le \eta, \ i = 1, 2\}$$

$$\ll_M \exp\left(CMU^{\delta_1+\delta_2}\log\log U\right)L_T(6M)$$

$$\times \frac{\max(1, |s_1 + s_2|)^d}{\max(1, |s_1|)^a \max(1, |s_2|)^b}$$

$$\ll_M \exp\left(CMU^{\delta_1+\delta_2}\log\log U\right).$$

$$(4.58)$$

It follows from the first line of this derivation that

$$\frac{\partial^m}{\partial s_2^m} \frac{\partial^j}{\partial s_1^j} D(s_1, s_2) \Big|_{\substack{s_1=0 \\ s_2=0}} \ll_m L_U(CM). \tag{4.59}$$

It follows using (4.56), (4.57) and the estimate we have just derived, (4.59), we obtain the bounds given in (4.48).

(7) We now bound I_1. By the estimates (4.54) and (4.58), we have

$$I_1 \ll_M \log^u R \int_{\mathcal{L}_2 \cup H_2} e^{CMU^{\delta_2} \log\log U} \frac{\log(|t_2| + 3)^{3M} \max(1, |s_2|^d)}{|s_2|^{v+d+1} \max(1 + |s_2|^b)} |R^{s_2}| |ds_2|. \tag{4.60}$$

By our assumptions, $b + v \geq 1$ on H_2, $|R^{s_2}| \ll \exp(c\sqrt{\log R})$ and on $\mathcal{L}_2 \cap H_2$ we also have $U^{\delta_2} \ll 1$. In addition, when $|s_2| \geq 1$ we can write

$$\frac{\max(1, |s_2|^d)}{|s_2|^{v+d+1} \max(1 + |s_2|^b)} \ll \frac{1}{|s_2|^{v+b+1}} \ll \frac{1}{|s_2|^2}.$$

Therefore, the contribution to the integral I_1 from the contour H_2 is

$$O_M\left(\frac{1}{V^2} L_R(7M/2 - 1/2) \exp\left(c\sqrt{\log R}\right)\right) \ll_M \exp\left(-c\sqrt{\log R}\right).$$

Next we see the integral on $\mathcal{L}_2$ is

$$O\left(L_R(2M) R^{-\frac{c}{16\log V}} \int_{-V}^{V} \log(|t_2| + 3)^{3M} \min(t^{-2}, L_V(3M)) \, dt\right)$$

$$\ll_M L_R(3M) R^{-\frac{c}{16\log V}}$$

$$\ll_M \exp\left(-c\sqrt{\log R}\right).$$

Therefore, we have $|I_1| \ll_M \exp\left(-c\sqrt{\log R}\right)$.

(8) Pattern matching and substitution enables us to derive the same bound as in Step (7) for I_2. Using (4.49) and the constraint $|s_1 + s_2| \gg c/\log V$ gives same bound for I_4.

(9) Next, we consider I_3, which is zero unless $d \geq 1$. Let

$$\mathcal{B}_i(s_2) := \binom{d-1}{i} \sum_{j=0}^{i} \binom{i}{j} \frac{\partial^{i-j}}{\partial s_1^{i-j}} D(s_1, s_2) \Big|_{s_1 = -s_2} \times \frac{(-1)^j (u+1) \cdots (u+j)}{(-1)^{u+j+1} s_2^{u+v+j+2}}, \tag{4.61}$$

and define for $0 \leq i \leq d - 1$,

$$\mathscr{C}_i := \frac{1}{2\pi i} \int_{L_2} \mathscr{B}_i(s_2) \, ds_2. \tag{4.62}$$

Then

$$\operatorname*{Res}_{s_1=-s_2} \frac{D(s_1,s_2) R^{s_1+s_2}}{s_1^{u+1} s_2^{v+1} (s_1+s_2)^d} = \lim_{s_1 \to -s_2} \frac{1}{(d-1)!} \frac{\partial^{d-1}}{\partial s_1^{d-1}} \frac{D(s_1,s_2) R^{s_1+s_2}}{s_1^{u+1} s_2^{v+1}}$$

$$= \frac{1}{(d-1)!} \sum_{i=0}^{d-1} \mathscr{B}_i(s_2) (\log R)^{d-1-i}. \tag{4.63}$$

By (4.53), (4.63) and (4.61), we can then write

$$I_3 = \frac{1}{(d-1)!} \sum_{i=0}^{d-1} \mathscr{C}_i \log^{d-1-i}(R). \tag{4.64}$$

Using the estimate (4.58) and (4.61), when s_2 is to the right of $\mathscr{L}$, we get

$$\mathscr{B}_i(s_2) \ll_M \exp(CMU^{|\delta_2|} \log\log U) \frac{\log^{4M}(|t_2|+3)}{|t_2|^{u+v+a+b+2} \max(1,|t_2|^i)}. \tag{4.65}$$

In the integral (4.62), first shift the contour L_2 to the imaginary axis together with a semicircle centre $s_2 = 0$ in the first and fourth quadrant and with radius $1/\log U$. Then extend this contour to the rest of the imaginary axis, introducing by the estimate (4.65) an error, which is $O_M \left(\exp\left(-c\sqrt{\log R}\right)\right)$. If we define the contour

$$\mathscr{L}' := \{s = it : |t| \geq 1/\log U\} \cup \{s = e^{i\vartheta}/\log U : -\pi/2 \leq \vartheta \leq \pi/2\} \tag{4.66}$$

and integrate up the axis from ∞, along $\mathscr{L}'$ to $+\infty$, we get for $d \geq 1$ and $0 \leq i \leq d - 1$

$$\mathscr{C}_i = \frac{1}{2\pi i} \int_{\mathscr{L}'} \mathscr{B}(s_2) \, ds_2 + O_M(\exp(-c\sqrt{\log R})). \tag{4.67}$$

(10) The contribution to the integral $\mathscr{C}_i$ from the integral along the imaginary axis and the semicircular contour is

$$O_M \left((\log U)^{u+v+i+a+b+1} \exp(CM \log\log U)\right) \ll_M L_U(C'M). \tag{4.68}$$

At last, all of the claims made in the statement of the lemma have been verified, so this completes the proof. $\qquad\qquad\qquad\qquad\qquad\qquad\qquad\qquad\square$

Remark: This lemma provides some serious challenges, and the reader might wish to first consider Lemma 4.20, which is part of the simplified proof set out in Section 4.8.

The next Lemma 4.9 derives an estimate for a sum involving products of the Λ_R. There are only five steps, but some important congruence counting sums in Step (2).

Recall we have defined in the statement of Lemma 4.8

$$T_R^*(a,b,d,u,v,h) := \frac{1}{(2\pi i)^2} \int_{(1)} \int_{(1)} \frac{D(s_1,s_2) R^{s_1+s_2}}{s_1^{u+1} s_2^{v+1} (s_1+s_2)^d} \, ds_1 \, ds_2.$$

Recall also we have defined

$$\Lambda_R(n,\mathcal{H},l) := \frac{1}{(k+l)!} \sum_{\substack{d|P(n) \\ d \leq R}} \mu(d) L_{R/d}(k+l),$$

and the functions $\mathcal{D}_j$ are those which occur in the statement of Lemma 4.8.

Also recall

$$\mathcal{H} = \mathcal{H}_1 \cup \mathcal{H}_2, \ k_1 = |\mathcal{H}_1|, \ k_2 = |\mathcal{H}_2|, \ r = |\mathcal{H}_1 \cap \mathcal{H}_2|, k = k_1 + k_2,$$
$$M = k + l_1 + l_2.$$

Lemma 4.9 *[72, proposition 4] Let $C > 0$ be a fixed constant and $h \ll R^C$. Then when $R, N \to \infty$, we have*

$$\sum_{n=1}^{N} \Lambda_R(n,\mathcal{H}_1,l_1) \Lambda_R(n,\mathcal{H}_2,l_2) = \frac{N}{(r+l_1+l_2)!} \binom{l_1+l_2}{l_1} \mathcal{G}(\mathcal{H}) L_R(r+l_1+l_2)$$

$$+ N \sum_{j=1}^{r+l_1+l_2} \mathcal{D}_j(l_1,l_2,\mathcal{H}_1,\mathcal{H}_2) L_R(r+l_1+l_2-j)$$

$$+ O_M\left(N \exp\left(-c\sqrt{\log R}\right)\right) + O(R^2 (3\log R)^{3k+M}),$$

$$(4.69)$$

where the $\mathcal{D}_j$ are functions not depending on R or N and which satisfy for some constants C_j, C_j' depending at most on M,

$$\mathcal{D}_j(l_1,l_2,\mathcal{H}_1,\mathcal{H}_2) \ll_M \log^{C_j}(U) \ll_M (\log\log(10h))^{C_j'}.$$

$$(4.70)$$

Proof **(1)** If either of the $k_i = 0$, then the proof is along the lines of Lemma 4.7, so we will assume $k_1 \geq 1$ and $k_2 \geq 1$. Recall the definition

$$\mathscr{S}_R := \mathscr{S}_R(N,\mathcal{H}_1,\mathcal{H}_2,l_1,l_2) := \sum_{n=1}^{N} \Lambda_R(n,\mathcal{H}_1,l_1) \Lambda_R(n,\mathcal{H}_2,l_2). \quad (4.71)$$

Expanding the definitions, we get

$$\mathscr{S}_R = \frac{1}{(k_1+l_1)!\,(k_2+l_2)!} \sum_{d,e\leq R} \mu(d)\mu(e)L_{R/d}(k_1+l_1)L_{R/e}(k_2+l_2)\left(\sum_{\substack{1\leq n\leq N \\ d\mid P_{\mathscr{H}_1}(n) \\ e\mid P_{\mathscr{H}_2}(n)}} 1\right). \tag{4.72}$$

(2) Consider the inner sum in (4.72). Let $g := (d,e)$, $d = d'g$, $e = e'g$ and $f := d'e'g$. Then since d and e are squarefree, d', e' and g are pairwise relatively prime. The n which satisfy $d \mid P_{\mathscr{H}_1}(n)$ and $e \mid P_{\mathscr{H}_2}(n)$ translate to $d' \mid P_{\mathscr{H}_1}(n)$, $e' \mid P_{\mathscr{H}_2}(n)$, $g \mid P_{\mathscr{H}_1}(n)$, $g \mid P_{\mathscr{H}_2}(n)$. We obtain $v_{d'}(\mathscr{H}_1)$ solutions $n \bmod d'$ and $v_{e'}(\mathscr{H}_2)$ solutions $n \bmod e'$. For $p \mid g$, we get

$$v_p(\mathscr{H}_1) + v_p(\mathscr{H}_2) - v_p(\mathscr{H}_1 \cup \mathscr{H}_2) =: \bar{v}_p(\mathscr{H}_1 \bar{\cap} \mathscr{H}_2) \tag{4.73}$$

solutions $n \bmod p$, and this can be extended to squarefree numbers by multiplicativity, so for the inner sum we get

$$\sum_{\substack{1\leq n\leq N \\ d\mid P_{\mathscr{H}_1}(n) \\ e\mid P_{\mathscr{H}_2}(n)}} 1 = \sum_{\substack{1\leq n\leq N \\ d\mid P_{\mathscr{H}_1}(n) \\ e\mid P_{\mathscr{H}_2}(n)}} v_{d'}(\mathscr{H}_1)v_{e'}(\mathscr{H}_2)\bar{v}_g(\mathscr{H}_1\bar{\cap}\mathscr{H}_2)\left(\frac{N}{\varphi(f)} + O(1)\right). \tag{4.74}$$

(3) Therefore, following Step (2) and using Lemma 4.5 to derive the error bound,

$$\mathscr{S}_R = \frac{N}{(k_1+l_1)!\,(k_2+l_2)!}$$

$$\times \sum_{\substack{d'g\leq R \\ e'g\leq R}} \frac{\mu(d')\mu(e')\mu(g)^2 v_{d'}(\mathscr{H}_1)v_{e'}(\mathscr{H}_2)\bar{v}_g((\mathscr{H}_1\bar{\cap}\mathscr{H}_2))}{\varphi(f)}$$

$$\times \left(\log \frac{R}{d'g}\right)^{k_1+l_1}\left(\log \frac{R}{e'g}\right)^{k_2+l_2}$$

$$+ O\left(L_R(M) \sum_{\substack{d'g\leq R \\ e'g\leq R \\ d',e',g\ \text{coprime}}} \mu(d')^2\mu(e')^2\mu(g)^2 v_{d'}(\mathscr{H}_1)v_{e'}(\mathscr{H}_2)\bar{v}_g((\mathscr{H}_1\bar{\cap}\mathscr{H}_2))\right)$$

$$= N\mathcal{T}_R(l_1,l_2,\mathscr{H}_1,\mathscr{H}_2) + O\left((\log R)^M \sum_{\substack{q\leq R^2 \\ \text{squarefree}}} \sum_{q=d'e'g} \tau_k(q)\right)$$

$$= N T_R(l_1, l_2, \mathcal{H}_1, \mathcal{H}_2) + O((\log R)^M \sum_{\substack{q \le R^2 \\ \text{squarefree}}} \tau_3(q) \tau_k(q)$$

$$= N T_R(l_1, l_2, \mathcal{H}_1, \mathcal{H}_2) + O((\log R)^M \sum_{\substack{q \le R^2 \\ \text{squarefree}}} \tau_{3k}(q))$$

$$= N T_R(l_1, l_2, \mathcal{H}_1, \mathcal{H}_2) + O(R^2 (3 \log R)^{3k+M}).$$

$$(4.75)$$

(4) Next, we consider the main term in (4.75). The integral (4.32) shows, if we define for $\Re s_1, \Re s_2 > 0$,

$$F(s_1, s_2) := \sum_{1 \le d', e', g < \infty} \frac{\mu(d') \mu(e') \mu(g)^2 v_{d'}(\mathcal{H}_1) v_{e'}(\mathcal{H}_2) \bar{v}_g(\mathcal{H}_1 \bar{\cap} \mathcal{H}_2)}{d'^{s_1+1} e'^{s_2+1} g^{s_1+s_2+1}}$$

$$= \prod_p \left(1 - \frac{v_p(\mathcal{H}_1)}{p^{s_1+1}} - \frac{v_p(\mathcal{H}_2)}{p^{s_2+1}} + \frac{\bar{v}_p(\mathcal{H}_1 \bar{\cap} \mathcal{H}_2)}{p^{s_1+s_2+1}} \right), \qquad (4.76)$$

that

$$T_R = \frac{1}{(2\pi i)^2} \int_{(1)} \int_{(1)} F(s_1, s_2) \frac{R^{s_1}}{s_1^{k_1+l_1+1}} \frac{R^{s_2}}{s_2^{k_2+l_2+1}} \, ds_1 \, ds_2. \qquad (4.77)$$

(5) Now $v_p(\mathcal{H}_1) = k_1$, $v_p(\mathcal{H}_2) = k_2$ and $v_p(\mathcal{H}_1 \cap \mathcal{H}_2) = r$. Separating out the dominant zeta Euler factors, we can define a function $G_{\mathcal{H}_1, \mathcal{H}_2}(s_1, s_2)$ by

$$F(s_1, s_2) = \frac{G_{\mathcal{H}_1, \mathcal{H}_2}(s_1, s_2) \zeta(1 + s_1 + s_2)^r}{\zeta(1 + s_1)^{k_1} \zeta(1 + s_2)^{k_2}}, \qquad (4.78)$$

where, using the Euler product representation for $\zeta(s)$,

$$G_{\mathcal{H}_1, \mathcal{H}_2}(s_1, s_2) = \prod_p \left(\frac{\left(1 - \frac{v_p(\mathcal{H}_1)}{p^{s_1+1}} - \frac{v_p(\mathcal{H}_2)}{p^{s_2+1}} + \frac{\bar{v}_p((\mathcal{H}_1 \bar{\cap} \mathcal{H}_2))}{p^{s_1+s_2+1}} \right) \left(1 - \frac{1}{p^{1+s_1+s_2}} \right)^r}{\left(1 - \frac{1}{p^{1+s_1}} \right)^{k_1} \left(1 - \frac{1}{p^{1+s_2}} \right)^{k_2}} \right).$$

$$(4.79)$$

This function is holomorphic and uniformly bounded in $\sigma_1, \sigma_2 > -\frac{1}{4} + \delta$ for given $\delta > 0$. In addition, it follows from (4.4), (4.1) and (4.73) that

$$G_{\mathcal{H}_1, \mathcal{H}_2}(0, 0) = \mathcal{G}(\mathcal{H}), \qquad (4.80)$$

and for s_1, s_2 on $\mathcal{L}$ or to the right of this contour, we have, for $\delta_i := -\min(0, \sigma_i)$ and $U := C k^2 \log(2h)$ as before,

$$G_{\mathcal{H}_1, \mathcal{H}_2}(s_1, s_2) \ll \exp(C k U^{\delta_1+\delta_2} \log\log U). \qquad (4.81)$$

(6) If we define an entire function $W(s) := s\zeta(1+s)$, then we would have

$$D(s_1, s_2) = G_{\mathcal{H}_1, \mathcal{H}_2}(s_1, s_2) \frac{W(s_1 + s_2)^r}{W(s_1)^{k_1} W(s_2)^{k_2}}, \tag{4.82}$$

and

$$T_R(l_1, l_2, \mathcal{H}_1, \mathcal{H}_2) = \frac{1}{(2\pi i)^2} \int_{(1)} \int_{(1)} D(s_1, s_2) \frac{R^{s_1 + s_2}}{s_1^{l_1+1} s_2^{l_2+1} (s_1 + s_2)^r} \, ds_1 \, ds_2. \tag{4.83}$$

We can then use the evaluation of this type of integral given by Lemma 4.8 to complete the proof of the lemma. □

Now we must attend to the so-called prime sums with products of the Λ_R. The proof follows a somewhat familiar pattern, but uses the Cauchy–Schwarz inequality and the generalized divisor function $\tau_k(q) := k^{\omega(q)}$ which we met in Section 4.3.

First recall the definitions $\theta(p) = \log p$ and $\theta(n) = 0$ if $n \in \mathbb{N}$ is not prime. $\mathcal{H} = \mathcal{H}_1 \cup \mathcal{H}_2$, $\mathcal{H}^0 = \mathcal{H} \cup \{h_0\}$ and $L_X(Y) = (\log X)^Y$ for $X > 0$. Also recall

$$\mathcal{H} = \mathcal{H}_1 \cup \mathcal{H}_2, \; k_1 = |\mathcal{H}_1|, \; k_2 = |\mathcal{H}_2|, \; r = |\mathcal{H}_1 \cap \mathcal{H}_2|, k = k_1 + k_2,$$
$$M = k + l_1 + l_2.$$

Lemma 4.10 *[72, proposition 5] Let $\mathcal{H}_1 \neq \emptyset$, $\mathcal{H}_2 \neq \emptyset$, $h \ll R$ and $h_0 \notin \mathcal{H}$. For all $A > 0$, there exists a B dependent at most on A and M such that if R and N are sufficiently large and*

$$R \ll_{M,A} \frac{N^{\frac{1}{4}}}{L_N(B)},$$

then

$$\sum_{n=1}^{N} \Lambda_R(n, \mathcal{H}_1, l_1) \Lambda_R(n, \mathcal{H}_2, l_2) \theta(n + h_0)$$

$$= \frac{1}{(r + l_1 + l_2)!} \binom{l_1 + l_2}{l_1} \mathcal{G}(\mathcal{H}^0) N L_R(r + l_1 + l_2)$$

$$+ N \sum_{j=1}^{r} \mathcal{D}_j(l_1, l_2, \mathcal{H}_1, \mathcal{H}_2, h_0) L_R(r + l_1 + l_2 - j)$$

$$+ O_{M,A}\left(\frac{N}{L_N(A)}\right), \tag{4.84}$$

where the $\mathcal{D}_j$ are the functions not depending on R or N, which appear in Lemma 4.10, and which satisfy for some constants C_j, C_j' depending at most on M,

$$\mathcal{D}_j(l_1, l_2, \mathcal{H}_1, \mathcal{H}_2, h_0) \ll_M L_U(C_j) \ll_M (\log\log(10h))^{C_j'}. \tag{4.85}$$

In case $h_0 \in \mathcal{H}_1 \cap \mathcal{H}_2$, we obtain the same asymptotic expression for the prime sum, except the main term should be replaced by

$$\frac{1}{(r + l_1 + l_2 + 1)!} \binom{l_1 + l_2 + 2}{l_1 + 1} \mathcal{G}(\mathcal{H}^0) N L_R (r + l_1 + l_2 + 1).$$

Proof **(1)** We may assume $k_1 \geq 1$ and $k_2 \geq 1$. Recall the definition

$$\widetilde{\mathcal{S}}_R := \widetilde{\mathcal{S}}_R(N, \mathcal{H}_1, \mathcal{H}_2, l_1, l_2, h_0) := \sum_{n=1}^{N} \Lambda_R(n, \mathcal{H}_1, l_1) \Lambda_R(n, \mathcal{H}_2, l_2) \theta(n + h_0).$$

$$(4.86)$$

Expanding the definitions, we get

$$\widetilde{\mathcal{S}}_R = \frac{1}{(k_1 + l_1)! \, (k_2 + l_2)!} \sum_{d, e \leq R} \mu(d) \mu(e) L_{R/d}(k_1 + l_1) L_{R/e}(k_2 + l_2)$$

$$\times \left(\sum_{\substack{1 \leq n \leq N \\ d | P_{\mathcal{H}_1}(n) \\ e | P_{\mathcal{H}_2}(n)}} \theta(n + h_0) \right). \quad (4.87)$$

(2) Consider the inner sum in (4.87). Let $g := (d, e)$, $d = d'g$, $e = e'g$ and $f = d'e'g$. Then since d and e are squarefree, d', e' and g are pairwise relatively prime and $f = [d, e]$. The n which satisfy $d \mid P_{\mathcal{H}_1}(n)$ and $e \mid P_{\mathcal{H}_2}(n)$ represent a set of residue classes modulo the least common modulo (LCM) $[d, e]$. If $n \equiv b \bmod f$ is one of these classes, then setting

$$m = n + h_0 \equiv b + h_0 \bmod f,$$

this residue class contributes to the inner sum a quantity

$$\sum_{\substack{1 + h_0 \leq m \leq N + h_0 \\ m \equiv b + h_0 \bmod f}} \theta(m) = \theta(N + h_0, f, b + h_0) - \theta(h_0, f, b + h_0)$$

$$= \frac{N}{\varphi(f)} + E(N, f, b + h_0) + O(h \log N), \quad (4.88)$$

where the leading term is to be replaced by zero in case $(b + h_0, f) \neq 1$ and E is defined in Lemma 4.6.

To count the number of classes for which $(b + h_0, f) = 1$, note that if $p \mid d'$, then $b \equiv -h_j \bmod p$ for some $h_j \in \mathcal{H}_1$ so $b + h_0 \equiv h_0 - h_j \bmod p$. Therefore, if h_0 is distinct modulo p from all the the $h_j \in \mathcal{H}_1$, then all $v_p(\mathcal{H}_1)$ classes satisfy the relatively prime condition; otherwise, $h_0 \equiv h_j \bmod p$ for some $j \in \mathcal{H}_1$ and just $v_p(\mathcal{H}_1) - 1$ classes have a nonzero main term.

Recall we have defined $\mathcal{H}^0 := \mathcal{H} \cup \{h_0\}$ and we define $v_p^*(\mathcal{H})$ to be either $v_p(\mathcal{H})$ or $v_p(\mathcal{H}) - 1$ to cover both possible cases as described in the previous paragraph. Extend this definition to $v_d^*(\mathcal{H})$ for squarefree d by multiplicativity. Finally, define $\mathcal{H}(p)$ to be the $v_p(\mathcal{H})$ set of congruence classes for $\mathcal{H}$ mod p and

$$\bar{v}_p^*(\mathcal{H}_1(p)\bar{\cap}\mathcal{H}_2(p)) := v_p^*(\mathcal{H}_1) + v_p^*(\mathcal{H}_2) - v_p^*(\mathcal{H}_1 \cup \mathcal{H}_2). \tag{4.89}$$

(3) In this step, the divisibility conditions $e \mid P_{\mathcal{H}_2}(n)$, $g \mid P_{\mathcal{H}_1}(n)$ and $g \mid P_{\mathcal{H}_2}(n)$ are dealt with in a similar manner to that employed in Lemma 4.9, together with the observations made in Step (2). Because $(a,q) > 1$ and $q \leq N$ imply $E(n,q,a) \ll \log N$, it follows that

$$\sum_{\substack{1 \leq n \leq N \\ d \mid P_{\mathcal{H}_1}(n) \\ e \mid P_{\mathcal{H}_2}(n)}} \theta(n + h_0) = v_{d'}^*(\mathcal{H}_1^0)v_{e'}^*(\mathcal{H}_2^0)\bar{v}_g^*((\mathcal{H}_1\bar{\cap}\mathcal{H}_2)^0)\frac{N}{\varphi(f)}$$

$$+ O\left(\tau_k(f)\left(\max_{b:\ (b,f)=1} |E(N,f,b)| + h \log N\right)\right). \tag{4.90}$$

Substituting this value into (4.14) to get

$$\widetilde{\mathscr{S}}_R = \frac{N}{(k_1+l_1)!\,(k_2+l_2)!}$$

$$\times \sum_{\substack{d'g \leq R \\ e'g \leq R}} \frac{\mu(d')\mu(e')\mu(g)^2 v_{d'}^*(\mathcal{H}_1^0)v_{e'}^*(\mathcal{H}_2^0)\bar{v}_g^*((\mathcal{H}_1\bar{\cap}\mathcal{H}_2)^0)}{\varphi(f)}$$

$$\times L_{R/(d'g)}(k_1+l_1)L_{R/(e'g)}(k_2+l_2)$$

$$+ O\left(L_R(M)\sum_{\substack{d'g \leq R \\ e'g \leq R}} \tau_k(d'e'g)E'(N,d'e'g)\right)$$

$$+ O\left(hR^2(3\log N)^{M+3k+1}\right),$$

where we used Lemma 4.5 for the last error term. This expression can now be written

$$\widetilde{\mathscr{S}}_R = N\widetilde{\mathscr{T}}_R(\mathcal{H}_1,\mathcal{H}_2,l_1,l_2,h_0) + O(L_R(M)\mathscr{E}_k(N))$$
$$+ O\left(hR^2(3\log N)^{M+3K+1}\right), \tag{4.91}$$

so $\widetilde{\mathscr{T}}_R(\mathcal{H}_1,\mathcal{H}_2,l_1,l_2,h_0)$ is the coefficient of N in the main term and

$$\mathscr{E}_k(N) := \sum_{\substack{d'g \leq R \\ e'g \leq R}} \tau_k(d'e'g)E'(N,d'e'g).$$

(4) Now we estimate the first error term in (4.91) using Lemma 4.5, Lemma 4.6, the estimate $q \leq N \implies E(N,q) \leq 2N/q$ and the Cauchy–Schwarz inequality to derive the estimate valid for $R^2 \ll \sqrt{N}/L_N(B)$ and uniform for $k \leq \sqrt{(\log N)/18}$:

$$|\mathcal{E}_k(N)| \leq \sum_{\substack{q \leq R^2 \\ \text{squarefree}}} \tau_k(q) \max_{b:(b,q)=1} |E(N,q,b)| \sum_{q=d'e'g} 1$$

$$= \sum_{\substack{q \leq R^2 \\ \text{squarefree}}} \tau_k(q)\tau_3(q)E(N,q)$$

$$\leq \sqrt{\sum_{\substack{q \leq R^2 \\ \text{squarefree}}} \frac{\tau_{3k}(q)^2}{q}} \sqrt{\sum_{q \leq R^2} q\,E(N,q)^2}$$

$$\leq \sqrt{L_N(9k^2)}\sqrt{2N \log N} \sqrt{\sum_{q \leq R^2} E(N,q)}$$

$$\ll N L_N((9k^2 + 1 - A)/2). \tag{4.92}$$

Changing the value of B, this implies that given any $A, M > 0$ there is a $B = B(A,M) > 0$ such that for $h \leq R \ll N^{\frac{1}{4}} \log^B N$, we have

$$\widetilde{\mathcal{F}}_R(N, \mathcal{H}_1, \mathcal{H}_2, l_1, l_2, h_0) = N\widetilde{T}_R(\mathcal{H}_1, \mathcal{H}_2, l_1, l_2, h_0) + O_M\left(\frac{N}{L_N(A)}\right). \tag{4.93}$$

(5) Next, we consider the main term in (4.91). Equation (4.32) shows, if we define for $\Re s_1, \Re s_2 > 0$,

$$F(s_1,s_2) := \sum_{1 \leq d', e', g < \infty} \frac{\mu(d')\mu(e')\mu(g)^2 v_{d'}^*(\mathcal{H}_1^0)v_{e'}^*(\mathcal{H}_2^0)\bar{v}_g^*((\mathcal{H}_1 \bar{\cap} \mathcal{H}_2)^0)}{\varphi(d')d'^{s_1}\varphi(e')e'^{s_2}\varphi(g)g^{s_1+s_2}}$$

$$= \prod_p \left(1 - \frac{v_p^*(\mathcal{H}_1^0)}{(p-1)p^{s_1}} - \frac{v_p^*(\mathcal{H}_2^0)}{(p-1)p^{s_2}} + \frac{\bar{v}_p^*((\mathcal{H}_1 \bar{\cap} \mathcal{H}_2)^0)}{(p-1)p^{s_1+s_2}}\right), \tag{4.94}$$

then

$$\widetilde{T}_R = \frac{1}{(2\pi i)^2} \int_{(1)} \int_{(1)} F(s_1,s_2) \frac{R^{s_1}}{s_1^{k_1+l_1+1}} \frac{R^{s_2}}{s_2^{k_2+l_2+1}} \, ds_1 \, ds_2. \tag{4.95}$$

(6) Assume now $h_0 \in \mathcal{H}_1 \cap \mathcal{H}_2$ (case (a)), and for $p > h$ get

$$v_p^*(\mathcal{H}_1^0) = k_1 - 1, \quad v_p^*(\mathcal{H}_2^0) = k_2 - 1, \quad \bar{v}_p^*((\mathcal{H}_1 \bar{\cap} \mathcal{H}_2)^0) = r - 1,$$

and can therefore define the function $G_{\mathcal{H}_1, \mathcal{H}_2}(s_1, s_2)$ by

$$F(s_1,s_2) = \frac{G_{\mathcal{H}_1, \mathcal{H}_2}(s_1,s_2)\zeta(1 + s_1 + s_2)^{r-1}}{\zeta(1+s_1)^{k_1-1}\zeta(1+s_2)^{k_2-1}}. \tag{4.96}$$

In case (b), we have $h_0 \notin \mathcal{H} = \mathcal{H}_1 \cup \mathcal{H}_2$, and when $p > h$, get

$$v_p^*(\mathcal{H}_1^0) = k_1, \quad v_p^*(\mathcal{H}_2^0) = k_2, \quad \bar{v}_p^*((\mathcal{H}_1 \cap \mathcal{H}_2)^0) = r,$$

and can therefore define the function $G_{\mathcal{H}_1, \mathcal{H}_2}(s_1, s_2)$ by

$$F(s_1, s_2) = \frac{G_{\mathcal{H}_1, \mathcal{H}_2}(s_1, s_2)\zeta(1 + s_1 + s_2)^r}{\zeta(1 + s_1)^{k_1}\zeta(1 + s_2)^{k_2}}. \tag{4.97}$$

In both cases, G is holomorphic and uniformly bounded in each half plane $\sigma_1, s_2 > c$ for any $c > -\frac{1}{4}$.

The same is true in the remaining case (c), wherein $h_0 \in \mathcal{H}_1$ but not in $\mathcal{H}_2$ with a suitably designed definition for $F(s_1, s_2)$, but we don't need this case, so the details are omitted.

(7) We now evaluate $G_{\mathcal{H}_1, \mathcal{H}_2}(0, 0)$. In case (a), using the Euler product for $\zeta(s)$, the definition of $\bar{v}_p$ (4.73), $\mathcal{H}^0 = \mathcal{H}$, and (4.94) and (4.97), and defining

$$\alpha(\mathcal{H}_1, \mathcal{H}_2, h_0) := k_1 + k_2 - r - 1 = k - r - 1,$$

we get

$$
\begin{aligned}
G_{\mathcal{H}_1, \mathcal{H}_2}(0, 0) &= \prod_p \left(1 - \frac{v_p(\mathcal{H}_1^0) + v_p(\mathcal{H}_2^0) - \bar{v}_p((\mathcal{H}_1 \cap \mathcal{H}_2)^0)}{p - 1}\right) \\
&\quad \times \left(1 - \frac{1}{p}\right)^{-\alpha(\mathcal{H}_1, \mathcal{H}_2, h_0)} \\
&= \prod_p \left(\frac{p - v_p(\mathcal{H})}{p - 1}\right)\left(1 - \frac{1}{p}\right)^{r - k + 1} \\
&= \prod_p \left(1 - \frac{v_p(\mathcal{H}^0)}{p}\right)\left(1 - \frac{1}{p}\right)^{-(k - r)} \\
&= \mathcal{G}(\mathcal{H}) = \mathcal{G}(\mathcal{H}^0). \tag{4.98}
\end{aligned}
$$

In case (b), using the Euler product for $\zeta(s)$, the definition of $\bar{v}_p$ (4.73), $v_p^*(\mathcal{H}^0) = v_p(\mathcal{H}^0) - 1$ and (4.94), if we define

$$\beta(\mathcal{H}_1, \mathcal{H}_2, h_0) := k_1 + k_2 - r = k - r,$$

we get

$$
\begin{aligned}
G_{\mathcal{H}_1,\mathcal{H}_2}(0,0) &= \prod_p \left(1 - \frac{v_p(\mathcal{H}_1^0) + v_p(\mathcal{H}_2^0) - \bar{v}_p((\mathcal{H}_1 \bar{\cap} \mathcal{H}_2)^0)}{p-1} \right) \\
&\qquad \times \left(1 - \frac{1}{p} \right)^{-\beta(\mathcal{H}_1,\mathcal{H}_2,h_0)} \\
&= \prod_p \left(\frac{p - v_p(\mathcal{H}^0)}{p-1} \right) \left(1 - \frac{1}{p} \right)^{r-k} \\
&= \prod_p \left(1 - \frac{v_p(\mathcal{H}^0)}{p} \right) \left(1 - \frac{1}{p} \right)^{-(k-r+1)} \mathscr{G}(\mathcal{H}^0). \qquad (4.99)
\end{aligned}
$$

(8) This final step follows the method of Lemma 4.9 to evaluate $\mathcal{T}_R$. The estimates for the lower-order terms and errors is essentially the same. For the main term, we use Lemma 4.8 and (4.97) to get

$$ a = k_1, \; b = k_2, \; d = r, \; u = l_1, \; v = l_2. $$

This completes the proof of the lemma. $\qquad\square$

The hard work has been done! Now we simplify the expression of two main lemmas to get Lemmas 4.11 and 4.12.

Lemma 4.11 *[72, proposition 1] Let as before*

$$ \mathcal{H} = \mathcal{H}_1 \cup \mathcal{H}_2, \; k_1 = |\mathcal{H}_1|, \; k_2 = |\mathcal{H}_2|, \; r = |\mathcal{H}_1 \cap \mathcal{H}_2|, k = k_1 + k_2, $$
$$ M = k + l_1 + l_2. $$

If $R \ll N^{\frac{1}{2}}/L_N(4M)$, and $h \le R^C$ for any fixed $C > 0$. Then as $R, N \to \infty$, we have

$$
\sum_{n=1}^{N} \Lambda_R(n,\mathcal{H}_1,l_1) \Lambda_R(n,\mathcal{H}_2,l_2) = N \binom{l_1+l_2}{l_1} \frac{L_R(r+l_1+l_2)}{(r+l_1+l_2)!} (\mathscr{G}(\mathcal{H}) + o_M(1)).
$$
$$ (4.100) $$

Proof This is an application of Lemma 4.9. $\qquad\square$

Recall we have defined $\mathcal{H}^0 = \mathcal{H} \cup \{h_0\}$ and $\mathcal{H} = \mathcal{H}_1 \cup \mathcal{H}_2$.

Lemma 4.12 *[72, proposition 2] Let $B(M) > 0$ be sufficiently large, $R \ll_M N^{\frac{1}{4}}/L_N(B(M))$ with $B(M)$ sufficiently large and positive and $h \le R$. Then as $R, N \to \infty$ we have in case (a) $h_0 \in \mathcal{H}_1 \cap \mathcal{H}_2$ or case (b) $h_0 \notin \mathcal{H}_1 \cup \mathcal{H}_2$, we have respectively for the sum*

$$
\sum_{n=1}^{N} \Lambda_R(n,\mathcal{H}_1,l_1) \Lambda_R(n,\mathcal{H}_2,l_2)\theta(n+h_0)
$$

the estimate

(a) $N\binom{l_1 + l_2 + 2}{l_1 + 1} \dfrac{L_R(r + l_1 + l_2 + 1)}{(r + l_1 + l_2 + 1)!}(\mathcal{G}(\mathcal{H}^0) + o_M(1))$,

(b) $N\binom{l_1 + l_2}{l_1} \dfrac{L_R(r + l_1 + l_2)}{(r + l_1 + l_2)!}(\mathcal{G}(\mathcal{H}^0) + o_M(1))$.

Proof   This follows directly from Lemma 4.10.   ☐

4.6 The Essential Theorem of Gallagher

Gallagher's proof of his theorem [67, 68] uses two properties of Stirling numbers of the second kind $\sigma(a, v)/v!$. These give a count of the number of different equivalence relations, with precisely v equivalence classes, for $1 \leq v \leq a$, that can be defined on a set with a distinct elements. Indeed, $\sigma(a, v)$ is the number of surjective functions from a set of a elements onto a set of v elements. The properties we need are derived in Lemma 4.13. The notation $\binom{b}{v}$ is the standard binomial coefficient.

Lemma 4.13   (i) $\sum_{v=1}^{b} \binom{b}{v}\sigma(a, v) = b^a$,

(ii) $\sum_{v=1}^{b} v\binom{b}{v}\sigma(a, v) = b^{a+1} - (b - 1)^a b$.

Proof   (i) Let $A = \{1, 2, 3, \ldots, a\}$ and $B = \{1, 2, 3, \ldots, b\}$. We count the functions which map from A to B:

$$b^a = \#\{f : A \to B\} = \sum_{v=1}^{b} \#\{f : A \to B : |f(A)| = v\}$$

$$= \sum_{v=1}^{b} \sum_{S \subset B : |S| = v} \#\{f : f(A) = S\}$$

$$= \sum_{v=1}^{b} \binom{b}{v}\sigma(a, v).$$

(ii) Note that for $v \leq b - 1$ we get

$$v\binom{b}{v} = b\binom{b-1}{v-1} = b\binom{b}{v} - b\binom{b-1}{v},$$

and $\binom{b-1}{b} = 0$. The result then follows substituting this identity in the result of Step (i) and summing.   ☐

Recall from (4.4) the definition of $\mathscr{G}(\mathscr{H})$:

$$\mathscr{G}(\mathscr{H}) := \prod_p \left(1 - \frac{1}{p}\right)^{-k} \left(1 - \frac{v_p(\mathscr{H})}{p}\right). \qquad (4.101)$$

Keeping $k \geq 2$ fixed and allowing $h \to \infty$, we next have the average result of Gallagher given by Theorem 4.14, which is crucial for the proof of Theorem 4.19.

Theorem 4.14 **(Gallagher)** *[67, 68, equation (3)] Let $k \in \mathbb{N}$. Then as $h \to \infty$, we have*

$$\sum_{\substack{\mathscr{H} \subset [1,h] \\ |\mathscr{H}| = k}} \mathscr{G}(\mathscr{H}) = (1 + o(1)) \frac{h^k}{k!} \quad \text{and} \quad \sum_{\substack{1 \leq h_1, \dots, h_k \leq h \\ h_i \text{ distinct}}} \mathscr{G}(\mathscr{H}) = (1 + o(1)) h^k,$$

where the sum in the second form is over all positive integer vectors $(h_1, \dots, h_k)$ with components bounded by h, and with no two components equal. In other words, the second form counts all reorderings of the elements of each $\mathscr{H}$, whereas the first counts only those with, say, increasing order of the h_j.

Proof **(1)** Let $\Delta = \Delta_{\mathscr{H}} := \prod_{1 \leq i < j \leq k} (h_i - h_j)$. Then $v_{\mathscr{H}}(p) < k$ if and only if $p \mid \Delta$. The pth Euler factor of $\mathscr{G}(\mathscr{H})$ can be written

$$1 + \frac{p^k - v_{\mathscr{H}}(p) p^{k-1} - (p-1)^k}{(p-1)^k} = 1 + f(p, v_{\mathscr{H}}(p)), \qquad (4.102)$$

where

$$v = k \implies f(p, v) \ll_k 1/(p-1)^2 \leq \frac{\eta}{(p-1)^2}$$

$$\text{and } v < k \implies f(p, v) \ll_k 1/(p-1) \leq \frac{\eta}{(p-1)}.$$

Thus the Euler product $\mathscr{G}(\mathscr{H})$ converges, and for all v with $1 \leq v \leq k$, we have

$$f(p, v) \ll_k \frac{1}{p-1}. \qquad (4.103)$$

If we define for squarefree q

$$f_{\mathscr{H}}(q) := \prod_{p \mid q} f(p, v_{\mathscr{H}}(p)).$$

Then we can write

$$\mathscr{G}(\mathscr{H}) = \sum_{q \text{ squarefree}} f_{\mathscr{H}}(q). \qquad (4.104)$$

(2) Next we estimate the tail of this sum, so let $x > 0$ be large and consider the sum over $q > x$. (Later we will make the choice $x = \sqrt{h}$). To simplify the inner sum in line two, we use a similar proof to that of Lemma 3.6 which gives

$$\sum_{n>x} \frac{\tau(n)}{\varphi(n)^2} \ll \frac{(\log x)^3}{x},$$

with the bound $n/\varphi(n) \ll \log\log n$, [90, theorem 317] and Lemma 4.15, which give

$$\tau(n) < 2^{\frac{(1+\epsilon)\log n}{\log\log n}} \text{ and } \omega(n) \ll \frac{\log n}{\log\log n}$$

respectively. For the outer sum in the following, we also use $d \mid \Delta \implies \tau(d) \ll \eta^\epsilon$. Note that the constant $\eta > 0$ depends at most on $k = |\mathcal{H}|$. We get for a given fixed $\mathcal{H}$

$$\sum_{x<q} |f_{\mathcal{H}}(q)| \leq \sum_{x<q} \frac{\mu(q)^2 \eta^{\omega(q)} \varphi((q,\Delta))}{\varphi(q)^2}$$

$$= \sum_{d\mid\Delta} \frac{\mu(d)^2 \eta^{\omega(d)}}{\varphi(d)} \sum_{\substack{e>x/d \\ (e,\Delta)=1}} \frac{\mu(e)^2 \eta^{\omega(e)}}{\varphi(e)^2}$$

$$\ll \sum_{d\mid\Delta} \frac{\mu(d)^2 \eta^{\omega(d)}}{\varphi(d)} \frac{d(\log x)^3}{x}$$

$$\ll \left(\sum_{d\mid\Delta} \frac{d}{\varphi(d)} \eta^{\log d/\log\log d}\right) \frac{(\log x)^3}{x}$$

$$\ll_{k,\epsilon} \frac{(xh)^\epsilon}{x}.$$

Hence

$$\sum_{\substack{1\leq h_1,\dots,h_k\leq h \\ h_i \text{ distinct}}} \mathcal{G}(\mathcal{H}) = \sum_{q\leq x} \sum_{\substack{1\leq h_1,\dots,h_k\leq h \\ h_i \text{ distinct}}} f_{\mathcal{H}}(q) + O_{k,\epsilon}\left(h^k \frac{(xh)^\epsilon}{x}\right). \qquad (4.105)$$

(3) Now we consider the inner sum of (4.105). Let $\beta_h(k,q,v)$ be the number of k-tuples of positive integers in $[1,h]$, which are not necessarily distinct, and which for each $p \mid q$ occupy exactly a given number $v(p)$ with $1 \leq v(p) \leq p$ of residue classes modulo p. In what follows, the outer sums over v are over all vectors $(v(p))_{p\mid q}$, where $v(p)$ is the number of residue classes modulo p occupied by the elements $h_1, \dots, h_k$. The inner sum of (4.105) can then be written

$$\sum_v \left(\prod_{p|q} f(p, v(p)) \right) (\beta_h(k, q, v) + O(h^{k-1})).$$

Using the Chinese Remainder Theorem, we get for $q \le h$,

$$\beta_h(k, q, v) = \left(\frac{h^k}{q^k} + O\left(\left(\frac{h}{q} \right)^{k-1} \right) \right) \prod_{p|q} \left(\frac{p}{v(p)} \right) \sigma(k, v(p)).$$

Thus the inner sum in (4.105) can be written

$$\frac{h^k}{q^k} A(q) + O\left(\left(\frac{h}{q} \right)^{k-1} B(q) \right) + O(h^{k-1} C(q)), \tag{4.106}$$

$$A(q) = \sum_v \prod_{p|q} f(p, v(p)) \left(\frac{p}{v(p)} \right) \sigma(k, v(p)) = \prod_{p|q} \left(\sum_{v=1}^{p} f(p, v) \binom{p}{v} \sigma(k, v) \right)$$

$$B(q) = \sum_v \prod_{p|q} |f(p, v(p))| \left(\frac{p}{v(p)} \right) \sigma(k, v(p)) = \prod_{p|q} \left(\sum_{v=1}^{p} |f(p, v)| \binom{p}{v} \sigma(k, v) \right)$$

$$C(q) = \sum_v \prod_{p|q} |f(p, v(p))| = \prod_{p|q} \left(\sum_{v=1}^{p} |f(p, v)| \right).$$

(4) Note that $A(1) = 1$ being the empty product. Next we **claim** that for $q > 1$ we have $A(q) = 0$. To see this, by (4.102) the pth Euler factor of $A(q)$, using (i) and (ii) of Lemma 4.13, is

$$(p-1)^k \left(p^k - (p-1)^k \right) \sum_{v=1}^{p} \binom{p}{v} \sigma(k, v) - p^{k-1} \sum_{v=1}^{p} v \binom{p}{v} \sigma(k, v) \right)$$

$$= (p-1)^k \left((p^k - (p-1)^k) p^k - p^{k-1} (p^{k+1} - p(p-1)^k) \right) = 0,$$

so, indeed, all of the Euler factors vanish.

For $B(q)$ and $C(q)$, we derive bounds, again for $q > 1$ squarefree. By the estimate (4.103), we have $f(a, v) \ll 1/(p-1)$. Hence, using (1) of Lemma 4.13, the pth Euler factor of $B(q)$ is $O(p^k/(p-1))$ and therefore $B(q) \le \eta^{\omega(q)} q^k / \varphi(q)$. Similarly for $C(q)$, the Euler factor is $O(p/(p-1))$, so $C(q) \le \eta^{\omega(q)} q / \varphi(q)$.

(5) We can now complete the derivation substituting (4.106) for the inner sum in (4.105). Now let $x = \sqrt{h}$ to get

$$\sum_{\substack{1\le h_1,\dots,h_k\le h \\ h_i \text{ distinct}}} \mathcal{G}(\mathcal{H}) = h^k + O\left(h^{k-1}\sum_{q\le x} \eta^{\omega(q)}\frac{q}{\varphi(q)}\right) + O\left(\frac{h^k(xh)^\epsilon}{x}\right)$$

$$= h^k + O\left(h^{k-1}x^{1+\epsilon} + \frac{h^k(hx)^\epsilon}{x}\right)$$

$$= h^k + O(h^{k-\frac{1}{2}+\epsilon}),$$

and the proof of the theorem is complete.   □

Kevin Ford gave a proof also, which is shorter but has a poorer error than that of Gallagher, namely $O(h^k/\log\log h)$, [50]. First we need a standard lemma.

Lemma 4.15   *There exists an absolute implied constant such that for all $n \ge 3$, we have*

$$\omega(n) \ll \frac{\log n}{\log\log n}.$$

Proof   **(1)** Since the function of n on the right of the estimate is increasing, we can assume n is squarefree. Let P_j be the jth prime and suppose that $n = p_1 \dots p_m$ with $p_1 < p_2 < \cdots < p_m$. Then $P_j \le p_j$ for all j with $1 \le j \le m$. Thus

$$n \ge \prod_{1\le j\le m} P_j \quad \Longrightarrow \quad \log n \ge \sum_{1\le j\le m} \log P_j = P_m(1+o(1)),$$

by the prime number theorem. But $P_m = (1+o(1))m\log m$, so $\log n \ge (1+o(1))m\log m$.

(2) Suppose $1 > \epsilon > 0$ is given. Then for m sufficiently large, we have, from the result of Step (1), $\log n \ge (1-\epsilon)m\log m$. If in addition $m > (\log n)^{1-\epsilon}$, we then get

$$\log n > (1-\epsilon)^2 m\log\log n \quad \Longrightarrow \quad m < \frac{1}{(1-\epsilon)^2}\frac{\log n}{\log\log n}.$$

If on the other hand $m \le (\log n)^{1-\epsilon}$, for all n sufficiently large we have

$$m \le (\log n)^{1-\epsilon} \le \frac{1}{(1-\epsilon)^2}\frac{\log n}{\log\log n}.$$

Note that m and n can be made simultaneously sufficiently large, so we get for all n, m sufficiently large

$$\omega(n) = m < \frac{1}{(1-\epsilon)^2}\frac{\log n}{\log\log n}.$$

Finally note that in the bounded ranges both sides of the estimate are bounded and nonzero, so we can choose the implied constant sufficiently large to ensure it is true in that range also, which completes the proof. $\qquad\square$

Recall the definition

$$\mathscr{G}(\mathscr{H}) = \prod_{p}\left(1 - \frac{v_p(\mathscr{H})}{p}\right)\left(1 - \frac{1}{p}\right)^{-k}.$$

Theorem 4.16 **(Ford)** *[50] Let $k \in \mathbb{N}$. Then as $h \to \infty$ we have*

$$\sum_{\substack{1\leq h_1,\ldots,h_k\leq h \\ h_i \text{ distinct}}} \mathscr{G}(\mathscr{H}) = (1 + o(1))h^k,$$

where the sum is over all positive integer vectors $(h_1, \ldots, h_k)$ with components bounded by h, and with no two components equal.

Proof **(1)** We may assume $h \geq 2$. Set $y = (\log h)/2$ and recall $\Delta := \prod_{1\leq i<j\leq k} |h_i - h_j|$ with

$$\mathscr{H} = (h_1, \ldots, h_k)$$

with the h_i distinct but not necessarily in increasing order.

First note that if $p \nmid \Delta$ then $v_p(\mathscr{H}) = k$. In addition, by Lemma 4.15, the number of primes dividing Δ, because the implied constants may depend on k, is

$$O\left(\frac{\log \Delta}{\log\log \Delta}\right) = O\left(\frac{\log h}{\log\log h}\right).$$

(2) It follows from Step (1) that

$$\prod_{y<p}\left(1 - \frac{v_p(\mathscr{H})}{p}\right)\left(1 - \frac{1}{p}\right)^{-k} = \prod_{\substack{y<p \\ p|\Delta}}\left(1 + O\left(\frac{1}{p}\right)\right)\prod_{\substack{y<p \\ p\nmid\Delta}}\left(1 + O\left(\frac{1}{p^2}\right)\right)$$

$$= 1 + O\left(\frac{\log h}{y\log\log h}\right)$$

$$= 1 + O\left(\frac{1}{\log\log h}\right).$$

Therefore, we can write

$$A \times B = \sum_{\substack{1 \le h_1,\ldots,h_k \le h \\ h_i \text{ distinct}}} \mathscr{G}(\mathscr{H}) \text{ where}$$

$$A = \left(1 + O\left(\frac{1}{\log\log h}\right)\right) \prod_{p \le y} \left(1 - \frac{1}{p}\right)^{-k}, \text{ and}$$

$$B = \sum_{\substack{1 \le h_1,\ldots,h_k \le h \\ h_j \text{ distinct}}} \prod_{p \le y} \left(1 - \frac{v_p(\mathscr{H})}{p}\right).$$

(3) Now the asymptotic relation of the lemma is an identity for $k = 1$, so using induction we can assume it is true for k replaced by $k - 1$. This implies, again since the implied constants can depend on k, we can write $B = B' + O(h^{k-1})$, were B' is the same sum as the lemma, but in it we relax the condition that the h_j be distinct.

Let $M = \prod_{p \le y} p$ so by the prime number theorem and our choice $y = (\log h)/2$ we have $M = \exp(y + o(y)) = h^{\frac{1}{2}+o(1)}$. By the Chinese Remainder Theorem, the product in B' is

$$\frac{1}{M} \prod_{p \le y}(p - v_p(\mathscr{H})) = \frac{1}{M} \times \#\left\{n : 0 \le n < M, \prod_{1 \le j \le k}(n + h_j, M) = 1\right\}.$$

It follows that

$$B' = \sum_{\substack{1 \le h_1,\ldots,h_k \le h}} \frac{1}{M} \sum_{0 \le n \le M-1} \prod_{j=1}^{k} \sum_{d_j|(n+h_j,M)} \mu(d_j)$$

$$= \frac{1}{M} \sum_{0 \le n < M} \sum_{d_j|M, \, 1 \le j \le k} \mu(d_1)\cdots\mu(d_k) \prod_{j=1}^{k}\left(\frac{h}{d_j} + O(1)\right)$$

$$= h^k \sum_{d_j|M, \, 1 \le j \le k} \frac{\mu(d_1)\cdots\mu(d_k)}{d_1\cdots d_k} + O\left(h^{k-1} \sum_{d_j|M, 1 \le j < k} \frac{1}{d_1\cdots d_{k-1}}\right)$$

$$= h^k \prod_{p \le y}\left(1 - \frac{1}{p}\right)^k + O\left(h^{k-1+o(1)}\right).$$

(4) Using the expressions we have derived for A, B and B', we get

$$A \times B = \left(1 + O\left(\frac{1}{\log\log h}\right)\right) \prod_{p \le y}\left(1 - \frac{1}{p}\right)^{-k}(B' + O(h^{k-1}))$$

$$= \left(1 + O\left(\frac{1}{\log\log h}\right)\right)\left(h^k + O(h^{k-1+o(1)})\right)$$

$$= h^k(1 + o(1)).$$

This completes the proof. $\square$

As well as Kevin Ford, Pintz took a careful look at Gallagher's result and arrived at a stronger expression containing more detail, [166]. First some definitions:

$$\mathcal{H} := (h_1, \ldots, h_k) \text{ distinct,}$$
$$\mathcal{H}' := \mathcal{H} \cup \{h\} \text{ with } h \notin \mathcal{H},$$
$$P(n) := (n + h_1) \cdots (n + h_k),$$
$$S_{\mathcal{H}}(h) := \frac{1}{h} \sum_{n=1}^{h} \frac{\mathcal{G}(\mathcal{H} \cup \{n\})}{\mathcal{G}(\mathcal{H})},$$
$$v_p := v_p(\mathcal{H}) \text{ and } v'_p := v_p(\mathcal{H}').$$

Note that Pintz uses $P(n) := (n - h_1) \cdots (n - h_k)$, which should make no essential difference to the asymptotic estimate.

Theorem 4.17 **(Pintz)** *[166] If $\epsilon > 0$ is sufficiently small, we have*

$$h \geq \exp(k^{1/\epsilon}) \implies S_{\mathcal{H}}(h) = 1 + O(\epsilon). \tag{4.107}$$

Proof **(1)** Let $h \geq 2$ be an integer, set $y = \log h$ and $M = \prod_{p \leq y} p$. Then we can write

$$\frac{\mathcal{G}(\mathcal{H}')}{\mathcal{G}(\mathcal{H})} = \prod_p \frac{\left(1 - \frac{v_p(\mathcal{H} \cup \{h\})}{p}\right)}{\left(1 - \frac{v_p(\mathcal{H})}{p}\right)\left(1 - \frac{1}{p}\right)} = A \times B \times C, \tag{4.108}$$

where respectively $A = A(h)$ is the product restricted to $p \leq y$, B to $p > y$ with $p \mid P(h)$ and C to $p > y$ with $p \nmid P(h)$.

(2) If $p \nmid P(h)$, then $v_p(\mathcal{H} \cup \{h\}) = v_p(\mathcal{H}) + 1$. Thus

$$C = \prod_{p > y} \left(1 + O\left(\frac{k}{p^2}\right)\right) = 1 + O\left(\frac{k}{y \log y}\right).$$

(3) Next we have

$$B = \prod_{\substack{p > y \\ p \mid P(h)}} \left(1 - \frac{1}{p}\right)^{-1} = \exp\left(O\left(\sum_{\substack{p > y \\ p \mid P(h)}} \frac{1}{p}\right)\right).$$

We have

$$\sum_{p \mid P(h)} \log p \leq \log |P(h)| \leq 2ky.$$

Thus the sum over $1/p$ will be a maximum when the primes p are taken to be as small as possible and satisfy the inequality $\sum \log p \leq 2ky$. Hence, by the prime number theorem, for y and hence h sufficiently large, we can take the maximum prime $p \leq 4ky$, so

$$\sum_{\substack{p > y \\ p|P(h)}} \frac{1}{p} \leq \sum_{y < p < 4ky} \frac{1}{p} \leq \log\left(\frac{\log(5ky)}{\log y}\right) \leq 2\epsilon.$$

Thus for fixed h, we have $B \times C = 1 + O(\epsilon)$.

(4) Now we consider the term A with $h \leq nM + r$, $0 \leq r < M$ and such that $n \to \infty$. The key observation is that A is periodic in h with period M, so the average A is the same as that taken over $h \in [1, M]$. If $p \leq y$, there are precisely $v_p(\mathcal{H})$ possible values modulo p for h with $v'_p = v_p$, and $p - v'_p$ with $v'_p = v_p + 1$. Therefore,

$$\frac{1}{M} \sum_{h=1}^{M} A(h) = \prod_{p|M} \frac{\left(\frac{v_p}{p}\left(1 - \frac{v_p}{p}\right) + \left(1 - \frac{v_p}{p}\right)\left(1 - \frac{v_p+1}{p}\right)\right)}{\left(1 - \frac{v_p}{p}\right)\left(1 - \frac{1}{p}\right)} = 1. \qquad (4.109)$$

(5) Using the result of Step (3) in (4.108) completes the proof. $\qquad\square$

We may use this theorem to obtain Gallager's theorem as a corollary using induction on k:

Corollary 4.18   *As $h \to \infty$, we have an alternative proof of Theorem 4.14, namely*

$$\sum_{\substack{1 \leq h_1, \ldots, \leq h_k \leq h \\ h_i \text{ distinct}}} \mathcal{G}(\mathcal{H}) \sim h^k.$$

Proof   For $k = 1$, both sides have the value h. Assuming the result holds for some $k \geq 1$, we can write

$$\sum_{\substack{1 \leq h_1, \ldots, h_k \leq h \\ h_i \text{ distinct}}} S_{\mathcal{H}}(h)\mathcal{G}(\mathcal{H}) = \sum_{\substack{1 \leq h_1, \ldots, h_k \leq h \\ h_i \text{ distinct} \\ 1 \leq n \leq h}} \mathcal{G}(\mathcal{H} \cup \{n\}),$$

which implies, using Theorem 4.17,

$$(1 + O(\epsilon)) \sum_{\substack{1 \leq h_1, \ldots, h_k \leq h \\ h_i \text{ distinct}}} \mathcal{G}(\mathcal{H}) = \frac{1}{h} \sum_{\substack{1 \leq h_1, \ldots, h_k \leq h \\ h_i \text{ distinct} \\ 1 \leq n \leq h}} \mathcal{G}(\mathcal{H} \cup \{n\}),$$

so therefore, using the inductive assumption,

$$h^k \sim \frac{1}{h} \sum_{\substack{1 \leq h_1, \ldots, h_k \leq h \\ h_i \text{ distinct}}} \mathcal{G}(\mathcal{H}),$$

which completes the proof. $\qquad\square$

4.7 The Main GPY Theorem

The argument is attributed by Goldston, Pintz and Yildirim to Granville and Soundararajan.

For $v \in \mathbb{N}$, define

$$\Delta_v := \liminf_{n \to \infty} \frac{p_{n+v} - p_n}{\log p_n} \geq 0.$$

A positive real number ϑ is the assumed level of distribution of rational primes and by Lemma 4.6 (the Bombieri–Vinogradov theorem) is valid unconditionally in the range $0 < \vartheta \leq \frac{1}{2}$.

Note that we use $\mathcal{H} = \mathcal{H}_1 = \mathcal{H}_2$, $k = k_1 = k_2$ and $l = l_1 = l_2$ in applying Lemmas 4.11 and 4.12 as promised.

> **Theorem 4.19** **(GMPY's prime gap estimate)** *[72, theorem 2] Let p_n be the nth prime. Then*
>
> $$\Delta_1 = \liminf_{n \to \infty} \frac{p_{n+1} - p_n}{\log p_n} = 0.$$

Proof **(1)** Let $l \geq 0$ and $\epsilon > 0$ be given and set $\mathcal{H}_k := (h_1, \ldots, h_k)$ with as usual

$$1 \leq h_1 < h_2 < h_k \leq h \leq R \ll \frac{\sqrt{N}}{L_N(B(M))}, \quad R, N \to \infty.$$

In Step (3), the variables h, R, l will be chosen as functions of N which can be arbitrarily large. The tuple size k should be regarded as fixed but will be chosen sufficiently large.

By Lemma 4.11 with $h_i \in \mathcal{H}_1 = \mathcal{H}_2$, we have

$$\sum_{n \leq N} \Lambda_R(n, \mathcal{H}_k, l)^2 \sim \frac{1}{(k + 2l)!} \binom{2l}{l} \mathcal{G}(\mathcal{H}_k) N L_R(k + 2l). \tag{4.110}$$

Using Lemma 4.12(a), we get for all $h_i \in \mathcal{H}_k$ and $R \ll N^{\vartheta/2-\epsilon}$ with $R, N \to \infty$,

$$\sum_{n \leq N} \Lambda_R(n, \mathcal{H}_k, l)^2 \theta(n + h_i) \sim \frac{1}{(k + 2l + 1)!} \binom{2l + 2}{l + 1} \mathcal{G}(\mathcal{H}_k) N L_R(k + 2l + 1). \tag{4.111}$$

(2) Now let $v \in \mathbb{N}$, and define $\mathcal{S} = \mathcal{S}(N, h, l)$ by

$$\mathcal{S} := \sum_{n=N+1}^{2N} \left(\sum_{1 \leq h_0 \leq h} \theta(n + h_0) - v \log(3N) \right) \sum_{\substack{(h_1, \ldots, h_k) \\ 1 \leq h_1, h_2, \ldots, h_k \leq h \\ h_i \text{ distinct}}} \Lambda_R(n, \mathcal{H}_k, l)^2. \tag{4.112}$$

Use Lemma 4.12(b), with $h_0 \notin \mathcal{H}_k = \mathcal{H}_1 = \mathcal{H}_2$, to write

$$\sum_{n=1}^{N} \Lambda_R(n, \mathcal{H}_k, l)^2 \theta(n + h_0) \sim \frac{1}{(k + 2l)!} \binom{2l}{l} \mathcal{G}(\mathcal{H}_k \cup \{h_0\}) N L_N(k + 2l).$$

(4.113)

Next, use (4.110), (4.111), (4.113) and Gallagher's theorem 4.14, splitting the sum over h_0 into $h_0 = h_i$ for some $1 \le i \le k$ and $h_0 \ne h_i$ for any i in the inner sum of $\mathcal{S}$, to derive

$$\mathcal{S} \sim \sum_{\substack{1 \le h_1, h_2, \dots, h_k \le h \\ \text{distinct}}} \left(\frac{k}{(k + 2l + 1)!} \binom{2l + 2}{l + 1} \mathcal{G}(\mathcal{H}_k) N L_R(k + 2l + 1) \right.$$

$$+ \sum_{\substack{1 \le h_0 \le h \\ h_i \ne h_0, \ 1 \le i \le k}} \frac{1}{(k + 2l)!} \binom{2l}{l} \mathcal{G}(\mathcal{H}_k \cup \{h_0\}) N L_R(k + 2l)$$

$$\left. - v \log(3N) \frac{1}{(k + 2l)!} \binom{2l}{l} \mathcal{G}(\mathcal{H}_k) N L_R(k + 2l) \right)$$

$$\sim \left(\frac{2k}{(k + 2l + 1)} \frac{2l + 1}{l + 1} (\log R) + h - v \log(3N) \right) \frac{1}{(k + 2l)!} \binom{2l}{l} N h^k L_R(k + 2l).$$

(4.114)

(3) We can now complete the proof. Let $0 < \vartheta \le \frac{1}{2}$ and $R = N^{\vartheta/2 - \epsilon}$. Given $\epsilon > 0$ and small depending on k but not N, let $h = 5\epsilon \log N$ and $l := \lfloor \sqrt{k}/2 \rfloor$. It follows from (4.114) that, as $N \to \infty$, if

$$\left(v(1 + o(1)) - \frac{2k}{k + 2l + 1} \frac{2l + 1}{l + 1} (\vartheta/2 - \epsilon) \right) \log N < h, \qquad (4.115)$$

there are at least $v + 1$ primes in some interval $(n, n + h]$ with $N < n \le 2N$. Choosing k sufficiently large, this simplifies to

$$\left(v - 2\vartheta + 4\epsilon + O\left(\frac{1}{\sqrt{k}} \right) \right) \log N < 5\epsilon \log N. \qquad (4.116)$$

This implies $0 \le \Delta_v \le \max(v - 2\vartheta, 0)$. If we let $v = 1$ and $\vartheta = 1/2$, we get $\Delta_1 = 0$, which completes the proof of the theorem. $\qquad \square$

4.8 The Simplified Proof

The reader will have observed that to prove Theorem 4.19, nothing like the complexity of the Lemmas 4.8 through 4.10 is used, only the case $\mathcal{H}_1 = \mathcal{H}_2$ and $l_1 = l_2$. GMPY anticipated this by publishing earlier in [70] a version of the proof which implements these simplifications. As far as this writer knows, Goldston and Yildirim first published in preprint form some preliminary ideas which led to an early draft of the *Annals of Mathematics* article [72]. They were joined by Motohashi, who was an expert on many aspects of the prime gaps problem, eventuating in the Japan Academy article [70], the simplified proof. It still depends on Gallagher's theorem, zeta estimates and the Bombieri–Vinogradov theorem. There is an interesting backstory to the relationship between the Japan Academy and *Annals* papers and the role of Yoichi Motohashi, part of which is related in [31].

Since the ideas are important for what follows in later developments, we give replacement statements and proofs for some of the lemmas in this section.

Some notation and definitions. The integers $k \geq 2$ and $l > 0$ are otherwise arbitrary but can be regarded as fixed, along with an admissible k-tuple $\mathcal{H}$. Implicit constants can depend on $k, l, \mathcal{H}$ at most. The real positive constant c can also depend on these and also take different values in different contexts. We also require distinct integers $1 \leq h_j \leq h$ so

$$\mathcal{H} = \{h_1, \dots, h_k\} \subset [1, h],$$

which we restrict to being admissible and regard as fixed.

We also let $\mathcal{V}(p)$ be the set of distinct residue classes among the $-h_j$ so $v_p(\mathcal{H}) = \#\mathcal{V}(p)$ and extend this definition for squarefree d, so we can write $n \in \mathcal{V}(d)$ provided $d \mid P(n) = (n + h_1) \cdots (n + h_k)$. Whenever we use the symbol d, subscripted or otherwise, it will be restricted to being a squarefree positive integer. Note that many of entities defined here depend on $\mathcal{H}$ through $\mathcal{V}(d)$, although this dependence is suppressed.

Regarding the sizes of parameters which tend to infinity, we need only assume for real $R > 0$,

$$\log N \ll h \ll \log N \ll \log R \leq \log N.$$

We detect primes as usual using the restricted von Mangoldt function and define for integer $a \geq 0$,

$$\lambda_R(d, a) := \begin{cases} \frac{1}{a!} \mu(d) L_{R/d}(a), & d \leq R, \\ 0, & R < d, \end{cases}$$

where, as before, we use $L_X(Y) := (\log X)^Y$, and then define

$$\widetilde{\Lambda}_R(n, \mathcal{H}, a) := \sum_{n \in \mathcal{V}(d)} \lambda_R(d, a) = \frac{1}{a!} \sum_{\substack{d \mid P(n) \\ d \leq R}} \mu(d) L_{R/d}(a).$$

It is a pity this notation differs from that used in the earlier sections of this chapter. We have retained the previous form and here use $\widetilde{\Lambda}_R$, keeping faith with the original, noting that

$$\Lambda_R(n, \mathcal{H}, l) = \widetilde{\Lambda}_R(n, \mathcal{H}, k + l).$$

First, we give an asymptotic expression for the sum over n of the $\widetilde{\Lambda}_R(n, \mathcal{H}, k+l)^2$ with a main term and error term. Note that the error is smaller than the main term, but only just!

Lemma 4.20   (Simplified proof nonprime sums) *[70, lemma 1] If $h \ll \log N$ with $R \le \sqrt{N}/L_N(C)$ for $C > 0$ sufficiently large, and k,l arbitrary and fixed integers with $k \ge 2$ and $l > 0$, then*

$$\sum_{N < n \le 2N} \widetilde{\Lambda}_R(n, \mathcal{H}, k + l)^2$$

$$= \frac{\mathcal{G}(\mathcal{H})}{(k + 2l)!} \binom{2l}{l} N L_R(k + 2l) + O\left(N L_N(k + 2l - 1)(\log\log N)^c\right).$$

Proof   **(1)** First, let

$$\mathcal{T} := \sum_{\substack{d,e \\ \text{squarefree}}} \frac{|\mathcal{V}([d,e])|}{[d,e]} \lambda_R(d, k + l)\lambda_R(e, k + l).$$

Expanding the square, we have

$$\sum_{N < n \le 2N} \widetilde{\Lambda}_R(n, \mathcal{H}, k + l)^2 = \sum_{\substack{d,e \\ \text{squarefree}}} \lambda_R(d, k + l)\lambda_R(e, k + l) \sum_{\substack{N < n \le 2N \\ n \in \mathcal{V}(d) \cap \mathcal{V}(e)}} 1$$

$$= \sum_{\substack{d,e \\ \text{squarefree}}} \lambda_R(d, k + l)\lambda_R(e, k + l) \sum_{\substack{N < n \le 2N \\ n \in \mathcal{V}([d,e])}} 1$$

$$= \sum_{\substack{d,e \\ \text{squarefree}}} \lambda_R(d, k + l)\lambda_R(e, k + l)|\mathcal{V}([d,e]) \cap (N, 2N]|$$

$$= \sum_{\substack{d,e \\ \text{squarefree}}} \lambda_R(d, k + l)\lambda_R(e, k + l)\left(\frac{N}{[d,e]} + O(1)\right) ||\mathcal{V}([d,e])|$$

$$= N\mathcal{T} + O\left(\left(\sum_d |\mathcal{V}(d)||\lambda_R(d, k + l)|\right)^2\right).$$

We also have $|\mathcal{V}(d)| \le k^{\omega(d)} \le \tau_k(d)$. Therefore, using Lemma 4.5, we can write

$$\sum_{N < n \le 2N} \widetilde{\Lambda}_R(n, \mathcal{H}, k + l)^2 = N\mathcal{T} + O\left(R^2 L_R(c)\right).$$

(2) If we define for $\Re s_1$, $\Re s_2$ sufficiently large

$$F(s_1, s_2) := \sum_{d,e} \mu(d)\mu(e)\frac{|\mathcal{V}([d,e])|}{[d,e]d^{s_1}e^{s_2}}$$

and note that for a given prime p and d, eth term, since the term is multiplicative and d, e squarefree, the contribution to this sum is zero when $p \nmid [d,e]$, is $-v_p(\mathcal{H})/p^{1+s_1}$ if $p \mid d$ and $p \nmid e$, is $-v_p(\mathcal{H})/p^{1+s_2}$ if $p \mid e$ and $p \nmid d$ and which overcompensates by $+v_p(\mathcal{H})/p^{1+s_1+s_2}$ when $p \mid (d,e)$, then we can write

$$F(s_1, s_2) = \prod_p \left(1 - \frac{v_p(\mathcal{H})}{p}\left(\frac{1}{p^{s_1}} + \frac{1}{p^{s_2}} - \frac{1}{p^{s_1+s_2}}\right)\right),$$

where the complex variables are such that the product converges absolutely. Alternatively, the reader might expand the Euler product and observe that it matches the expansion of $F(s_1, s_2)$.

In addition, following the method of Lemma 4.7 Step (2), we can write for $a \geq 1$ and $d \leq R$,

$$\lambda_R(d, a) = \frac{\mu(d)}{2\pi i} \int_{(1)} \left(\frac{R}{d}\right)^s \frac{ds}{s^{a+1}}.$$

The definitions of F and $\mathcal{T}$ from Step (1) then enable us to write

$$\mathcal{T} = \frac{1}{(2\pi i)^2}\int_{(1)}\int_{(1)} F(s_1, s_2)\frac{R^{s_2+s_2}}{(s_1 s_2)^{k+l+1}} ds_1 ds_2.$$

(3) Note that $v_p(\mathcal{H}) = k$ when $p > h$, so we define, with $\zeta(s)$ Riemann's zeta function,

$$G(s_1, s_2) := F(s_1, s_2)\left(\frac{\zeta(s_1+1)\zeta(s_2+1)}{\zeta(s_1+s_2+1)}\right)^k.$$

Considering the Euler product on the right, we see that this function is analytic and bounded for $\Re s_1, \Re s_2 > -c$. It follows directly from the definitions, taking the limit as $s_1, s_2 \to 0$, that it is such that

$$G(0,0) = \prod_p \left(1 - \frac{v_p(\mathcal{H})}{p}\right)\left(1 - \frac{1}{p}\right)^{-k} = \mathcal{G}(\mathcal{H}).$$

Since $\mathcal{H}$ is admissible, we have $G(0,0) \neq 0$. In addition, in the half plane

$$\sigma := \min(\Re s_1, \Re s_2, 0) \geq -c > -\frac{1}{4}$$

$$\implies -\sigma = \max(-\Re s_1, -\Re s_2, 0) \leq c < \frac{1}{4},$$

using the Euler product expansion for $G(s_1, s_2)$, because $v_p(\mathcal{H}) = k$ when $p > h$, the corresponding Euler factor is uniformly bounded. Taking the logarithm of the factor where $p \leq h$ and using the a priori bounds

$$\log N \ll h \ll \log N \ll \log R \leq \log N, \tag{4.117}$$

we get an upper bound estimate for the logarithm of that factor in the strip $0 \leq -\sigma < \frac{1}{4}$ for some real constants a_1, a_2, a_3,

$$\sum_{p \leq h} \log\left(1 + \frac{a_1}{p^{s_1+1}} + \frac{a_2}{p^{s_2+1}} + \frac{a_3}{p^{s_1+s_2+1}} + O\left(p^{-2+\epsilon}\right)\right).$$

Since $1/|p^{s_i+1}| \leq h^{-\sigma}/p$, the bound for the logarithm is

$$O\left(h^{-2\sigma} \sum_{p \leq h} \frac{1}{p}\right).$$

Since $\sigma \leq 0$, this leads to the estimate for some $c > 0$

$$G(s_1, s_2) \ll \exp\left(\frac{c \log\log\log N}{L_N(2\sigma)}\right). \tag{4.118}$$

(4) In this step, we modify the contours $s_1 = 1 + it$, $s_2 = 1 + it$ to first truncate them and then simplify the integrand. A representation of each contour used in the proof is included in Figure 4.7.

Using Steps (2) and (3), we can write

$$\mathcal{T} = \frac{1}{(2\pi i)^2} \int_{(1)} \int_{(1)} G(s_1, s_2) \left(\frac{\zeta(s_1 + s_2 + 1)}{\zeta(s_1 + 1)\zeta(s_2 + 1)}\right)^k \frac{R^{s_2+s_2}}{(s_1 s_2)^{k+l+1}} \, ds_1 ds_2.$$

Let $U := \exp(\sqrt{\log N})$ and shift the contours for s_1 and s_2 to the vertical lines in $\mathbb{C}$ through $c_0/\log U$ and $c_0/(2 \log U)$ respectively, where c_0 is sufficiently small so $\zeta(s)$ is nonzero on these vertical lines, truncated to the bounded intervals $|t| \leq U$, $|t| \leq U/2$, denoting them by $\mathcal{L}_1$, $\mathcal{L}_2$ respectively. Using the bounds (4.117) and (4.118) and Lemma 4.2(a),(b) or [207, section 3.11, equation 3.11.8], [4, theorem 13.4] for the zeta bounds, gives since $-2\sigma < \frac{1}{2}$

$$\mathcal{T} = \frac{1}{(2\pi i)^2} \int_{\mathcal{L}_2} \int_{\mathcal{L}_1} G(s_1, s_2) \left(\frac{\zeta(s_1 + s_2 + 1)}{\zeta(s_1 + 1)\zeta(s_2 + 1)}\right)^k$$

$$\times \frac{R^{s_2+s_2}}{(s_1 s_2)^{k+l+1}} \, ds_1 ds_2 + O\left(\exp\left(-c\sqrt{\log N}\right)\right).$$

Now we shift the contour for s_1 to $\mathcal{L}_3$, which is the line $-c_0/\log U + it$ with $|t| \leq U$. When making this move, we pass over singularities of the integrand at

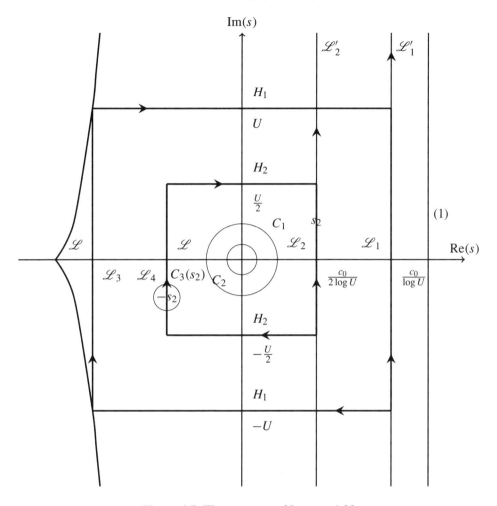

Figure 4.7 The contours of Lemma 4.20.

$s_1 = 0$, a pole of order $l + 1$ and $s_1 = -s_2$ a pole of order k. Denoting the integrand by I so

$$I = I(s_1, s_2) := G(s_1, s_2) \left(\frac{\zeta(s_1 + s_2 + 1)}{\zeta(s_1 + 1)\zeta(s_2 + 1)} \right)^k \frac{R^{s_2 + s_2}}{(s_1 s_2)^{k+l+1}},$$

and using the estimate (4.118), this gives

$$\mathcal{T} = \frac{1}{2\pi i} \int_{\mathscr{L}_3} (\text{Res}(I, s_1 = 0) + \text{Res}(I, s_1 = -s_2)) \, ds_2 + O\left(\exp\left(-c\sqrt{\log N} \right) \right).$$

(5) We show in this step that the second residue is small. Regarding $s_2 \in \mathcal{L}_2$ as fixed, and defining a circular contour

$$C(s_2) := \{s_1 : |s_1 + s_2| = 1/\log N\},$$

we can write

$$\mathrm{Res}(I, s_1 = -s_2) = \frac{1}{2\pi i} \int_{C(s_2)} G(s_1, s_2) \left(\frac{\zeta(s_1 + s_2 + 1)}{\zeta(s_1 + 1)\zeta(s_2 + 1)} \right)^k \frac{R^{s_2 + s_2}}{(s_1 s_2)^{k+l+1}} \, ds_1.$$

On the circle $C(s_2)$, we have the estimates, for the first using (4.118),

$$G(s_1, s_2) \ll (\log\log N)^c, \quad \zeta(s_1 + s_2 + 1) \ll \log N, \quad \text{and } R^{s_1 + s_2} \ll 1.$$

In addition, because $|s_2| \ll |s_1| \ll |s_2|$ we can write for the domain which is the region inside and on $C(s_2)$

$$\frac{1}{s_1 \zeta(s_1 + 1)} \ll \frac{\log(|s_2| + 2)}{|s_2| + 1} \quad \text{and} \quad \frac{1}{s_2 \zeta(s_2 + 1)} \ll \frac{\log(|s_2| + 2)}{|s_2| + 1}.$$

Therefore,

$$\mathrm{Res}(I, s_1 = -s_2) \ll L_N (k - 1)(\log\log N)^c \left(\frac{\log(|s_2| + 2)}{|s_2| + 1} \right)^{2k} \frac{1}{|s_2|^{2l+2}}.$$

Using this estimate in the result of Step (4), writing $|s_2|^2 = t^2 + \epsilon^2$ with $\epsilon = c_0/\sqrt{\log N}$, the integral estimate,

$$\int_0^U \frac{dt}{(t^2 + \epsilon^2)^{l+1}} \ll (\log N)^{l + \frac{1}{2}},$$

obtained by splitting the integral at ϵ and replacing the first integrand by $1/\epsilon^{2l+2}$ and the second by $1/t^{2l+2}$, and then and bounding the third factor, we get

$$\mathcal{T} = \frac{1}{2\pi i} \int_{\mathcal{L}_2} \mathrm{Res}(I, s_1 = 0) \, ds_2 + O \left(L_N \left(k + l - \tfrac{1}{2} \right) (\log\log N)^c \right).$$

(6) We now complete the evaluation of the integral for $\mathcal{T}$ by first computing the remaining residue $\mathrm{Res}(I, s_1 = 0)$. First we define a function Z, which is analytic in a neighbourhood of $(0, 0) \in \mathbb{C}^2$, by

$$Z(s_1, s_2) := G(s_1, s_2) \left(\frac{(s_1 + s_2)\zeta(s_1 + s_2 + 1)}{s_1 s_2 \zeta(s_1 + 1)\zeta(s_2 + 1)} \right)^k.$$

Then, since the pole of I at $s_1 = 0$ has order $l + 1$, we get

$$\mathrm{Res}(I, s_1 = 0) = \frac{R^{s_2}}{l! \, s_2^{l+1}} \left(\frac{\partial}{\partial s_1} \right)^l \frac{Z(s_1, s_2) R^{s_1}}{(s_1 + s_2)^k} \Bigg|_{s_1 = 0}.$$

Using this expression in the result of Step (5) and shifting the integration contour to $\mathcal{L}_4$, which is the line segment

$$\mathcal{L}_4 = \left\{ s \in \mathbb{C} : s = -c_0/(2 \log U) + it, \ |t| \leq \frac{U}{2} \right\},$$

and using the same bounds as in Step (5), it follows that the integral along horizontal line segments produced is $O(\exp(-c\sqrt{\log N}))$. Thus, from the result of Step (5), if C_1 and C_2 are circles centre 0 with radii ρ and 2ρ respectively, with $\rho > 0$ sufficiently small so Z is analytic inside and on the circles, we can write using $s = s_1$ and $s_2 = \xi s$, $Z(0,0) = \mathcal{G}(\mathcal{H})$, with C_3 the circular contour with centre zero and radius 2, and using the estimate (4.118) to bound G again,

$$\mathcal{T} = \mathrm{Res}(\mathrm{Res}(I, s_1 = 0), s_2 = 0) + O(L_N(k+l))$$

$$= \frac{1}{(2\pi i)^2} \int_{C_2} \left(\int_{C_1} \frac{Z(s_1, s_2) R^{s_1 + s_2}}{(s_1 + s_2)^k (s_1 s_2)^{l+1}} \, ds_1 \right) ds_2 + O(L_N(k+l))$$

$$= \frac{1}{(2\pi i)^2} \int_{C_3} \left(\int_{C_1} \frac{Z(s, s\xi) R^{s(1+\xi)}}{(\xi + 1)^k \xi^{l+1} s^{k+2l+1}} \, ds \right) d\xi + O(L_N(k+l))$$

$$= \frac{\mathcal{G}(\mathcal{H}) L_R(k+2l)}{2\pi i (k+2l)!} \int_{C_3} \frac{(\xi + 1)^{2l}}{\xi^{l+1}} \, d\xi + O\left(L_N(k+2l-1)(\log\log N)^c \right).$$

Carrying out the integration over ξ, and using the result of Step (1), we get

$$\sum_{N < n \leq 2N} \widetilde{\Lambda}_R(n, \mathcal{H}, k+l)^2$$

$$= \frac{\mathcal{G}(\mathcal{H})}{(k+2l)!} \binom{2l}{l} N L_R(k+2l) + O\left(N L_N(k+2l-1)(\log\log N)^c \right),$$

which completes the proof of the lemma. $\qquad\square$

Following a well-trodden path, we now make the corresponding derivation of a good asymptotic estimate for the prime sums.

Recall $\theta(n) := \log n$ when n is prime and 0 otherwise. As well as the sum of Lemma 4.20, we also need to estimate

$$\sum_{N < n \leq 2N} \theta(n+h) \widetilde{\Lambda}_R(n, \mathcal{H}, k+l)^2$$

for any natural number $h_0 \leq h$. On the face of it, this is significantly more difficult than the sum of Lemma 4.20. It is here that we employ the Bombieri–Vinogradov theorem by way of the prime *level of distribution* ϑ. To obtain an unconditional result we need $\vartheta < \frac{1}{2}$, but we retain this as a parameter in the range $0 < \vartheta < 1$. The details are given in Step (2) of Lemma 4.21.

Recall we say that the (rational) primes have a **level of distribution** ϑ for a given ϑ with $0 < \vartheta < 1$ provided that for any fixed real $A > 0$ and $\epsilon > 0$, if we define

$$\theta^*(y,q,a) := \sum_{\substack{y < n \le 2y \\ n \equiv a \bmod q}} \theta(n),$$

then we assume for all $A > 0$,

$$\sum_{1 \le q \le x^{\vartheta-\epsilon}} \max_{y \le x} \max_{a:(a,q)=1} \left| \theta^*(y,q,a) - \frac{y}{\varphi(q)} \right| \ll_{A,\epsilon} \frac{x}{L_x(A)}.$$

These simplified Lemmas 4.20 and 4.21 are then fed into a proof analogous to that of Theorem 4.19 to complete the proof of the prime gaps result.

Lemma 4.21 (**Simplified proof prime sums**) *[70, lemma 2] Assume the primes have a level of distribution ϑ for some fixed ϑ with $0 < \vartheta < 1$. Assume also $h \ll \log N$ and k,l are arbitrary fixed integers with $k \ge 2$ and $l > 0$, and that $R \le N^{\vartheta/2}$. Define*

$$\mathscr{S} := \sum_{N < n \le 2N} \theta(n + h_0) \widetilde{\Lambda}_R(n, \mathscr{H}, k + l)^2.$$

If $h_0 \notin \mathscr{H}$, we have

$$\mathscr{S} = \frac{\mathscr{G}(\mathscr{H} \cup \{h_0\})}{(k+2l)!} \binom{2l}{l} N L_R(k + 2l) + O\left(N L_N(k + 2l - 1)(\log\log N)^c\right).$$

If however we have $h_0 \in \mathscr{H}$, we get

$$\mathscr{S} = \frac{\mathscr{G}(\mathscr{H})}{(k+2l+1)!} \binom{2l+2}{l+1} N L_R(k + 2l + 1) + O\left(N L_N(k + 2l)(\log\log N)^c\right).$$

Proof (1) First assume $h_0 \notin \mathscr{H}$. Let $\delta(x) = 1$ if and only if $x = 1$ and 0 otherwise, and note that $\theta^*(N, [d, e, b + h_0]) = 0$ if $b + h_0$ and $[d, e]$ are not coprime, and N is sufficiently large. Expand the square as in Lemma 4.20 and define

$$T^* := \sum_{\substack{d,e \\ \text{squarefree}}} \frac{\lambda_R(d, k + l)\lambda_R(e, k + l)}{\varphi([d,e])} \sum_{b \in \mathscr{V}([d,e])} \delta((b + h_0, [d,e])),$$

where $\delta(1) = 1$ and $\delta(n) = 0$ for $n > 1$, to get

$$\mathscr{S} = \sum_{N < n \le 2N} \theta(n + h_0) \widetilde{\Lambda}_R(n, \mathscr{H}, k + l)^2$$

$$= \sum_{\substack{d,e \\ \text{squarefree}}} \lambda_R(d, k + l)\lambda_R(e, k + l) \sum_{n \in \mathscr{V}([d,e])} \delta((b + h_0, [d,e]))\theta^*(N, [d,e], n + h_0)$$

$$+ O(R^2 (\log N)^c) = N T^* + O\left(\frac{N}{L_N(A/3)}\right),$$

where the second error term will be derived in Step (2). The first error term arises in a similar manner to that of Lemma 4.20 Step (1).

(2) For the error term in the estimate which is the result of Step (1), since $0 < \vartheta < 1$ and $R \le N^{\vartheta/2}$, we have

$$R^2 (\log N)^c \ll \frac{N}{L_N(A/3)}.$$

For the terms with

$$|\mathcal{V}([d,e])| \le \tau_k([d,e]) < L_N(A/2),$$

since d, e are squarefree we have $\#\{(d,e) : [d,e] = n\} = \tau_3(n)$, so their subsum will be less than $L_N (A \log 3/(2 \log k))$, and the terms in $\mathcal{T}^*$ follow from the level of distribution assumption. For the remaining terms, the sum, provided A is chosen sufficiently large, is

$$\ll NL_R(2k + 2l) \log N \sum_{d,e \le R} \frac{\tau_k([d,e])|\mathcal{V}([d,e])|}{L_N(A/2)[d,e]}$$

$$\ll \frac{N}{L_N(A/3)}.$$

(3) In this step, we derive an integral representation for $\mathcal{T}^*$. We **claim** the inner sum in the definition of $\mathcal{T}^*$ can be written

$$\prod_{p|[d,e]} \left(\sum_{b \in \mathcal{V}(p)} \delta((b + h_0, p)) \right) = \prod_{p|[d,e]} (v_p(\mathcal{H} \cup \{h_0\}) - 1).$$

To see this, note that $\delta((b + h_0, p)) = 0$ when and only when

$$-h_0 \in \mathcal{V}(p) \quad \Longleftrightarrow \quad v_p(\mathcal{H}) = v_p(\mathcal{H} \cup \{h_0\}).$$

In other words, $\delta((b + h_0, p)) = 1 \quad \Longleftrightarrow \quad v_p(\mathcal{H} \cup \{h_0\}) = \varphi(\mathcal{H})$. Thus, in analogy to Step (2) of Lemma 4.20, we have

$$\mathcal{T}^* = \frac{1}{(2\pi i)^2} \int_{(1)} \int_{(1)} F^*(s_1, s_2) \frac{R^{s_2+s_2}}{(s_1 s_2)^{k+l+1}} \, ds_1 ds_2,$$

where, since in addition $\varphi(p) = p - 1$,

$$F^*(s_1, s_2) := \prod_p \left(1 - \frac{v_p(\mathcal{H} \cup \{h_0\}) - 1}{p - 1} \left(\frac{1}{p^{s_1}} + \frac{1}{p^{s_2}} - \frac{1}{p^{s_1+s_2}} \right) \right).$$

(4) If $p > h$, since $h_0 \notin \mathcal{H}$ we have $v_p(\mathcal{H} \cup \{h_0\}) = k + 1$. Provided $\mathcal{H} \cup \{h_0\}$ is admissible, the computation of T^* is in direct analogy to that carried out in Lemma 4.20, replacing $\mathcal{H}$ by $\mathcal{H} \cup \{h_0\}$ and using the result of Step (3), leading to

$$T^* = \frac{\mathcal{G}(\mathcal{H} \cup \{h_0\})}{(k+2l)!} \binom{2l}{l} L_R(k + 2l) + O\left(L_N(k + 2l - 1)(\log\log N)^c\right).$$

$$(4.119)$$

If however $\mathcal{H} \cup \{h_0\}$ is not admissible, then for some prime p we have $v_p(\mathcal{H} \cup \{h_0\}) = p$ so, substituting in the Euler product from Step (3), it vanishes at either $s_1 = 0$ or $s_2 = 0$, to an order which equals the number of primes p such that $v_p(\mathcal{H} \cup \{h_0\}) = p$. Necessarily we must have $p \leq k + 1$. Going through the derivation of Lemma 4.20 again using this constraint, and $\mathcal{G}(\mathcal{H} \cup \{h_0\} = 0$, the formula (4.119) remains true.

(5) Now assume $h_0 \in \mathcal{H}$. If $\theta(n + h_0) \neq 0$, then $n + h_0$ is prime. Therefore, because $R < N$, the factor $n + h$ of $P_{\mathcal{H} \cup \{h_0\}}(n)$ has no divisors d with $d \leq R$, so can be disregarded when evaluating $\widetilde{\Lambda}_R(n, \mathcal{H}, k + l)$. In other words, in this situation we have

$$\mathcal{S} := \sum_{N < n \leq 2N} \theta(n + h_0)\widetilde{\Lambda}_R(n, \mathcal{H} \setminus \{h_0\}, k + l)^2$$

$$= \sum_{N < n \leq 2N} \theta(n + h_0)\widetilde{\Lambda}_R(n, \mathcal{H} \setminus \{h_0\}, (k - 1) + (l + 1))^2. \qquad (4.120)$$

So the sum will be the same as that found when $h_0 \notin \mathcal{H}$ in Step (7) but with k replaced by $k - 1$ and l by $l + 1$, leading to

$$\mathcal{S} = \frac{\mathcal{G}(\mathcal{H})}{(k + 2l + 1)!} \binom{2l + 2}{l + 1} N L_R(k + 2l + 1) + O\left(N L_N(k + 2l)(\log\log N)^c\right).$$

This completes the proof. □

We now combine these two lemmas to give the equivalent of GPY's Theorem 4.19. The proof is almost the same as the proof of that theorem, but is included for completeness and readability.

Theorem 4.22 (GMPY's simplified prime gaps) *[70, theorem 0] If p_m is the mth prime with $2 = p_1$, then*

$$\liminf_{m \to \infty} \frac{p_{m+1} - p_m}{\log p_m} = 0.$$

Proof Let $\epsilon > 0$ be given, $R = N^{\vartheta/2}$ and $h = 2\epsilon \log N$. Define

$$\mathscr{T} := \sum_{\substack{\mathscr{H} \subset [1,h] \\ |\mathscr{H}|=k}} \sum_{N < n \leq 2N} \left(\left(\sum_{0 \leq h_0 \leq h} \theta(n + h_0) \right) - \log(3N) \right) \times \widetilde{\Lambda}_R(n, \mathscr{H}, k + l)^2.$$

Then by Lemma 4.20 and Gallagher's Theorem 4.14, we have

$$\mathscr{T} = \sum_{\substack{\mathscr{H} \subset [1,h] \\ |\mathscr{H}|=k}} \sum_{N < n \leq 2N} \sum_{\substack{h_0 \leq h \\ h_0 \notin \mathscr{H}}} \theta(n + h_0) \widetilde{\Lambda}_R(n, \mathscr{H}, k + l)^2$$

$$+ \sum_{\substack{\mathscr{H} \subset [1,h] \\ |\mathscr{H}|=k}} \sum_{N < n \leq 2N} \sum_{\substack{h_0 \leq h \\ h_0 \in \mathscr{H}}} \theta(n + h_0) \widetilde{\Lambda}_R(n, \mathscr{H}, k + l)^2$$

$$- \frac{N}{(k + 2l)!} \binom{2l}{l} h^k \log(N) L_R(k + 2l) + o(N h^k L_N(k + 2l + 1)).$$

By Lemma 4.21, we then get

$$\mathscr{T} \sim \frac{N}{(k + 2l)!} \binom{2l}{l} h^{k+1} L_R(k + 2l) + \frac{kN}{(k + 2l + 1)!} \binom{2l + 2}{l + 1} h^k L_R(k + 2l + 1)$$

$$- \frac{N}{(k + 2l)!} \binom{2l}{l} h^k \log(N) L_R(k + 2l)$$

$$= \left(h + \frac{k}{k + 2l + 1} \times \frac{2(2l + 1)}{l + 1} \log(R) - \log(N) \right)$$

$$\times \frac{N}{(k + 2l)!} \binom{2l}{l} h^k L_R(k + 2l).$$

Therefore, since $R = N^{\vartheta/2}$ we have both Lemmas 4.20 and 4.21, and $\mathscr{T}$ will be positive provided that for given $\epsilon > 0$ and h sufficiently large, we have

$$\frac{h}{\log N} > 1 + \epsilon - \frac{k}{k + 2l + 1} \times \frac{2(2l + 1)}{l + 1} \times \frac{\vartheta}{2}.$$

Choosing as we may $l = \lfloor \sqrt{k} \rfloor$ and given $h = 2\epsilon \log N$, this condition is implied by

$$\epsilon > 1 - 2\vartheta,$$

which is true for all ϑ smaller than and sufficiently close to $\frac{1}{2}$, which we are able achieve using the Bombieri–Vinogradov theorem. Therefore, with all N sufficiently large there is a least one integer $n \in (N, 2N]$ such that

$$\left(\sum_{h_0 \leq h} \theta(n + h_0) \right) - \log(3N) > 0.$$

Thus there is a subinterval of length h in $(N, 2N + h]$ which contains at least two primes, so

$$\min_{N < p_m \le 2N+h} (p_{m+1} - p_m) \le h \le 2\epsilon \log(p_m).$$

Thus

$$\liminf_{m \to \infty} \frac{p_{m+1} - p_m}{\log p_m} = 0.$$

This completes the simplified proof. □

4.9 GPY's Conditional Bounded Gaps Theorem

Here we go back to the more general Lemmas 4.11 and 4.12 and use their greater flexibility by employing a range of l values to obtain an eigenvalue problem which is then minimized. The level of distribution of primes can be arbitrarily close to 1, i.e., we are assuming EH[1]. Note that Manyard making this same assumption obtained a better result, i.e., $p_{n+1} - p_n \le 12$. Polymath8b assumed a stronger form of EH to get $p_{n+1} - p_n \le 6$. But the reader must be warned. Making any improvement on EH[$\frac{1}{2}$] is regarded as being extremely hard.

Even though Theorem 4.23 is the only application provided by GPY for the double k-tuple results, they may be useful in a wider range of circumstances, justifying the considerable effort to prove and understand them. For example, this could be in studies where the more detailed combinatorics of, say, special classes of admissible tuples are developed.

> **Theorem 4.23** (**GPY bounded gaps assuming EH**) *[72, theorem 1] Assume the prime numbers have a level of distribution $\frac{1}{2} < \vartheta < 1$. Then there is an effective constant c_ϑ such that any admissible k-tuple with $k \ge c_\vartheta$ contains at least two primes infinitely often. In particular, if $\vartheta \ge 0.971$, then we can choose $c_\vartheta = 6$. Using $\mathcal{H} = \{0, 4, 6, 10, 12, 16\}$ and assuming EH[ϑ], there would be an infinite number of integers n such that $p_{n+1} - p_n \le 16$.*

Proof (**1**) The proof follows a familiar pattern, but uses a vector of real parameters $a_l := L_R(-l)b_l$, characterizing $\mathbf{b} = (b_l)$ as an eigenvector. Let $R := N^{\vartheta/2}$. Then we can derive, using Lemmas 4.11 and 4.12 with weights

$$w(n) := \left(\sum_{l=0}^{L} a_l \Lambda_R(n, \mathcal{H}_k, l) \right)^2,$$

where $\mathcal{H}_k$ is an admissible k-tuple, regarded as fixed, by first expanding the square and changing the order of summation,

$$
\mathcal{S} := \sum_{N < n \le 2N} \left(\sum_{i=1}^{k} \theta(n + h_i) - \log(3N) \right) \left(\sum_{l=0}^{L} a_l \Lambda_R(n, \mathcal{H}_k, l) \right)^2
$$

$$
= \sum_{N < n \le 2N} \left(\sum_{i=1}^{k} \theta(n + h_i) \right) \sum_{\substack{0 \le l_1 \le L \\ 0 \le l_2 \le L}} a_{l_1} a_{l_2} \Lambda_R(n, \mathcal{H}_k, l_1) \Lambda_R(n, \mathcal{H}_k, l_2)
$$

$$
\sim \sum_{0 \le l_1, l_2 \le L} a_{l_1} a_{l_2} \left(k \binom{l_1 + l_2 + 2}{l_1 + 1} \frac{L_R(k + l_1 + l_2 + 1)}{(k + l_1 + l_2 + 1)!} \mathcal{G}(\mathcal{H}_k) N \right)
$$

$$
- \sum_{0 \le l_1, l_2 \le L} a_{l_1} a_{l_2} \left(\binom{l_1 + l_2}{l_1} \frac{L_R(k + l_1 + l_2)}{(k + l_1 + l_2)!} \mathcal{G}(\mathcal{H}_k) N \log(3N) \right)
$$

$$
\sim \sum_{0 \le l_1, l_2 \le L} a_{l_1} a_{l_2} A_{l_1, l_2},
$$

where

$$
A_{l_1, l_2} := \binom{l_1 + l_2}{l_1} \mathcal{G}(\mathcal{H}_k) N \frac{L_R(k + l_1 + l_2)}{(k + l_1 + l_2)!}
$$
$$
\times \left(\frac{k(l_1 + l_2 + 2)(l_1 + l_2 + 1)}{(l_1 + 1)(l_2 + 1)(k + l_1 + l_2 + 1)} \log(R) - \log(3N) \right).
$$

(2) Dividing out the common factor and defining an $(L + 1) \times (L + 1)$ real matrix by $B := A/(\mathcal{G}(\mathcal{H}_k) N L_R(k + 1))$, we now define

$$
\mathcal{S}' := \frac{\mathcal{S}}{\mathcal{G}(\mathcal{H}_k) N L_R(k + 1)}
$$

$$
\sim \sum_{0 \le l_1, l_2 \le L} b_{l_1} b_{l_2} \binom{l_1 + l_2}{l_1} \frac{1}{(k + l_1 + l_2)!}
$$

$$
\times \left(\frac{k(l_1 + l_2 + 2)(l_1 + l_2 + 1)}{(l_1 + 1)(l_2 + 1)(k + l_1 + l_2 + 1)} - \frac{2}{\vartheta} \right)
$$

$$
= \mathbf{b}^T B \mathbf{b}.
$$

Let $\mathbf{b}$ be an eigenvector of B corresponding to the eigenvalue λ. Since

$$
\mathbf{b}^T B \mathbf{b} = \lambda \sum_{l=0}^{L} |b_l|^2,
$$

the real number $\mathscr{S}'$, hence $\mathscr{S}$, will be positive provided $\lambda > 0$. We computed the following values set out in Table 4.1 using the PGpack function GPYEH, which returns the maximum eigenvalue. Note, however, that these results are subject to numerical errors, so should be regarded as only indicative. The values of ϑ were typically 40 places of precision or more, i.e., in Mathematica $\vartheta = 0.990\,40$ was the value used in line 1 of the table. The last line used 200 places of precision. The gap widths were taken from Engelsma's tables [43] and the rightmost column is the order of magnitude of the maximum eigenvalue, which of course is positive.

Table 4.1 *Prime gaps of GPY using EH.*

ϑ	L	k	gap	$\log \lambda_{max}$
0.990	1	6	16	−6
0.970	2	6	16	−7
0.950	1	7	26	−7
0.900	2	8	26	−9
0.850	2	11	36	−11
0.800	2	12	42	−15
0.750	2	16	60	−21
0.700	4	22	90	−33
0.650	4	35	158	−53
0.620	4	50	246	−80
0.615	4	54	270	−87
0.600	6	65	336	−114
0.580	9	105	600	−186
0.550	9	193	1204	−401
0.520	7	1783	≤ 14950	−5072

Taking the first row of the table with $k = 6$, $L = 1$, $b_0 = 1$, $b_1 =: b$, so B is 2×2, writing the asymptotic approximation to $\mathscr{S}'$ as a quadratic in b and completing the square, we can derive with

$$\alpha := \frac{9\vartheta - 8}{2(1 - \vartheta)} \text{ and } \beta := \frac{15\vartheta^2 - 64\vartheta + 48}{4(1 - \vartheta)^2},$$

the expression

$$\mathscr{S}' \sim -\frac{4(1 - \vartheta)}{8!\,\vartheta} \left((b - \alpha)^2 + \beta \right).$$

At $b = \alpha$, we have

$$\mathscr{S}' \sim -\frac{15\vartheta^2 - 64\vartheta + 48}{8!\,\vartheta\,(1 - \vartheta)}.$$

This expression will be positive if $0 < \vartheta < 1$ and ϑ is strictly between the two roots of the numerator, which is true if

$$4\frac{(8 - \sqrt{19})}{15} < 0.97097 \le \vartheta < 1.$$

Therefore, for ϑ in this range, there are at least two primes in any 6-tuple $\mathscr{H}_6$. This completes the proof. $\square$

4.10 End Notes

The work of GMPY was celebrated throughout the number theory world. To see why, the reader need only consult again the timeline Table 1.1, where one way of describing their result is that they showed the steadily reducing constant in the upper bound could be replaced by any $\epsilon > 0$. Their work is seminal, in that many of the new ideas they introduced, such as using admissible tuples, testing the method using the Elliott–Halberstam conjecture and their optimization step, have found their way into other developments and applications.

So in the early 2000s the prime gaps game was in play. Motohashi had earlier provided deeper insights into the scope of Bombieri–Vinogradof-style estimates by showing the arithmetic functions which could have an equivalent role to Λ or θ had an inductive property [146] with respect to Dirichlet multiplication. In addition, he also had revealed a bilinear structure in the error arising from Selberg-style sieves [148]. Both were to play a crucial role in the breakthroughs of Zhang and Polymath8a.

Finally, and most significantly, Motohashi and Pintz showed in 2008 [151] that the sum over all moduli in Bombieri–Vinogradov was not essential for the GPY method or sieve (as it has come to be known) to work, only a sum over those which are squarefree and smooth. Applying this to Theorem 4.23, this shows any advance over one half in the level of prime distributions, with squarefree and smooth moduli, of course, would give rise to prime gaps. The specific contribution of Zhang, described in part in the following chapter as published in 2014 [215] with Polymath8a's enhancements in Chapter 8, was to show that with squarefree and smooth moduli one could unconditionally extend Bombieri–Vinogradov by a small amount, indeed as far as

$$\frac{1}{2} + \frac{1}{1168}.$$

The GPY function GPYEH relates to the work of this chapter. See the manual entry in Appendix I.

5

The Astounding Result of Yitang Zhang

5.1 Introduction

The story of how Yitang Zhang (Figure 5.1) came to be the first to complete a proof that there are an infinite number of consecutive primes with the gap between them less than a fixed bound is best told in his own words, even though memory is not always an accurate guide to past events. We have paraphrased some of these thoughts here from various reports of interviews. There have been many in the popular press and newsletters such as the write-ups in *Quanta* and *Nautilus* magazines [120, 188], and even a full-length documentary, *Counting from Infinity* [34].

Zhang reflected on his early schooling in China: "During the Cultural Revolution it was difficult to find a person with a college education, indeed even to find a book." Indeed, he did not attend school but taught himself from books. Following his admission to Peking University, he did extremely well and was a top student.

In the 1980s, Zhang moved to the USA and undertook study towards a Ph.D. under Professor Moh, working on the Jacobian conjecture, and graduating in 1991 from Purdue University. Evidently his supervisor was not particularly impressed with his work there, and left him to fend for himself in the job market, where his strengths and potential were not recognized. For Moh's view on his relationship with Zhang, see [141].

Zhang had a quiet and shy disposition, which did not help as far as professional progress was concerned. But he certainly applied his energies to mathematics research: "I was born for math, but for many years the situation was not easy, but I didn't give up. I just kept going, kept pushing. Curiosity was of first rank importance – it's what makes mathematics an indispensable part of my life."

Sometimes Zhang lived in his car, and in 1992 worked in a friend's Subway restaurant. Seven years working at odd jobs sustained his life. In 1999, at age forty-four, Zhang was appointed to an adjunct professor position at the University of New Hampshire, teaching first-year calculus. His mind was on issues in number theory, and claims that around 2009 he had started to think about twin primes. He thought that problems such as this were interesting to every mathematician because they addressed essential problems of the theory of numbers. Problems must be difficult and regarded as being important by the whole mathematical community.

He says he thought about bounded gaps between primes for three years, stimulated by the work of Goldston, Motohashi, Pintz and Yildirim. However, the depth of his knowledge of related concepts and theories is such that his thinking on this problem almost certainly would have been much longer. Now he is considering several problems in number theory, including the distribution of the zeros of zeta and L-functions. As for twin primes: "That is not an easy problem. I didn't find a certain way to do it."

He may be quiet, but certainly does not lack when it comes to self-belief: "When I was very young I imagined there would be a day that I would solve a major math problem. I'm self-confident."

Figure 5.1 Yitang Zhang (1955–). Courtesy of Yitang Zhang.

Once the paper on bounded gaps between primes was submitted to the *Annals of Mathematics*, it caused quite a stir amongst the leading number theorists asked to look at the work. Refereeing, apart from anonymous final reports, is normally top secret, but in this case the process has been comprehensively documented by John Friedlander [59]. That there should be a fixed, finite, explicit bound such that an infinite number of consecutive primes are distant apart less than this bound, would have been regarded by many who have thought about the problem as being true, but bordering on the impossible to prove with current knowledge and tools. (But not however Goldston, Motohash, Pintz or Yildirim, for example, who had already in the first decade of the century made significant advances towards a full proof.) Hence the excitement, and hence the flowering of further work resulting from the inspiring achievements, including the better results of of Polymath8a, Maynard, Polymath8b, and as a sample, the enhancements and applications listed in [82, section 12], or discussed in Chapter 9.

Recognition, awards and fame have come thick and fast: Zhang has been awarded the Ostrowski Prize, the Frank Nelson Cole Prize, and the Rolf Schock Prize amongst others, including prestigious speaking invitations. One feels he would prefer many aspects of his quiet mathematical life before "bounded gaps".

As far as this writer can judge, Zhang's inspiration came most directly from the results and methods of Motohashi and Pintz (MP) [151] , together with those of Iwaniec [115] and Motohashi [148], and as already noted from Goldston, Motohashi, Pintz and Yildirim's [70, 72]. That an improvement of Bombieri–Vinogradof would solve the problem was already pointed to by GMPY and GPY in [70, 72], and by MP in [151] wherein details were given on how "smoothing" would give the sought-after prime gaps result, given a suitable improvement to Bombieri–Vinogradov.

To clarify this, Zhang's principal contribution to the complete proof of bounded gaps was his variation on Bombieri–Vinogradov's estimate, and this depended critically on the work of many mathematicians. The list is long and includes Birch, Bombieri, Deligne, Deshouillers, Fouvry, Friedlander, Iwaniec, Heath-Brown, Motohashi, Kuznetsov and Weil. The same applies to the work of Polymath8a reported in Chapter 8, wherein Zhang's work is refined and improved. Further details, including references, are given in the end note for that chapter.

To quote Andrew Granville [83, page 336], referring to Zhang's preprint of his *Annals of Mathematics* paper [215]:

It is to the credit of the mathematical community that this preprint, with such extraordinary claims by someone far removed from the usual suspects, was quickly accepted as correct (as opposed to the many other claims of great advances on famous open problems, which the experts rapidly dismiss). But there is good reason for that: Zhang followed the ideas of Goldston, Pintz and Yildirim [72] circulating in 2005, brilliantly developing techniques from the important papers of Bombieri, Friedlander and Iwaniec [15, 16, 18, 61] and even

Deligne [36], writing his proof as if he had spent a lifetime immersed in these deep and difficult techniques.

5.2 Summary of Zhang's Method

We saw in Chapter 4 that the assumption EH[ϑ] for some $\vartheta > \frac{1}{2}$ was sufficient to demonstrate bounded gaps between primes. See Theorem 4.23. In making his wonderful breakthrough, Zhang showed two things. First, the full strength of EH[ϑ] for some $\vartheta > \frac{1}{2}$ was not needed. In particular, instead of summing over all $q \leq N^{\vartheta}$ one could restrict the q values of the moduli to being smooth integers with prime factors all being less than N^{ϖ} with a particular value $\varpi = 1/1168$. This part of the argument was given in an almost equivalent form by Motohashi and Pintz in [151, page 309], as confirmed by Terence Tao [202, page 2] and enshrined in the Polymath8 expression MPZ[ϖ]. Then, in a tour de force of advanced methods, Zhang showed that this restricted EH[ϑ] was unconditionally true for $\vartheta = \frac{1}{2} + 2\varpi$.

To reiterate, the approach taken by Zhang had already been suggested by Motohashi and Pintz [151], i.e., that a "smoothed" version of the GPY sieve would enable bounded gaps to be proved, and the GPY paper [72], with its theorem 4.23, reinforced the utility of this road map direction. But no one before Zhang believed it to be possible to carry this through to an unconditional proof using techniques which were available at the time.

In more detail, Zhang restricted his attention to integers with all prime factors in $I = (1, N^{\varpi}]$ for the particular small $\varpi = 1/1168$ mentioned earlier. He defined S_I to be the set of these smooth integers with all prime factors in I. Let $P(n) = (n + h_1) \cdots (n + h_k)$ and $(a, q) = 1$, and define the error

$$E(N, q, a) := \left| \sum_{\substack{N < n \leq N \\ n \equiv a \bmod q}} \chi_{\mathbb{P}}(n) - \frac{1}{\varphi(q)} \sum_{\substack{N < n \leq 2N \\ (n, q) = 1}} \chi_{\mathbb{P}}(n) \right|.$$

Then, using the notation of Polymath8a, the key definition is as follows: we say MPZ[ϖ] is true if for any fixed $A > 0$ we have

$$\sum_{\substack{q \leq N^{\frac{1}{2} + 2\varpi} \\ q \in S_I}} \sum_{\substack{a: 1 \leq a \leq q \\ (a, q) = 1 \\ P(a) \equiv 0 \bmod q}} E(N, q, a) \ll_A \frac{N}{L_N(A)}.$$

Note that on the face of it EH[$\frac{1}{2} + 2\varpi$], is stronger than MPZ[ϖ], and this has been shown to be the case. Here by EH[δ] we mean the assumption that we could replace $q \leq x^{\frac{1}{2} - \epsilon}$ by $q < x^{\delta - \epsilon}$ in the range of moduli in Bombieri–Vinogradov's estimate from Section 1.9.

Zhang showed [215, theorem 2] that MPZ[ϖ] is true for all ϖ with $0 < \varpi < 1/1168$. Others had already experimented successfully with finding ways of moving ϑ, the level of distribution of the rational primes, beyond the $\frac{1}{2}$ barrier, namely Bombieri, Fouvry, Friedlander and Iwaniec, in their various joint paper combinations, [18, 54, 61]. Zhang used their ideas, but significantly extended them to obtain sufficient flexibility to attain bounded gaps.

Motohashi and Pintz had already shown [151], and Zhang gave a closely related proof [215], that if MPZ[ϖ] is true for any $\varpi > 0$, then DHL[k, 2] is true for some explicitly computable k. Recall DHL[k, 2] means for any admissible k-tuple $\mathcal{H}$ there exist an infinite number of translates $n + \mathcal{H}$ containing at least two primes.

To prove this, we need to go through the GPY argument with weights $w(n)$, but instead of summing over $n \in (N, 2N]$ we restrict the sums to $n \in S_I \cap (N, 2N]$. The detail of Zhang's argument is set out in this chapter, based on how he presented it in [215]. Note that the "hard-wiring" of particular constant values such as $\varpi = 1/1168$, is somewhat unusual. In this regard, Zhang in his paper makes the enigmatic remark after the statement of his theorem 1:

This result is, of course, not optimal. The condition $k \geq 3.5 \times 10^6$ is also crude and there are certain ways to relax it. To replace the constant 7×10^7 by a value as small as possible is an open problem that will not be discussed in this paper. [215, page 1123]

In fact, Zhang proves that for an infinite number of $n \in \mathbb{N}$ we have

$$p_{n+1} - p_n \leq 7 \times 10^7,$$

which immediately cries out for improvements, namely reducing the size of the right-hand side. These improvements came thick and fast, but none as thorough and far-reaching as those of Polymath8a, described in part in Chapter 8.

We don't cover all of Zhang's work in this chapter or the entire book, but simply quote and use his theorem 2. We cover just sufficient of Polymath8a's improvements in Chapter 8, giving the details including an extensive set of appendices giving background concepts and results with their proofs, to lower the bound so for infinitely many $n \in \mathbb{N}$ we have

$$p_{n+1} - p_n \leq 14950.$$

In both Zhang's and Polymath8a's approaches, ultimately, applications of Deligne's work on the Weil conjectures for varieties over finite fields are used, so rely on l-adic cohomology to be completely understood. In not going that far, we will derive a bound which is a little greater than Polymath8a, but significantly better than Zhang's original. All of these bounds, however, as we will see in Chapter 6, are weaker than that of Maynard (600) and Polymath8b (246), which will be derived in Chapters 6 and 7, respectively.

5.3 Notation

For the most part, we have kept to Zhang's notation from [215], but have made some changes to simplify the formulas. So let $\mathcal{H} = \{h_1, \ldots, h_k\}$ be an admissible k-tuple. We set

$$P(n) := (n + h_1) \cdots (n + h_k),$$

$$R = x^{\frac{1}{4} + \varpi}, \ \varpi = 1/1168,$$

$$R_1 = x^{\varpi},$$

$$\mathscr{P}_1 = \prod_{p < R_1} p,$$

$$R_0 = \exp((\log x)^{1/k}),$$

$$\mathscr{P}_0 = \prod_{p \leq R_0} p,$$

$$\eta = 1 + \frac{1}{(\log x)^A},$$

$$k = 3.5 \times 10^6, \ l = 180,$$

$$L_A(B) := (\log A)^B,$$

$$g(d) := \begin{cases} \frac{1}{(k+l)!} L_{R/d}(k+l), & d < R, \\ 0, & d \geq R. \end{cases}$$

$$\lambda(n) := \sum_{d \mid (P(n), \mathscr{P}_1)} \mu(d) g(d),$$

$$S_1 := \sum_{x \leq n < 2x} \lambda(n)^2,$$

$$S_2 := \sum_{x \leq n < 2x} \left(\sum_{j=1}^{k} \theta(n + h_j) \right) \lambda(n)^2,$$

where for $n \in \mathbb{P}$, $\theta(n) := \log(n)$ and $\theta(n) = 0$ otherwise.

Note that Zhang expects $R_0 < R_1$, and for this we need at least $x > e^{1168}$, so his results are truly asymptotic!

5.4 Chapter Summary

In Section 5.5, we note the variations on the Bombieri–Vinogradov estimates of Zhang and those of Polymath. Section 5.6 gives Zhang's preliminary lemmas, essential to the argument set out in the three sections which follow it. Note that Lemma 5.1 is imported almost directly from Chapter 4.

Recall the definition from Section 1.6: $\mathscr{H}$ is an admissible k-tuple and

$$\mathscr{G}(\mathscr{H}) := \prod_p \left(1 - \frac{v_p(\mathscr{H})}{p}\right)\left(1 - \frac{1}{p}\right)^{-k}.$$

In Section 5.7, we derive an upper bound for the sum S_1:

Lemma 5.6   (Zhang's upper bound) *[215, section 4] Let $\mathscr{H}$ be an admissible k-tuple. As $x \to \infty$, we have the upper bound*

$$S_1 \le \frac{1 + \kappa_1}{(k + 2l)!}\binom{2l}{l}x\mathscr{G}(\mathscr{H})(\log R)^{k+2l} + o(x(\log x)^{k+2l}),$$

where the parameter κ_1 depends on ϖ, k and l, and is very small.

In Section 5.8, we derive a lower bound for the sum S_2:

Lemma 5.9   (Zhang's lower bound) *[215, section 5] Let $\mathscr{H}$ be an admissible k-tuple. As $x \to \infty$, we have the lower bound*

$$S_2 \ge \frac{k(1 - \kappa_2)}{(k + 2l + 1)!}\binom{2l + 2}{l + 1}x\mathscr{G}(\mathscr{H})(\log R)^{k+2l+1} + o\left(x(\log x)^{k+2l+1}\right),$$

where the parameter κ_2 depends on the same quantities as κ_1 and is small.

Section 5.9 is devoted to proving Zhang's main theorem, assuming his variation on the Bombieri–Vinogradov estimates. As usual p_n is the nth rational prime with $p_1 = 2$ and $\mathscr{H}_k$ is a k-tuple.

Theorem 5.10   (Zhang's prime gap) *[215, theorem 1] If $\mathscr{H}_k$ is admissible with $k = 3.5 \times 10^6$. Then there are infinitely many positive integers n such that the translates $n + \mathscr{H}_k$ contain at least two primes. From this it follows that*

$$\liminf_{n \to \infty}(p_{n+1} - p_n) \le 7 \times 10^7.$$

The proof of Zhang's theorem 1 depends on his theorem 2, which is described in Section 5.5. This and other dependencies are portrayed in Figure 5.2. In the final Section 5.10 "End Notes", we give some context around Zhang's discovery and point the way forward to the chapters which follow.

5.5 Variations on the Bombieri–Vinogradov Estimates

The estimate of Bombieri and Vinogradov is discussed in Section 1.9. Here we give a brief outline of Zhang and Polymath's variations wherein the range of moduli is extended but they are restricted to being smooth. We give in Section 5.8 the briefest of outlines of Zhang's theorem 2 (our Theorem 5.8). There is a much more complete discussion of this topic in Chapter 8.

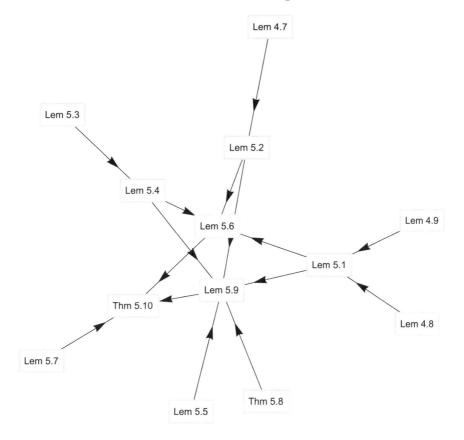

Figure 5.2 Chapter 5 dependencies.

Let $\gamma : \mathbb{N} \to \mathbb{C}$ be an arithmetic function, let $a \in \mathbb{Z}$ and let $q \in \mathbb{N}$ with $(a,q) = 1$. Define the **signed discrepancy** Δ by

$$\Delta(\gamma, q, a) := \sum_{\substack{x < n \le 2x \\ n \equiv a \bmod q}} \gamma(n) - \frac{1}{\varphi(q)} \sum_{\substack{x < n \le 2x \\ (n,q)=1}} \gamma(n).$$

Let $k \ge 2$, and for $1 \le i \le k$ and for squarefree q, let

$$\mathscr{C}_i(q) := \{a : 1 \le a \le q, \ (a,q) = 1, \ P(a - h_i) \equiv 0 \bmod q\}.$$

For example, if $\mathscr{H} = \{0, 2, 6\}$, then $\mathscr{C}_1(15) = \{4, 13\}$, $\mathscr{C}_2(15) = \{2, 11\}$ and $\mathscr{C}_3(15) = \{1, 4\}$.

As usual, let

$$\theta(x, q, a) := \sum_{\substack{n \le x \\ n \in \mathbb{P} \\ n \equiv a \bmod q}} \log n.$$

Then an alternative presentation of Zhang's main theorem to that given in the introduction to this chapter is as follows:

Theorem 5.8 (Zhang's theorem) *[215, theorem 2] Let $\varpi = 1/1168$, $R = x^{\frac{1}{4}+\varpi}$, and $\mathscr{P}_1 = \prod_{p<x^\varpi} p$. Then for $1 \le i \le k$ and all $A > 0$, we have*

$$\sum_{\substack{d<R^2 \\ d|\mathscr{P}_1}} \sum_{a\in\mathscr{C}_i(q)} |\Delta(\theta,q,a)| \ll \frac{x}{(\log x)^A}.$$

In comparison, we state Polymath8a's theorem, which has a wider range but somewhat different conditions:

Theorem (Polymath8a's theorem) *[174, theorem 1.1] Let $\theta := 1/2 + 7/300$ and let $\epsilon > 0$ and $A \ge 1$ be fixed and real. For each prime p, let a_p be an integer representing a fixed residue class with $(a_p, p) = 1$. For each squarefree $q \ge 1$, by the Chinese remainder theorem, let a_q be an integer representing the unique invertible residue class modulo q such that for all primes $p \mid q$ we have $a_q \equiv a_p$ mod p. Then there exists a $\delta = \delta_\epsilon > 0$, depending only on ϵ, such that for all $x \ge 1$ we have*

$$\sum_{\substack{q\le x^{\theta-\epsilon} \\ x^\delta-smooth \\ squarefree}} \left| \theta(x,q,a_q) - \frac{x}{\varphi(q)} \right| \ll_{A,\epsilon,\delta} \frac{x}{(\log x)^A},$$

where the implied constant is independent of the residue classes a_p.

Polymath extended Zhang in a number of ways. Perhaps the most comprehensive is based on the concept of "multiple dense divisibility", a generalization of smoothness. This is fully described in Chapter 8, and used to get one of Polymath's very significant improvements of Zhang's upper bound 7×10^7.

5.6 Preliminary Lemmas

The first preliminary lemma is a major import from the work of GMPY set out in Chapter 4. First we need some definitions.

$\mathscr{H}_k$ is a fixed admissible k-tuple. Define

$$\rho_1(p) = v_p(\mathscr{H}) := |\{h_i \bmod p : 1 \le i \le k\}| < p,$$
$$\rho_2(p) := v_p(\mathscr{H}) - 1,$$

and extend ρ_1, ρ_2 multiplicatively to all squarefree numbers. In particular, $\rho_1(p)$ is the number of distinct residue classes modulo p of the elements of $\mathscr{H}_k$. For numbers which are not squarefree, the values of the two functions are defined to be zero. This implies sums of multiples of these function values can be regarded as being over squarefree numbers. Next recall

$$g(d) := \begin{cases} \frac{1}{(k+l)!} \left(\log \frac{R}{d} \right)^{k+l}, & d < R, \\ 0, & d \geq R. \end{cases}$$

Thus sums of multiples of g values are finite. Also define

$$T_1^* := \sum_{d_0, d_1, d_2 \in \mathbb{N}} \frac{\mu(d_1 d_2)\rho_1(d_0 d_1 d_2)}{d_0 d_1 d_2} g(d_0 d_1) g(d_0 d_2),$$

$$T_2^* := \sum_{d_0, d_1, d_2 \in \mathbb{N}} \frac{\mu(d_1 d_2)\rho_2(d_0 d_1 d_2)}{\varphi(d_0 d_1 d_2)} g(d_0 d_1) g(d_0 d_2),$$

and recall the definition from Section 1.10

$$\mathscr{G}(\mathscr{H}) := \prod_p \left(1 - \frac{1}{p} \right)^{-k} \left(1 - \frac{v_p(\mathscr{H})}{p} \right).$$

Lemma 5.1 *[215, lemma 1] Using the given notations, we can write*

$$T_1^* = \frac{1}{(k+2l)!} \binom{2l}{l} \mathscr{G}(\mathscr{H})(\log R)^{k+2l} + o((\log x)^{k+2l}),$$

$$T_2^* = \frac{1}{(k+2l+1)!} \binom{2l+2}{l+1} \mathscr{G}(\mathscr{H})(\log R)^{k+2l+1} + o((\log x)^{k+2l+1}).$$

Proof The quantities on the right are the same as those of (4.75) and (4.91) respectively in case $\mathscr{H}_1 = \mathscr{H}_2 = \mathscr{H}$, $k_1 = k_2 = k$ and $l_1 = l_2 = l$, so the result follows from Lemma 4.8 Step (3) and 4.9 Step (3) in Chapter 4 with as defined before $r = |\mathscr{H}_1 \cap \mathscr{H}_2| = k$ and $M = k + l_1 + l_2 = k + 2l$, and with $T_1^* = T_R$ and $T_2^* = \widetilde{T_R}$. □

Next define the finite sums

$$\mathscr{A}_1(q) := \sum_{\substack{n \in \mathbb{N} \\ (n,q)=1}} \frac{\mu(n)\rho_1(n)}{n} g(qn),$$

$$\mathscr{A}_2(q) := \sum_{\substack{n \in \mathbb{N} \\ (n,q)=1}} \frac{\mu(n)\rho_2(n)}{\varphi(n)} g(qn),$$

where, as before, $\vartheta_1(q)$, $\vartheta_2(q)$, $\rho_1(q)$, $\rho_2(q)$ are multiplicative functions supported on squarefree integers such that

$$\rho_1(p) := v_p(\mathcal{H}), \quad \rho_2(p) := v_p(\mathcal{H}) - 1,$$

$$\vartheta_1(p) := \left(1 - \frac{\rho_1(p)}{p}\right)^{-1}, \quad \text{and } \vartheta_2(p) := \left(1 - \frac{\rho_2(p)}{p-1}\right)^{-1}.$$

Recall also the assignments

$$R = x^{\frac{1}{4}+\varpi},$$

$$\varpi = 1/1168, \text{ and}$$

$$R_0 = \exp((\log x)^{1/k}) = \exp(L_x(1/k)).$$

To find tractable expressions for the sums $\mathcal{A}_1(q)$ and $\mathcal{A}_2(q)$, we also need to build on the work of GMPY set out in Chapter 4, in particular from Lemma 4.7. Note the explicit dependence of the sums on the parameter l, with the k dependence implicit in $\mathcal{G}(\mathcal{H})$.

Lemma 5.2 *[215, lemma 2] If q is such that $1 \le q < R$ and is squarefree, then*

$$\mathcal{A}_1(q) = \frac{\vartheta_1(q)}{l!}\mathcal{G}(\mathcal{H})\left(\log \frac{R}{q}\right)^l + O\left((\log x)^{l-1+\epsilon}\right),$$

$$\mathcal{A}_2(q) = \frac{\vartheta_2(q)}{(l+1)!}\mathcal{G}(\mathcal{H})\left(\log \frac{R}{q}\right)^{l+1} + O\left((\log x)^{l+\epsilon}\right).$$

Proof **(1)** Because $\rho_1(p) \le k = \tau_k(p)$, we get $\rho_1(q) \le \tau_k(q)$ for all squarefree q and can write

$$\mathcal{A}_1(q) \ll 1 + \left(\log \frac{R}{q}\right)^{2k+l}.$$

Because of this, we **claim** we can assume $R/q > \exp(\log^2 R_0)$. To see this, if $R/q \le \exp(\log^2 R_0)$, then

$$\left(\log \frac{R}{q}\right)^{2k+l} \le (\log x)^{4+l/k} \le (\log x)^{l-1+\epsilon},$$

so $\mathcal{A}_1(q) = O((\log x)^{l-1+\epsilon})$ and the estimate for $\mathcal{A}_1(q)$ is satisfied.

Now let $\sigma > 0$ and $s \in \mathbb{C}$ be given by $s = \sigma + it$. Then if we define

$$\vartheta_1(q,s) := \prod_{p|q}\left(1 - \frac{v_p(\mathcal{H})}{p^{1+s}}\right)^{-1} \text{ and } G_1(s) := \prod_{p\in\mathbb{P}}\left(1 - \frac{v_p(\mathcal{H})}{p^{1+s}}\right)\left(1 - \frac{1}{p^{s+1}}\right)^{-k},$$

we have

$$\sum_{\substack{n\in\mathbb{N}\\(n,q)=1}} \frac{\mu(n)\rho_1(n)}{n^{1+s}} = \vartheta_1(q,s)G_1(s)\zeta(1+s)^{-k}.$$

Thus, using the integral formula of Lemma 4.7 Step (2), we get

$$\mathscr{A}_1(q) = \frac{1}{2\pi i} \oint_{(1/(\log x)^2)} \frac{\vartheta_1(q,s)G_1(s)}{\zeta(1+s)^k} \frac{(R/q)^s}{s^{k+l+1}}\, ds.$$

(2) Because we have $v_p(\mathscr{H}) = k$ for all $p > h_k$, the function $G_1(s)$ is holomorphic and bounded in the half plane $\sigma \geq -1/3$. Split the line of integration into a bounded part with $|t| \leq R_0$ and an unbounded part $|t| > R_0$. Using the standard zero free region for $\zeta(s)$, we can move the bounded part of the contour of integration for some constant $c > 0$ to

$$\left\{ s : \sigma = -\frac{c}{\log R_0},\ |t| \leq R_0 \right\}$$

and get

$$\mathscr{A}_1(q) = \frac{1}{2\pi i} \oint_{|s|=1/(\log x)^2} \frac{\vartheta_1(q,s)G_1(s)}{\zeta(1+s)^k} \frac{(R/q)^s}{s^{k+l+1}}\, ds + O\left(\frac{1}{(\log x)^A}\right).$$

(3) We next derive bounds for $\vartheta_1(q)$ and $\vartheta_1(q,s)$. For $p > 2k$, we have $-\log(1 - k/p) < -2k\log(1 - 1/p)$. The estimate of Rosser and Schoenfeld gives

$$\frac{n}{\varphi(n)} < e^\gamma \log\log(n) + \frac{5}{2\log\log(n)},\ n \geq n_0,$$

where n_0 is an absolute constant [184, theorem 15] or [21, theorem 5.25]. Using these estimates as $x \to \infty$, we get

$$\vartheta_1(q) = \prod_{p|q}\left(1 - \frac{\rho_1(p)}{p}\right)^{-1} \ll_k \prod_{\substack{p|q\\p>k}}\left(1 - \frac{k}{p}\right)^{-1}$$

$$\ll_k \prod_{\substack{p|q\\p>2k}}\left(1 - \frac{1}{p}\right)^{-k} \ll_k \left(\frac{q}{\varphi(q)}\right)^k$$

$$\ll_k (\log\log q)^k \ll_k (\log\log x)^k \ll (\log x)^{o_k(1)}.$$

When $|s| \leq 1/(\log x)^2$, we get by the Mean Value Theorem,

$$\prod_{p|q}\left(1 - \frac{\rho_1(p)}{p}\right)^{-1} = \prod_{p|q}\left(1 - \frac{\rho_1(p)}{p^{1+1/\log x}}\right)^{-1} + O_k\left(\frac{1}{\log x}\right),$$

so $\vartheta_1(q,s) \ll (\log\log x)^{o_k(1)}$ also. Next we see that it follows from the definitions that $\vartheta_1(q,0) = \vartheta_1(q)$ and that we can write

$$\vartheta_1(q,s) - \vartheta_1(q) = \vartheta_1(q,s)\vartheta_1(q)\sum_{n|q} \frac{\mu(n)\rho_1(n)}{n}\left(1 - \frac{1}{n^s}\right).$$

But since $|s| \le 1/(\log x)^2$, we have

$$\left|1 - \frac{1}{n^s}\right| \le \frac{2}{(\log x)^2}\exp(-\Re s \log n)\log n \ll \frac{1}{\log x}.$$

Therefore,

$$\vartheta_1(q,s) - \vartheta_1(q) \ll \frac{1}{(\log x)^{1-\epsilon}}.$$

In addition, using Cauchy's integral formula in the disk $|z| \le 1/(\log x)^2$, we get for $|s| < 1/(\log x)^2$,

$$G_1(s) - \mathcal{G}(\mathcal{H}) = G_1(s) - G_1(0) = \frac{s}{2\pi i}\oint_{|z|=1/(\log x)^2} \frac{G_1(z)}{z(z-s)}\,dz \ll \frac{1}{(\log x)^2}.$$

Therefore, using the result of Step (1), and noting that

$$\frac{1}{\zeta(1+s)^k} = s^k + g(s)$$

for some function $g(s)$ holomorphic in a neighbourhood of 0, we get

$$\mathscr{A}_1(q) - \frac{1}{2\pi i}\vartheta_1(d)\mathcal{G}(\mathcal{H})\oint_{|s|=1/(\log x)^2} \frac{(R/q)^s}{s^{l+1}}\,ds$$

$$= \frac{1}{2\pi i}\oint_{|s|=1/(\log x)^2} \frac{\vartheta_1(q,s)G_1(s)}{\zeta(1+s)^k}\frac{(R/q)^s}{s^{k+l+1}}\,ds$$

$$- \frac{1}{2\pi i}\vartheta_1(d)\mathcal{G}(\mathcal{H})\oint_{|s|=1/(\log x)^2} \frac{(R/q)^s}{s^{l+1}}\,ds \ll (\log x)^{l-1+\epsilon}.$$

The first result of the lemma follows from this using Lemma 4.7 Step (2) again.

(4) The proof of the second result is similar. To begin, let

$$\vartheta_2(q,s) := \prod_{p|q}\left(1 - \frac{v_p(\mathcal{H})-1}{(p-1)p^s}\right)^{-1}, \quad \text{and}$$

$$G_2(s) := \prod_{p|q}\left(1 - \frac{v_p(\mathcal{H})-1}{(p-1)p^s}\right)\left(1 - \frac{1}{p^{s+1}}\right)^{1-k},$$

so

$$\mathcal{A}_2(q) = \frac{1}{2\pi i} \int_{|s|=1/(\log x)^2} \frac{\vartheta_2(q,s)G_2(s)}{\zeta(1+s)^{k-1}} \frac{(R/q)^s}{s^{k+l+1}} \, ds,$$

and proceed as in Steps (2) and (3) to complete the proof. $\qquad\square$

We need Perron's famous formula for the partial sum of the Dirichlet coefficients. We don't use a general form that includes estimates for the error when the
leading integral term is truncated such as [144, corollary 5.3], but do the truncation
explicitly in Step (1).

Lemma 5.3 **(Perron's formula)** *[4, theorem 11.18] Let the Dirichlet series*

$$f(s) = \sum_{n=1}^{\infty} \frac{a_n}{n^s}$$

*converge absolutely for $\sigma > \sigma_a$, where σ_a is the abscissae of absolute convergence.
Let $\sigma_0 > \max(0, \sigma_a)$ and $x > 0$. Then if $x \notin \mathbb{N}$, we have*

$$\sum_{n \le x} a_n = \frac{1}{2\pi i} \int_{(\sigma_0)} f(s) \frac{x^s}{s} \, ds.$$

*In case $x \in \mathbb{N}$, if the final term a_x is counted with weight one half the integral formula
remains valid.*

Recall the definitions $\rho_1(p) = v_p(\mathcal{H})$, $\vartheta_1(p) = (1 - v_p(\mathcal{H})/p)^{-1}$, $\rho_2(p) = v_p(\mathcal{H}) - 1$, and $\vartheta_2(p) = (1 - (v_p(\mathcal{H}) - 1)/(p-1))^{-1}$, extended multiplicatively
and supported on squarefree positive integers.

Perron's formula is applied in Lemma 5.4 to derive expressions for sums over q
of terms related to those appearing in Lemma 5.2.

Lemma 5.4 *[215, lemma 3] As $x \to \infty$, we have the estimates for the given sums
over squarefree numbers*

$$\sum_{q < x^{\frac{1}{4}}} \frac{\rho_1(q)\vartheta_1(q)}{q} = \frac{(1+4\varpi)^{-k}}{k!} \frac{(\log R)^k}{\mathcal{G}(\mathcal{H})} + O\left((\log x)^{k-1}\right),$$

$$\sum_{q < x^{\frac{1}{4}}} \frac{\rho_2(q)\vartheta_2(q)}{q} = \frac{(1+4\varpi)^{1-k}}{(k-1)!} \frac{(\log R)^{k-1}}{\mathcal{G}(\mathcal{H})} + O\left((\log x)^{k-2}\right).$$

Proof (1) To begin, observe $\vartheta_1(p)/p = 1/(p - v_p(\mathcal{H}))$. If we define for $\sigma > 0$

$$B_1(s) := \prod_p \left(1 + \frac{v_p(\mathcal{H})}{(p - v_p(\mathcal{H}))p^s}\right) \left(1 - \frac{1}{p^{1+s}}\right)^k,$$

then

$$\sum_{n=1}^{\infty} \frac{\rho_1(n)\vartheta_1(n)}{n^{1+s}} = B_1(s)\zeta(1+s)^k.$$

Therefore, using Perron's formula from Lemma 5.3 with $\sigma_a = 0, \sigma_0 = 1/(\log x)$, and the estimates on the vertical line of integration

$$|B_1(s)| \ll 1,$$
$$|\zeta(s+1)| \ll \log x,$$
$$\left| \int_T^{\infty} \frac{e^{it}}{t} \, dt \right| \ll \frac{1}{T},$$

we get

$$\mathscr{S} := \sum_{n \le x^{\frac{1}{4}}} \frac{\rho_1(n)\vartheta_1(n)}{n^1} = \frac{1}{2\pi i} \int_{1/(\log x)-iR_0}^{1/(\log x)+iR_0} \frac{B_1(s)\zeta(1+s)^k x^{s/4}}{s} \, ds$$

$$+ O\left(\frac{(\log x)^k}{R_0} \right).$$

(2) Since $B_1(s)$ is holomorphic and bounded in the half plane $\sigma \ge -1/3$, if we move the path of integration to the vertical line segment

$$[-1/3 - iR_0, \, -1/3 + iR_0], \quad R_0 = \exp(L_x(1/k)),$$

the formula for the sum $\mathscr{S}$ becomes

$$\mathscr{S} = \frac{1}{2\pi i} \oint_{|s|=1/(\log x)} \frac{B_1(s)\zeta(1+s)^k x^{s/4}}{s} \, ds + O\left(\frac{(\log x)^B}{R_0} \right).$$

Zhang's argument here was found to be somewhat mysterious, so we give an alternative. First we expand the exponential and zeta functions in Laurent series

$$\frac{1}{2\pi i} \oint_{|s|=1/(\log x)} \frac{\zeta(1+s)^k x^{s/4}}{s} \, ds$$

$$= \frac{1}{2\pi i} \oint_{|s|=1/(\log x)} \sum_{n=0}^{\infty} \frac{\left(\frac{\log x}{4}\right)^n}{n!} \left(\frac{1}{s}\right) \left(\frac{1}{s} + \gamma + a_1 s + \cdots\right)^k \, ds$$

$$= \frac{1}{k!} \left(\frac{\log x}{4}\right)^k + O((\log x)^{k-1}).$$

On the disk $|s| \leq 1/(\log x)$, we can write $B_1(s) = B_1(0) + sf(s)$ for some holomorphic function $f(s)$, and using Laurent series again

$$\frac{1}{2\pi i} \oint_{|s|=1/(\log x)} sf(s) \frac{\zeta(1+s)^k x^{s/4}}{s} \, ds$$

$$= \frac{1}{2\pi i} \oint_{|s|=1/(\log x)} f(s)\zeta(1+s)^k x^{s/4} \, ds \ll (\log x)^{k-1}.$$

In addition,

$$B_1(0) = \prod_p \left(1 + \frac{v_p(\mathcal{H})}{(p - v_p(\mathcal{H}))} \right) \left(1 - \frac{1}{p} \right)^k = \frac{1}{\mathcal{G}(\mathcal{H})}.$$

It follows that

$$\frac{1}{2\pi i} \oint_{|s|=1/(\log x)} \sum_{n=0}^{\infty} B_1(s) \frac{\zeta(1+s)^k x^{s/4}}{s} \, ds = \frac{1}{k!} \left(\frac{\log x}{4} \right)^k + O((\log x)^{k-1}),$$

and therefore

$$\mathcal{S} = \frac{1}{k! \, \mathcal{G}(\mathcal{H})} \left(\frac{1}{4} \log x \right)^k + O((\log x)^{k-1}).$$

Finally, since $R = x^{\frac{1}{4}+\varpi}$ implies $(\log x)/4 = (\log R)/(1 + 4\varpi)$, a substitution gives the first formula of the lemma.

(3) To prove the second formula, proceed in a similar manner. To begin, define

$$B_2(s) := \prod_p \left(1 + \frac{v_p(\mathcal{H}) - 1}{(p - v_p(\mathcal{H}))p^s} \right) \left(1 - \frac{1}{p^{1+s}} \right)^{k-1} \implies B_2(0) = \frac{1}{\mathcal{G}(\mathcal{H})},$$

and thus

$$\sum_{n=1}^{\infty} \frac{\rho_2(n)\vartheta_2(n)}{\varphi(n)n^s} = B_2(s)\zeta(1+s)^{k-1}.$$

The proof is completed in an analogous manner to Steps (1) and (2). $\square$

Finally we need a simple consequence of the Chinese Remainder Theorem useful for counting the sets $\mathcal{C}_i(q)$.

Recall the definition for squarefree q:

$$\mathcal{C}_i(q) := \{n : 1 \leq n \leq q, \ (n,q) = 1, \ P(n - h_i) \equiv 0 \bmod q\},$$

where $P(n) := (n + h_1) \cdots (n + h_k)$.

Lemma 5.5   *[215, lemma 5] Let* $1 \leq i \leq k$ *and* q_1 *and* q_2 *be squarefree with* $(q_1, q_2) = 1$. *There is a bijection*

$$\mathscr{C}_i(q_1 q_2) \to \mathscr{C}_i(q_1) \times \mathscr{C}_i(q_2)$$

where the integer $c \to (a, b)$ *is such that* $c \bmod q_1 q_2$ *is a common solution to* $c \equiv a \bmod q_1$ *and* $c \equiv b \bmod q_2$.

Proof   Because $(q_1, q_2) = 1$, the bijection between sets satisfying the first two conditions of the sets $\mathscr{C}_i$ is a consequence of the Chinese Remainder Theorem. If the set on the left satisfies the third condition, so do both sets on the right using this bijection, and vice versa.   □

For example, if $\mathscr{H} = \{0, 2, 6\}$, then

$$\mathscr{C}_2(3 \cdot 5) = \{4, 13\},$$
$$\mathscr{C}_2(7 \cdot 11) = \{2, 24, 51, 73\},$$
$$\mathscr{C}_2(3 \cdot 5 \cdot 7 \cdot 11) = \{3, 101, 227, 332, 821, 926, 1052, 1151\}.$$

Of course, one need only check the sizes of these sets to verify the existence of a bijection.

5.7 Upper Bound for the Sum S_1

The following is the first of Zhang's fundamental lemmas. In spite of its length, it is the easier of the two, providing an upper bound for S_1. The second Lemma 5.9, which provides a lower bound for S_2, requires Zhang's extension of Bombieri–Vinogradov's theorem, which we state but do not prove in detail, giving a brief overview of Zhang's proof.

Recall the assignments $R = x^{\frac{1}{4}+\varpi}$, $R_1 = x^\varpi$, $\mathscr{P}_1 = \prod_{p<R_1} p$, $R_0 = \exp((\log x)^{1/k})$ and $\delta_1 = 1/(1+4\varpi)^k$. Also recall the definitions:

$$g(q) := \begin{cases} \frac{1}{(k+l)!} \left(\log \frac{R}{q} \right)^{k+l}, & q < R, \\ 0, & q \geq R, \end{cases}$$

$$\lambda(n) := \sum_{q|(P(n), \mathscr{P}_1)} \mu(q) g(q),$$

$$T_1 := \sum_{\substack{d,e \in \mathbb{N} \\ d,e | \mathscr{P}_1}} \frac{\mu(d) g(d) \mu(e) g(e)}{[d,e]} \rho_1([d,e]),$$

$$T_1^* := \sum_{h,d,e \in \mathbb{N}} \frac{\mu(de) \rho_1(hde)}{hde} g(hd) g(he),$$

$$S_1 := \sum_{x \leq n < 2x} \lambda(n)^2.$$

If q is such that $1 \le q < R$, and is squarefree, then we have derived in Lemma 5.2:

$$\mathscr{A}_1(q) = \frac{\vartheta_1(q)}{l!}\mathscr{G}(\mathscr{H})\left(\log\frac{R}{q}\right)^l + O(L_x(l-1+\epsilon)), \text{ and}$$

$$\mathscr{A}_2(q) = \frac{\vartheta_2(q)}{(l+1)!}\mathscr{G}(\mathscr{H})\left(\log\frac{R}{q}\right)^{l+1} + O(L_x(l+\epsilon)).$$

We also have the important function κ_1, used in Lemma 5.6 and estimated in Lemma 5.7:

$$\delta_1 := (1+4\varpi)^{-k},$$

$$\delta_2 := 1 + \sum_{n=1}^{292}\frac{((\log 293)k)^n}{n!},$$

$$\kappa_1 := \delta_1(1+\delta_2^2+(\log 293)k)\binom{k+2l}{k}.$$

Finally recall we have defined $\vartheta_1(q)$, $\vartheta_2(q)$, $\rho_1(q)$ and $\rho_2(q)$ to be multiplicative functions supported by squarefree integers such that

$$\rho_1(p) := v_p(\mathscr{H}), \quad \rho_2(p) := v_p(\mathscr{H}) - 1,$$

$$\vartheta_1(p) := \left(1 - \frac{\rho_1(p)}{p}\right)^{-1},$$

$$\vartheta_2(p) := \left(1 - \frac{\rho_2(p)}{p-1}\right)^{-1}.$$

Having assembled these definitions and results, we are now in a position to give the proof of the first of Zhang's fundamental lemmas, the upper bound for S_1. It takes seven steps to complete, but is intrinsically the easier of the two penultimate inputs to Zhang's prime gap theorem.

Lemma 5.6 (Zhang's upper bound) *[215, section 4] As $x \to \infty$, we have the upper bound*

$$S_1 \le \frac{1+\kappa_1}{(k+2l)!}\binom{2l}{l}x\mathscr{G}(\mathscr{H})(\log R)^{k+2l} + o(x(\log x)^{k+2l}).$$

Proof **(1)** First, since $\rho_1(n)$ is zero if n is not squarefree, if $h = (d,e)$ and we replace d/h by d and e/h by e, we get

$$T_1 := \sum_{h,d,e|\mathscr{P}_1}\frac{\mu(de)\rho_1(hde)}{hde}g(hd)g(he).$$

(2) Next, interchanging the order of summation in the definition of S_1 gives

$$S_1 = \sum_{d,e|\mathscr{P}_1} \mu(d)g(d)\mu(e)g(e) \sum_{\substack{x<n\leq 2x \\ P(n)\equiv 0 \bmod [d,e]}} 1.$$

If q is a squarefree integer, by the Chinese Remainder Theorem, there are precisely $\rho_1(q)$ distinct residue classes modulo q such that $P(n) \equiv 0 \bmod q$ if and only if n is in one of those classes. Thus the inner sum is equal to

$$\rho_1\left([d,e]\left(\frac{x}{[d,e]} + O(1)\right)\right) = \frac{\rho_1([d,e])x}{[d,e]} + O(\rho_1([d,e])).$$

For the error term, we can then write the following, using $\rho_1(q) \ll \tau_k(q)$ and the divisor sum bound of (4.26), and noting that the number of integer pairs (d,e) with $f = [d,e]$ is $\tau_3([d,e])$. We derive, using Cauchy–Schwarz,

$$\left| \sum_{d,e|\mathscr{P}_1} \rho_1([d,e])\ \mu(d)\mu(e)g(d)g(e) \right| \leq (\log x)^{2k+2l} \sum_{\substack{d,e|\mathscr{P}_1 \\ d,e<R}} \tau_k([d,e])$$

$$\leq (\log x)^{2k+2l} \sum_{\substack{f<R^2 \\ \text{squarefree}}} \tau_3(f)\tau_k(f)$$

$$\leq (\log x)^{2k+2l} \left(\sum_{f<R^2} \tau_3(f) \right)^{\frac{1}{2}} \times \left(\sum_{f<R^2} \tau_k(f) \right)^{\frac{1}{2}}$$

$$\leq (\log x)^{2k+2l} \left(R^2(\log x)^{B_1} \right)^{\frac{1}{2}} \times \left(R^2(\log x)^{B_2} \right)^{\frac{1}{2}}$$

$$\leq R^{2+\epsilon}.$$

Therefore,

$$S_1 = \mathcal{T}_1 x + O(R^{2+\epsilon}).$$

(3) We now find an expression for the difference $|\mathcal{T}_1 - \mathcal{T}_1^*|$ in terms of three sums which we then bound in successive steps. Considering $\mathcal{T}_1^*$ and splitting the sum over h at $x^{\frac{1}{4}}$, defining

$$\Sigma_1 := \sum_{h\leq x^{\frac{1}{4}}} \sum_{d,e\in\mathbb{N}} \frac{\mu(de)\rho_1(hde)}{hde} g(hd)g(he),$$

$$\Sigma_1' := \sum_{x^{\frac{1}{4}}<h<R} \sum_{d,e\in\mathbb{N}} \frac{\mu(de)\rho_1(hde)}{hde} g(hd)g(he).$$

Since $g(d)$ is supported on $[1, R)$, we get $\mathcal{T}_1^* = \Sigma_1 + \Sigma_1'$.

Note that if $h > x^{\frac{1}{4}}$, $hd < R$, $he < R$ and de are squarefree, then the conditions $d, e \mid \mathscr{P}_1$ are automatically satisfied. Thus, if we define

$$\Sigma_2 := \sum_{\substack{h \le x^{\frac{1}{4}} \\ h \mid \mathscr{P}_1}} \sum_{d, e \mid \mathscr{P}_1} \frac{\mu(de)\rho_1(hde)}{hde} g(hd)g(he),$$

$$\Sigma_2' := \sum_{\substack{x^{\frac{1}{4}} < h < R \\ h \mid \mathscr{P}_1}} \sum_{d, e \in \mathbb{N}} \frac{\mu(de)\rho_1(hde)}{hde} g(hd)g(he),$$

$$\Sigma_3 := \sum_{\substack{x^{\frac{1}{4}} < h < R \\ h \nmid \mathscr{P}_1}} \sum_{d, e \in \mathbb{N}} \frac{\mu(de)\rho_1(hde)}{hde} g(hd)g(he),$$

then $\mathcal{T}_1 = \Sigma_2 + \Sigma_2'$ and

$$|\mathcal{T}_1 - \mathcal{T}_1^*| = |\Sigma_1 - \Sigma_2 + \Sigma_1' - \Sigma_2'| \le |\Sigma_1| + |\Sigma_2| + |\Sigma_1' - \Sigma_2'|$$
$$\le |\Sigma_1| + |\Sigma_2| + |\Sigma_3|.$$

In the following three steps, we estimate each of these right-hand-side terms.
(4) We first simplify the inner sum of Σ_1, with h fixed, recalling the definition

$$\mathscr{A}_1(q) := \sum_{\substack{n \in \mathbb{N} \\ (n, q) = 1}} \frac{\mu(n)\rho_1(n)}{n} g(qn),$$

we derive, setting $d = qd'$ and $e = qe'$ and recalling d and e are squarefree,

$$\sum_{\substack{d, e \in \mathbb{N} \\ (h, d) = (h, e) = 1}} \frac{\mu(d)\mu(e)\rho_1(d)\rho_1(e)}{de} g(hd)g(he) \sum_{q \mid (d, e)} \mu(q)$$

$$= \sum_{\substack{q \in \mathbb{N} \\ (q, h) = 1}} \frac{\mu(q)\mu(d')\mu(e')\rho_1(qd')\rho_1(qe')}{d'e'q^2} g(hd'q)g(he'q)$$

$$= \sum_{\substack{q \in \mathbb{N} \\ (q, h) = 1}} \frac{\mu(q)\rho_1(q)^2}{q^2} \mathscr{A}_1(hq)^2.$$

Therefore, we can write, where Σ_1 is defined in Step (3),

$$\Sigma_1 := \sum_{h \le x^{\frac{1}{4}}} \sum_{(q, h) = 1} \frac{\rho_1(h)\mu(q)\rho_1(q)^2}{hq^2} \mathscr{A}_1(hq)^2.$$

The terms in this sum with $q \geq R_0$ contribute $O(L_x(B)/R_0)$. Hence if we substitute $hq = d$ and define for squarefree d,

$$\vartheta^*(d) := \sum_{\substack{hq=d \\ h<x^{\frac{1}{4}} \\ q<R_0}} \frac{\mu(q)\rho_1(q)}{q}.$$

With this definition, we get

$$\Sigma_1 = \sum_{d<x^{\frac{1}{4}}R_0} \frac{\rho_1(d)\vartheta^*(d)}{d} \mathscr{A}_1(d)^2 + O\left(\frac{L_x(B)}{R_0}\right). \tag{5.1}$$

(5) We have seen in Lemma 5.2 that

$$\mathscr{A}_1(q) = \frac{\vartheta_1(q)}{l!} \mathscr{G}(\mathscr{H}) \left(\log \frac{R}{q}\right)^l + O(L_x(l-1+\epsilon)).$$

Recall we have seen in Step (3) of Lemma 5.2 that

$$\vartheta_1(d) := \sum_{q|d} \frac{\mu(q)\rho_1(q)}{q} \ll (\log x)^{o_k(1)}.$$

These estimates imply $\mathscr{A}_1(q) \ll (\log x)^l (\log x)^{o_k(1)}$.

We **claim** also

$$\sum_{x^{\frac{1}{4}} \leq q < x^{\frac{1}{4}}R_0} \frac{\rho_1(q)}{q} \ll L_x(k+1/k-1).$$

To see this first note by Lemma 4.5, we have

$$\sum_{\substack{q \leq x \\ \text{squarefree}}} \frac{\tau_k(q)}{q} \leq (k+\log x)^k,$$

and then note that $(\log(x^{\frac{1}{4}}R_0)+k)^k \ll_k L_x(k-1+1/k)$ to complete the derivation.

These two estimates imply the contribution to the sum we have derived for Σ_1 from terms with $x^{\frac{1}{4}} \leq q < x^{\frac{1}{4}}R_0$, is $o(L_x(k+2l))$. In addition, with d squarefree, because

$$\sum_{q|d} \frac{\mu(q)\rho_1(q)}{q} = \frac{1}{\vartheta_1(d)},$$

for $q < x^{\frac{1}{4}}$ we can write

$$\vartheta^*(q) = \frac{1}{\vartheta_1(q)} + O\left(\frac{\tau_k(q)}{R_0}\right). \tag{5.2}$$

Thus, using the expression for $\mathscr{A}_1(q)$ from Lemma 5.2 given earlier, we have for a sufficiently large constant B (depending on l),

$$\vartheta^*(q)\mathscr{A}_1(q)^2 = \frac{1}{(l!)^2}\mathscr{G}(\mathcal{H})^2 \vartheta_1(q)\left(\log\frac{R}{q}\right)^{2l}$$
$$+ O(L_x(2l - 1 + \epsilon)) + O\left(\tau_k(q)\frac{L_x(B)}{R_0}\right).$$

Substituting into the expression derived for Σ_1, (5.1), then gives

$$\Sigma_1 = \frac{1}{(l!)^2}\mathscr{G}(\mathcal{H})^2 \sum_{q \le x^{\frac{1}{4}}} \frac{\rho_1(q)\vartheta_1(q)}{q}\left(\log\frac{R}{q}\right)^{2l} + o(L_x(k + 2l)).$$

Replacing the sum with the expression derived in Lemma 5.4, using a sum similar to Lemma 8.24 (iv), and using $\delta_1 := (1 + 4\varpi)^{-k}$, then gives

$$|\Sigma_1| \le \frac{\delta_1}{k!\,(l!)^2}\mathscr{G}(\mathcal{H})(\log R)^{k+2l} + o(L_x(k + 2l)),$$

which is Zhang's final estimate for Σ_1.

(6) To estimate Σ_2, we first define

$$\mathscr{A}_1^*(d) := \sum_{\substack{n:(n,d)=1 \\ n|\mathscr{P}_1}} \frac{\mu(n)\rho_1(n)g(dn)}{n},$$

so we can write using the same approach as in Step (4) for $\mathscr{A}_1(q)$

$$\Sigma_2 = \sum_{\substack{h \le x^{\frac{1}{4}} \\ h|\mathscr{P}_1}} \sum_{\substack{(q,h)=1 \\ q|\mathscr{P}_1}} \frac{\rho_1(h)\mu(q)\rho_1(q)^2}{hq^2}\mathscr{A}_1^*(hq)^2.$$

Again using a similar approach to that taken in Step (4), where we defined and estimated $\vartheta^*(d)$, we then derive

$$\Sigma_2 = \sum_{\substack{d < x^{\frac{1}{4}} \\ d|\mathscr{P}_1}} \frac{\rho_1(d)\vartheta^*(d)}{d}\mathscr{A}_1^*(d)^2 + O\left(\frac{L_x(B)}{R_0}\right). \tag{5.3}$$

Now recall $R = x^{\frac{1}{4}+\varpi}$, $R_1 = x^{\varpi}$ and let $\mathscr{P}^* = \prod_{R_1 \le p < R} p$. Assume $d \mid \mathscr{P}_1 = \prod_{p<R_1} p$. Zhang as in Step (4) suggests a derivation using Möbius inversion, but we use a different approach. We derive letting $n = qe$ to get the second line

$$\mathscr{A}_1^*(d) = \sum_{(n,d)=1} \frac{\mu(n)\rho_1(n)g(dn)}{n} \sum_{q|(n,\mathscr{P}^*)} \mu(q)$$

$$= \sum_{q|\mathscr{P}^*} \sum_{(qe,d)=1} \frac{\mu(e)\rho_1(q)\rho_1(e)}{qe} g(dqe)$$

$$= \sum_{q|\mathscr{P}^*} \frac{\rho_1(q)}{q} \sum_{(e,dq)=1} \frac{\mu(e)\rho_1(e)}{e} g(dqe)$$

$$= \sum_{q|\mathscr{P}^*} \frac{\rho_1(q)}{q} \mathscr{A}_1(dq).$$

If $q \mid \mathscr{P}^*$ and $q < R$, we have

$$\vartheta_1(q) = 1 + O((\log x)/R_1),$$

and then by Lemma 5.2 we can write

$$|\mathscr{A}_1^*(d)| \le \frac{\mathscr{G}(\mathscr{H})\vartheta_1(d)}{l!} \left(\log \frac{R}{d}\right)^l \sum_{\substack{q|\mathscr{P}^* \\ q<R}} \frac{\rho_1(q)}{q} + O(L_x(l-1+\epsilon)).$$

If $q \mid \mathscr{P}^*$ and $q < R$, then we **claim** q has at most 292 prime factors. To see this, let $q = p_1 \cdots p_l$ so $R_1^l \le q < R$, which implies $x^{\varpi l} < x^{\frac{1}{4}+\varpi}$ and thus $l < \frac{1}{4}/\varpi + 1 = 293$. Now,

$$\sum_{p \le x} \frac{1}{p} = \log\log x + B + O(1/\log x),$$

so

$$\sum_{R_1 \le p < R} \frac{1}{p} = \log 293 + O(L_x(-1)).$$

Consider squarefree $q < R$ with precisely $1 \le v \le l$ prime factors p_j with $R_1 \le p_j < D$. Then summing over all such q, we have

$$\sum_{\substack{q|\mathscr{P}^* \\ q<R, \ \omega(q)=v}} \frac{v_q(\mathscr{H})}{q} \le \frac{1}{v!} \prod_{\substack{\Pi_j p_j<R \\ p_j \text{ distinct}}} \left(\sum_{\substack{p_j|\mathscr{P}^* \\ 1\le j\le v}} \frac{v_{p_j}(\mathscr{H})}{p_j} \right)^v \le \frac{1}{v!} \left(\sum_{\substack{p_j|\mathscr{P}^* \\ 1\le j\le v \\ p_j \text{ distinct}}} \frac{k}{p_j} \right)^v$$

$$\le \frac{k^v}{v!} \left(\sum_{\substack{p_j|\mathscr{P}^* \\ 1\le j\le v \\ p_j \text{ distinct}}} \frac{1}{p_j} \right)^v \le \frac{k^v (\log 293)^v}{v!}.$$

This implies, assigning

$$\delta_2 = 1 + \sum_{n=1}^{292} \frac{(\log(293)k)^n}{n!},$$

that

$$\sum_{\substack{q|\mathscr{P}^* \\ q<R}} \frac{\rho_1(q)}{q} \leq \delta_2 + O(L_x(-1)).$$

Next we substitute this bound in the right-hand side of the estimate derived for $|\mathscr{A}_1^*(d)|$, to get

$$|\mathscr{A}_1^*(d)| \leq \frac{\delta_2 \mathscr{G}(\mathscr{H})\vartheta_1(d)}{l!} \left(\log \frac{R}{d}\right)^l + O(L_x(l-1+\epsilon)).$$

Finally in this step, substituting the bound we have derived in the expression we have derived for Σ_2, (5.3), using (5.2), recalling $\delta_1 = (1+4\varpi)^{-k}$ and applying Lemma 5.4 to get the second inequality, then gives

$$|\Sigma_2| \leq \frac{\delta_2^2 \mathscr{G}(\mathscr{H})^2}{(l!)^2} \sum_{d<x^{\frac{1}{4}}} \frac{\rho_1(d)\vartheta_1(d)}{d} \left(\log \frac{R}{d}\right)^{2l} + o(L_x(k+2l))$$

$$\leq \frac{\delta_1 \delta_2^2}{k!\,(l!)^2} \mathscr{G}(\mathscr{H})L_R(k+2l) + o(L_x(k+2l)),$$

which completes the derivation of the bound for Σ_2.

(7) For Σ_3, we follow Zhang's familiar process. First, define

$$\widetilde{\vartheta}(d) := \sum_{\substack{hq=d \\ x^{\frac{1}{4}}<h \\ h\nmid\mathscr{P}_1}} \frac{\mu(q)\rho_1(q)}{q}.$$

This definition enables us to derive, in a similar manner as we did in Step (5) for Σ_2,

$$\Sigma_3 = \sum_{x^{\frac{1}{4}}<d<R} \frac{\rho_1(d)\widetilde{\vartheta}(d)}{d} \mathscr{A}_1(d)^2. \tag{5.4}$$

Using the two estimates from Step (4), namely $\mathscr{A}_1(d) \ll (\log x)^l (\log\log x)^B$ and

$$\sum_{x^{\frac{1}{4}}\leq d<x^{\frac{1}{4}}R_0} \frac{\rho_1(d)}{d} \ll L_x(k+1/k-1),$$

we find the terms in this sum with $x^{\frac{1}{4}} < d < x^{\frac{1}{4}} R_0$ contribute $o((\log x)^{k+2l})$. The upper range $x^{\frac{1}{4}} R_0 < d < R$ is more difficult. Assume d satisfies these bounds and is squarefree with $d \nmid \mathscr{P}_1$. If an integer $h \mid d$ and $x^{\frac{1}{4}} < h$ and $h \mid \mathscr{P}_1$, then

$$\frac{d}{h} < \frac{R}{x^{\frac{1}{4}}} = R_1 \implies \frac{d}{h} \mid \mathscr{P}_1 \implies d \mid \mathscr{P}_1,$$

which is false. Therefore, $h \nmid \mathscr{P}_1$. Thus, using a sum from Step (4), namely

$$\sum_{q \mid d} \frac{\mu(q)\rho_1(q)}{q} = \frac{1}{\vartheta_1(d)},$$

we get

$$\widetilde{\vartheta}(d) = \sum_{\substack{hq=d \\ x^{\frac{1}{4}} < h}} \frac{\mu(q)\rho_1(q)}{q} = \frac{1}{\vartheta_1(d)} + O\left(\frac{\tau_{k+1}(d)}{R_0}\right).$$

Using Lemma 5.4 then gives

$$\widetilde{\vartheta}(d)\mathscr{A}_1(d)^2 = \frac{\mathscr{G}(\mathscr{H})^2\vartheta_1(d)}{(l!)^2}\left(\log\frac{R}{d}\right)^{2l} + O\left(\tau_{k+1}(d)\frac{L_x(B)}{R_0}\right)$$
$$+ O(L_x(2l - 1 + \epsilon)).$$

Substituting this in the expression we have derived for Σ_3, i.e., (5.4), gives

$$\Sigma_3 = \frac{\mathscr{G}(\mathscr{H})^2}{(l!)^2} \sum_{\substack{x^{\frac{1}{4}} R_0 < d < R \\ d \nmid \mathscr{P}_1}} \frac{\rho_1(d)\vartheta_1(d)}{d}\left(\log\frac{R}{d}\right)^{2l} + o(L_x(k + 2l)).$$

Using the definition for $\vartheta(q)$ and the sum for $1/p$ given in Step (6) together with Lemma 5.4 enables us to derive the bounds

$$\sum_{\substack{x^{\frac{1}{4}} < d < R \\ d \nmid \mathscr{P}_1}} \frac{\rho_1(d)\vartheta_1(d)}{d} \leq \sum_{d < R} \frac{\rho_1(d)\vartheta_1(d)}{d} \sum_{p \mid (d, \mathscr{P}^*)} 1$$

$$\leq \sum_{R_1 \leq p < R} \frac{\rho_1(p)\vartheta_1(p)}{p} \sum_{d < R/p} \frac{\rho_1(d)\vartheta_1(d)}{d}$$

$$\leq \frac{\delta_1(\log 293)}{(k - 1)!\,\mathscr{G}(\mathscr{H})} L_R(k) + o(L_x(k)).$$

Substituting this bound in the expression we have derived for Σ_3 gives, finally,

$$|\Sigma_3| \leq \frac{\delta_1(\log 293)}{(k - 1)!\,(l!)^2}\mathscr{G}(\mathscr{H})L_R(k + 2l) + o(L_x(k + 2l)).$$

(8) By Steps (3) through (7), if we define

$$\kappa_1 := \delta_1(1 + \delta_2^2 + (\log 293)k)\binom{k + 2l}{k}$$

and use the identity

$$\frac{1}{k!\,(l!)^2} = \frac{1}{(k + 2l)!}\binom{k + 2l}{k}\binom{2l}{l},$$

after simplifying we can write

$$|\mathcal{T}_1 - \mathcal{T}_1^*| \le |\Sigma_1| + |\Sigma_2| + |\Sigma_3| \le \frac{\kappa_1}{(k + 2l)!}\binom{2l}{l}\mathscr{G}(\mathscr{H})L_R(k + 2l))$$

$$+ o(L_x(k + 2l)).$$

Using Lemma 5.1, this shows

$$\mathcal{T}_1 \le \frac{1 + \kappa_1}{(k + 2l)!}\binom{2l}{l}\mathscr{G}(\mathscr{H})L_R(k + 2l) + o(L_x(k + 2l)).$$

Finally, using this bound and the estimate $S_1 = x\mathcal{T}_1 + O(R^{2+\epsilon})$ from Step (2), we get

$$S_1 \le \frac{1 + \kappa_1}{(k + 2l)!}\binom{2l}{l}x\mathscr{G}(\mathscr{H})L_R(k + 2l) + o(xL_x(k + 2l)),$$

which completes the proof.   □

The parameter κ_1 defined in Step (8) of Lemma 5.6 is very small, and we show this in two ways, both of which involve a computation. First from Lemma 5.6 we extract the definitions

$$\delta_1 := (1 + 4\varpi)^{-k},$$

$$\delta_2 := 1 + \sum_{n=1}^{292}\frac{((\log 293)k)^n}{n!},$$

$$\kappa_1 := \delta_1(1 + \delta_2^2 + (\log 293)k)\binom{k + 2l}{k}.$$

We also have the assignments of Zhang, $\varpi = 1/1168$, $k = 3.5 \times 10^6$, $l = 180$. These lead to the numerical evaluation $\log \kappa_1 < -1228 < -1200$, via the **PGpack** function KappaZhang or the following derivation.

Lemma 5.7 *[215, section 4] We have the upper bound* $\kappa_1 < \exp(-1200)$.

Proof From the definition of δ_2, a computation to verify the first inequality and using $\sqrt{2\pi n}(n/e)^n < n!$ with $n = 292$ for the second, we get

$$1 + \delta_2^2 + (\log 293)k < 2\left(\frac{(\log 293)k)^{292}}{292!}\right)^2 < \frac{(185100)^{584}}{292\pi}.$$

Since $k = 3.5 \times 10^6$ and $l = 180$, we also get with $n = 180$,

$$\binom{k+2l}{k} < \frac{2k^{2l}}{(2l)!} < \frac{26500^{360}}{\sqrt{180\pi}}.$$

Using these bounds, from the definition of κ_1 we compute

$$\log \kappa_1 < -3.5 \times 10^6 \log\left(\frac{293}{292}\right) + 584\log(185100) + 360\log(26500) < -1200.$$

This completes the proof.   □

5.8 Lower Bound for the Sum S_2

Next we derive the crucial fundamental Lemma 5.9. The proof is relatively short, but it uses a wide range of inputs from what has gone before in this chapter. Zhang's variation of Bombieri–Vinogradov, Theorem 5.8, is used in Step (3). The small parameter κ_2 appears in the lower bound, mopping up a lot of the detailed variation.

Recall the definitions: $\mathcal{H} = \{h_1, \ldots, h_k\}$ is an admissible k-tuple,

$$S_2 := \sum_{x \le n < 2x}\left(\sum_{i=1}^{k}\theta(n+h_i)\right)\lambda(n)^2,$$

$$g(y) := \frac{1}{(k+l)!}\left(\log\frac{R}{y}\right)^{k+l}, \quad \text{if } y < R \text{ else } 0,$$

$$\lambda(n) := \sum_{q|(P(n),\mathscr{P}_1)}\mu(q)g(q),$$

$$P(n) := \prod_{i=1}^{k}(n+h_i),$$

$$\mathscr{C}_i(q) := \{a : 1 \le a \le q, \ (a,q) = 1, \ P(a-h_i) \equiv 0 \bmod q\}, \ 1 \le i \le k,$$

where $\varpi = 1/1168$, $R = x^{\frac{1}{4}+\varpi}$, $\mathscr{P} = \prod_{p<R}p$, $R_1 = x^{\varpi}$, $\mathscr{P}_1 = \prod_{p<R_1}p$.

Finally, following Polymath8a, the expression MPZ$[\varpi,\delta]$ for $0 < \varpi < \frac{1}{4}$ and $0 < \delta < \frac{1}{4} + \varpi$ means

Definition 5.1 **(Motohashi–Pintz–Zhang property)** Let $\varpi, \delta > 0$ be real numbers and $P_\Omega := \prod_{p \in \Omega} p$. We say MPZ$[\varpi, \delta]$ is true if $\Omega \subset [1, x^\delta]$ is a subinterval and $Q \prec^o x^{\frac{1}{2}+2\varpi}$, where $f(x) \prec^o g(x)$ means $f(x) \leq x^{o(1)} g(x)$ as $x \to \infty$, a mod P_Ω is a primitive residue class and $A \geq 1$ is a given real number, then with an implied constant depending only on A, ϖ and δ, we have

$$\sum_{\substack{q \leq Q \\ q \in S_\Omega}} \left| \sum_{\substack{x \leq n < 2x \\ n \equiv a \bmod q}} \Lambda(n) - \frac{1}{\varphi(q)} \sum_{\substack{x \leq n < 2x \\ (n,q)=1}} \Lambda(n) \right| \ll \frac{x}{(\log x)^A}, \qquad (5.5)$$

where S_Ω is the set of all squarefree numbers being products of primes in Ω.

We also need the signed discrepancy $\Delta(\gamma, q, a)$, with γ an arithmetic function, $q > 0$ an integer and a an integer with $(a, q) = 1$. This measures the difference between values of $\gamma(n)$ over a particular congruence class compared with the average over all classes coprime to the modulus. Then define as before for $(a, q) = 1$,

$$\Delta(\gamma, q, a) := \sum_{\substack{x \leq n < 2x \\ n \equiv a \bmod q}} \gamma(n) - \frac{1}{\varphi(q)} \sum_{\substack{x \leq n < 2x \\ (n,q)=1}} \gamma(n).$$

We need Zhang's remarkable theorem, which we state and give an outline of the proof. In Chapter 8, we give the proof of a restricted form of Polymath8a's variation. Zhang's Theorem 5.8 is used in Step (3) of Lemma 5.9.

Theorem 5.8 *[215, theorem 2]* **(Zhang's Bombieri–Vinogradov variation)** *Let $k \geq 2$. For $1 \leq i \leq k$ and all $A > 0$, we have*

$$\sum_{\substack{q < R^2 \\ q | \mathcal{P}_1}} \sum_{a \in \mathscr{C}_i(q)} |\Delta(\theta, q, a)| \ll \frac{x}{(\log x)^A}.$$

This implies that if $\varpi = \delta = 1/1168$, then MPZ$[\varpi, \delta]$ is true.

Next we give a brief outline of Zhang's proof. Using the combinatorial identity of Heath-Brown Lemma 8.21, he first decomposed $\Lambda(n)$ (a suitable approximation for $\theta(n)$) into a finite sum of Dirichlet convolutions $\gamma(n)$ of differing forms, and estimated each form separately. They were in two groups labelled Type I/II and Type III, which we will describe using the following assumptions $(A_1) - (A_5)$.

First, define $x_1 = x^{3/8+8\varpi}$ and $x_2 = x^{\frac{1}{2} - 4\varpi}$ so $x^{\frac{1}{4}} < x_1 < x_2 < \sqrt{x}$.
Type I/II forms have $\gamma = \alpha \star \beta$, where $\eta = 1 + 1/(\log x)^{2A}$ and:

(A_1) α has support in $[M, \eta^{j_1} M)$ with $j_1 \leq 19$ and $\alpha(m) \leq \tau_{j_1}(m) \log x$,

(A_2) β has support in $[N, \eta^{j_2} N)$ with $j_2 \leq 19$ and $\beta(n) \leq \tau_{j_2}(n) \log x$,

and such that $x_1 < N < 2\sqrt{x}$.

(A_3) $j_1 + j_2 \leq 20$, $[MN, \eta^{20}MN) \subset [x, 2x)$.

In addition, under (A_2) we have the assumption for any fixed q, r, a with $(a, r) = 1$,

$$\sum_{n \equiv a \bmod r} \beta(n) - \frac{1}{\varphi(r)} \sum_{(n, qr) = 1} \beta(n) \ll \tau_{20}(q) N / (\log x)^{200A}.$$

Then under Type I, N is in the lower range and under Type II the upper range:

Type I : $x_1 < N \le x_2$,

Type II : $x_2 < N < 2\sqrt{x}$.

The arithmetic functions γ are denoted by Zhang as being Type III if they can be written

$$\gamma = \alpha \star \mathcal{X}_{N_1} \star \mathcal{X}_{N_1} \star \mathcal{X}_{N_2} \star \mathcal{X}_{N_3},$$

where $\mathcal{X}_N$ is the characteristic function of $[N, \eta N) \cap \mathbb{Z}$, with α satisfying (A_1) and $j_1 \le 17$ and such that N_1, N_2, N_3 satisfy (A_4) and (A_5), where

(A_4) $N_3 \le N_2 \le N_1$, $M N_1 \le x_1$,

(A_5) $[M N_1 N_2 N_3, \eta^{20} M N_1 N_2 N_3) \subset [x, 2x)$.

Following this decomposition into types, for Types I/II Zhang used the dispersion method of Linnik [127] (see Appendix F) used by Friedlander–Iwaniec in [61] and then by Heath-Brown in [97]. Then with a factorization $d = qr$, he reduced the problem to estimating certain incomplete Kloosterman sums, which he does using a variant of Weil's Kloosterman sum bound.

The estimation of Type III sums is much more difficult, and Zhang has recourse to an exponential sum estimate of Bombieri and Birch found in an appendix to the Friedlander–Iwaniec work [61]. This estimate depends on Deligne's results [36]. It is insufficient to complete the proof, which required Zhang to exploit the factorization $d = qr$ yet again.

This outline is a pale representation of Zhang's work and the interested reader is encouraged to consult the original [215]. The theorem will be revisited in Chapter 8, reporting part of the work of Polymath8a, with a more general approach based on the concept dense divisibility.

Next recall the definitions from Section 5.7,

$$\delta_1 := \frac{1}{(1 + 4\varpi)^k},$$

$$\delta_2 := 1 + \sum_{n=1}^{292} \frac{((\log 293)k)^n}{n!}, \text{ and then define}$$

$$\kappa_2 := \delta_1 (1 + 4\varpi)(1 + \delta_2^2 + (\log 293)k) \binom{k + 2l + 1}{k - 1},$$

and from Section 1.6

$$\mathcal{G}(\mathcal{H}) := \prod_p \left(1 - \frac{\upsilon_p(\mathcal{H})}{p}\right)\left(1 - \frac{1}{p}\right)^{-k}.$$

We can now give the second of the fundamental lemmas of Zhang.

Lemma 5.9   (Zhang's lower bound) *[215, section 5] As $x \to \infty$ we have the lower bound*

$$S_2 \geq \frac{k(1 - \kappa_2)}{(k + 2l + 1)!}\binom{2l + 2}{l + 1}x\mathcal{G}(\mathcal{H})(\log R)^{k+2l+1} + o\left(x(\log x)^{k+2l+1}\right).$$

Proof   **(1)** First, note that given $\epsilon > 0$, since the number of primes in any interval of fixed length is $O(1/\log x)$ and $\mathcal{H}$ is fixed, we have

$$S_2 = \sum_{1 \leq i \leq k}\sum_{x \leq n < 2x} \theta(n)\lambda(n - h_i)^2 + O(x^\epsilon).$$

Now fix $i \in [1, k] \cap \mathbb{N}$ and as usual interchange the order of summation in the inner sum of this expression to get

$$\sum_{x \leq n < 2x} \theta(n)\lambda(n - h_i)^2 = \sum_{\substack{d, e \in \mathbb{N} \\ d, e | \mathscr{P}_1}} \mu(d)g(d)\mu(e)g(e)\left(\sum_{\substack{x \leq n < 2x \\ P(n - h_i) \equiv 0 \bmod [d, e]}} \theta(n)\right).$$

(2) In the sum derived in Step (1) we can assume d and e are squarefree and consider the inner sum over n. Note that the conditions

$$P(n - h_i) \equiv 0 \bmod d \text{ and } (n, d) = 1 \iff \exists c \in \mathscr{C}_i(d) \text{ with } n \equiv c \bmod d.$$

In addition, for any prime p, $|\mathscr{C}_i(p)|$ is the number of distinct residue classes of the $h_i - h_j$ for $h_i - h_j \not\equiv 0 \bmod p$. Thus, using Lemma 5.5, since $\rho_2(d)$ is defined multiplicatively on squarefree integers,

$$|\mathscr{C}_i(p)| = \upsilon_p(\mathcal{H}) - 1 = \rho_2(p) \implies |\mathscr{C}_i(d)| = \rho_2(d).$$

Therefore, using the definition of signed discrepancy, we can write

$$\sum_{\substack{x \leq n < 2x \\ P(n - h_i) \equiv 0 \bmod [d, e]}} \theta(n) = \sum_{c \in \mathscr{C}_i([d, e])}\sum_{\substack{x \leq n < 2x \\ n \equiv c \bmod [d, e]}} \theta(n)$$

$$= \frac{\rho_2([d, e])}{\varphi([d, e])}\left(\sum_{x \leq n < 2x} \theta(n)\right) + \sum_{a \in \mathscr{C}_i([d, e])} \Delta(\theta, [d, e], a).$$

(3) Next, note that because

$$\tau_3(n) := \#\{(a, b, c) : abc = q\}$$

and d, e are squarefree, we have

$$\#\{(d,e) : [d,e] = q\} = \tau_3(q).$$

Define

$$T_2 := \sum_{d,e \mid \mathcal{P}_1} \frac{\mu(d)g(d)\mu(e)g(e)}{\varphi([d,e])} \rho_2([d,e])$$

and also

$$\mathcal{E}_i := \sum_{\substack{q < R^2 \\ q \mid \mathcal{P}_1}} \tau_3(q) \sum_{a \in \mathcal{C}_i(q)} |\Delta(\theta, q, a)|.$$

It follows from Steps (1) and (2) that with i fixed,

$$\sum_{x \le n < 2x} \theta(n)\lambda(n - h_i)^2 = T_2 \sum_{x \le n < 2x} \theta(n) + O(|\mathcal{E}_i|(\log R)^{2k+2l}).$$

(4) In this step, we show that the error is sufficiently small. First, let

$$E(x,q) := \sup_{\substack{(a,q)=1 \\ 1 \le a \le q}} |\Delta(\theta, q, a)|.$$

By the prime number theorem and the estimate $q/\varphi(q) \ll \log\log q$, we have

$$E(x,q) \ll \frac{x \log x}{q} + \frac{x}{\varphi(q)} \ll \frac{x \log x}{q}.$$

Since q is squarefree, we also have $\tau_3(q) = 3^{\Omega(q)}$ and $|\mathcal{C}_i(q)| = \rho_2(q) \le k^{\Omega(q)}$. If we let $a = 3k$ and note that

$$\sum_{\substack{q < x \\ \text{squarefree}}} \frac{a^{\Omega(q)}}{q} \le \prod_{p < x}\left(1 + \frac{a}{p}\right) \le \left(\prod_{p < x}\left(1 + \frac{1}{p}\right)\right)^a \ll (\log x)^a,$$

then

$$\sum_{\substack{q < R^2 \\ q \mid \mathcal{P}_1}} a^{2\Omega(q)} E(x,q) \ll (x \log x) \sum_{\substack{q < R^2 \\ q \mid \mathcal{P}_1}} \frac{a^{2\Omega(q)}}{q} \ll x(\log x)^{1+a^2}.$$

Therefore, using Cauchy–Schwarz and Theorem 5.8, and setting $B = 1+a^2$, we get

$$|\mathcal{E}_i| \le \sum_{\substack{q<R^2 \\ q|\mathscr{P}_1}} \tau_3(q)\tau_k(q)E(x,q)$$

$$\le \left(\sum_{\substack{q<R^2 \\ q|\mathscr{P}_1}} a^{2\Omega(q)} E(x,q) \right)^{\frac{1}{2}} \times \left(\sum_{\substack{q<R^2 \\ q|\mathscr{P}_1}} E(x,q) \right)^{\frac{1}{2}}$$

$$\ll \sqrt{x(\log x)^B} \times \sqrt{\frac{x}{(\log x)^A}} = \frac{x}{(\log x)^{(A-B)/2}}.$$

It follows that

$$|\mathcal{E}_i|(\log x)^{2k+2l} \ll x/(\log x)^A.$$

Using this bound, Steps (1) through (3) and the prime number theorem, we get

$$S_2 = k \cdot \mathcal{T}_2 \cdot x + O\left(\frac{x}{L_x(A)} \right).$$

(5) We are now able to complete the proof. Let the GCD $h = (d,e)$ and rename d,e as d/h and e/h respectively to rewrite $\mathcal{T}_2$ as

$$\mathcal{T}_2 = \sum_{h,d,e|\mathscr{P}_1} \frac{\mu(de)\rho_2(hde)}{\varphi(hde)} g(hd)g(he).$$

Also we have as before (see Section 5.6)

$$\mathcal{T}_2^* := \sum_{h,d,e\in\mathbb{N}} \frac{\mu(de)\rho_2(hde)}{\varphi(hde)} g(hd)g(he).$$

Also recall we have defined a parameter κ_2, which like κ_1 is extremely small, by

$$\kappa_2 := \delta_1(1+4\varpi)(1+\delta_2^2 + (\log 293)k)\binom{k+2l+1}{k-1}.$$

We next follow Steps (1) through (7) of Lemma 5.6 making appropriate amendments and use the second parts of Lemmas 5.2 and 5.4 to get

$$|\mathcal{T}_2 - \mathcal{T}_2^*| < \frac{\kappa_2}{(k+2l+1)!}\binom{2l+2}{l+1}\mathcal{G}(\mathcal{H})L_R(k+2l+1) + o\left(L_x(k+2l+1)\right).$$

By Lemma 5.1, we also have

$$\mathcal{T}_2^* = \frac{1}{(k+2l+1)!}\binom{2l+2}{l+1}\mathcal{G}(\mathcal{H})(\log R)^{k+2l+1} + o\left(L_x(k+2l+1)\right).$$

Therefore,

$$T_2 \geq \frac{1 - \kappa_2}{(k + 2l + 1)!} \binom{2l + 2}{l + 1} \mathscr{G}(\mathscr{H}) L_R(k + 2l + 1) + o\left(L_x(k + 2l + 1)\right).$$

This bound, together with the result of Step (4), implies

$$S_2 \geq \frac{k(1 - \kappa_2)}{(k + 2l + 1)!} \binom{2l + 2}{l + 1} \mathscr{G}(\mathscr{H}) x L_R(k + 2l + 1) + o\left(x L_x(k + 2l + 1)\right).$$

This completes the proof.   □

Remark: In [215, section 2] on page 1125, Zhang makes an observation regarding the error terms based on the $\mathscr{E}_i$. He states that this may be treated using the "well factorable function" idea of Iwaniec. However, as the reader can judge from Step (4) of the proof of Lemma 5.9, this is not needed to bound the error, neither also are the alternative strategies of Polymath, but these could give better insights into what is needed in this step.

5.9 Zhang's Prime Gap Result

The proof of Zhang's main result uses only the two fundamental lemmas, Lemmas 5.6 and 5.9 and the bounds we have derived for the κ's, so it is very short.

Recall the assignments $k = 3.5 \times 10^6$, $l = 180$, $R = x^{\frac{1}{4} + \varpi}$ and

$$\kappa_1 := \delta_1(1 + \delta_2^2 + (\log 293)k) \binom{k + 2l}{k},$$

$$\kappa_1 := \delta_1(1 + 4\varpi)(1 + \delta_2^2 + (\log 293)k) \binom{k + 2l + 1}{k - 1}.$$

Theorem 5.10   (Zhang's prime gap) [215, theorem 1] *If $\mathscr{H}_k$ is admissible with $k = 3.5 \times 10^6$, then there are infinitely many positive integers n such that the translates $n + \mathscr{H}_k$ contain at least two primes. From this, it follows that*

$$\liminf_{n \to \infty}(p_{n+1} - p_n) \leq 7 \times 10^7.$$

Proof **(1)** Using Lemmas 5.6, 5.9 and the identity

$$\log x = \frac{4 \log R}{1 + 4\varpi},$$

and defining the parameter

$$\omega := \frac{k(1 - \kappa_2)}{k + 2l + 1} \binom{2l + 2}{l + 1} - \frac{4(1 + \kappa_1)}{(1 + 4\varpi)(k + 2l)!} \binom{2l}{l}$$

$$= \frac{1}{(k + 2l)!} \binom{2l}{l} \left(\frac{2(2l + 1)}{l + 1} \frac{k(1 - \kappa_2)}{k + 2l + 1} - \frac{4(1 + \kappa_1)}{1 + 4\varpi} \right),$$

we get

$$S_2 - \log(3x)S_1 \geq \omega \mathcal{G}(\mathcal{H})(\log R)^{k+2l+1} + o\left(x(\log x)^{k+2l+1}\right)$$
$$= \omega \mathcal{G}(\mathcal{H})(\log R)^{k+2l+1}(1 + o(1)). \tag{5.6}$$

Now a simple manipulation leads to

$$\frac{\kappa_2}{\kappa_1} = \frac{k(k + 2l + 1)(1 + 4\omega)}{(2l + 1)(2l + 2)} < 9.41 \times 10^7 < 10^8.$$

By Lemma 5.7, we have shown $\kappa_1 < \exp(-1200)$, so κ_2 is also very small. Since

$$\frac{2(2l + 1)k}{(l + 1)(k + 2l + 1)} = \frac{2527 \times 10^6}{633565341} > 3.988 > 3.987 > \frac{1168}{293} = \frac{4}{1 + 4\omega},$$

it follows that $\omega > 0$.

(2) We next note that if a set $\mathcal{H}$ consists of k distinct primes all greater than k, then, because necessarily the zero class is missing for each prime modulus $p \leq k$, it is admissible. We have

$$\pi(7 \times 10^7) - \pi(3.5 \times 10^6) = 3867984 > 3.5 \times 10^6$$

so we can find such an $\mathcal{H}$ choosing the first $k = 3.5 \times 10^6$ primes greater than k. Using this $\mathcal{H}$ in the estimate (5.6), we see there is a translate $n + \mathcal{H}$ in $[x, 2x)$ which contains at least two primes. Thus

$$\liminf_{n \to \infty} p_{n+1} - p_n \leq h_k - h_1 \leq 7 \times 10^7.$$

This completes the proof.   □

5.10 End Notes

The reader may recall the quote from Zhang [215, page 1123] "This result ($p_{n+1} - p_n \leq 7 \times 10^7$ is, of course, not optimal".

A slightly improved Step (2) in Theorem 5.10: we have

$$\pi(6 \times 10^7) - \pi(3 \times 10^6) = 3345299 > 3 \times 10^6$$

so we can find such an $\mathcal{H}$ choosing the first $k = 3 \times 10^6$ primes greater than k. Using this $\mathcal{H}$ in the estimate (5.6), we see there is a translate $n + \mathcal{H}$ in $[x, 2x)$ which contains at least two primes. Thus

$$\liminf_{n \to \infty} p_{n+1} - p_n \leq h_k - h_1 \leq 6 \times 10^7.$$

Of course, this was not the end of the matter. For example, choosing $k = 3 \times 10^4$, $l = 30$ and $\varpi = 1/10$ gives a positive lower bound for ω, namely its leading coefficient. But this value of ϖ is too much to expect. We get positive ω for

$k = 3 \times 10^6$, $l = 130$ and $\varpi = 1/1000$. The reader is invited to experiment with the **PGpack** functions KappaZhang and OmegaZhang.

To this writer's knowledge, the first published improvement to Zhang was by Tim Trudgian [208], who showed the gap could be reduced by a simple modification to the original argument to

$$59874594 < 6 \times 10^7.$$

A more comprehensive analysis of Zhang's approach conducted by János Pintz [167] demonstrated that the arguments would lead to MPZ$[\varpi, \delta]$ whenever

$$828\varpi + 172\delta < 1.$$

This enabled him to find bounded gap sizes of not more than 2530338 with $k = 181000$.

Terence Tao [202] also analysed Zhang's argument and suggested many ways in which the bound 7×10^7 could be improved, for example by tightening bounds in different places, especially those for κ_1 and κ_2, however retaining Zhang's variation of Bombieri–Vinogradov. This was in part to set the stage for the comprehensive work of Polymath8a, before the focus shifted to the Maynard/Tao approach. This "note" [202] is an excellent example of how to critically read a mathematics paper, and it is strongly recommended the reader consider the note alongside the work of Zhang.

Looking ahead in the book, Polymath8a made significant improvements to Zhang's method and results, and these are reported in part in Chapter 8. Meanwhile, Maynard and Tao were moving the subject forward at an even more rapid pace. Maynard was able to reduce the gap size to 600 without using the deep ideas which were needed for example for Zhang's Theorem 5.8, i.e., his variation on Bombieri–Vinogradov's theorem, or the improvements of Polymath8a. Maynard's work on prime gaps is the subject of Chapter 6. Simultaneously and initially independently, Tao and Polymath8b (reported in Chapter 7) were also establishing smaller gap bounds using a completely fresh approach.

The PGpack functions KappaZhang and OmegaZhang relate to the work of this chapter. See the manual entries in Appendix I.

6

Maynard's Radical Simplification

6.1 Introduction

In this chapter, we report on the work of James Maynard (Figure 6.1). We don't completely neglect the work of Terence Tao, who independently arrived at comparable results using a closely related method in many respects. Like Zhang, Maynard studied the work of GPY and GMPY closely, but adopted and developed an old idea of Selberg for the sieve weights [194, page 245]. Recall $P(n) = (n + h_1) \cdots (n + h_k)$. Instead of the now classical choice,

$$w(n) := \left(\sum_{\substack{d \mid P(n) \\ d \leq R}} \lambda_d \right)^2,$$

he developed the properties and then used multidimensional/multidivisor weights

$$w(n) := \left(\sum_{\substack{d, \ldots, d_k \\ \forall j \; d_j \mid n + h_j \\ d_1 \cdots d_k < R}} \lambda_{d_1, \ldots, d_k} \right)^2.$$

He considered this choice of weights gave added flexibility in applying the sieve, and would result in "arbitrarily large" improvements, provided k was sufficiently large. Note that Goldston and Yildirim [71] used similar weights, originally it appears suggested by Selberg. Although technically more complicated, these weights enabled him to not only reduce the prime gap to 600, without using any variation or extension to Bombieri–Vinogradov's theorem, but also to obtain a bound, parameterized in terms of m, for the general gap $p_{m+n} - p_n$ for any $m \geq 1$! In other words, for each $m \in \mathbb{N}$, there is a bounded interval I_m with an infinite number of translates each containing at least $m + 1$ primes. In addition, he obtained bounded gaps between primes under any level of distribution of the primes ϑ with $0 < \vartheta < 1$.

Maynard was born in Chelmsford, England (about 50 kilometres northeast of London) in 1987 and attended King Edward VI Grammar School. He completed his Master's degree at nearby Cambridge in 2009 and his Ph.D. from Oxford in 2013. His supervisor was the great number theorist Roger Heath-Brown and his thesis entitled "Topics in Analytic Number Theory". In his thesis, he made significant advances regarding the value of the explicit constant in the Brun–Titchmarsh inequality [135]. Among other discoveries, he showed that there are infinitely many n such that $[n, n + 90]$ contains two primes and an almost prime with at most four prime factors, assuming numbers with up to four prime factors have a "level of distribution" at least 0.99.

Figure 6.1 James Maynard (1987–). Courtesy of James Maynard and the American Mathematical Society.

He is the youngest by far of the mathematicians whose work is reported in this book. He has received the Sastra Ramanujan Prize, the Whitehead Prize and the EMS Prize, and presented an invited lecture at the International Congress of Mathematicians (ICM) 2018 in Rio de Janeiro.

Terence Tao, independently and at about the same time as James Maynard, also decided to study the bounded gaps between primes problem using a multidimensional version of Selberg's sieve. His method was based on the Fourier transform and Riemann zeta function, and a description of the method is included in the Polymath8b article [175, lemma 30, theorems 19, 20]. See also Chapter 7 in this book where the method is reported in detail. Sometimes people refer to one or more of the results in this chapter as the "Maynard/Tao" theorem(s). There is a very nice overview of Maynard's work and the very positive interactions between Tao and Maynard, fostered by Andrew Granville, in Vicky Neal's book [154, chapter 13].

6.2 Definitions

Let $N \in \mathbb{N}$ be sufficiently large. In what follows, we will have $N \to \infty$. Let

$$D_0 := \text{logloglog } N \text{ and } W := \prod_{p \leq D_0} p,$$

so W is a product of very small primes at scale N. The constant R will be a small (certainly less than $1/2$) power of N. The integer k has $k \geq 2$, and could be quite large but should be considered fixed.

Let F be a fixed real-valued piecewise differentiable function supported on the simplex

$$\mathcal{R}_k := \left\{ (x_1, \ldots, x_k) \in [0,1]^k : \sum_{i=1}^{k} x_i \leq 1 \right\},$$

and such that $\int_{\mathcal{R}_k} F^2 > 0$. The set of all such functions will be denoted $\mathcal{S}_k$. The precise definition of piecewise differentiable in this context is not important – the functions used finally will be explicit.

When $(\prod_{i=1}^{k} d_i, W) = 1$, define the sieve lambdas $\lambda_{\mathbf{d}}$ by

$$\lambda_{d_1, \ldots, d_k} := \left(\prod_{i=1}^{k} \mu(d_i) d_i \right) \sum_{\substack{r_1, \ldots, r_k \\ \forall i \ d_i | r_i \\ \forall i \ (r_i, W)=1}} \frac{\mu(\prod_{i=1}^{k} r_i)^2}{\prod_{i=1}^{k} \varphi(r_i)} F \left(\frac{\log r_1}{\log R}, \ldots, \frac{\log r_k}{\log R} \right).$$

Let $d = \prod_{1 \leq i \leq k} d_i$. When $(\prod_{i=1}^{k} d_i, W) \neq 1$, set $\lambda_{d_1, \ldots, d_k} = 0$. Note also from the definition that $\lambda_{\mathbf{d}} = 0$ whenever any of the d_i is not squarefree, and we apply the additional constraint that $(d_i, d_j) = 1$ for all $i \neq j$ (i.e., $\mu(d)^2 = 1$). Note also that because F has compact support, there are at most a finite number of terms in these sums, and because the support of F is in the simplex $\mathcal{R}_k$ we get $\prod_{i=1}^{k} r_i \leq R$ is necessary for the corresponding term in the sum for $\lambda_{\mathbf{d}}$ to be nonzero.

Maynard offers extensive motivation for this choice of λ's, which are fundamental to his method. He also implies that their general shape is rather insensitive to the leading coefficients, apart from the $\mu(d_i)$ of course.

We say the primes have **a level of distribution** θ if for any given $A > 0$ and $\epsilon > 0$,

$$\sum_{q \leq x^{\theta - \epsilon}} \max_{(a,q)=1} \left| \pi(x,q,a) - \frac{\pi(x)}{\varphi(q)} \right| \ll_{A,\epsilon} \frac{x}{(\log x)^A}.$$

Then, by the Bombieri–Vinogradov theorem, unconditionally, we can choose any $\theta \leq \frac{1}{2}$. (Note this definition is common but different from that used by some other authors, so the expression "level of distribution" should be used with care.)

We can now define the two integral forms, and quantities in terms of those forms, which are fundamental to the expression of Maynard's results:

$$I_k(F) := \int_0^1 \cdots \int_0^1 F(t_1, \ldots, t_k)^2 \, dt_1 \ldots dt_k,$$

$$J_k^{(m)}(F) := \int_0^1 \cdots \int_0^1 \left(\int_0^1 F(t_1, \ldots, t_k) \, dt_m \right)^2 dt_1 \ldots dt_{m-1} dt_{m+1} \ldots dt_k.$$

Then let

$$M_k := \sup_{F \in \mathscr{S}_k} \frac{\sum_{m=1}^k J_k^{(m)}(F)}{I_k(F)} \quad \text{and} \quad r_k := \left\lceil \frac{\theta M_k}{2} \right\rceil.$$

Maynard also makes a preliminary restriction of the integers n which appear in sums so that all lie in a fixed nonzero congruence class v_0 modulo W. This ensures that the integers in $n + \mathscr{H}$ do not have small prime divisors – these would further complicate formulas. Then, for admissible $\mathscr{H}$ we have for $1 \leq i \leq k$, $(v_0 + h_i, W) = 1$: to construct such a v_0, choose for each $p \mid W$ a missing class representative a_p and use the Chinese Remainder Theorem to construct $v_0 \equiv -a_p \mod p$ for each such p. Then if $p \mid W$ and for some i we had $p \mid v_0 + h_i$, we would have $p \mid -a_p + h_i$ so $h_i \equiv a_p \mod p$, which is false. It follows that if $n \equiv v_0 \mod W$, then $(n + h_i, W) = 1$ also, so each $n + h_i$ will have no prime divisors less than or equal to $\log\log\log N$. This technique is sometimes referred to as the "W trick".

Using v_0, we define four sums. These are the subject of the fundamental lemmas, which are stated and proved in Section 6.6:

$$W = \prod_{p \leq \log\log\log N} p,$$

$$S_1 := \sum_{\substack{N \leq n < 2N \\ n \equiv v_0 \bmod W}} \left(\sum_{\forall i \; d_i \mid n + h_i} \lambda_{d_1, \ldots, d_k} \right)^2, \tag{6.1}$$

$$S_2 := \sum_{\substack{N \le n < 2N \\ n \equiv v_0 \bmod W}} \left(\sum_{i=1}^{k} \chi_{\mathbb{P}}(n + h_i) \right) \times \left(\sum_{\forall i \; d_i | n + h_i} \lambda_{d_1, \dots, d_k} \right)^2, \qquad (6.2)$$

$$S_2^{(m)} := \sum_{\substack{N \le n < 2N \\ n \equiv v_0 \bmod W}} \chi_{\mathbb{P}}(n + h_m) \left(\sum_{\forall i \; d_i | n + h_i} \lambda_{d_1, \dots, d_k} \right)^2, \qquad (6.3)$$

$$w(n) := \left(\sum_{\substack{d_i | n + h_i \\ n \equiv v_0 \bmod W}} \lambda(d_1, \dots, d_k) \right)^2. \qquad (6.4)$$

As a direct consequence, we have $S_2 = \sum_{m=1}^{k} S_2^{(m)}$.

We use the notation $L_X(Y) = (\log X)^Y$ for X, Y real and $X > 0$.

6.3 Chapter Summary

The three Sections 6.4, 6.5 and 6.6 are all devoted to Maynard's development of properties of the multidimensional Selberg sieve. This culminates in the summary Lemma 6.8, which gives asymptotic expressions for S_1 and S_2. We see this in Figure 6.2. In Figure 6.3, we see how the main results are derived from Lemma 6.8.

Lemma 6.8 *[138, proposition 4.1] If F is such that $I_k(F) \ne 0$ and for all m with $1 \le m \le k$, $J_k^{(m)}(F) \ne 0$, $R = N^{\theta/2 - \epsilon}$ for some small $\epsilon > 0$ and level of distribution of primes θ, then*

$$S_1 = \frac{\varphi(W)^k N L_R(k)(1 + o(1))}{W^{k+1}} I_k(F),$$

$$S_2 = \frac{\varphi(W)^k N L_R(k + 1)(1 + o(1))}{W^{k+1} \log N} \sum_{m=1}^{k} J_k^{(m)}(F),$$

$$S := S_2 - \rho S_1 = \sum_{\substack{N \le n < 2N \\ n \equiv v_0 \bmod W}} \left(\left(\sum_{i=1}^{k} \chi_{\mathbb{P}}(n + h_i) \right) - \rho \right) w_n.$$

Here we see how Maynard links with the other contributors. If $S > 0$ for some $\rho \ge 1$, then there is an $n \in [N, 2N)$ with

$$\sum_{i=1}^{k} \chi_{\mathbb{P}}(n + h_i) > \rho \ge 1,$$

so at least two of the $n + h_i$ are prime and their distance apart will be not greater than the width of the k-tuple. In Lemma 6.9, we see ρ is chosen just slightly less than $\theta M_k/2$.

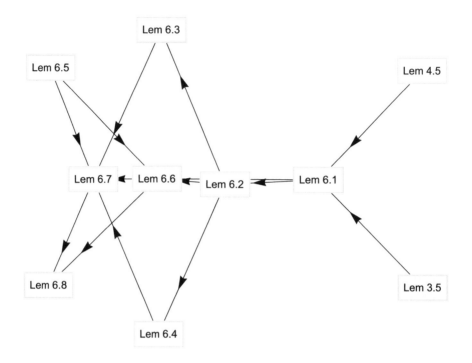

Figure 6.2 Selberg sieve lemmas for Chapter 6.

The reader will note from the flow diagram of dependencies, Figure 6.2, the determining role played by Lemma 6.5. As proved in this chapter, it is a compacted form of the proof of a result taken from the first few chapters of [88]. It is used by Maynard in an essential way and does not appear in the work of GPY, GMPY, Zhang or Polymath on prime gaps.

The final fundamental result is really a theorem, since some of Maynard's main results flow from it as virtual corollaries:

Lemma 6.11   (lower bounds for M_k) *[138, proposition 4.3] For $k \in \mathbb{N}$, we have*

(i) $M_5 > 2$,
(ii) $M_{105} > 4$,
(iii) *for all k sufficiently large $M_k > \log k - 2\log\log k - 1$.*

We show that Maynard's choice of the constant -2 in (iii) is optimal for lower bounds of the form $\log k - \alpha \log\log k$.

In Section 6.7, a number of useful integration formulas are derived. These find their way into improving the efficiency of computations, many of which are incorporated in the PGpack function RayleighQuotientMaynard.

Section 6.9 gives detailed proofs of Maynard's main theorems on small prime gaps. The first is an improvement of the theorem of Polymath8a of 2013, which gave an upper bound of 4680, following a series of improvements to Zhang's 70×10^6:

Theorem 6.15 (**Maynard's prime gap**) *[138, theorem 1.3]*

$$\liminf_{n \to \infty}(p_{n+1} - p_n) \le 600.$$

As discussed, many authors use the Elliott–Halberstam conjecture to see just how far their methods could presumably take us, assuming of course we found some way to turn the conjecture into a theorem. To repeat, this is most unlikely to be achieved in the foreseeable future!

Theorem 6.16 (**Maynard's conditional prime gap**) *[138, theorem 1.4] Assume the Elliott–Halberstam conjecture in the form that the primes have a level of distribution θ for all $0 < \theta < 1$. Then*

$$(i) \liminf_{n \to \infty}(p_{n+1} - p_n) \le 12, \ and \ (ii) \liminf_{n \to \infty}(p_{n+2} - p_n) \le 600.$$

The next result of Maynard is quite remarkable. Not only does it include bounded gaps between primes but shows that we get bounded gaps for an arbitrary preassigned finite number of primes.

Theorem 6.17 (**Maynard's consecutive primes gap**) *[138, theorem 1.1] For all $m \in \mathbb{N}$, there is an absolute constant such that*

$$\liminf_{n \to \infty}(p_{n+m} - p_n) \ll m^3 e^{4m}.$$

We show that the constants 3 and 4 in Maynard's choice of upper bound are optimal amongst bounds of the given form.

Finally, using great combinatorial counting, Maynard showed that prime m-tuples have positive relative density.

Theorem 6.18 (**Maynard's positive relative density**) *[138, theorem 1.2] Let $m \in \mathbb{N}$ be fixed and $r > m$ sufficiently large depending on m, and let $\mathscr{A} = \{a_1, a_2, \ldots, a_r\}$ be any given set of **any** r distinct integers. Then*

$$\frac{\#\{\{h_1, \ldots, h_m\} \subset \mathscr{A} : all \ of \ n + h_j \ are \ prime \ for \ infinitely \ many \ n\}}{\#\{\{h_1, \ldots, h_m\} \subset \mathscr{A}\}} \gg_m 1,$$

where the implied constant depends on m but not on $\mathscr{A}$.

Recall that the flow graph Figure 6.3 gives the main interrelationships of Maynard's results. Note also the central role played by Lemma 6.11 stated earlier in this section.

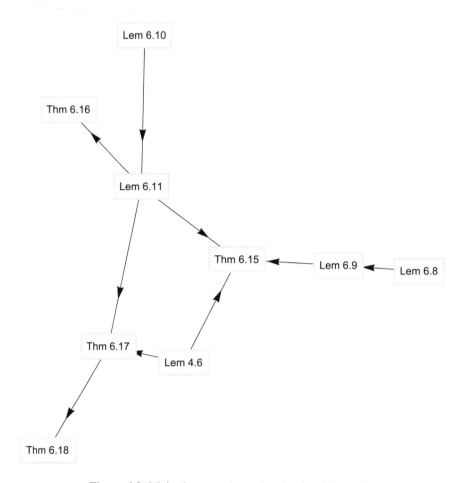

Figure 6.3 Main theorem dependencies for Chapter 6.

6.4 Selberg's Sieve Lemmas

We begin with two lemmas providing estimates for some arithmetic sums. Recall from Chapter 4 the definition $\tau_k(u) := k^{\omega(u)}$ for u squarefree.

Lemma 6.1 *For fixed $k, W \in \mathbb{N}$, we have as $x \to \infty$*

$$(i) \quad \sum_{\substack{u \le x \\ \text{squarefree}}} \frac{\tau_k(u)}{\varphi(u)} \ll (\log x)^k,$$

$$(ii) \quad \sum_{\substack{u \leq x \\ \text{squarefree} \\ (u, W)=1}} \frac{1}{\varphi(u)} \ll \frac{\varphi(W) \log x}{W}.$$

Proof (i) We have (compare Lemma 4.5)

$$\sum_{\substack{u \leq x \\ \text{squarefree}}} \frac{\tau_k(u)}{\varphi(u)} = \sum_{\substack{u \leq x \\ \text{squarefree}}} \frac{k^{\omega(u)}}{\varphi(u)}$$

$$\leq \sum_{u \leq x} \prod_{p|u} \left(1 + \frac{k}{p-1}\right)$$

$$\leq \sum_{u \leq x} \prod_{p|u} \left(1 + \frac{1}{p-1}\right)^k$$

$$= \sum_{u \leq x} \frac{u^k}{\varphi(u)^k} \ll \sum_{u \leq x} (\log\log u)^k$$

$$\ll \left(\sum_{u \leq x} \log\log u\right)^k \ll (\log x)^k.$$

(ii) Using Lemma 3.5 (see also [142]), we get

$$\sum_{\substack{n \leq x \\ (n, W)=1}} \frac{\mu(n)^2}{\varphi(n)} = \sum_{n \leq x} \frac{\mu(n)^2}{\varphi(n)} \sum_{d|(n, W)} \mu(d)$$

$$= \sum_{\substack{d \leq x \\ e \leq x/d \\ d|W}} \frac{\mu(de)^2 \mu(d)}{\varphi(d)\varphi(e)}$$

$$\leq \sum_{d|W} \frac{\mu(d)}{\varphi(d)} \sum_{e \leq x} \frac{\mu(e)^2}{\varphi(e)}$$

$$\leq \frac{\varphi(W)}{W} \sum_{e \leq x} \frac{1}{\varphi(e)}$$

$$\ll \frac{\varphi(W)}{W} (\log x). \qquad \square$$

The next two lemmas derive good expressions for S_1 and $S_2^{(m)}$ and go to the heart of Maynard's method. They are not to be taken lightly and require seven and six steps respectively.

Recall the definition of the sieve weights. Let $d = \prod_{i=1}^{k} d_i$. When $(d, W) = 1$ and $\mu(d)^2 = 1$ (so $(d_i, d_j) = 1$ for all $i \neq j$), we set

$$\lambda_{d_1,\ldots,d_k} := \left(\prod_{i=1}^{k} \mu(d_i)d_i \right) \sum_{\substack{r_1,\ldots,r_k \\ \forall i \; d_i|r_i \\ \forall i \; (r_i, W)=1}} \frac{\mu(\prod_{i=1}^{k} r_i)^2}{\prod_{i=1}^{k} \varphi(r_i)} F\left(\frac{\log r_1}{\log R}, \ldots, \frac{\log r_k}{\log R} \right).$$

When $(\prod_{i=1}^{k} d_i, W) \neq 1$, or the d_i are not coprime, then we set $\lambda_{d_1,\ldots,d_k} = 0$.

Now this looks quite complicated, but the heart of the method is in the function F, which is defined later in such a way that $F(x) = 0$ for all x other than those in the simplex $0 \leq x_j$ and $x_1 + \cdots + x_k \leq 1$, as well as ensuring that the r_j are not divisible by small primes, these constraints make derivations simpler later.

Lemma 6.2 *[138, lemma 5.1] Let θ be the level of distribution of primes, $R = N^{\theta/2-\epsilon}$, $R^2 \leq N^{1-2\epsilon}$, $W \ll N^\epsilon$ and $D_0 = \log\log\log N$. Define for $r_i \in \mathbb{N}$,*

$$y_{r_1,\ldots,r_k} := \left(\prod_{i=1}^{k} \mu(r_i)\varphi(r_i) \right) \sum_{\substack{d_1,\ldots,d_k \\ \forall i \; r_i|d_i}} \frac{\lambda_{d_1,\ldots,d_k}}{\prod_{i=1}^{k} d_i}$$

and $y_{max} := \sup_{r_1,\ldots,r_k} |y_{r_1,\ldots,r_k}|$. Then

$$S_1 = \frac{N}{W} \sum_{r_1,\ldots,r_k} \frac{y_{r_1,\ldots,r_k}^2}{\prod_{i=1}^{k} \varphi(r_i)} + O\left(\frac{y_{max}^2 N L_R(k)}{W D_0} \right).$$

Proof **(1)** Expanding the square in the definition of S_1 and interchanging the order of summation gives

$$S_1 = \sum_{\substack{N \leq n < 2N \\ n \equiv v_0 \bmod W}} \left(\sum_{\forall i \; d_i|n+h_i} \lambda_{d_1,\ldots,d_k} \right)^2 = \sum_{\substack{d_1,\ldots,d_k \\ e_1,\ldots,e_k}} \lambda_{d_1,\ldots,d_k} \lambda_{e_1,\ldots,e_k} \sum_{\substack{N \leq n < 2N \\ n \equiv v_0 \bmod W \\ \forall i \; [d_i,e_i]|n+h_i}} 1.$$

The Chinese Remainder Theorem, when the integers $W, [d_1, e_1], \ldots, [d_k, e_k]$ are pairwise coprime, shows that the inner sum in this expression can be written as a sum over a single residue class modulo

$$q := W \prod_{i=1}^{k} [d_i, e_i],$$

and in this case the inner sum will be $N/q + O(1)$. If these integers are not coprime, because of the way the weights have been defined, we **claim** the inner sum is zero. To see this, first note from the definition of the weights that $(W, d_i) = (W, e_i) = 1$ so $(W, [d_i, e_i]) = 1$ for all i.

If for a distinct pair of indices $([d_i, e_i], [d_j, e_j]) \neq 1$, then since $(d_i, d_j) = (e_i, e_j) = 1$, we have for some prime p, $p \mid (d_i, e_j)$ or $p \mid (d_j, e_i)$. Suppose the former is true. Then $p \mid (n + h_i, n + h_j)$ so $p \mid h_i - h_j$. But for N sufficiently large, we then have $p \mid W$ also, so $(W, d_j) \neq 1$, which is false. Thus each of the $W, [d_i, e_i], 1 \leq i \leq k$ are coprime.

Thus, if $\sum^{\flat}$ means the vectors of integers over which the sum is being taken are restricted to having coprime components and corresponding components, we get

$$S_1 = \frac{N}{W} \sum_{\substack{d_1,\ldots,d_k \\ e_1,\ldots,e_k}}^{\flat} \frac{\lambda_{d_1,\ldots,d_k} \lambda_{e_1,\ldots,e_k}}{\prod_{i=1}^{k}[d_i, e_i]} + O\left(\sum_{\substack{d_1,\ldots,d_k \\ e_1,\ldots,e_k}}^{\flat} |\lambda_{d_1,\ldots,d_k} \lambda_{e_1,\ldots,e_k}| \right).$$

Define $\lambda_{max} = \sup_{d_1,\ldots,d_k} |\lambda_{d_1,\ldots,d_k}|$. Because, from the support of F, $\lambda_{d_1,\ldots,d_k}$ is zero when $\prod_{i=1}^{k} d_i \geq R$, using (4.26), the error term can be expressed as

$$O\left(\lambda_{max}^2 \left(\sum_{d<R} \tau_k(d) \right)^2 \right) = O\left(\lambda_{max}^2 R^2 (\log^{2k} R) \right).$$

(2) Next, we note that because

$$n = \sum_{d|n} \varphi(d) \implies \frac{1}{[d_i, e_i]} = \frac{1}{d_i e_i} \sum_{u_i | (d_i, e_i)} \varphi(u_i),$$

we can rewrite the main term in the expression for S_1 as

$$\frac{N}{W} \sum_{u_1,\ldots,u_k} \prod_{i=1}^{k} \varphi(u_i) \sum_{\substack{d_1,\ldots,d_k \\ e_1,\ldots,e_k \\ \forall i \ u_i | (e_i, d_i)}}^{\flat} \frac{\lambda_{d_1,\ldots,d_k} \lambda_{e_1,\ldots,e_k}}{\prod_{i=1}^{k} d_i \prod_{i=1}^{k} e_i}.$$

Because $\lambda_{d_1,\ldots,d_k}$ is zero unless $(d_i, d_j) = 1$ for all $i \neq j$ and $(W, d_i) = 1$ for all i, we automatically satisfy $(W, [d_i, e_i]) = 1$ for the summation, as we have seen in Step (1). We may also drop the requirement that the variables d_i be pairwise coprime and the variables e_i be pairwise coprime. So all we need to insist on explicitly to obtain $W, [d_1, e_1], \ldots, [d_k, e_k]$ are all coprime is that we have $(d_i, e_j) = 1$ for all $i \neq j$.

(3) Because $\sum_{d|n} \mu(d) = 0$ if $n > 1$ and $\mu(1) = 1$, we can rewrite the main terms as

$$\frac{N}{W} \sum_{u_1,\ldots,u_k} \prod_{i=1}^{k} \varphi(u_i) \sum_{s_{1,2},\ldots,s_{k,k-1}} \left(\prod_{\substack{1 \leq i, j \leq k \\ i \neq j}} \mu(s_{i,j}) \right) \sum_{\substack{d_1,\ldots,d_k \\ e_1,\ldots,e_k \\ \forall i \ u_i | (e_i, d_i) \\ \forall i \neq j \ s_{i,j} | (e_i, d_j)}} \frac{\lambda_{d_1,\ldots,d_k} \lambda_{e_1,\ldots,e_k}}{\prod_{i=1}^{k} d_i \prod_{i=1}^{k} e_i}.$$

$$(6.5)$$

Note that we can constrain the $s_{i,j}$, the divisors of (e_i, d_j), to be coprime to the u_i because the terms when this is false, say $p \mid (s_{i,j}, u_i)$, since $s_{i,j} \mid (e_i, d_j)$ gives $p \mid (d_i, d_j)$ so $(d_i, d_j) \neq 1$. Thus such terms make a zero contribution to the sum. The same applies to the $s_{i,j}$ in that we can insist $(s_{i,j}, s_{i,a}) = (s_{i,j}, s_{b,j}) = 1$ for all $a \neq j$ and $b \neq i$. Sums satisfying all these constraints are denoted $\sum^{\sharp}$.

(4) Next, we introduce the key change of variables which will simplify the sums and their estimations:

$$y_{r_1,\dots,r_k} := \prod_{i=1}^{k} \mu(r_i) \varphi(r_i) \sum_{\substack{d_1,\dots,d_k \\ \forall i \ r_i \mid d_i}} \frac{\lambda_{d_1,\dots,d_k}}{\prod_{i=1}^{k} d_i}. \tag{6.6}$$

Note that this shows the y variables are supported on r_i such that $r = \prod_{i=1}^{k} r_i$ is squarefree. A straightforward derivation shows we can express the λ variables in terms of the y variables. To see this, if we are given d_i such that $\prod_{i=1}^{k} d_i$ is squarefree, then

$$\sum_{\substack{r_1,\dots,r_k \\ \forall i \ d_i \mid r_i}} \frac{y_{r_1,\dots,r_k}}{\prod_{i=1}^{k} \varphi(r_i)} = \sum_{\substack{r_1,\dots,r_k \\ \forall i \ d_i \mid r_i}} \prod_{i=1}^{k} \mu(r_i) \sum_{\substack{e_1,\dots,e_k \\ \forall i \ r_i \mid e_i}} \frac{\lambda_{e_1,\dots,e_k}}{\prod_{i=1}^{k} e_i}$$

$$= \sum_{\substack{e_1,\dots,e_k}} \frac{\lambda_{e_1,\dots,e_k}}{\prod_{i=1}^{k} e_i} \sum_{\substack{r_1,\dots,r_k \\ \forall i \ d_i \mid r_i \\ \forall i \ r_i \mid e_i}} \prod_{i=1}^{k} \mu(r_i)$$

$$= \sum_{\substack{e_1,\dots,e_k}} \frac{\lambda_{e_1,\dots,e_k}}{\prod_{i=1}^{k} e_i} \prod_{i=1}^{k} \mu(d_i) \sum_{\substack{r_1,\dots,r_k \\ \forall i \ d_i \mid r_i \\ \forall i \ r_i/d_i \mid e_i/d_i}} \mu(r_i/d_i)$$

$$= \frac{\lambda_{d_1,\dots,d_k}}{\prod_{i=1}^{k} d_i \mu(d_i)},$$

since the internal sum in the penultimate line is 0 unless $e_i = d_i$ for all i. It thus follows that any assignment of the $y_{r_1,\dots,r_k}$ with $r = \prod_{i=1}^{k} r_i$ squarefree and satisfying $(r, W) = 1$ and $r < R$ will, using this formula, give a suitable set of values for the $\lambda_{d_1,\dots,d_k}$.

(5) In this step, we rewrite the error terms using y_{max}. Recall we have defined $y_{max} := \sup_{r_1,\dots,r_k} |y_{r_1,\dots,r_k}|$ in the lemma statement. Because when d is squarefree, we have

$$\frac{d}{\varphi(d)} = \prod_{p \mid d} \frac{p}{p-1} = \prod_{p \mid d} \left(1 + \frac{1}{\varphi(p)}\right) = \sum_{e \mid d} \frac{1}{\varphi(e)},$$

defining $r' := \prod_{i=1}^{k} r_i/d_i$, and using $u := dr'$ and $\tau_k(r') \leq \tau_k(dr')$ we get from the change of variables equation derived in Step (4), changing the order of summation, using r and d are squarefree to get $(r',d) = 1$ and using Lemma 6.1(a) and the independence of the sum over u of the d_i to get the last line,

$$
\lambda_{max} \leq \sup_{\substack{d_1,\ldots,d_k \\ \prod_{i=1}^{k} d_i \text{ squarefree}}} y_{max} \prod_{i=1}^{k} d_i \sum_{\substack{r_1,\ldots,r_k \\ \forall i \ d_i|r_i \\ \prod_{i=1}^{k} r_i < R \\ \prod_{i=1}^{k} d_i \text{ squarefree}}} \prod_{i=1}^{k} \frac{\mu(r_i)^2}{\varphi(r_i)}
$$

$$
\leq y_{max} \sup_{\substack{d_1,\ldots,d_k \\ \prod_{i=1}^{k} d_i \text{ squarefree}}} \prod_{i=1}^{k} \frac{d_i}{\varphi(d_i)} \sum_{\substack{r' < R/\prod_{i=1}^{k} d_i \\ (r', \prod_{i=1}^{k} d_i)=1}} \frac{\mu(r')^2 \tau_k(r')}{\varphi(r')}
$$

$$
\leq y_{max} \sup_{d|\prod_{i=1}^{k} d_i} \sum \frac{\mu(d)^2}{\varphi(d)} \sum_{\substack{r' < R/\prod_{i=1}^{k} d_i \\ (r', \prod_{i=1}^{k} d_i)=1}} \frac{\mu(r')^2 \tau_k(dr')}{\varphi(r')}
$$

$$
\leq y_{max} \sum_{u < R} \frac{\mu(u)^2 \tau_k(u)}{\varphi(u)}
$$

$$
\ll y_{max} L_R(k).
$$

Therefore,

$$
\lambda_{max} \ll y_{max} L_R(k). \tag{6.7}
$$

This enables us to estimate the error term derived in Step (1) as

$$
O\left(\lambda_{max}^2 R^2 L_R(2k)\right) = O\left(y_{max}^2 R^2 L_R(4k)\right).
$$

(6) Next, we rewrite the main term as a function of the y variables. Using (6.6) in (6.5) and the revised error term from Step (5), setting

$$
a_j := u_j \prod_{i \neq j} s_{j,i} \quad \text{and} \quad b_j := u_j \prod_{i \neq j} s_{i,j}
$$

and requiring the $s_{i,j}$ to be coprime with the other factors of a_j and b_j, we get

$$
S_1 = \frac{N}{W} \sum_{u_1,\ldots,u_k} \prod_{i=1}^{k} \varphi(u_i) \sum_{s_{1,2},\ldots,s_{k,k-1}}^{\sharp} \left(\prod_{\substack{1 \leq i,j \leq k \\ i \neq j}} \mu(s_{i,j}) \right) \left(\prod_{i=1}^{k} \frac{\mu(a_i)\mu(b_i)}{\varphi(a_i)\varphi(b_i)} \right)
$$

$$
\times y_{a_1,\ldots,a_k} y_{b_1,\ldots,b_k} + O\left(y_{max}^2 R^2 L_R(4k)\right).
$$

And then, since we can assume a_j, b_j are squarefree and μ is multiplicative, we can write, referring to Step (3),

$$\mu(a_j) = \mu(u_j) \prod_{i \neq j} \mu(s_{j,i}) \text{ and } \mu(b_j) = \mu(u_j) \prod_{i \neq j} \mu(s_{i,j}),$$

so

$$S_1 = \frac{N}{W} \sum_{u_1,\ldots,u_k} \prod_{i=1}^{k} \frac{\mu(u_i)^2}{\varphi(u_i)} \sum_{s_{1,2},\ldots,s_{k,k-1}}^{\sharp} \prod_{\substack{1 \leq i, j \leq k \\ i \neq j}} \frac{\mu(s_{i,j})}{\varphi(s_{i,j})^2} y_{a_1,\ldots,a_k} y_{b_1,\ldots,b_k}$$

$$+ O\left(y_{max}^2 R^2 L_R(4k)\right).$$

(7) In this last step, we see that, because there is zero contribution when $(s_{i,j}, W) \neq 1$, we need only consider the $s_{i,j} > D_0$ or $s_{i,j} = 1$. In case $s_{i,j} > D_0$, using Lemma 6.1 (b) and $\varphi(n) \geq \sqrt{n}$ for composite n and summing over primes separately, the main term contributes

$$O\left(\frac{y_{max}^2 N}{W} \left(\sum_{\substack{u < R \\ (u, W) = 1}} \frac{\mu(u)^2}{\varphi(u)}\right)^k \left(\sum_{s_{i,j} > D_0} \frac{\mu(s_{i,j})^2}{\varphi(s_{i,j})^2}\right) \left(\sum_{1 \leq s} \frac{\mu(s)^2}{\varphi(s)^2}\right)^{k^2-k-1}\right)$$

$$\ll \frac{y_{max}^2 \varphi(W)^k N L_R(k)}{W^{k+1} D_0}.$$

Therefore, taking only the terms for the main term with $s_{i,j} = 1$ for all $i \neq j$, and using Lemma 6.1, gives

$$S_1 = \frac{N}{W} \sum_{u_1,\ldots,u_k} \frac{y_{u_1,\ldots,u_k}^2}{\prod_{i=1}^{k} \varphi(u_i)} + O\left(\frac{y_{max}^2 \varphi(W)^k N L_R(k)}{W^{k+1} D_0}\right) + O\left(y_{max}^2 R^2 L_R(4k)\right).$$

Finally, since we have $R^2 = N^{\theta-2\epsilon} \leq N^{1-2\epsilon}$, $W \ll N^\epsilon$ and $D_0 = \log\log\log N$, the error is dominated by the first error terms and we get

$$S_1 = \frac{N}{W} \sum_{u_1,\ldots,u_k} \frac{y_{u_1,\ldots,u_k}}{\prod_{i=1}^{k} \varphi(u_i)} + O\left(\frac{y_{max}^2 N L_R(k)}{W D_0}\right),$$

which completes the proof.   □

The derivation of an asymptotic expression for $S_2^{(m)}$ in terms of the y variables is similar once Step (1) is established. A novel feature is the use of $g(p) = p - 2$ rather than $\varphi(p) = p - 1$. Recall the definitions where N is large:

$$D_0 := \log\log\log N,$$

$$W := \prod_{p \leq D_0} p,$$

$$S_2^{(m)} := \sum_{\substack{N \leq n < 2N \\ n \equiv v_0 \bmod W}} \chi_{\mathbb{P}}(n + h_m) \left(\sum_{\forall i \ d_i | n + h_i} \lambda_{d_1,\dots,d_k} \right)^2.$$

Lemma 6.3 *[138, lemma 5.2] Let* $R^2 = N^{\theta-2\epsilon} \leq N^{1-2\epsilon}$, $W \ll N^\epsilon$ *and* $D_0 = \log\log\log N$. *Define a completely multiplicative function* $g \colon \mathbb{N} \to \mathbb{Z}$ *on primes by* $g(p) = p - 2$, *and let* $1 \leq m \leq k$. *Set for* $r_i \in \mathbb{N}$,

$$y_{r_1,\dots,r_k}^{(m)} := \left(\prod_{i=1}^k \mu(r_i) g(r_i) \right) \sum_{\substack{d_1,\dots,d_k \\ \forall\, i\ r_i | d_i \\ d_m = 1}} \frac{\lambda_{d_1,\dots,d_k}}{\prod_{i=1}^k \varphi(d_i)}.$$

Let $y_{max}^{(m)} := \sup_{r_1,\dots,r_k} |y_{r_1,\dots,r_k}^{(m)}|$. *Then for fixed real* $A > 0$, *we have the estimate*

$$S_2^{(m)} = \frac{N}{\varphi(W) \log N} \sum_{r_1,\dots,r_k} \frac{(y_{r_1,\dots,r_k}^{(m)})^2}{\prod_{i=1}^k g(r_i)} + O\left(\frac{(y_{max}^{(m)})^2 \varphi(W)^{k-2} N L_N(k-2)}{W^{k-1} D_0} \right)$$

$$+ O\left(\frac{y_{max}^2 N}{L_N(A)} \right).$$

Proof **(1)** First, consider the definition of $S_2^{(m)}$. Expand and interchange the order of summation to get

$$S_2^{(m)} = \sum_{\substack{d_1,\dots,d_k \\ e_1,\dots,e_k}} \lambda_{d_1,\dots,d_k} \lambda_{e_1,\dots,e_k} \sum_{\substack{N \leq n < 2N \\ n \equiv v_0 \bmod W \\ \forall i\ [d_i,e_i] | n + h_i}} \chi_{\mathbb{P}}(n + h_m).$$

The proof follows a similar path to that of Lemma 6.2. If the numbers $W, [d_1, e_1], \dots$ $[d_k, e_k]$ are pairwise coprime, then, using the Chinese Remainder Theorem, the inner sum in this expression can be written as a sum over a residue class modulo $q := W \prod_{i=1}^k [d_i, e_i]$. We **claim** the integer $n + h_m$ is in a class, represented by a say, coprime to this modulus if and only if $d_m = e_m = 1$. To see this, on the one hand, since $[d_m, h_m] \mid n + h_m$, we have $d_m \mid n + h_m$ and $d_m \mid q$. Thus since $n + h_m \equiv a \bmod q$, we have $d_m \mid a$. But $(a, q) = 1$ so $d_m = 1$. Similarly, $e_m = 1$.

On the other hand, if $d_m = e_m = 1$, since $n \equiv v_0 \bmod W$ and $p \mid (n + h_m, W)$, we would get $v_0 \equiv -a_p \bmod p$, which is forbidden. Therefore, $(v_0 + h_m, W) = 1$. If finally $p \mid (n + h_m, W)$, then $(n + h_m, [d_j, e_j])$ for some $j \neq m$. Then we would get $p \mid (n + h_j, n + h_m)$, so $p \mid h_j - h_m \neq 0$, so $p \mid W$, which is false. This completes the proof of the claim.

Next, we define an expression of the Bombieri–Vinogradov type.

$$E(N,q) := 1 + \sup_{a:(a,q)=1} \left| \sum_{\substack{N \le n < 2N \\ n \equiv a \bmod q}} \chi_{\mathbb{P}}(n) - \frac{1}{\varphi(q)} \sum_{N \le n < 2N} \chi_{\mathbb{P}}(n) \right|, \text{ and}$$

$$X_N := \sum_{N \le n < 2N} \chi_{\mathbb{P}}(n),$$

the inner sum will contribute

$$\frac{X_N}{\varphi(q)} + O(E(N,q)).$$

If a pair of $W, [d_1, e_1], \ldots [d_k, e_k]$ had a common factor or $d_m e_m \ne 1$, then the contribution of the corresponding term to the inner sum would be zero. Therefore, if $\flat$ indicates the sum is restricted to terms where $W, [d_1, e_1], \ldots, [d_k, e_k]$ are coprime,

$$S_2^{(m)} = \frac{X_N}{\varphi(W)} \sum_{\substack{d_1, \ldots, d_k \\ e_1, \ldots, e_k \\ e_m = d_m = 1}}^{\flat} \frac{\lambda_{d_1, \ldots, d_k} \lambda_{e_1, \ldots, e_k}}{\prod_{i=1}^{k} \varphi([d_i, e_i])} + O\left(\sum_{\substack{d_1, \ldots, d_k \\ e_1, \ldots, e_k}} |\lambda_{d_1, \ldots, d_k} \lambda_{e_1, \ldots, e_k}| E(N,q) \right).$$

(2) Now we consider the error term. Because $\lambda_{d_1, \ldots, d_k} = 0$ for $\mu^2(d) = 0$ or $\mu^2(d_i) = 0$ and $\prod r_i \le R$, we need only consider squarefree q with $q < R^2 W$. Given a squarefree q, the number of values of the d and e variables such that

$$q = W \prod_{i=1}^{k} [d_i, e_i]$$

is at most $\tau_{3k}(q)$. By (6.7), we have $\lambda_{max} \le y_{max} L_R(k)$. Therefore, the error is

$$O\left(y_{max}^2 L_R(2k) \sum_{q < R^2 W} \mu(q)^2 \tau_{3k}(q) E(N,q) \right).$$

Using the Cauchy–Schwarz inequality, the simple bound $E(N,q) \ll N/\varphi(q)$, the rewrite $\tau_{3k}(q) = (3k)^{\omega(q)}$, the estimate

$$\prod_{p \le x} \left(1 + \frac{3k}{p} \right) \ll (\log x)^{3k}$$

and the level of distribution of the primes θ, for fixed $k \in \mathbb{N}$ and $A > 0$, we get the error bound

$$O\left(y_{max}^2 L_R(2k)\left(\sum_{q < R^2 W} \mu(q)^2 \tau_{3k}^2(q)\frac{N}{\varphi(q)}\right)^{\frac{1}{2}}\left(\sum_{q < R^2 W} \mu(r)^2 E(N,q)\right)^{\frac{1}{2}}\right) \ll \frac{y_{max}^2 N}{L_N(A)}.$$

$$(6.8)$$

(3) Now consider the main term. We use a similar method to that of Step (6) of Lemma 6.2 and rewrite the expression by replacing the conditions $(d_i, e_j) = 1$ by $\sum_{s_{i,j}|(d_i,e_j)} \mu(s_{i,j})$ and using the same restrictions as in that step. We also replace the $1/\varphi([d_i, e_i])$ terms using a completely multiplicative function defined on primes by $g(p) := p - 2$, so that for squarefree d_i, e_i we have (see Step (2) of Lemma 6.2 or expand each term explicitly)

$$\frac{1}{\varphi([e_i, d_i])} = \frac{1}{\varphi(e_i)\varphi(d_i)}\sum_{u_i|(d_i,e_i)} g(u_i).$$

With these replacements, the main term becomes

$$\frac{X_N}{\varphi(W)}\sum_{u_1,\dots,u_k}\left(\prod_{i=1}^k g(u_i)\right)\sum_{s_{1,2},\dots,s_{k,k-1}}^{\sharp}\left(\prod_{1 \le i,j \le k} \mu(s_{i,j})\right)\sum_{\substack{d_1,\dots,d_k \\ e_1,\dots,e_k \\ \forall i\ u_i|(d_i,e_i) \\ \forall i \ne j\ s_{i,j}|(d_i,e_j) \\ d_m=e_m=1}} \frac{\lambda_{d_1,\dots,d_k}\lambda_{e_1,\dots,e_k}}{\prod_{i=1}^k \varphi(d_i)\varphi(e_i)}.$$

Notice that in this expression the d and e variables are now separate.

(4) Next, recall the definition given in the hypothesis of the lemma:

$$y_{r_1,\dots,r_k}^{(m)} := \left(\prod_{i=1}^k \mu(r_i)g(r_i)\right)\sum_{\substack{d_1,\dots,d_k \\ \forall\ i\ r_i|d_i \\ d_m=1}} \frac{\lambda_{d_1,\dots,d_k}}{\prod_{i=1}^k \varphi(d_i)}.$$

Note that because in the summation $d_m = 1$, we have $y_{r_1,\dots,r_k}^{(m)} \ne 0$ implies $r_m = 1$. Substituting in the result of Step (3), with a_j and b_j defined as in Step (6) of Lemma 6.2, and observing the terms with a_j or b_j not squarefree make a zero contribution, we obtain for the main term,

$$\frac{X_N}{\varphi(W)}\sum_{u_1,\dots,u_k}\left(\prod_{i=1}^k \frac{\mu(u_i)^2}{g(u_i)}\right)\sum_{s_{1,2},\dots,s_{k,k-1}}^{\sharp}\left(\prod_{\substack{1 \le i,j \le k \\ i \ne j}} \frac{\mu(s_{i,j})}{g(s_{i,j})^2}\right) y_{a_1,\dots,a_k}^{(m)} y_{b_1,\dots,b_k}^{(m)}.$$

(5) The contribution from terms with $s_{i,j} \neq 1$ is

$$O\left(\frac{(y_{max}^{(m)})^2 N}{\varphi(W)\log N}\left(\sum_{\substack{u<R\\(u,W)=1}}\frac{\mu(u)^2}{g(u)}\right)^{k-1}\left(\sum_{s\in\mathbb{N}}\frac{\mu(s)^2}{g(s)^2}\right)^{k^2-k-1}\sum_{s_{i,j}>D_0}\frac{\mu(s_{i,j})^2}{g(s_{i,j})^2}\right)$$

$$\ll \frac{(y_{max}^{(m)})^2 N\varphi(W)^{k-2}L_R(k-1)}{W^{k-1}D_0\log N}.$$

(6) It follows from Steps (3) through (5) that we can write

$$S_2^{(m)} = \frac{X_N}{\varphi(W)}\sum_{u_1,\dots,u_k}\frac{(y_{u_1,\dots,u_k}^{(m)})^2}{\prod_{i=1}^k g(u_i)} + O\left(\frac{(y_{max}^{(m)})^2 N\varphi(W)^{k-2}L_R(k-2)}{W^{k-1}D_0}\right)$$

$$+ O\left(\frac{y_{max}^2 N}{L_N(A)}\right).$$

By the prime number theorem, we have

$$X_N := \sum_{N\leq n<2N}\chi_{\mathbb{P}}(n) = \frac{N}{\log N} + O\left(\frac{N}{\log^2 N}\right),$$

and the error in this expression, using $p-2 \geq (p-1)/2$ for $p \geq 3$ and Lemma 6.1, contributes

$$O\left(\frac{(y_{max}^{(m)})^2 N}{\varphi(W)L_N(2)}\left(\sum_{\substack{u<R\\(u,W)=1}}\frac{\mu(u)^2}{g(u)}\right)^{k-1}\right) \ll \frac{(y_{max}^{(m)})^2 N\varphi(W)^{k-2}L_R(k-3)}{W^{k-1}}$$

$$\ll \frac{(y_{max}^{(m)})^2 N\varphi(W)^{k-2}L_R(k-2)}{W^{k-1}D_0},$$

so, finally,

$$S_2^{(m)} = \frac{N}{\varphi(W)\log N}\sum_{r_1,\dots,r_k}\frac{(y_{r_1,\dots,r_k}^{(m)})^2}{\prod_{i=1}^k g(r_i)} + O\left(\frac{(y_{max}^{(m)})^2\varphi(W)^{k-2}NL_N(k-2)}{W^{k-1}D_0}\right)$$

$$+ O\left(\frac{y_{max}^2 N}{L_N(A)}\right),$$

which completes the proof.   □

Recall again the definitions:

$$S_1 := \sum_{\substack{N \le n < 2N \\ n \equiv v_0 \bmod W}} \left(\sum_{\forall i \ d_i \mid n + h_i} \lambda_{d_1, \ldots, d_k} \right)^2,$$

$$y^{(m)}_{r_1, \ldots, r_k} := \left(\prod_{i=1}^{k} \mu(r_i) g(r_i) \right) \sum_{\substack{d_1, \ldots, d_k \\ \forall i \ r_i \mid d_i \\ d_m = 1}} \frac{\lambda_{d_1, \ldots, d_k}}{\prod_{i=1}^{k} \varphi(d_i)},$$

$$g(p) := p - 2 \text{ is completely multiplicative.}$$

Recall also the variable R is a small fixed power of N which is less than $1/4$, and that $\lambda_{d_1, \ldots, d_k} = 0$ for $d_1 \cdots d_k \ge R$.

Next we derive an asymptotic expression for the $y^{(m)}$ variables in terms of the y variables, which later will be set to be values of the functions F.

Lemma 6.4 *[138, lemma 5.3] When $r_m = 1$ we have*

$$y^{(m)}_{r_1, \ldots, r_k} = \sum_{a_m \in \mathbb{N}} \frac{y_{r_1, \ldots, r_{m-1} a_m, r_{m+1}, \ldots, r_k}}{\varphi(a_m)} + O\left(\frac{y_{max} \varphi(W) \log R}{W D_0} \right).$$

Proof **(1)** Whenever r_m appears in this proof, we will assume its value is 1. We begin by substituting the expression

$$\sum_{\substack{r_1, \ldots, r_k \\ \forall i \ d_i \mid r_i}} \frac{y_{r_1, \ldots, r_k}}{\prod_{i=1}^{k} \varphi(r_i)} = \frac{\lambda_{d_1, \ldots, d_k}}{\prod_{i=1}^{k} d_i \mu(d_i)},$$

derived in Step (4) of Lemma 6.2, into the definition of $y^{(m)}_{r_1, \ldots, r_k}$ to get

$$y^{(m)}_{r_1, \ldots, r_k} = \left(\prod_{i=1}^{k} \mu(r_i) g(r_i) \right) \sum_{\substack{d_1, \ldots, d_k \\ \forall i \ r_i \mid d_i \\ d_m = 1}} \left(\prod_{i=1}^{k} \frac{\mu(d_i) d_i}{\varphi(d_i)} \right) \sum_{\substack{a_1, \ldots, a_k \\ \forall i \ d_i \mid a_i}} \frac{y_{a_1, \ldots, a_k}}{\prod_{i=1}^{k} \varphi(a_i)}.$$

Next, interchange the two summations in this expression to get

$$y^{(m)}_{r_1, \ldots, r_k} = \left(\prod_{i=1}^{k} \mu(r_i) g(r_i) \right) \sum_{\substack{a_1, \ldots, a_k \\ \forall i \ r_i \mid a_i}} \frac{y_{a_1, \ldots, a_k}}{\prod_{i=1}^{k} \varphi(a_i)} \sum_{\substack{d_1, \ldots, d_k \\ \forall i \ d_i \mid a_i \\ \forall i \ r_i \mid d_i \\ d_m = 1}} \left(\prod_{i=1}^{k} \frac{\mu(d_i) d_i}{\varphi(d_i)} \right).$$

Then we perform the sum over the d variables by setting $d'_i = d_i / r_i$ and $a'_i = a_i / r_i$, interchanging the product and sum, and using $\sum_{d \mid n} \mu(d) f(d) = \prod_{p \mid n} (1 - f(p))$, which is valid for the multiplicative $f(n) = n / \varphi(n)$, we get

$$
y^{(m)}_{r_1,\dots,r_k} = \left(\prod_{i=1}^{k}\mu(r_i)g(r_i)\right) \sum_{\substack{a_1,\dots,a_k \\ \forall i\; r_i|a_i}} \frac{y_{a_1,\dots,a_k}}{\prod_{i=1}^{k}\varphi(a_i)} \prod_{i\neq m}\frac{\mu(a_i)r_i}{\varphi(a_i)}.
$$

(2) Each of the a variables are squarefree and they are coprime with $(a_j, W) = 1$ and with $y_{a_1,\dots,a_k} = 0$ otherwise, so since $r_j \mid a_j$, either $a_j = r_j$ or $a_j > D_0 r_j$. If $j \neq m$ the contribution to this sum of all the $a_j \neq r_j$ is thus, noting the subscript in the first sum has been amended - (see [138, equation 5.31]),

$$
O\left(y_{max}\left(\prod_{i=1}^{k}g(r_i)r_i\right)\left(\sum_{\substack{a_j>D_0 r_j \\ r_j|a_j}}\frac{\mu(a_j)^2}{\varphi(a_j)^2}\right)\left(\sum_{\substack{a_m<R \\ (a_m,W)=1}}\frac{\mu(a_m)^2}{\varphi(a_m)}\right)\prod_{\substack{1\le i\le k \\ i\neq j,\,m}}\left(\sum_{r_i|a_i}\frac{\mu(a_i)^2}{\varphi(a_i)^2}\right)\right)
$$

$$
\ll \left(\prod_{i=1}^{k}\frac{g(r_i)r_i}{\varphi(r_i)^2}\right)\frac{y_{max}\varphi(W)\log R}{W D_0}
$$

$$
\ll \frac{y_{max}\varphi(W)\log R}{W D_0}.
$$

To get the last line, we note that

$$
\frac{g(p)p}{\varphi(p)^2} = \frac{p(p-2)}{(p-1)^2} = \frac{1-\frac{2}{p}}{\left(1-\frac{1}{p}\right)^2} = 1 + O\left(\frac{1}{p^2}\right).
$$

Thus, since nonzero terms must have the r variables coprime to $W = \prod_{p\le D_0}p$, the leading factor may be replaced by $1 + O(1/D_0)$.

(3) Therefore, the main contribution to the expression is when for all $j \neq m$ we have $a_j = r_j$, so

$$
y^{(m)}_{r_1,\dots,r_k} = \left(\prod_{i=1}^{k}\frac{g(r_i)r_i}{\varphi(r_i)^2}\right)\sum_{a_m\in\mathbb{N}}\frac{y_{r_1,\dots,r_{m-1},a_m,r_{m+1},\dots,r_k}}{\varphi(a_m)} + O\left(\frac{y_{max}\varphi(W)\log R}{W D_0}\right),
$$

which completes the proof.   □

6.5 Other Preliminary Lemmas

The following major sieve lemma, expressing a particular sum as an integral plus an error term, is essential to Maynard's approach and is used in Lemmas 6.6 and 6.6. It is built on a result of Goldston, Graham, Pintz and Yildirim [77, lemmas 4, 5], which in turn comes from Halberstam and Richert's lemmas 5.3 and 5.4 in chapter 4 of their book [88]. The lemma was not used by Polymath. We have made some simplifications, but retained the parameter κ, the so-called **sieve dimension** or **sieve**

density, even though in Maynard's applications only the value $\kappa = 1$ is used, so the underlying sieve is "linear" – see Section 2.8. Note also that $1 = \kappa \neq k \geq 2$.

First we need a set of results with some based on Merten's theorem. Here we have $2 \leq w \leq z$, γ is Euler's constant, (D) is proved in Lemma 2.21, (F) is a standard integration formula and implied constants are absolute:

$$(A) \sum_{w \leq p < z} \frac{\log p}{p} = \log\left(\frac{z}{w}\right) + O(1),$$

$$(B) \sum_{w \leq p < z} \frac{1}{p} = \log\frac{\log z}{\log w} + O\left(\frac{1}{\log w}\right),$$

$$(C) \; V(z) := \prod_{p < z}\left(1 - \frac{1}{p}\right) = \frac{1}{e^\gamma \log z}\left(1 + O\left(\frac{1}{\log z}\right)\right),$$

$$(D) \; \frac{V(w)}{V(z)} = \frac{\log z}{\log w}\left(1 + O\left(\frac{1}{\log w}\right)\right),$$

$$(E) \sum_{p | n} \frac{\log p}{p} \ll \log\log n,$$

$$(F) \int_1^\infty \frac{(\log y)^{\lambda-1}}{y^{\sigma+1}} \, dy = \frac{1}{\sigma^\lambda}\int_0^\infty e^{-t}t^{\lambda-1} \, dt = \frac{\Gamma(\lambda)}{\sigma^\lambda}, \; \lambda, \sigma > 0.$$

Lemma 6.5 **(Halberstam–Richert–GGPY)** *Let κ, A_1, A_2, L be positive real constants, and γ a multiplicative function which satisfies bounds given as follows, for $2 \leq w \leq z$, which we call the "first constraint" and "second constraint" respectively:*

$$0 \leq \frac{\gamma(p)}{p} \leq 1 - \frac{1}{A_1} \; \text{ and } \; -L \leq \sum_{w \leq p \leq z} \frac{\gamma(p)\log p}{p} - \frac{\kappa \log z}{w} \leq A_2.$$

Let h be a completely multiplicative function with $h(p) := \gamma(p)/(p - \gamma(p))$ for $p \in \mathbb{P}$. Let $F : [0, 1] \to \mathbb{R}$ be piecewise differentiable and define

$$F_{max} := \sup\{|F(t)| + |F'(t)| : 0 \leq t \leq 1\}.$$

Finally define

$$\mathscr{G} := \prod_p\left(1 - \frac{\gamma(p)}{p}\right)^{-1}\left(1 - \frac{1}{p}\right)^\kappa.$$

Then

$$\sum_{d < z} \mu(d)^2 h(d) F\left(\frac{\log d}{\log z}\right) = \mathscr{G}\frac{(\log z)^\kappa}{\Gamma(\kappa)}\int_0^1 F(x)x^{\kappa-1} \, dx$$

$$+ O_{A_1, A_2, \kappa}\left(\mathscr{G}L F_{max}(\log z)^{\kappa-1}\right).$$

Proof (**1**) Since by the second constraint, we have

$$\sum_{w \le p < z} \frac{\gamma(p) \log p}{p} \le \frac{\kappa \log z}{w} + A_2,$$

using Abel's theorem [4, theorem 4.2], we get

$$\sum_{w \le p < z} \frac{\gamma(p)}{p} \le \frac{\kappa \log(z/w) + A_2}{\log z} + \int_w^z \frac{\kappa \log(t/w) + A_2}{t \log^2 t} \, dt$$

$$= \kappa \log \left(\frac{\log z}{\log w} \right) + \frac{A_2}{\log w}.$$

Similarly,

$$-L + \frac{\kappa \log z}{w} \le \sum_{w \le p < z} \frac{\gamma(p) \log p}{p}$$

implies

$$-\frac{L}{\log w} + \kappa \log \left(\frac{\log z}{\log w} \right) \le -\frac{L}{\log w} + \kappa \frac{\log z}{\log w} \le \sum_{w \le p < z} \frac{\gamma(p)}{p}.$$

Therefore,

$$-\frac{L}{\log w} \le \sum_{w \le p < z} \frac{\gamma(p)}{p} - \kappa \log \left(\frac{\log z}{\log w} \right) \le \frac{A_2}{\log w}.$$

Note also that using Abel's theorem again, a similar derivation shows that

$$\sum_{w \le p < z} \frac{\gamma(p) \log p}{p} \le \frac{1}{\log w} \left(\kappa + \frac{A_2}{\log w} \right).$$

(**2**) Next we **claim**

$$\sum_{w \le p < z} h(p) \le \kappa \log \left(\frac{\log z}{\log w} \right) + O \left(\frac{1}{\log w} \right).$$

To see this, first note that using $w = p$, $z = p + \epsilon$ and allowing $\epsilon \to 0+$ we get from the second constraint,

$$\frac{\gamma(p) \log p}{p} \le A_1.$$

Thus, since $h(p) \le A_1 \gamma(p)/p$, we can write

$$\sum_{w \le p < z} \frac{\gamma(p) h(p)}{p} \le A_1 A_2 \sum_{w \le p < z} \frac{\gamma(p)}{p \log p} \le \frac{A_1 A_2 \gamma(p)}{p \log p} \left(\kappa + \frac{A_2}{\log w} \right).$$

But we have

$$h(p) = \frac{1}{1 - \gamma(p)/p} - 1 \implies h(p) = \frac{\gamma(p)}{p} + \frac{\gamma(p)}{p} h(p),$$

so using the results from Step (1) we get

$$\sum_{w \leq p < z} h(p) \leq \kappa \log \left(\frac{\log z}{\log w} \right) + \frac{A_2}{\log w} + \frac{A_1 A_2}{\log w} \left(\kappa + \frac{A_2}{\log w} \right),$$

which completes the proof of the claim.

(3) It follows from Step (2) that

$$\sum_{w \leq p < z} h(p) - \kappa \sum_{w \leq p < z} \frac{1}{p} \ll \frac{1}{\log w}.$$

On the other hand, because $h(p) \geq \gamma(p)/p$ and result (B) listed at the start of this section, the first inequality derived in Step (1) shows that

$$\sum_{w \leq p < z} h(p) - \kappa \sum_{w \leq p < z} \frac{1}{p} \gg \frac{L}{\log w}.$$

(4) Now let $P(z) := \prod_{p < z} p$, $p \mid P(z)$ be a fixed prime, and define

$$G(z) := \sum_{\substack{d < z \\ d \mid P(z)}} h(d) = \sum_{d < z} \mu(d)^2 h(d), \text{ and}$$

$$G_p(z) := \sum_{\substack{d < z \\ d \mid P(z) \\ (d, p) = 1}} h(d).$$

Then

$$G(z) = \sum_{\substack{d < z \\ d \mid P(z) \\ (d, p) = 1}} h(d) + h(p) \sum_{\substack{m < z/p \\ m \mid P(z) \\ (m, p) = 1}} h(m) = G_p(z) + h(p) G_p(z/p).$$

Multiplying this identity by $1 - \gamma(p)/p$ gives

$$\left(1 - \frac{\gamma(p)}{p} \right) G(z) = G_p(z) - \frac{\gamma(p)}{p} \left(G_p(z) - G_p(z/p) \right).$$

Replace z by z/p in this expression and manipulate to get

$$G_p(z/p) = \left(1 - \frac{\gamma(p)}{p} \right) G(z/p) + \frac{\gamma(p)}{p} \left(G_p(z/p) - G_p(z/p^2) \right).$$

(5) Next, using the result of Step (4) to derive the fourth line that follows, we get

$$
\sum_{\substack{d<z \\ d|P(z)}} h(d) \log d = \sum_{\substack{d<z \\ d|P(z)}} h(d) \sum_{p|d} \log(p)
$$

$$
= \sum_{p<z} h(p) \log(p) \sum_{\substack{m<z/p \\ m|P(z) \\ (m,p)=1}} h(m)
$$

$$
= \sum_{p<z} h(p) \log(p) G_p(z/p)
$$

$$
= \sum_{p<z} \frac{\gamma(p)}{p} \log(p) \sum_{\substack{d<z/p \\ d|P(z/p) \\ (d,p)=1}} h(d)
$$

$$
+ \sum_{p<z} \frac{h(p)\gamma(p)}{p} \log(p) \sum_{\substack{z/p^2 \le d < z/p \\ d|P(z) \\ (d,p)=1}} h(d)
$$

$$
= \sum_{\substack{d<z \\ d|P(z)}} h(d) \sum_{\substack{p<z/d \\ (d,p)=1}} \frac{\gamma(p)}{p} \log(p)
$$

$$
+ \sum_{\substack{d<z \\ d|P(z)}} h(d) \sum_{\substack{\sqrt{z/d} \le p < z/d \\ p \nmid d}} \frac{h(p)\gamma(p)}{p} \log(p).
$$

(6) We will next simplify this expression. For the inner sum in the first term of the expression derived in Step (5), we use the second constraint, which gives

$$
\sum_{1<p<y} \frac{\gamma(p) \log p}{p} = \kappa \log y + O(L).
$$

For the inner sum in the second term, we use the inequality $(\gamma(p)/p) \log p$ derived in Step (2) and the bound for the sum of the $h(p)$ also derived in Step (2) to get

$$
\sum_{\substack{\sqrt{z/d} \le p < z/d \\ p \nmid d}} \frac{h(p)\gamma(p)}{p} \log p \ll 1.
$$

Hence

$$
\sum_{\substack{d<z \\ d|P(z)}} h(d) \log d = \sum_{\substack{d<z \\ d|P(z)}} h(d) \left(\kappa \left(\log \frac{z}{d} \right) + O(LG(z)) \right).
$$

Adding the second term on the left to each side in the following, we get

$$\sum_{\substack{d<z \\ d|P(z)}} h(d) \log d + \sum_{\substack{d<z \\ d|P(z)}} h(d) \log\left(\frac{z}{d}\right) = (\kappa + 1) \sum_{\substack{d<z \\ d|P(z)}} h(d) \log\left(\frac{z}{d}\right) + O(LG(z)).$$

If we then define

$$T(z) := \int_1^z \frac{G(t)}{t}\, dt = \sum_{\substack{d<z \\ d|P(z)}} h(d) \log\left(\frac{z}{d}\right),$$

we get for some function $r(z) \ll L/\log z$,

$$G(z) \log z = (\kappa + 1)T(z) + G(z)(\log z)r(z).$$

Assuming we have z sufficiently large, we can assume in what follows that $|r(y)| \le \frac{1}{2}$ for all $y \ge z$ and write

$$G(z) = \frac{1}{1 - r(z)} \frac{\kappa + 1}{\log z} T(z).$$

(7) Next we define

$$E(y) := \log\left(\frac{\kappa + 1}{(\log y)^{\kappa+1}} T(y)\right) = \log(\kappa + 1) - (\kappa + 1)\log\log y + \log(T(y)).$$

This implies that the derivative

$$E'(y) = -\frac{\kappa + 1}{y \log y} + \frac{G(y)}{yT(y)} = -\frac{\kappa + 1}{y \log y} + \frac{1}{1 - r(y)} \frac{\kappa + 1}{y \log y}$$

$$= \frac{r(y)}{1 - r(y)} \frac{\kappa + 1}{y \log y} \ll \frac{L}{y \log^2 y}.$$

Thus the integral $\int_z^\infty E'(y)\, dy$ converges and there is a constant c_2 such that

$$\exp(E(z)) = \frac{\kappa + 1}{(\log z)^{\kappa+1}} T(z)$$

$$= c_2 \exp\left(-\int_z^\infty E'(y)\, dy\right)$$

$$= c_2\left(1 + O\left(\frac{L}{\log z}\right)\right) \implies$$

$$T(z) = \frac{c_2 (\log z)^{\kappa+1}}{\kappa + 1}\left(1 + O\left(\frac{L}{\log z}\right)\right).$$

In addition, we can write

$$\frac{1}{1 - r(z)} = 1 + \frac{r(z)}{1 - r(z)} = 1 + O\left(\frac{L}{\log z}\right).$$

Using this, the expression we have derived for $T(z)$ and the result of Step (6) gives

$$G(z) = c_2(\log z)^\kappa \left(1 + O\left(\frac{L}{\log z}\right)\right).$$

(8) In this step, we evaluate the constant c_2. From Step (7), we get for $y \to \infty$,

$$G(y) = c_2(\log y)^\kappa + O(c_2 L(\log y)^{\kappa-1}).$$

We can therefore write with real $s > 0$, using Abel's theorem [4, theorem 4.2], the integral identity result (F), and $1 + h(p) = 1/(1 - \gamma(p)/p)$,

$$
\begin{aligned}
\prod_p \left(1 + \frac{h(p)}{p^s}\right) &= \sum_{n=1}^\infty \frac{\mu(n)^2 h(n)}{n^s} \\
&= s \int_1^\infty \frac{G(y)}{y^{s+1}} \, dy \\
&= s \int_1^\infty \frac{c_2(\log y)^\kappa + O(c_2 L(\log y)^{\kappa-1}}{y^{s+1}} \, dy \\
&= \frac{c_2 \Gamma(\kappa+1)}{s^\kappa} + O\left(\frac{c_2 L}{s^{\kappa-1}}\right).
\end{aligned}
$$

Hence, recalling the definition of $\mathscr{G}$ from the statement of the lemma and using $\lim_{s\to 0+} s\zeta(s+1) = 1$, we get

$$
\begin{aligned}
c_2 &= \frac{1}{\Gamma(\kappa+1)} \lim_{s\to 0+} s^\kappa \prod_p \left(1 + \frac{h(p)}{p^s}\right) \\
&= \frac{1}{\Gamma(\kappa+1)} \lim_{s\to 0+} \prod_p \left(1 + \frac{h(p)}{p^s}\right)\left(1 - \frac{1}{p^{s+1}}\right)^\kappa \\
&= \frac{1}{\Gamma(\kappa+1)} \lim_{s\to 0+} \prod_p \left(1 - \frac{\gamma(p)}{p}\right)^{-1}\left(1 - \frac{1}{p^1}\right)^\kappa = \frac{\mathscr{G}}{\Gamma(\kappa+1)},
\end{aligned}
$$

so $c_2 = \mathscr{G}/\Gamma(\kappa+1)$.

(9) Next we write using the definition of $G(z)$ from Step (4), together with Steps (7) and (8),

$$G(u) := \sum_{d<u} \mu(d)^2 h(d) = \mathscr{G}\frac{(\log u)^\kappa}{\Gamma(\kappa+1)} + H(u).$$

Thus, using the Riemann–Stieltjes integral (see for example, [22, appendix C]),

$$\sum_{d<z} \mu(d)^2 h(d) F\left(\frac{\log d}{\log z}\right) = \int_{1-}^z F\left(\frac{\log u}{\log z}\right) dG(u).$$

Therefore, we can express the right-hand side as

$$\int_{1-}^{z} F\left(\frac{\log u}{\log z}\right) d\left(\mathscr{G}\frac{(\log u)^{\kappa}}{\Gamma(\kappa+1)}\right) + \int_{1-}^{z} F\left(\frac{\log u}{\log z}\right) dH(u).$$

(10) Now make the substitution $u = z^x$ in the first integral to get

$$\int_{1}^{z} F\left(\frac{\log u}{\log z}\right) d\left(\mathscr{G}\frac{(\log u)^{\kappa}}{\Gamma(\kappa+1)}\right) = \frac{\mathscr{G}(\log z)^{\kappa}}{\Gamma(\kappa)} \int_{0}^{1} F(x) x^{\kappa-1} \, dx.$$

Then integrate by parts in the second integral and use the bound obtained in Step (7), namely $H(u) \ll \mathscr{G}L(\log 2u)^{\kappa-1}$, to get

$$\int_{1-}^{z} F\left(\frac{\log u}{\log z}\right) dH(u) = F\left(\frac{\log u}{\log z}\right) H(u)\bigg|_{1-}^{z} - \int_{1}^{z} H(u) F'\left(\frac{\log u}{\log z}\right) \frac{1}{u \log z} \, du$$

$$= O\left(\mathscr{G}L F_{max}(\log z)^{\kappa-1}\right).$$

Therefore,

$$\sum_{d<z} \mu(d)^2 h(d) F\left(\frac{\log d}{\log z}\right) = \mathscr{G}\frac{(\log z)^{\kappa}}{\Gamma(\kappa)} \int_{0}^{1} F(x) x^{\kappa-1} \, dx$$

$$+ O_{A_1, A_2, \kappa}\left(\mathscr{G}L F_{max}(\log z)^{\kappa-1}\right).$$

This completes the proof. $\qquad\square$

Remark: Note again that in each of Maynard's applications of Lemma 6.5, namely Lemmas 6.6 and 6.7, we need only the special case $\kappa = 1$. In that case, we can rewrite the integral in the result of Lemma 6.5 as

$$\sum_{d<z} \mu(d)^2 h(d) F\left(\frac{\log d}{\log z}\right) = \mathscr{G}(\log z) \int_{0}^{1} F(x) \, dx + O_{A_1, A_2}(\mathscr{G}L F_{max}).$$

Next, recall the definitions

$$I_k(F) := \int_{0}^{1} \cdots \int_{0}^{1} F(t_1, \ldots, t_k)^2 dt_1, \ldots, dt_k,$$

and

$$S_1 := \sum_{\substack{N \le n < 2N \\ n \equiv v_0 \bmod W}} \left(\sum_{\forall i \, d_i | n + h_i} \lambda_{d_1, \ldots, d_k}\right)^2.$$

In what follows, we use the following definition for the y variables:

$$y_{r_1, \ldots, r_k} := F\left(\frac{\log r_1}{\log R}, \ldots, \frac{\log r_k}{\log R}\right), \tag{6.9}$$

or zero when $(W, r_1 \cdots r_k) \neq 1$ or $r_1 \cdots r_k$ is not squarefree. Because the support of F is contained in the standard unit simplex, we can also assume $y_{r_1,\ldots,r_k} = 0$ when $r_1 \cdots r_k > R$. Note that can obtain a valid set of the sieve $\lambda_{d_1,\ldots,d_k}$ corresponding to these y variables - see Lemma 6.2.

Lemma 6.6   *[138, lemma 6.2] Let*

$$F_{max} := \sup_{(t_1,\ldots,t_k) \in [0,1]^k} |F(t_1,\ldots,t_k)| + \sum_{i=1}^{k} \left| \frac{\partial F}{\partial t_i} (t_1,\ldots,t_k) \right|,$$

then

$$S_1 = \frac{N\varphi(W)^k L_R(k)}{W^{k+1}} I_k(F) + O\left(\frac{F_{max}^2 N\varphi(W)^k L_R(k)}{W^{k+1} D_0} \right).$$

Proof   **(1)** By Lemma 6.2, we have

$$S_1 = \frac{N}{W} \sum_{u_1,\ldots,u_k} \frac{y_{u_1,\ldots,u_k}^2}{\prod_{i=1}^{k} \varphi(u_i)} + O\left(\frac{y_{max}^2 N L_R(k)}{W D_0} \right).$$

We substitute into this expression the particular choice we are making for the y variables, namely

$$y_{r_1,\ldots,r_k} := F\left(\frac{\log r_1}{\log R}, \ldots, \frac{\log r_k}{\log R} \right),$$

and zero when $(W, r_1 \cdots r_k) \neq 1$ or $r_1 \cdots r_k$ is not squarefree, to get from Step (7) of the proof of Lemma 6.2:

$$S_1 = \frac{N}{W} \sum_{\substack{u_1,\ldots,u_k \\ \forall i \neq j \ (u_i, u_j)=1 \\ \forall i \ (u_i, W)=1 \\ \text{squarefree}}} \left(\prod_{i=1}^{k} \frac{\mu(u_i)^2}{\varphi(u_i)} \right) F\left(\frac{\log u_1}{\log R}, \ldots, \frac{\log u_k}{\log R} \right)^2$$

$$+ O\left(\frac{F_{max}^2 \varphi(W)^k N L_R(k)}{W^{k+1} D_0} \right).$$

(2) Since $W := \prod_{p \leq D_0} p$, any two integers a, b with $(a,b) \neq 1$ and $(a, W) = (b, W) = 1$ must have a common prime factor larger than D_0. Thus if we remove the requirement $(u_i, u_j) = 1$ from the sum from Step (1) and use Lemma 6.1 (b), exchanging the inner sum and product and setting $u_i = Pu$, where $p \mid P$ and P

is the product of primes in common with other u_j so $\varphi(u) \leq (p-1)^2 \varphi(u_i)$, we introduce an error, which is for $k \geq 2$,

$$O\left(\frac{F_{max}^2 N}{W} \sum_{p > D_0} \sum_{\substack{u_1 \cdots u_k < R \\ \forall i \neq j \ p|(u_i, u_j) \\ \forall i \ (u_i, W)=1}} \prod_{i=1}^{k} \frac{\mu(u_i)^2}{\varphi(u_i)}\right) \ll \frac{F_{max}^2 N}{W} \sum_{p > D_0} \frac{1}{(p-1)^2} \left(\sum_{\substack{u < R \\ (u, W)=1}} \frac{\mu(u)^2}{\varphi(u)}\right)^k$$

$$\ll \frac{F_{max}^2 N \varphi(W)^k L_R(k)}{W^{k+1} D_0},$$

which can be absorbed into the error from Step (1).

(3) Therefore, by Step (3) we can write the main term as the sum

$$\frac{N}{W} \sum_{\substack{u_1, \ldots, u_k \\ \forall i \ (u_i, W)=1}} \left(\prod_{i=1}^{k} \frac{\mu(u_i)^2}{\varphi(u_i)}\right) F\left(\frac{\log u_1}{\log R}, \ldots, \frac{\log u_k}{\log R}\right)^2.$$

We now use the result of Lemma 6.5 to the sum over each u_i variable in turn, with $\kappa = 1$, choosing for all of these applications the constants A_1, A_2 and L such that

$$\gamma(p) = \begin{cases} 1, & p \nmid W \\ 0, & p \mid W \end{cases} \quad \text{and} \quad L \ll 1 + \sum_{p|W} \frac{\log p}{p} \ll \log D_0.$$

We also choose $w = D_0$, $z = R$ and get $\mathcal{G} = \varphi(W)/W$.

Then, using the result of Step (2) for the error, this gives

$$\sum_{\substack{u_1, \ldots, u_k \\ \forall i \ (u_i, W)=1}} \left(\prod_{i=1}^{k} \frac{\mu(u_i)^2}{\varphi(u_i)}\right) F\left(\frac{\log u_1}{\log R}, \ldots, \frac{\log u_k}{\log R}\right)^2 = \frac{\varphi(W)^k L_R(k)}{W^k} I_k(F)$$

$$+ O\left(\frac{F_{max}^2 \varphi(W)^k L_R(k-1)}{W^k} \log D_0\right).$$

Combining these expressions with that derived in Step (1), and using

$$\frac{\log N}{\log\log\log N} \gg \log\log\log\log N,$$

completes the proof. $\qquad\qquad\qquad\qquad\qquad\qquad\qquad\qquad\qquad \square$

Now we must apply the same type of analysis to the more difficult sum for $S_2^{(m)}$.
Recall the definitions, for $1 \leq m \leq k$,

$$S_2^{(m)} := \sum_{\substack{N \leq n < 2N \\ n \equiv v_0 \bmod W}} \chi_{\mathbb{P}}(n + h_m) \left(\sum_{\forall i \; d_i \mid n + h_i} \lambda_{d_1, \dots, d_k} \right)^2,$$

$$J_k^{(m)}(F) := \int_0^1 \cdots \int_0^1 \left(\int_0^1 F(t_1, \dots, t_k) \, dt_m \right)^2 dt_1 \dots dt_{m-1} dt_{m+1} \dots dt_k,$$

$$F_{max} := \sup_{(t_1, \dots, t_k) \in [0,1]^k} \left(|F(t_1, \dots, t_k)| + \sum_{i=1}^k \left| \frac{\partial F}{\partial t_i}(t_1, \dots, t_k) \right| \right),$$

$$y_{r_1, \dots, r_k}^{(m)} := \left(\prod_{i=1}^k \mu(r_i) g(r_i) \right) \sum_{\substack{d_1, \dots, d_k \\ \forall i \; r_i \mid d_i \\ d_m = 1}} \frac{\lambda_{d_1, \dots, d_k}}{\prod_{i=1}^k \varphi(d_i)},$$

$$y_{max}^{(m)} := \sup_{r_1, \dots, r_k} |y_{r_1, \dots, r_k}^{(m)}|,$$

$g(p) := p - 2$ and completely multiplicative.

Lemma 6.7   *[138, lemma 6.3] We have the estimate*

$$S_2^{(m)} = \frac{N\varphi(W)^k L_R(k+1)}{W^{k+1} \log N} J_k^{(m)}(F) + O\left(\frac{F_{max}^2 N \varphi(W)^k L_R(k)}{W^{k+1} D_0} \right).$$

Proof   The proof is similar to that of Lemma 6.6 and takes five steps.
(1) We begin by estimating $y_{r_1, \dots, r_k}^{(m)}$, which is given by Lemma 6.4 as

$$y_{r_1, \dots, r_k}^{(m)} = \sum_{a_m \in \mathbb{N}} \frac{y_{r_1, \dots, r_{m-1} a_m, r_{m+1}, \dots, r_k}}{\varphi(a_m)} + O\left(\frac{y_{max} \varphi(W) \log R}{W D_0} \right)$$

when $r_m = 1$, $(r, W) = 1$, $\mu(r)^2 = 1$ for $r = r_1 \cdots r_k$, $r \leq R$, and otherwise
$y_{r_1, \dots, r_k}^{(m)} = 0$. Substitute the value we have chosen for the y variables, namely

$$y_{r_1, \dots, r_k} := F\left(\frac{\log r_1}{\log R}, \dots, \frac{\log r_k}{\log R} \right),$$

in the nonzero case, and use Lemma 6.1(b) for the error, recalling $r_m = 1$ in the
product, to get

$$y_{r_1, \dots, r_k}^{(m)} = \sum_{\substack{u \in \mathbb{N} \\ (u, W \prod_{i=1}^k r_i) = 1}} \frac{\mu(u)^2}{\varphi(u)} F\left(\frac{\log r_1}{\log R}, \dots, \frac{r_{m-1}}{\log R}, \frac{\log u}{\log R}, \frac{\log r_{m+1}}{\log R}, \dots, \frac{\log r_k}{\log R} \right)$$

$$+ O\left(\frac{F_{max} \varphi(W) \log R}{W D_0} \right) \implies y_{max}^{(m)} \ll \frac{\varphi(W) F_{max} \log R}{W}.$$

(2) Next we estimate the sum over u in the main term of the expression from Step (1). To apply Lemma 6.5, define

$$\gamma(p) = \begin{cases} 1, & p \nmid W \prod_{i=1}^{k} r_i, \\ 0, & \text{otherwise}, \end{cases}$$

so

$$\mathscr{G} = \prod_{p} \left(1 - \frac{\gamma(p)}{p} \right)^{-1} \left(1 - \frac{1}{p} \right) = \frac{\varphi(W) \prod_{i=1}^{k} \varphi(r_i)}{W} \frac{\prod_{i=1}^{k} \varphi(r_i)}{\prod_{i=1}^{k} r_i}.$$

Using [4, theorem 4.10] and noting there are not more than $O(\log R)$ primes dividing $\prod r_i$,

$$L \ll 1 + \sum_{\substack{p \mid W \prod_{i=1}^{k} r_i}} \frac{\log p}{p} \ll \sum_{p < \log R} \frac{\log p}{p} + \sum_{\substack{p \mid W \prod_{i=1}^{k} r_i \\ p > \log R}} \frac{\log\log R}{\log R} \ll \log\log N.$$

Recall we have also chosen $z = R$. Also recall

$$F_{r_1,\dots,r_k}^{(m)} := \int_0^1 F\left(\frac{\log r_1}{\log R}, \dots, t_m, \frac{\log r_{m+1}}{\log R}, \dots, \frac{\log r_k}{\log R} \right) dt_m.$$

Applying Lemma 6.5, these choices enable us to write

$$y_{r_1,\dots,r_k}^{(m)} = \log(R) \frac{\varphi(W)}{W} \left(\prod_{i=1}^{k} \frac{\varphi(r_i)}{r_i} \right) F_{r_1,\dots,r_k}^{(m)} + O\left(\frac{F_{max}\varphi(W)\log R}{W D_0} \right).$$

If $r_m \neq 1$ or $(r, W) \neq 1$ or $\mu(r) = 0$ or $r > R$, then $y_{r_1,\dots,r_k}^{(m)} = 0$. Squaring and simplifying the cross term, we get

$$(y_{r_1,\dots,r_k}^{(m)})^2 = \log^2(R) \frac{\varphi(W)^2}{W^2} \left(\prod_{i=1}^{k} \frac{\varphi(r_i)^2}{r_i^2} \right) (F_{r_1,\dots,r_k}^{(m)})^2 + O\left(\frac{F_{max}^2 \varphi(W)^2 \log^2 R}{W^2 D_0} \right).$$

(3) From Lemma 6.3 with $g(p) = p - 2$ completely multiplicative, we can write

$$S_2^{(m)} = \frac{N}{\varphi(W)\log N} \sum_{r_1,\dots,r_k} \frac{(y_{r_1,\dots,r_k}^{(m)})^2}{\prod_{i=1}^{k} g(r_i)} + O\left(\frac{(y_{max}^{(m)})^2 \varphi(W)^{k-2} N L_N (k-2)}{W^{k-1} D_0} \right)$$

$$+ O\left(\frac{y_{max}^2 N}{L_N(A)} \right).$$

Substituting the values for the $y^{(m)}$ variables squared from Step (2), using Step (2) to bound the $y^{(m)}_{max}$ and choosing as before A sufficiently large in terms of k, we get

$$S_2^{(m)} = \frac{\varphi(W) N L_R(2)}{W^2 \log N} \sum_{\substack{r_1,\ldots,r_k \\ \forall i \ (r_i, W)=1 \\ \forall i \neq j \ (r_i, r_j)=1 \\ r_m=1}} \left(\prod_{i=1}^{k} \frac{\mu(r_i)^2 \varphi(r_i)^2}{r_i^2 g(r_i)} \right) (F_{r_1,\ldots,r_k}^{(m)})^2$$

$$+ O \left(\frac{F_{max}^2 \varphi(W)^k N L_R(k)}{W^{k+1} D_0} \right).$$

(4) Next remove the coprime condition $(r_i, r_j) = 1$ in the same manner as in Step (2) of Lemma 6.6. Because for $p > 2$,

$$\frac{\varphi(p)^4}{p^4 g(p)^2} = \frac{1}{p^2} + O\left(\frac{1}{p^3}\right) \quad \text{and} \quad \frac{\varphi(p)^2}{p^2 g(p)} \leq \frac{1}{\varphi(p)},$$

the error this gives is of order

$$\frac{F_{max}^2 (\log R)^2 N \varphi(W)}{W^2 \log N} \left(\sum_{p > D_0} \frac{\varphi(p)^4}{p^4 g(p)^2} \right) \left(\sum_{\substack{r < R \\ (r, W)=1}} \frac{\mu(r)^2 \varphi(r)^2}{r^2 g(r)} \right)^{k-1}$$

$$\ll \frac{F_{max}^2 N \varphi(W)^k L_N(k)}{W^{k+1} D_0}.$$

(5) Finally, we estimate the sum which occurs in the leading term (recall $r_m = 1$) of $S_2^{(m)}$ in Step (3). Note that for all $p \in \mathbb{P}$ with $p > 2$, we have $\varphi(p)^2/(p^2 h(p) \leq 1/\varphi(p)$ and using Lemma 6.1(b). The sum is

$$\sum_{\substack{r_1,\ldots,r_{m-1},r_{m+1},\ldots,r_k \\ \forall i \ (r_i, W)=1}} \left(\prod_{\substack{1 \leq i \leq k \\ i \neq j}} \frac{\mu(r_i)^2 \varphi(r_i)^2}{r_i^2 g(r_i)} \right) (F_{r_1,\ldots,r_k}^{(m)})^2.$$

We proceed by summing over each value of i in turn using Lemma 6.5 with the new definitions:

$$\gamma(p) = \begin{cases} 1 - \frac{p^2 - 3p + 1}{p^3 - p^2 - 2p + 1}, & p \nmid W, \\ 0, & p \mid W. \end{cases}$$

Note that this clever choice of $\gamma(p)$ gives for $p \nmid W$,

$$h(p) := \frac{\gamma(p)}{p - \gamma(p)} = \frac{(p-1)^2}{p^2(p-2)} = \frac{\varphi(p)^2}{p^2 g(p)},$$

which is precisely what we need to apply the integral formula of Lemma 6.5. We also choose L, so

$$L \ll 1 + \sum_{p|W} \frac{\log p}{p} \ll \log D_0.$$

This then gives

$$S_2^{(m)} = \frac{\varphi(W)^k N L_R(k+1)}{W^{k+1} \log N} J_k^{(m)}(F) + O\left(\frac{F_{max}^2 N \varphi(W)^k L_N(k)}{W^{k+1} D_0}\right),$$

and completes the proof. $\qquad\qquad\qquad\qquad\qquad\qquad\qquad\qquad\qquad\qquad\qquad$ $\square$

6.6 Fundamental Lemmas

We now give Maynard's fundamental lemmas from which his main results flow. First recall the setting:

The function F is a fixed piecewise differentiable function supported on the simplex

$$\mathscr{R}_k := \{(x_1, \ldots, x_k) \in [0, 1]^k : \sum_{i=1}^{k} x_i \leq 1\},$$

and when $(\prod_{i=1}^{k} d_i, W) = 1$ (where $W = \prod_{p \leq D_0} p$, $D_0 = \logloglog N$), we have defined sieve λ's by

$$\lambda_{d_1,\ldots,d_k} := \left(\prod_{i=1}^{k} \mu(d_i) d_i\right) \sum_{\substack{r_1,\ldots,r_k \\ \forall i\ d_i|r_i \\ \forall i\ (r_i, W)=1}} \frac{\mu(\prod_{i=1}^{k} r_i)^2}{\prod_{i=1}^{k} \varphi(r_i)} F\left(\frac{\log r_1}{\log R}, \ldots, \frac{\log r_k}{\log R}\right).$$

If $(\prod_{i=1}^{k} d_i, W) \neq 1$, then $\lambda_{d_1,\ldots,d_k} := 0$. In addition, for N sufficiently large, we have $\mu(d)^2 = 1$, where $d := \prod_{i=1}^{k} d_i$.

Recall also, for θ the level of distribution of rational primes, that

$$I_k(F) := \int_0^1 \cdots \int_0^1 F(t_1, \ldots, t_k)^2\, dt_1 \ldots dt_k,$$

$$J_k^{(m)}(F) := \int_0^1 \cdots \int_0^1 \left(\int_0^1 F(t_1, \ldots, t_k)\, dt_m\right)^2 dt_1 \ldots dt_{m-1} dt_{m+1} \ldots dt_k,$$

$$M_k := \sup_{F \in \mathscr{F}_k} \frac{\sum_{m=1}^{k} J_k^{(m)}(F)}{I_k(F)},$$

$$r_k := \left\lceil \frac{\theta M_k}{2} \right\rceil,$$

$$w_n := \left(\sum_{\forall i\ d_i|n+h_i} \lambda_{d_1,\ldots,d_k}\right)^2.$$

Finally, in this recapitulation, the main objects of this investigation are the sums:

$$S_1 := \sum_{\substack{N \le n < 2N \\ n \equiv v_0 \bmod W}} \left(\sum_{\forall i \ d_i | n + h_i} \lambda_{d_1, \ldots, d_k} \right)^2,$$

$$S_2^{(m)} := \sum_{\substack{N \le n < 2N \\ n \equiv v_0 \bmod W}} \chi_{\mathbb{P}}(n + h_m) \left(\sum_{\forall i \ d_i | n + h_i} \lambda_{d_1, \ldots, d_k} \right)^2,$$

$$S_2 = \sum_{m=1}^{k} S_2^{(m)},$$

$$S := S_2 - \rho S_1 = \sum_{\substack{N \le n < 2N \\ n \equiv v_0 \bmod W}} \left(\left(\sum_{i=1}^{k} \chi_{\mathbb{P}}(n + h_i) \right) - \rho \right) w_n.$$

Lemma 6.8 *[138, proposition 4.1] If F is such that $I_k(F) \ne 0$ and for all m with $1 \le m \le k$, $J_k^{(m)}(F) \ne 0$, $R = N^{\theta/2 - \epsilon}$ for some small $\epsilon > 0$, then as $N \to \infty$*

$$S_1 = \frac{\varphi(W)^k N L_R(k)(1 + o(1))}{W^{k+1}} I_k(F),$$

$$S_2 = \frac{\varphi(W)^k N L_R(k+1)(1 + o(1))}{W^{k+1} \log N} \sum_{m=1}^{k} J_k^{(m)}(F).$$

Proof This is a summary of the results of Lemmas 6.6 and 6.7.   □

Note that the ratio of the errors in the expressions in Lemmas 6.6 and 6.7 to the main terms are in each case $O(1/D_0) \ll 1/\log\log\log N$). Thus, in a sense, the proof only just works! However, we can now leave the hard work of deriving estimates using the multidimensional sieve behind, and build the applications.

First we give a general condition on prime gaps in terms of the supremum M_k:

Lemma 6.9 (Maynard's lemma) *[138, proposition 4.2] Let $\mathcal{H} = \{h_1, \ldots, h_k\}$, with $0 \le h_1 < \cdots < h_k$, be a fixed admissible set of distinct integers with $k \ge 2$. Let the primes $\mathbb{P}$ have level of distribution θ for some θ with $0 < \theta < 1$. Let $\mathcal{S}_k$ be the set of all nonnegative piecewise differentiable functions F on $[0, 1]^k$ as defined previously and such that $I_k(F) \ne 0$ and $\prod_{m=1}^{k} J_k^{(m)}(F) \ne 0$, so both are positive. (Using $F \equiv 1$, the constant function, shows that $M_k > 0$.)*
  Then there are infinitely many integers $n \in \mathbb{N}$ such that for each such n at least $r_k := \lceil \theta M_k / 2 \rceil$ of the $n + h_i$, $1 \le i \le k$ are prime, and so

$$\liminf_{n \to \infty} p_{n+r_k-1} - p_n \le h_k - h_1.$$

Proof Let $\rho > 0$ be a given real parameter to be chosen later, S_1 and S_2 be as in Lemma 6.8 and $S := S_2 - \rho S_1$. Let $R = N^{\theta/2 - \epsilon}$ with N large and $\epsilon > 0$ small. By the definition of M_k, there exists $F_0 \in \mathcal{S}_k$ such that

$$\sum_{m=1}^{k} J_k^{(m)}(F_0) > (M_k - 2\epsilon) I_k(F_0).$$

By Lemma 6.8, there is a $\lambda_{d_1,\dots,d_k}$ with

$$S = S_2 - \rho S_1$$

$$= \frac{\varphi(W)^k N(L_R(k))}{W^{k+1}} \left(\frac{\log R}{\log N} \sum_{j=1}^{k} J_k^{(m)}(F_0) - \rho I_k(F_0) \right) (1 + o(1))$$

$$> \frac{\varphi(W)^k N(L_R(k)) I_k(F_0)}{W^{k+1}} \left((\theta/2 - \epsilon)(M_k - 2\epsilon) - \rho \right). \tag{6.10}$$

Next define $\rho = \theta M_k/2 - 2\delta$ with $\delta > 0$ sufficiently small in terms of ϵ so that both $\rho > 0$ and $S > 0$. For example, we could choose $\delta = \sqrt{\epsilon}$ with ϵ sufficiently small so

$$\frac{2}{\sqrt{\epsilon}} > M_k + \theta.$$

Therefore, at least one term in the sum over n in the definition of S must be positive. Since the weights w_n are nonnegative, this implies for each N sufficiently large there is at least one $n \in (N, 2N]$ such that at least $\lfloor \rho + 1 \rfloor$ of the $n + h_i$ are prime. Because when $\delta > 0$ is sufficiently small

$$\lfloor \rho + 1 \rfloor = \lceil \theta M_k/2 \rceil = r_k,$$

we get

$$\liminf_{n \to \infty} p_{n+r_k-1} - p_n \le h_k - h_1.$$

This completes the proof. $\qquad\square$

The following lemma is a standard result in optimization. It is used in Lemma 6.11 and in Maynard's algorithm described in Section 6.8.

Lemma 6.10 **(Rayleigh quotient)** *[138, lemma 7.3] Let A, B be real symmetric positive definite $n \times n$ matrices. Then the ratio*

$$\rho(x) := \frac{x^T A x}{x^T B x}$$

is a maximum for x an eigenvector of $B^{-1}A$ corresponding to its largest eigenvalue, and the maximum value is this largest eigenvalue.

Proof Since $\rho(\lambda x) = \rho(x)$ for all $x \ne 0$ and $\lambda \ne 0$, we can assume $x^T B x = 1$. Using Lagrange multipliers, the maximum of $x^T A x$ subject to the constraint $x^T B x = 1$ occurs at a stationary point of

$$L(x, \lambda) := x^T A x - \lambda(x^T B x - 1),$$

which satisfies for $1 \leq i \leq n$,

$$\frac{\partial L}{\partial x_i} = (((2A - 2\lambda B)x)_i = 0.$$

Thus, since B is invertible, when

$$B^{-1}Ax = \lambda x \implies \lambda = \frac{x^T A x}{x^T B x},$$

so the maximum value of $\rho(x)$ corresponds to the largest eigenvalue. $\qquad\square$

Maynard's lemma, in the context of gaps between consecutive primes, suggests we should seek values of k such that $r_k > 1$. Since $r_k := \lceil \theta M_k / 2 \rceil$, this leads to a relationship with the level of distribution of primes θ and lower bounds for M_k. Maynard's first and most elementary exploration of this is Lemma 6.11 which is given next.

Figure 6.4 is a plot of the approximate values of M_k computed using Maynard's algorithm implemented in PGpack as `RayleighQuotientMaynard` or `rqm`. These values are the result of optimizing over a finite-dimensional class of functions which will be described later, so they should be somewhat less than the best possible values for the M_k.

Recall the definitions:

$$I_k(F) := \int_0^1 \cdots \int_0^1 F(t_1, \ldots, t_k)^2 \, dt_1 \ldots dt_k,$$

$$J_k^{(m)}(F) := \int_0^1 \cdots \int_0^1 \left(\int_0^1 F(t_1, \ldots, t_k) \, dt_m \right)^2 dt_1 \ldots dt_{m-1} dt_{m+1} \ldots dt_k,$$

$$M_k := \sup_{F \in \mathscr{F}_k} \frac{\sum_{m=1}^k J_k^{(m)}(F)}{I_k(F)}.$$

We also define two symmetric polynomials:

$$P_1(t) := t_1 + \cdots + t_k,$$
$$P_2(t) := t_1^2 + \cdots + t_k^2.$$

Lemma 6.11 *[138, proposition 4.3] For $k \in \mathbb{N}$, we have*

(i) $M_5 > 2$,
(ii) $M_{105} > 4$,
(iii) *for all k sufficiently large, $M_k > \log k - 2 \log\log k - 1$.*

Proof (i) Let $k = 5$ and, guided perhaps by trial and error (see the aside later), define

$$P := (1 - P_1)P_2 + \frac{7}{10}(1 - P_1)^2 + \frac{1}{14}P_2 - \frac{3}{14}(1 - P_1),$$

and set $F = P$ for $t_1 + \cdots t_k \le 1$ and $F = 0$ otherwise. Using the PGpack functions $\mathtt{ComputeI}$ and $\mathtt{ComputeJ}$, we get

$$M_5 \ge \frac{\sum_{m=1}^{k} J_k^{(m)}(F)}{I_k(F)} = \frac{1417255}{708216} > 2.0015 > 2.$$

This completes the proof of part (i). As an aside, this choice of Maynard's for F is rather miraculous, and could have resulted from a very extensive search. In Figure 6.4, we have varied the coefficients $(7/10, 1/14, -3/14)$ used in the definition of P about these values along a line

$$\left(\frac{7}{10} - \frac{t}{1000}, \frac{1}{14} - \frac{t}{1000}, -\frac{3}{14} + \frac{t}{1000} \right)$$

for $-10 \le t \le 10$ to show that Maynard's choice of coefficients is very close to (at least a local) optimum. Note that the increments $t/1000$ are very small.

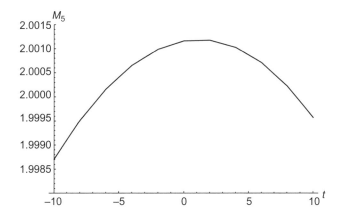

Figure 6.4 M_5 vs t for $-10 \le t \le 10$.

(ii) Let $k = 105$ and let P be a linear combination of all 42 polynomials $(1 - P_1)^b P_2^c$ with $0 \le b + 2c \le 11$. Let F be defined in terms of P. Then using the PGpack function $\mathtt{RayleighQuotientMaynard}$ (or $\mathtt{rqm}$), which encodes in a lexically scoped manner the function of Maynard, to first compute the coefficients of the quadratic forms B and A corresponding to $I_k(F)$ and $\sum_{m=1}^{k} J_k^{(m)}(F)$ respectively, and then the largest eigenvalue λ of $B^{-1}A$, we find using the Rayleigh quotient Lemma 6.10,

$$M_{105} \ge \lambda > 4.002 > 4.$$

(iii) This is the most difficult part and takes six steps. We give the proof of Maynard, which is related to that of Tao.

(1) Let $k > T > 0$ and let $g : [0, \infty) \to \mathbb{R}$ be piecewise differentiable with support in $[0, T]$. The key definition is the symmetric piecewise differentiable F defined by

$$F(t) := \begin{cases} \prod_{i=1}^k g(kt_i), & \text{if } \sum_{i=1}^k t_i < 1, \\ 0, & \text{otherwise.} \end{cases}$$

Since F is symmetric in the t_i, $J_k^{(m)}(F)$ is the same for all $1 \le m \le k$, so in this and the following steps, we use the notation $J_k := J_k^{(m)}$.

(2) Now observe that if the centre of mass integral expression for a line of density $g(x)^2$ satisfies

$$\frac{\int_0^\infty u g(u)^2 \, du}{\int_0^\infty g(u)^2 \, du} < 1,$$

then when k is large, the error removing the constraint $\sum_{i=1}^k t_i < 1$ in the definition of F is expected to be small, since the main contribution to the integrals defining I_k and J_k would come from the regions near the centre of mass. Thus when k is large, we will extend some integral upper limits to infinity.

Let $\gamma := \int_0^\infty g(u)^2 \, du$ and consider piecewise differentiable functions g which are not the zero function, so $\gamma > 0$. Then, using our chosen shape for F, we can write

$$I_k(F) = \int_0^1 \cdots \int_0^1 F(t_1, \ldots, t_k)^2 \, dt_1 \ldots dt_k \le \left(\int_0^\infty g(kt)^2 \, dt \right)^k = \frac{\gamma^k}{k^k}. \quad (6.11)$$

(3) In this step, we find a lower bound for $J_k(F)$. To do this, we restrict the outer integral in its definition to the region

$$\left\{ t \in [0,1]^k : \sum_{i=2}^k t_i < 1 - \frac{T}{k} \right\}.$$

Because the support of $g(kt)$ is in $[0, T/k]$ (see Step (1)), we can then write

$$J_k(F) \ge \int_{\substack{t_2,\ldots,t_k \ge 0 \\ \sum_{i=2}^k t_i < 1 - \frac{T}{k}}} \cdots \int \left(\int_0^{T/k} \left(\prod_{i=1}^k g(kt_i) \right) dt_1 \right)^2 dt_2 \ldots dt_k. \quad (6.12)$$

Next let

$$J_k'(F) := \int \cdots \int_{t_2,\ldots,t_k \geq 0} \left(\int_0^{T/k} \left(\prod_{i=1}^k g(kt_i) \right) dt_1 \right)^2 dt_2 \ldots dt_k$$

$$= \left(\int_0^\infty g(kt_1)\, dt_1 \right)^2 \left(\int_0^\infty g(kt)^2\, dt \right)^{k-1}$$

$$= \frac{\gamma^{k-1}}{k^{k+1}} \left(\int_0^\infty g(u)\, du \right)^2,$$

$$\tag{6.13}$$

and the error

$$E_k'(F) := \int \cdots \int_{\substack{t_2,\ldots,t_k \geq 0 \\ \sum_{i=2}^k t_i \geq 1 - \frac{T}{k}}} \left(\int_0^{T/k} \left(\prod_{i=1}^k g(kt_i) \right) dt_1 \right)^2 dt_2 \ldots dt_k$$

$$= \frac{1}{k^{k+1}} \left(\int_0^\infty g(u)\, du \right)^2 \int \cdots \int_{\substack{u_2,\ldots,u_k \geq 0 \\ \sum_{i=2}^k u_i \geq k - T}} \prod_{i=2}^k g(u_i)^2\, du_2 \ldots du_k.$$

$$\tag{6.14}$$

Thus, using (6.12) we can write $J_k(F) \geq J_k'(F) - E_k'(F)$.

(4) In this step, we derive a bound for the error $E_k'(F)$ in terms of k, μ, T and g. In Step (5), we show how to choose the parameters so that the error is small. For large k, we expect the bound for $E_k'(F)$ derived in Step (3) to be small if the centre of mass with density $g(x)^2$, is less than

$$\frac{k-T}{k-1} = 1 - \frac{T}{k} + o\left(\frac{1}{k} \right) \sim 1 - \frac{T}{k}.$$

Indeed, the final choice for T (see Step (6)) gives $T \ll k/\log k$. To this end, we constrain $g(x)$ so that

$$\mu := \frac{\int_0^\infty u g(u)^2\, du}{\int_0^\infty g(u)^2\, du} < 1 - \frac{T}{k}. \tag{6.15}$$

Let $g(x)$ be such that $\eta := (k-T)/(k-1) - \mu > 0$, and note that if

$$\sum_{i=2}^k u_i > k - T \implies \sum_{i=2}^k u_i > (k-1)\mu \implies 1 \leq \frac{1}{\eta^2} \left(\frac{1}{k-1} \left(\sum_{i=2}^k u_i \right) - \mu \right)^2.$$

We obtain an upper bound for $E'_k(F)$ if we multiply the integrand for the multiple integral factor of (6.14) by $\frac{1}{\eta^2}\left(\frac{1}{k-1}\sum_{i=2}^{k}u_i - \mu\right)^2$ and then remove the constraint $\sum_{i=2}^{k}u_i \geq k - T$, giving

$$E'_k(F) \leq \frac{1}{\eta^2 k^{k+1}}\left(\int_0^{\infty} g(u)\,du\right)^2$$

$$\times \int_0^{\infty}\cdots\int_0^{\infty}\left(\frac{\sum_{i=2}^{k}u_i}{k-1} - \mu\right)^2\left(\prod_{i=2}^{k}g(u_i)^2\right)du_2\ldots du_k.$$

Now we perform an operation which comes close to the dispersion method of Linnik – see Appendix G. Expanding the square containing μ, we get, for terms not of the form u_i^2, using $\mu\gamma = \int_0^{\infty}ug(u)^2\,du$, miraculously,

$$\int_0^{\infty}\cdots\int_0^{\infty}\left(\frac{2\sum_{2\leq i<j\leq k}u_iu_j}{(k-1)^2} - 2\mu\frac{\sum_{i=2}^{k}u_i}{k-1} + \mu^2\right)\left(\prod_{i=2}^{k}g(u_i)^2\right)du_2\ldots du_k$$

$$= -\frac{\mu^2\gamma^{k-1}}{k-1}.$$

Since $g(x) = 0$ for $x \notin [0, T]$ we have $u_j^2 g(u_j)^2 \leq Tu_j g(u_j)^2$, so for $2 \leq j \leq k$,

$$\int_0^{\infty}\cdots\int_0^{\infty}u_j^2\left(\prod_{i=2}^{k}g(u_i)^2\right)du_2\ldots du_k \leq T\gamma^{k-2}\int_0^{\infty}u_j g(u_j)^2\,du_j = \mu T\gamma^{k-1}.$$

Therefore, since there are $k - 1$ of these,

$$E'_k(F) \leq \frac{1}{\eta^2 k^{k+1}}\left(\int_0^{\infty}g(u)\,du\right)^2\left(\frac{\mu T\gamma^{k-1}}{k-1} - \frac{\mu^2\gamma^{k-1}}{k-1}\right)$$

$$\leq \frac{\mu T\gamma^{k-1}}{\eta^2 k^{k+1}(k-1)}\left(\int_0^{\infty}g(u)\,du\right)^2, \tag{6.16}$$

which is the upper bound we are seeking.

(5) Recall from (6.15) that we have $\mu < 1$ and can say

$$\eta = \frac{k-T}{k-1} - \mu > \left(\frac{k}{k-1}\right)\left(\frac{k-T}{k} - \mu\right) > \sqrt{\frac{k}{k-1}}\left(\frac{k-T}{k} - \mu\right),$$

which implies $k(1 - T/k - \mu)^2 \leq (k-1)\eta^2$. Combining these with the equations and estimates (6.11), (6.12), (6.13) and (6.16) and using $\mu < 1$ to replace μT in the numerator with T then gives

$$\frac{k J_k(F)}{I_k(F)} \geq \frac{\left(\int_0^\infty g(u) \, du\right)^2}{\int_0^\infty g(u)^2 \, du} \left(1 - \frac{T}{k(1 - T/k - \mu)^2}\right). \tag{6.17}$$

(6) We now find a maximum value for the lower bound obtained in Step (5) by maximizing $\int_0^T g(u) \, du$ subject to $\mu = \int_0^T u g(u)^2 \, du$ and $\gamma = \int_0^T g(u)^2 \, du$ for given μ and γ. By the method of Lagrange multipliers, we wish to find a function g and multipliers α, β to maximize

$$L(g, \alpha, \beta) := \int_0^T g(u) \, du - \alpha \left(\int_0^T g(u)^2 \, du - \gamma\right) - \beta \left(\int_0^T u g(u)^2 \, du - \mu\right).$$

The Euler–Lagrange necessary condition

$$\frac{\partial L}{\partial g} - \frac{d}{du} \frac{\partial L}{\partial g'} = 0,$$

since $g'(u)$ does not occur explicitly in L, gives for all $t \in [0, T]$,

$$\frac{\partial}{\partial g} \left(g(t) - \alpha g(t)^2 - \beta t g(t)^2\right) = 0 \implies g(t) = \frac{1}{2\alpha + 2\beta t}, \quad 0 \leq t \leq T$$

is one solution. Because we can multiply g by any positive constant and not change the ratio on the left in the estimate (6.17), we are able to consider the form $g(t) = 1/(1 + Bt)$, $B > 0$. Using this form for $g(u)$, we get

$$\int_0^T g(u) \, du = \frac{\log(1 + BT)}{B},$$

$$\gamma = \int_0^T g(u)^2 \, du = \frac{1}{B} \left(1 - \frac{1}{1 + BT}\right),$$

$$\mu \times \gamma = \int_0^T u g(u)^2 \, du = \frac{1}{B^2} \left(\log(1 + BT) - 1 + \frac{1}{1 + BT}\right).$$

Choosing T so $1 + BT = e^B$ gives $\int_0^T g(u) \, du = 1$, and with a bit of work we find

$$\mu = \frac{1}{1 - e^{-B}} - \frac{1}{B} \text{ and } T \leq \frac{e^B}{B}, \text{ which implies } \frac{1}{B}\left(1 - \frac{B}{e^B - 1} - \frac{e^B}{k}\right) \leq 1 - \frac{T}{k} - \mu. \tag{6.18}$$

Using these integrals and inequalities in the lower bound of the estimate (6.17) then gives

$$\frac{k J_k(F)}{I_k(F)} \geq \frac{B}{1 - e^{-B}} \left(1 - \frac{T}{k(1 - T/k - \mu)^2}\right)$$

$$\geq B \left(1 - \frac{B e^B}{k(1 - B/(e^B - 1) - e^B/k)^2}\right).$$

Now choose $B = \log k - 2 \log\log k > 0$, with k sufficiently large ($k \geq 8$ suffices), and using the inequality (6.18), we get

$$1 - \frac{k}{T} - \mu \geq \frac{1}{B}\left(1 - \frac{(\log k)^3}{k} - \frac{1}{(\log k)^2}\right) > 0.$$

Therefore, with these choices the required constraint $\mu < 1 - k/T$, which is (6.15), has been satisfied. Then using the given assignment $B = \log k - 2 \log\log k$, we get for k sufficiently large ($k \geq 37$ should suffice),

$$M_k \geq \frac{k J_k(F)}{I_k(F)} \geq B\left(1 - \frac{Be^B}{k\left(1 - \frac{B}{e^B-1} - \frac{e^B}{k}\right)^2}\right).$$

Note that since $B > 0$, we have

$$\frac{B}{e^B - 1} - \frac{B}{e^B} \leq \frac{1}{e^B}$$

and that

$$B\left(1 - \frac{Be^B}{k\left(1 - \frac{B}{e^B-1} - \frac{e^B}{k}\right)^2}\right) \geq B - 1$$

$$\iff Be^{B/2} \leq \sqrt{k}\left(1 - \frac{B}{e^B} + O(e^{-B}) + O(e^B/k)\right).$$

Thus using $B = \log k - 2 \log\log k$ and simplifying, we obtain a lower bound $B - 1$ provided

$$\frac{2 \log\log k}{\log k} \gg O\left(\frac{(\log k)^3}{k}\right),$$

which is true for all k sufficiently large ($k \geq 37$ should suffice). This completes the proof. $\qquad\square$

Remarks: Exploring a variation of the choice of B in Lemma 6.11(iii), let $B = \log k - \alpha \log\log k$ for $\alpha > 0$. Then to achieve $M_k \geq B - 1$, we would need with $B \to \infty$,

$$B\left(1 - \frac{Be^B}{k\left(1 - \frac{B}{e^B} - \frac{e^B}{k} + O(e^{-B})\right)^2}\right) \geq B - 1.$$

A manipulation shows this is equivalent to

$$1 \geq (\log k)^{1-\alpha/2}(1 + o(1)),$$

so necessarily we have $\alpha \geq 2$, and Maynard's choice is optimal for B of the given form. Similarly, for any choice $B = \beta \log k$ for $0 < \beta$ we must have $\beta < 1$ to meet the constraint (6.15). Also any choice $1 + BT = e^{\lambda T}$ with $\lambda \neq 1$ fails for the same reason.

Asymptotically, it appears the constant on the right should be -1 rather than -2.

This fundamentally simple but intricate clever derivation is very influential – see for example Theorem 6.17, Maynard's consecutive prime gaps theorem, in Section 6.9. It has improvement potential.

6.7 Integration Formulas

In this section, we derive some integration formulas which are used in Maynard's algorithm described in the following Section 6.8. Then in the culminating Section 6.9, the fundamental lemmas and Maynard's algorithm are combined with data on explicit admissible sets of integers to derive the main results. First we define some symmetric polynomials which will be integrated over the standard simplex.

$$\mathscr{R}_k := \left\{ x \in [0,1]^k : \sum_{i=1}^{k} x_i \leq 1 \right\},$$

$$\mathscr{S}_k := \{ F : [0,1]^k \rightarrow \mathbb{R} \mid \text{piecewise differentiable and supported on } \mathscr{R}_k \},$$

$$P_j(t) := \sum_{i=1}^{k} t_i^j, \ j \in \mathbb{N},$$

$$\binom{x}{r} := \frac{x(x-1)\cdots(x-r+1)}{r!}, \ x \in \mathbb{R}, \ r \in \mathbb{Z}, \ r \geq 0,$$

where we call $P_j(t)$ the jth symmetric (power sum) polynomial and $\binom{x}{r} \in \mathbb{Q}[x]$.

We now give three lemmas where exact formulas for the integrals over the standard simplex of multinomials in symmetric polynomials are derived. These are valuable in that they can be precomputed as exact polynomial expressions in a particular variable and reused, speeding up computations. The first is the well-known beta function integral. The second provides a higher-dimensional version, and the third gives evaluations for $I_k(F)$ and $J_k^{(m)}(F)$ for a particular class of functions F with support in $\mathscr{R}_k$.

Lemma 6.12 *For all integers $a,b \geq 0$, we have*

$$\int_0^1 t^a(1-t)^b \, dt = \frac{a!\,b!}{(a+b+1)!}.$$

Proof This is by induction on a, say. The cases $a = 0$ or $b = 0$ are immediate, and the inductive step follows using integration by parts. $\qquad \square$

That lemma is now used to derive a more general expression.

Lemma 6.13 (**general beta-function integral**) *[138, lemma 7.1] For $j \geq 1$, $k \geq 2$ and $a, b \geq 0$, let*

$$G_{b,j}(x) := b! \sum_{r=1}^{b} \binom{x}{r} \sum_{\substack{b_1,\dots,b_r \geq 1 \\ \sum_{i=1}^{r} b_i = b}} \prod_{i=1}^{r} \frac{(jb_i)!}{b_i!}$$

be a polynomial in $\mathbb{Q}[x]$. Then for all $k \geq 2$, we have

$$\int \cdots \int_{\mathscr{R}_k} (1 - P_1)^a P_j^b \, dt_1 \dots dt_k = \frac{a!}{(k + jb + a)!} G_{b,j}(k).$$

Proof (**1**) Consider first integration with respect to t_1 making the substitution

$$v := \frac{t_1}{1 - \sum_{i=2}^{k} t_i}.$$

This gives, using Lemma 6.12, for $a_i \geq 0$,

$$\int_0^{1 - \sum_{i=2}^{k} t_i} \left(1 - \sum_{i=1}^{k} t_i\right)^a \left(\prod_{i=1}^{k} t_i^{a_i}\right) dt_1 = \left(\prod_{i=2}^{k} t_i^{a_i}\right) \left(1 - \sum_{i=2}^{k} t_i\right)^{a+a_1+1}$$

$$\times \int_0^1 (1 - v)^a v^{a_1} \, dv$$

$$= \frac{a! \, a_1!}{(a + a_1 + 1)!} \left(\prod_{i=2}^{k} t_i^{a_i}\right) \left(1 - \sum_{i=2}^{k} t_i\right)^{a+a_1+1}.$$

Using this expression, which is part of the case $k = 1$, the integral

$$\int \cdots \int_{\mathscr{R}_k} \left(1 - \sum_{i=1}^{k} t_i\right)^a \left(\prod_{i=1}^{k} t_i^{a_i}\right) dt_1 \dots dt_k = \frac{a! \prod_{i=1}^{k} a_i!}{(k + a + \sum_{i=1}^{k} a_i)!}, \qquad (6.19)$$

then follows by induction on k. For example, the second application would use

$$v := \frac{t_2}{1 - \sum_{i=3}^{k} t_i}$$

and

$$\frac{a! \, a_1!}{(a + a_1 + 1)!} \times \int_0^1 v^{a_2} (1 - v)^{a+a_1+1} \, dv = \frac{a! \, a_1! \, a_2!}{(a + a_1 + a_2 + 2)!}.$$

(2) Expanding using the multinomial theorem, we have

$$P_j^b = \sum_{\substack{b_1,\ldots,b_k \geq 0 \\ \sum_{i=1}^k b_i = b}} \frac{b!}{\prod_{i=1}^k b_i!} \prod_{i=1}^k t_i^{jb_i}.$$

Using this with (6.19) then gives

$$\int_{\mathcal{R}_k} \cdots \int (1 - P_1)^a P_j^b \, dt_1 \ldots dt_k = \frac{b! \, a!}{(k+a+jb)!} \sum_{\substack{b_1,\ldots,b_k \geq 0 \\ \sum_{i=1}^k b_i = b}} \prod_{i=1}^k \frac{(jb_i)!}{b_i!}.$$

(3) Finally, for each r with $1 \leq r \leq k$, there are $\binom{k}{r}$ ways of choosing r of the $b_1, \ldots, b_k$ to be nonzero, so we can rewrite the final sum on the right as

$$\sum_{\substack{b_1,\ldots,b_k \geq 0 \\ \sum_{i=1}^k b_i = b}} \prod_{i=1}^k \frac{(jb_i)!}{b_i!} = \sum_{r=1}^b \binom{k}{r} \sum_{\substack{b_1,\ldots,b_r \geq 1 \\ \sum_{i=1}^r b_i = b}} \prod_{i=1}^r \frac{(jb_i)!}{b_i!}.$$

This completes the proof. $\qquad\qquad\qquad\qquad\qquad\qquad\qquad\qquad\qquad\qquad\qquad$ □

Now let $P(t_1, \ldots, t_k)$ be a multinomial and set for $t \in \mathcal{R}_k$, $F(t) = P(t_1, \ldots, t_k)$. If $t \in [0,1]^k \setminus \mathcal{R}_k$, let $F(t) = 0$, so $F \in \mathscr{S}_k$. Recall the definitions

$$P_1(t) := t_1 + \cdots + t_k \text{ and } P_2(t) := t_1^2 + \cdots + t_k^2.$$

The following expressions for evaluating the fundamental objects $I_k(F)$ and $J_k^{(m)}(F)$ might be regarded by some as extreme. However, for computer evaluation purposes, they are quite superior to numerical integration for small values of the bound d.

Lemma 6.14 (**explicit functional evaluations**) *[138, lemma 72] Define the multinomial $P := \sum_{i=1}^d a_i (1 - P_1)^{b_i} P_2^{c_i}$, where $d \geq 0$, the $a_i \in \mathbb{R}$ and $b_i, c_i, c_i' \in \mathbb{Z}$ satisfy $b_i, c_i, c_i' \geq 0$. Let for $t \in \mathcal{R}_k$, $F(t) = P(t_1, \ldots, t_k)$. If $t \in [0,1]^k \setminus \mathcal{R}_k$, let $F(t) = 0$, so $F \in \mathscr{S}_k$. Define*

$$\gamma(b_i, b_j, c_i, c_j, c_i', c_j')$$
$$:= \frac{b_i! \, b_j! \, (2c_i - 2c_i')! \, (2c_j - 2c_j')! \, (b_i + b_j + 2c_i + 2c_j - 2c_i' - 2c_j' + 2)!}{(b_i + 2c_i - 2c_i' + 1)! \, (b_j + 2c_j - 2c_j' + 1)!}.$$

Then for $k \geq 2$, we get

$$I_k(F) = \sum_{1 \leq i, j \leq d} a_i a_j \frac{(b_i + b_j)! \, G_{c_i + c_j, 2}(k)}{(k + b_i + b_j + 2c_i + 2c_j)!} \quad and$$

$$J_k^{(m)}(F) = \sum_{1 \leq i, j \leq d} a_i a_j \sum_{c_i'=0}^{c_i} \sum_{c_j'=0}^{c_j} \binom{c_i}{c_i'} \binom{c_j}{c_j'} \frac{\gamma(b_i, b_j, c_i, c_j, c_i', c_j') G_{c_i'+c_j', 2}(k-1)}{(k + b_i + b_j + 2c_i + 2c_j + 1)!},$$

where $G_{b,j}(x)$ is defined in the statement of Lemma 6.13.

Proof **(1)** The calculation of $I_k(F)$ is straightforward. Using Lemma 6.13, we have

$$I_k(F) = \int \cdots \int_{\mathcal{R}_k} P^2 \, dt_1 \ldots dt_k$$

$$= \sum_{1 \leq i, j \leq d} a_i a_j \int \cdots \int_{\mathcal{R}_k} (1 - P_1)^{b_i + b_j} P_2^{c_i + c_j} \, dt_1 \ldots dt_k$$

$$= \sum_{1 \leq i, j \leq d} a_i a_j \frac{(b_i + b_j)! \, G_{c_i + c_j, 2}(k)}{(k + b_i + b_j + 2c_i + 2c_j)!}.$$

(2) Recall

$$J_k^{(m)} := \int_0^1 \cdots \int_0^1 \left(\int_0^1 F(t_1, \ldots, t_k) \, dt_m \right)^2 dt_1 \ldots dt_{m-1} dt_{m+1} \ldots dt_k.$$

To calculate the $J_k^{(m)}(F)$, note that the symmetry of F implies $J_k^{(m)}(F)$ is independent of m, so we need calculate only $J_k^{(1)}(F)$. To do this, first let $P_1' = P_1 - t_1$ and $P_2' = P_2 - t_1^2$. Then, using Lemma 6.12,

$$\int_0^{1-P_1'} (1 - P_1)^b P_2^c \, dt_1 = \sum_{c'=0}^{c} \binom{c}{c'} \left(\sum_{i=2}^{k} t_i^2 \right)^{c'} \int_0^{1-P_1'} \left(1 - \sum_{i=1}^{k} t_i \right)^b t_1^{2c-2c'} \, dt_1$$

$$= \sum_{c'=0}^{c} \binom{c}{c'} (P_2')^{c'} (1 - P_1')^{b+2c-2c'+1}$$

$$\times \int_0^1 (1 - u)^b u^{2c-2c'} \, du$$

$$= \sum_{c'=0}^{c} \binom{c}{c'} (P_2')^{c'} (1 - P_1')^{b+2c-2c'+1} \frac{b! \, (2c - 2c')!}{(b + 2c - 2c' + 1)!}.$$

Squaring, we get

$$\left(\int_0^1 F(t)\,dt_1\right)^2 = \left(\sum_{i=1}^d a_i \int_0^{1-\sum_{j=2}^k t_j} (1-P_1)^{b_i} P_2^{c_i}\,dt_1\right)^2$$

$$= \sum_{1\le i,\,j\le d} a_i a_j \sum_{c_i'=0}^{c_i}\sum_{c_j'=0}^{c_j} \binom{c_i}{c_i'}\binom{c_j}{c_j'}(1-P_1')^{b_i+b_j+2c_i+2c_j-2c_i'-2c_j'+2}$$

$$\times (P_2')^{c_i'+c_j'} \frac{b_i!\,b_j!\,(2c_i-2c_i')!\,(2c_j-2c_j')!}{(b_i+2c_i-2c_i'+1)!\,(b_j+2c_j-2c_j'+1)!}.$$

Using Lemma 6.13 gives

$$\int_{\mathcal{R}_{k-1}}\!\!\cdots\int (1-P_1')^b(P_2')^c\,dt_2\ldots dt_k = \frac{b!\,G_{c,2}(k-1)}{(k+b+2c-1)!}.$$

Combining these equations completes the derivation. □

6.8 Maynard's Algorithm

In this section, we give a descriptive breakdown of Maynard's algorithm to compute a lower bound for M_k which was used for Lemma 6.11(ii) and is implemented in PGpack as `RayleighQuotientMaynard`.

(1) Set the integers $n \ge 1$ and $k \ge 2$ and the precision parameters $\{d_1, d_2, d_3\}$.

(2) Calculate all pairs of nonnegative integers (u, v) such that $u + 2v \le 2n + 1$ and store them in an array $A = ((u_j, v_j) : 1 \le j \le N)$.

(3) Form a polynomial with symbolic coefficients $a[j]$

$$F(x_1, \ldots, x_k) = \sum_{j=1}^N a[j](1-P_1)^{u_j} \times P_2^{v_j},$$

where $P_1 = x_1 + \cdots + x_k$ and $P_2 = x_1^2 + \cdots + x_k^2$ are unevaluated.

(4) Compute a matrix B corresponding to the $I(F)$ integral, which is a quadratic form in the $a[j]$, using the integral formulas of Lemmas 6.12 through 6.14.

(5) Similarly, compute a matrix A corresponding to the $J_1(F)$ integral. Both matrices will be symmetric and positive definite.

(6) Let $C = Inverse[B].A$, where "." represents matrix multiplication. This matrix will have rational number entries.

(7) Take a floating point form of C to precision d_1, Compute the eigenvector corresponding to the maximum eigenvalue, and then find a rational number approximation to this vector to precision d_2 and call it x, regarded as a column vector.

(8) Calculate the rational number (Lemma 6.10)

$$r := k\frac{x^T.A.x}{x^T.B.x}.$$

(9) Return a floating point approximation to r to precision d_3.

   Remarks: We verified Maynard's resulting prime gap of 600 using this approach, but spent little time attempting to improve it. This was because Polymath had obtained a better result as reported in Chapter 7. They used a richer set of polynomials and perturbed the standard simplex to do so. However, we did use Steps (6) through (9) of Maynard's algorithm as one option for `RayleighQuotient Polymath` in order that we might estimate the (slow) progress of the Polymath approach, since Maynard's algorithm appeared to give more precise information on the approximate optimal value of M_k,

6.9 Main Theorems

The first main theorem shows that an infinite number of consecutive pairs of primes are distant apart at most 600. This improved Polymath8a's bound of 4,680, which had been derived using a modified Zhang approach described in Chapter 8. It demonstrates the value of using a multidimensional sieve and optimization. Limits to this approach were derived by Polymath8b and are set out in Chapter 7.

   Recall we have defined the level of distribution of primes to be θ if it is the supremum of values $\theta - \epsilon$, which can be used as the range of moduli $q \leq x^{\theta-\epsilon}$ in the Bombieri–Vinogradov theorem. Thus unconditionally we can choose any $\theta \leq 1/2$.

   Theorem 6.15   **(Maynard's prime gap)** *[138, theorem 1.3]*

$$\liminf_{n\to\infty}(p_{n+1} - p_n) \leq 600.$$

Proof   Let $k = 105$. By Lemma 6.11(ii) we get $M_{105} > 4$. By Bombieri–Vinogradov's estimate Lemma 4.6, for every $\epsilon > 0$ the primes have a level of distribution $\theta = \frac{1}{2} - \epsilon$. Thus we can choose ϵ so that $\theta M_{105}/2 > 1$. Hence, in the notation of Lemma 6.9, $r_k = \lceil \theta M_k/2 \rceil \geq 2$. Therefore, for any admissible set $\mathcal{H} = \{h_1, \ldots, h_k\}$ with h_i in increasing order and with 105 elements, we have

$$\liminf_{n\to\infty}(p_{n+1} - p_n) \leq h_k - h_1.$$

Using the admissible set (which has minimal width for a given k) from the database of Engelsma [43],

$$\mathcal{H}_{105} = \{0, 10, 12, 24, 28, 30, 34, 42, 48, 52, 54, 64, 70, 72, 78, 82, 90, 94,$$
$$100, 112, 114, 118, 120, 124, 132, 138, 148, 154, 168, 174, 178, 180, 184, 190,$$
$$192, 202, 204, 208, 220, 222, 232, 234, 250, 252, 258, 262, 264, 268, 280, 288,$$
$$294, 300, 310, 322, 324, 328, 330, 334, 342, 352, 358, 360, 364, 372, 378, 384,$$
$$390, 394, 400, 402, 408, 412, 418, 420, 430, 432, 442, 444, 450, 454, 462, 468,$$
$$472, 478, 484, 490, 492, 498, 504, 510, 528, 532, 534, 538, 544, 558, 562, 570,$$
$$574, 580, 582, 588, 594, 598, 600\},$$

we get $h_{105} - h_1 = H(105) = 600$. This completes the proof $\qquad\square$

In Theorem 6.16, we show that even assuming a level of distribution of primes as large as possible, we cannot get close to demonstrating conditionally the twin prime conjecture using these methods.

Theorem 6.16 (**Maynard's conditional prime gap**) *[138, theorem 1.4] Assume the Elliott–Halberstam conjecture in the form that the primes are assumed to have a level of distribution θ for all $0 < \theta < 1$. Then*

$$(i)\ \liminf_{n\to\infty}(p_{n+1} - p_n) \leq 12\ and$$

$$(ii)\ \liminf_{n\to\infty}(p_{n+2} - p_n) \leq 600.$$

Proof For any ϵ in the range $0 < \epsilon < 1$, we may assume, by the Elliott–Halberstam conjecture, the primes have a level of distribution $\theta = 1 - \epsilon$.

(i) By Lemma 6.11(i), we have $M_5 > 2$, so we can choose ϵ sufficiently small so that $\theta M_5 > 2 \implies r_5 = \lceil \theta M_5/2 \rceil \geq 2$. Hence, using Lemma 6.9 and choosing $\mathcal{H}_5 = \{0, 4, 6, 10, 12\}$, we get

$$\liminf_{n\to\infty}(p_{n+1} - p_n) \leq 12.$$

(ii) By Lemma 6.11(ii), we have $M_{105} > 4$, so we can choose ϵ sufficiently small so that $\theta M_{105} > 4 \implies r_{105} = \lceil \theta M_{105}/2 \rceil \geq 3$. Hence, using Lemma 6.9 again, choosing $\mathcal{H}_{105}$ as in Theorem 6.15, we get

$$\liminf_{n\to\infty}(p_{n+2} - p_n) \leq 600.$$

This completes the proof.

In Table 6.1, BV represents the application of Bombieri–Vinogradov theorem Lemma 4.6, and EH the assumption the level of distribution of primes θ is true for all $0 < \theta < 1$. The PGpack function used to make this table was

Table 6.1 *Values of r_k and prime gaps computed using Maynard's algorithm.*

k	d	r_k BV	r_k EH	$H(k)$
2	5	1	1	2
3	5	1	1	6
4	5	1	1	8
5	5	1	1	12
6	5	1	2	16
⋮	⋮	⋮	⋮	⋮
100	6	1	2	558
101	6	1	2	572
102	6	2	3	576
103	6	2	3	578
104	6	2	3	590
105	6	2	3	600

`RayleighQuotientMaynard` with arguments k and d, the degree bound. Here the reader will recall also

$$r_k := \left\lceil \frac{\theta M_k}{2} \right\rceil \text{ and } \liminf_{n \to \infty} p_{n+r_k-1} - p_n \le H(k),$$

where $H(k)$ is the minimal width of an admissible k-tuple.

In Figure 6.5, we plot the lower bound for M_k computed using `Rayleigh QuotientMaynard` against k, using the same range as in Table 6.1.

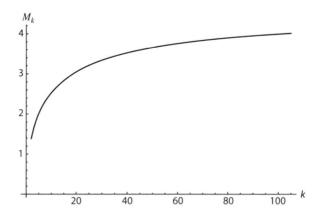

Figure 6.5 M_k vs k for $2 \le k \le 105$.

The next theorem represents a significant breakthrough result. It shows we have bounded gaps for arbitrary sets of consecutive primes, with the upper bound for the gap size depending on the number of primes. The upper bound is not regarded as best possible. But see the remarks following the proof.

Theorem 6.17 **(Maynard's consecutive primes gap)** *[138, theorem 1.1] For all* $m \in \mathbb{N}$, *there is an absolute constant such that*

$$\liminf_{n \to \infty}(p_{n+m} - p_n) \ll m^3 e^{4m}.$$

Proof Using Bombieri–Vinogradov's theorem, Lemma 4.6, we can assume the primes have a level of distribution $\theta = \frac{1}{2} - \epsilon$. By Lemma 6.11(iii), for k sufficiently large we have

$$\frac{\theta M_k}{2} \geq \left(\frac{1}{4} - \frac{\epsilon}{2}\right)(\log k - 2\log\log k - 2). \tag{6.20}$$

Setting $\epsilon = 1/k$ and $C = e^3$, if $k = \lceil Cm^2 \exp(4m) \rceil$ then we **claim** $\theta M_k/2 > m$. To see this, we derive successively

$$k \sim Cm^2 e^{4m}$$

$$\log k \sim 4m + 2\log m + \log C$$

$$= 4m\left(1 + \frac{2\log m + \log C}{4m}\right) \text{ and}$$

$$\log\log k \sim \log(4m) + \log\left(1 + \frac{2\log m + \log C}{4m}\right)$$

$$< \log(4m) + \frac{2\log m + \log C}{4m}.$$

Therefore,

$$\frac{\theta M_k}{2} \geq \left(\frac{1}{4} - \frac{\epsilon}{2}\right)(\log k - 2\log\log k - 2)$$

$$> \frac{4m + 2\log m + \log C}{4} - \frac{1}{2}\log(4m) - \frac{2\log m + \log C}{8m} + o(1)$$

$$= m + \frac{1}{4}\log C - \frac{1}{2}\log 4 + o(1) > m,$$

where C and the implied constants are absolute.

It follows that for any admissible $\mathcal{H}_k$, for infinitely many integers n at least $m+1$ of the $n + h_i$ (the i's depending on n) must be prime. Choose the prime k-tuple

$$\mathcal{H}_k := \{p_{\pi(k)+1}, \ldots, p_{\pi(k)+k}\},$$

which is such that no element is a multiple of a prime less than or equal to k, and for any prime $p > k$ we have $\upsilon_p(\mathcal{H}_k) \leq k < p$. Therefore, $\mathcal{H}_k$ is admissible and, in addition,

$$\text{width}(\mathcal{H}_k) = p_{\pi(k)+k} - p_{\pi(k)+1} \ll k \log k.$$

Hence, since $k = \lceil Cm^2 \exp(4m) \rceil$ we get

$$\liminf_{n \to \infty}(p_{n+m} - p_n) \ll k \log k \ll m^3 \exp(4m).$$

This completes the proof.   □

Remarks: In the final estimate, the first implied constant is almost certainly less than 2 and the second less than 50. Taking $k = \lceil Cm^\beta \exp(\gamma m) \rceil$, using (6.20), gives $\theta M_k > m$ provided $k = \lceil Cm^\beta \exp(\gamma m) \rceil$ with $\beta \geq 2$ and $\gamma \geq 4$. Thus again, using his method, Maynard's choice is optimal.

The next great result, Maynard's Theorem 6.18, shows that for any given fixed $m \geq 2$, a positive proportion of all m-tuples, appropriately counted, satisfies the prime m-tuples conjecture.

> **Theorem 6.18** (**Maynard's positive relative density**) *[138, theorem 1.2] Let $m \in \mathbb{N}$ be fixed and $r > m$ sufficiently large depending on m, and let $\mathcal{A} = \{a_1, a_2, \ldots, a_r\}$ be any given set of **any** r distinct integers. Then*
>
> $$\frac{\#\{\{h_1, \ldots, h_m\} \subset \mathcal{A}\} : \text{all of } n + h_j \text{ are prime for infinitely many } n\}}{\#\{\{h_1, \ldots, h_m\} \subset \mathcal{A}\}} \gg_m 1,$$
>
> *where the implied constant depends on m but not on $\mathcal{A}$.*

Proof **(1)** Following the proof of Theorem 6.17, for a given m let $k = \lceil Cm^2 \exp(4m) \rceil$. If $\mathcal{H} = \{h_1, \ldots, h_k\}$ is admissible, by Theorem 6.17, since the number of subsets of $\mathcal{H}$ of size m is finite, and there exists a fixed subset $\{h'_1, \ldots, h'_m\} \subset \mathcal{H}$ such that for an infinite set of positive integers n, the integers $n + h'_i$ are all prime for $1 \leq i \leq m$.

(2) Now let $\mathcal{A}_2$ be the set derived from $\mathcal{A} = \{a_1, \ldots, a_r\}$ by removing for each prime $p = 2, 3, \ldots \leq k$ all elements of $\mathcal{A}$ which are in one of the residue classes modulo p and which has the minimal number of elements among the modulo p residue classes. Then, because k depends on at most m,

$$s = s_{\mathcal{A}} := \#\mathcal{A}_2 \geq r \prod_{p \leq k}(1 - 1/p) \gg_m r.$$

In addition, because it cannot cover all residue classes modulo p for any prime $p \leq k$, every subset of $\mathcal{A}_2$ having size k must be admissible. Because we can choose r sufficiently large and $s = \#\mathcal{A}_2$, we can assume $s > k$.

(3) There are $\binom{s}{k}$ distinct subsets $\mathcal{H} \subset \mathcal{A}_2$ of size k, and we have seen in Step (2) that they are all admissible. Therefore, by Step (1) each contains a subset $\mathcal{H}' = \{h'_1, \ldots, h'_m\} \subset \mathcal{A}_2$ which satisfies $n + h'_i$ is prime for an infinite number of positive integers n and for each such n for all $1 \leq i \leq m$ (i.e., a subset which satisfies the prime m-tuples conjecture). But any given admissible set $\mathcal{H} \subset \mathcal{A}_2$ of size m is contained in $\binom{s-m}{k-m}$ sets $\mathcal{H}' \subset \mathcal{A}_2$ of size k. Therefore, since for each $\mathcal{H}$ there is at least one $\mathcal{H}'$ and for each $\mathcal{H}'$ there are $\binom{s-m}{k-m}$ sets $\mathcal{H}$, we have

$$\#\{\text{sets } \mathcal{H}' \text{ satisfying the prime } m\text{-tuples conjecture}\} \times \binom{s-m}{k-m} \geq \#\{\text{sets } \mathcal{H}\},$$

therefore, there are at least

$$\frac{\binom{s}{k}}{\binom{s-m}{k-m}} = \frac{s!\,(k-m)!}{(s-m)!\,k!} \gg_m \frac{s!}{(s-m)!} \gg_m s^m \gg_m r^m \qquad (6.21)$$

admissible sets $\mathcal{H}' \subset \mathcal{A}_2$ which satisfy the prime m-tuples conjecture. Because there are $\binom{r}{m} \leq r^m$ subsets of $\mathcal{A}$ of size m, we get, since r and k depend on m,

$$\frac{\#\{\{h_1, \ldots, h_m\} \subset \mathcal{A} : \text{each } n + h_j \text{ is prime for infinitely many } n\}}{\#\{\{h_1, \ldots, h_m\} \subset \mathcal{A}\}}$$

$$\gg_m \frac{r^m}{r^m} = 1.$$

This completes the proof. $\qquad\qquad\qquad\qquad\qquad\qquad\qquad\qquad\qquad\qquad\qquad\qquad$ □

Remarks: (1) In the estimate (6.21), note that k, r, s depend on m as does the implied constant as $\mathcal{A}$ varies over all subsets with r elements. It would be a useful exercise to find an explicit positive lower bound for the proportion as a function of m, which would require an explicit estimate for r and s as functions of m.

(2) The count in the denominator of the ratio is *all* subsets of $\mathcal{A}_2$ of size m. Thus in terms of the given measure, the proportion of subsets which satisfy the prime m-tuples conjecture is a positive proportion of all subsets of size m.

6.10 End Notes

Maynard made significant further advances and applications of these results and ideas, some of which are reported in several sections in Chapter 9. Polymath8a, on becoming aware of Maynard's work, appears to have ceased work refining and enhancing Zhang's approach, and under the name Polymath8b, and using Tao's ideas on how to treat the multidimensional form of Selberg's sieve, began to finesse the work of Maynard/Tao, arriving in a short space of time with a significant reduction of the best prime gap bound. This very reduced account is possibly only somewhat close to what actually happened! Since 2014, the Polymath8b gap size of 246, reported in detail in the following Chapter 7, has not apparently been superseded.

The PGpack functions ComputeI, ComputeJ, IntegrateRk and RayleighQuotient-Maynard relate to the work of this chapter. See the manual entries in Appendix I.

7

Polymath's Refinements of Maynard's Results

7.1 Introduction

Polymath8a had already made substantial improvements to Zhang's work, which was reported in Chapter 5. Some of their improvements are reported in Chapter 8. They might well have continued in this direction had it not been for the advent of Maynard's discoveries (as reported in Chapter 6), appearing first on the preprint server ArXiv, for example. A new Polymath subproject, Polymath8b, was formed, led again by Terence Tao (Figure 7.1), who had been working independently along similar lines to Maynard, but with a more analytical approach reported in this chapter. Again substantial improvements were made very quickly. For example, Polymath8b showed there are limits to Maynard's method. In particular, when using a standard simplex, getting a smaller bound than $p_{n+1} - p_n \leq 270$ for an infinite set of $n \in \mathbb{N}$ would require a significantly new idea. They then perturbed the simplex to derive $p_{n+1} - p_n \leq 246$, but did not show this was optimal for such simplices.

It is an impossible task to even hint at the impact Terence Tao has had on mathematics, let alone number theory, and this will not be attempted here. He was born in South Australia in 1975 and is now a faculty member at UCLA. His 1996 Princeton Ph.D. was supervised by Ellias Stein, and had the title "Three Regularity Results in Harmonic Analysis." The award of the Fields Medal is only one of a whole host of prizes and awards. He has been compared with David Hilbert, but mathematics is now a much deeper and broader subject than in that "Polymath's" day, so maybe Hilbert should be compared with Tao!

Tao was awarded a Fields Medal at the 2006 ICM at age thirty-one. The citation read: "For his contributions to partial differential equations, combinatorics, harmonic analysis and additive number theory". An extract from the American Mathematical Society describes his achievements up to that time (Notices of the AMS, October 2006):

Figure 7.1 Terence Tao (1975–). Photo: Steve Jennings/ Stringer/Getty Images.

Terence Tao is a supreme problem-solver whose spectacular work has had an impact across several mathematical areas. He combines sheer technical power, an other-worldly ingenuity for hitting upon new ideas, and a startlingly natural point of view that leaves other mathematicians wondering, "Why didn't anyone see that before?" At 31 years of age, Tao has written over eighty research papers, with over thirty collaborators, and his interests range over a wide swath of mathematics, including harmonic analysis, nonlinear partial differential equations, and combinatorics. "I work in a number of areas, but I don't view them as being disconnected," he said in an interview published in the Clay Mathematics Institute Annual Report. "I tend to view mathematics as a unified subject and am particularly happy when I get the opportunity to work on a project that involves several fields at once."

In this chapter, we do not report on all of Polymath8b's discoveries, but take a relatively short route through Tao/Polymath8b's derivation of the multidimensional Selberg sieve lemmas, the perturbation technique for Maynard's standard simplex, and the algorithm Polymath developed to derive the bound 246. We have included the derivation of the bounds which show the methods are limited, the analytic bounds and the lower bound of Ignace Bogaert using the Krylov basis method. We have included a new bound which is suitable for small values of the perturbation parameter, and which could be useful for future developments of Polymath8b's ideas.

PGpack has a lexically scoped (that is, all variables have as local a scope as possible, protecting the user from the curse of global variables) implementation of Polymath8b's algorithm for deriving their famous bound, and the author's implementation of the Krylov method. The latter is very computer memory intensive (Ignace Bogaert reported using 64 GBytes of RAM). Using months of CPU time and a combination of Polymath's and Maynard's approach, the author was able to verify but not improve on $p_{n+1} - p_n \leq 246$.

In the "End Note" Section 7.11, we summarize some of Polymath8b's additional results. For details, the reader might well consult the original article and trace alternative routes through the proofs. They used a "batched" approach to proof description, which this writer found particularly challenging to follow.

7.2 Definitions

We begin with some basic definitions: x is a suitably large real variable.

$$w := \log\log\log x,$$

$$W := \prod_{p \leq w} p,$$

$$B := \frac{\varphi(W)}{W}(\log x),$$

$$\theta(n) := \log n \text{ for } n \in \mathbb{P} \text{ and } 0 \text{ otherwise.}$$

Let $k \geq 2$ and $m \in \mathbb{N}$ be integers and $F \colon [0,\infty)^k \to \mathbb{R}$ a compactly supported square integrable function. Let $f \colon [0,\infty) \to \mathbb{R}$ be smooth and compactly supported on $(0,\infty)$. Define the functionals

$$I(F) := \int_{[0,\infty)^k} F(t_1, \ldots, t_k)^2 \, dt_1 \ldots dt_k,$$

$$J_i(F) := \int_{[0,\infty)^{k-1}} \left(\int_0^\infty F(t_1, \ldots, t_k) dt_i \right)^2 dt_1 \ldots dt_{i-1} dt_{i+1} \ldots dt_k, \ 1 \leq i \leq k,$$

$$\lambda_f(n) := \sum_{d \mid n} \mu(d) f \left(\frac{\log d}{\log x} \right).$$

Let $\mathscr{L}_k$ be all square integrable functions F supported on the simplex

$$\mathscr{R}_k := \left\{ (t_1, \ldots, t_k) \in [0,1]^k : t_1 + \cdots + t_k \leq 1 \right\}$$

and not equivalent in Lebesgue measure to the zero function. Let

$$M_k := \sup_{F \in \mathscr{L}_k} \frac{\sum_{m=1}^{k} J_m(F)}{I(F)}.$$

If $f : [0, \infty) \to \mathbb{R}$ is a function with compact support, then we define

$$S(f) := \sup\{x \geq 0 : f(x) \neq 0\}.$$

The expression DHL[k, j] (Dickson–Hardy–Littlewood) means that for any given admissible k-tuple $\mathscr{H}$ and an infinite number of positive integers n each translate $n + \mathscr{H}$ contains at least j prime values.

If $0 < \theta \leq 1$, the expression EH[θ] (Elliott–Halberstam) means that for all $\epsilon > 0$, any given $A > 0$, and $Q \leq x^{\theta - \epsilon}$, we have

$$\sum_{\substack{q \leq Q}} \sup_{\substack{1 \leq a \leq q \\ (a,q)=1}} \left| \sum_{\substack{x \leq n \leq 2x \\ n \equiv a \bmod q}} \Lambda(n) - \frac{1}{\varphi(q)} \sum_{\substack{x \leq n \leq 2x \\ (n,q)=1}} \Lambda(n) \right| \ll_{A,\epsilon} \frac{x}{(\log x)^A}.$$

7.3 Chapter Summary

The results proved in this chapter are independent of those of Maynard. Even "Maynard's lemma", Theorem 7.6, has an alternative proof to the corresponding result of Maynard given in Chapter 6. In this section, we use the definitions given in Section 7.2 without note, but they are frequently repeated in the body of the chapter.

Section 7.4 lays the groundwork, providing the Selberg sieve structure needed for the main results. We begin with a lemma, which provides a practical criterion for obtaining an infinite number of primes in k-tuples with fixed but unspecified nonnegative weights $v(n)$:

Lemma 7.1 (**Criterion for an infinitude of primes in k-tuples**) *[175, lemma 18]*
Let $k \geq 2$ be an integer and $\rho > 0$ real. Suppose that for each admissible k-tuple $\mathscr{H} = \{h_1, \ldots, h_k\}$ and n in the residue class a modulo W such that $(n + h_i, W) = 1$ for all $1 \leq i \leq k$, there is a nonnegative weight function $v \colon \mathbb{N} \to [0, \infty)$ and constants $\alpha > 0$, $\beta_1, \ldots, \beta_k \geq 0$ such that as $x \to \infty$ we have an upper bound

$$\sum_{\substack{x \leq n < 2x \\ n \equiv a \bmod W}} v(n) \leq (\alpha + o(1)) \frac{x}{W B^k}$$

and a lower bound, for all i with $1 \leq i \leq k$,

$$\sum_{\substack{x \leq n < 2x \\ n \equiv a \bmod W}} v(n) \theta(n + h_i) \geq (\beta_i + o(1)) \frac{x}{\varphi(W) B^{k-1}},$$

and, in addition, there exists $m \in \mathbb{N}$ such that $\beta_1 + \cdots + \beta_k > \alpha m$. Then there are infinitely many translates $n + \mathscr{H}$ which contain at least $m + 1$ primes.

Note that Polymath8b don't require asymptotic expressions for the "non-prime sums" and "prime sums", but only upper and lower bounds respectively. This novel flexibility is useful, at least for one of the proofs.

When carrying out optimizations for Maynard's method, we can restrict functions to being symmetric. This is demonstrated in Lemma 7.2 – this property was used implicitly by Maynard as reported in Chapter 6.

> **Lemma 7.2 (Optimization over symmetric functions)** *[175, lemma 41] Let $\mathscr{L}_k$ be the set of all square integrable functions on $\mathscr{R}_k$ which are not equivalent in Lebesgue measure to the zero function, extended with value zero on $[0,\infty)^k \setminus \mathscr{R}_k$. Then*
>
> $$M_k = \sup_{F \in \mathscr{L}_k} \frac{k J_1(F)}{I(F)}.$$

The main multidimensional Selberg sieve lemma is then demonstrated:

> **Lemma 7.3 (Multidimensional Selberg sieve approximation)** *[175, lemma 30] Let $k \geq 1$ and $N \in \mathbb{N}$ satisfy $(N, W) = 1$ and be such that $\log N \ll (\log x)^{O(1)}$. Let*
>
> $$F_j, G_j : [0,\infty) \to \mathbb{R},\ 1 \leq j \leq k$$
>
> *be smooth real functions which are compactly supported in $(0,\infty)$. Let*
>
> $$c := \prod_{j=1}^{k} \int_0^\infty F_j'(t) G_j'(t)\, dt.$$
>
> *Then as $x \to \infty$, we have*
>
> $$\sum_{\substack{d,e \in \mathbb{N}^k \\ [d_1,e_1],\dots,[d_k,e_k],W,N \\ coprime}} \prod_{j=1}^{k} \frac{\mu(d_j)\mu(e_j) F_j(\log_x d_j) G_j(\log_x e_j)}{[d_j, e_j]} = (c + o(1)) \frac{N^k}{B^k \varphi(N)^k},$$
>
> *where as usual $\log_x y = \log y / \log x$.*

The Selberg sieve lemma is used to approximate both so-called "prime sums" and "nonprime sums":

> **Lemma 7.4 (Approximating prime sums)** *[175, theorem 19i] Let $\mathscr{H} = \{h_1, \dots, h_k\}$ be a fixed admissible k-tuple with $k \geq 2$ and let $a \bmod W$ be such that $(a + h_i, W) = 1$ for all $1 \leq i \leq k$. Let $1 \leq i_0 \leq k$ be a fixed index and suppose that for $1 \leq i \leq k, i \neq i_0$, we are given*
>
> $$F_i, G_i : [0,\infty) \to \mathbb{R},$$

which are smooth, compactly supported functions on $(0, \infty)$. Assume also that there is a θ in $(0, 1)$ such that $EH[\theta]$ is true, and

$$\sum_{\substack{1 \le i \le k \\ i \ne i_0}} (S(F_i) + S(G_i)) < \theta.$$

Define

$$\beta_{i_0} := \prod_{\substack{1 \le i \le k \\ i \ne i_0}} \int_0^1 F_i'(t) G_i'(t) \, dt.$$

Then as $x \to \infty$, we have

$$\sum_{\substack{x \le n < 2x \\ n \equiv a \bmod W}} \theta(n + h_{i_0}) \prod_{\substack{1 \le i \le k \\ i \ne i_0}} \lambda_{F_i}(n + h_i) \lambda_{G_i}(n + h_i) = (\beta_{i_0} + o(1)) \frac{x}{B^{k-1} \varphi(W)}.$$

Lemma 7.5 (Approximating nonprime sums) *[175, theorem 20(i)] Let $\mathcal{H} = \{h_1, \ldots, h_k\}$ be a fixed admissible k-tuple with $k \ge 2$ and let a mod W be such that $(a + h_i, W) = 1$ for all $1 \le i \le k$. Suppose that for $1 \le i \le k$, we have functions*

$$F_i, G_i : [0, \infty) \to \mathbb{R},$$

which are smooth and compactly supported. Assume also there are uniformly bounded support upper bounds $S(F_i), S(G_i)$ satisfying

$$\sum_{i=1}^{k} (S(F_i) + S(G_i)) < 1,$$

and let, as in Lemma 7.3, $c := \prod_{i=1}^{k} \int_0^1 F_i'(t) G_i'(t) \, dt$. Then

$$\sum_{\substack{x \le n < 2x \\ n \equiv a \bmod W}} \prod_{i=1}^{k} \lambda_{F_i}(n + h_i) \lambda_{G_i}(n + h_i) = (c + o(1)) \frac{x}{B^k W}.$$

We also need the so-called Maynard's lemma, first proved by Maynard, but in this chapter we give Polymath's alternative proof. It is a pivotal step, revealing the relationship between approximations to M_k and prime gaps.

Lemma 7.6 (Maynard's lemma) *[175, theorem 22] Assume there is a fixed θ with $0 < \theta < 1$ such that EH[θ] is true and such that $\theta M_k > 2m$. Then DHL[k, m + 1] is true.*

Following these preparations we can give the proof of Theorem 7.7, namely $p_{n+1} - p_n \le 270$, using Polymath's algorithm for M_k which is set out in Section 7.5.

In Section 7.6, lower and upper bounds for M_k are derived using corollaries to Lemma 7.8. The first corollary gives a precise value for M_k as an eigenvalue, assuming the corresponding eigenfunction exists:

Corollary 7.9 **(upper bound for M_k)** *[175, corollary 35] Let $k \geq 2$ and suppose there is a positive measurable function $F : \mathscr{R}_k \to (0, \infty)$ which satisfies for some $\lambda > 0$ the equation*

$$\lambda F(t) = \sum_{i=1}^{k} \int_0^\infty F(t_1, \ldots, t_{i-1}, s, t_{i+1}, \ldots, t_k) \, ds,$$

where we have extended F to all of $[0, \infty)^k$ by giving it the value zero outside of $\mathscr{R}_k$. Then $\lambda = M_k$.

In case $k = 2$, we obtain an analytic value for M_2. In Corollary 7.10, we show $M_2 = 1/(1 - W(1/e))$, where W, for this corollary, is Lambert's function, i.e., the unique positive solution to $x = W(x) \exp(W(x))$. A tantalizing problem is to find analytic values for higher values of k.

Then we give Polymath's analytic upper bound for M_k:

Corollary 7.11 *[175, corollary 37] For all $k \geq 2$, we have $M_k \leq k \log(k)/(k-1)$.*

On the face of it, this could allow prime gaps to be derived for $k \geq 51$.

In Sections 7.7 and 7.8, we derive numerical lower bounds for M_k using the so-called Krylov basis method. This requires the inductive derivation of a set of symmetric polynomials and the symbolic calculation of integrals of monomials with coefficients in symbolic k, implemented in PGpack. The method is memory intensive. It is included since it is not without value for future variations and applications.

In Section 7.9, another degree of freedom is introduced by perturbing the standard simplex $\mathscr{R}_k$ in a simple manner. Polymath found an improved lower bound for a related functional $M_{k,\epsilon}$, which is defined in the statement of Theorem 7.14. They derived a sufficient condition for DHL[$k, m+1$] having the same form as Lemma 7.6, but with computations allowing smaller values of k, and thus smaller prime gaps.

Theorem 7.14 **(Sieving on an enlarged simplex)** *[175, theorem 26] Let $k \geq 2$ and $m \in \mathbb{N}$ be integers, $F : [0, \infty)^k \to \mathbb{R}$ a compactly supported square integrable function and let $\epsilon > 0$ be given. Define the functionals for $1 \leq i \leq k$,*

$$I(F) := \int_{[0, \infty)^k} F(t_1, \ldots, t_k)^2 \, dt_1 \ldots dt_k,$$

$$J_{i, 1-\epsilon}(F) := \int_{(1-\epsilon)\mathscr{R}_{k-1}} \left(\int_0^\infty F(t_1, \ldots, t_k) dt_i \right)^2 dt_1 \ldots dt_{i-1} dt_{i+1} \ldots dt_k.$$

Let $\mathscr{L}_{k, \epsilon}$ be all square integrable functions supported on the simplex

$$(1+\epsilon)\mathscr{R}_k := \left\{ (t_1, \ldots, t_k) \in [0, \infty)^k : t_1 + \cdots + t_k \leq 1 + \epsilon \right\}$$

and not equivalent in Lebesgue measure to the zero function, and

$$M_{k,\epsilon} := \sup_{F \in \mathcal{L}_{k,\epsilon}} \frac{\sum_{i=1}^{k} J_{i,1-\epsilon}(F)}{I(F)}.$$

Let the primes have level of distribution θ with $0 < \theta < \frac{1}{2}$ and suppose also that $1 + \epsilon < 1/\theta$. If

$$\theta M_{k,\epsilon} > 2m \quad \text{then} \quad \text{DHL}[k, m+1].$$

Application of Polymath's algorithm then enables us to show $M_{50,1/27} > 4.0018$ and thus DHL[50, 2], and therefore $p_{n+1} - p_n \le 246$. This is described in Theorems 7.15 and 7.16.

In Section 7.10, some limits to this approach for $M_{k,\epsilon}$ are derived. Similar to the upper bound derived for M_k, we have $M_{k,\epsilon} \le k \log(2k-1)/(k-1)$ valid for $k \ge 2$ and $0 \le \epsilon < 1$ as derived in Theorem 7.17. This upper bound obstruction (i.e., we would never have $M_{k,\epsilon} > 4$ if $M_{k,\epsilon} \le 4$) would on the face of it enable values of k as low as 24 – not that this value was achieved numerically. In addition, we derived a new bound which depends on ϵ and tends to the bound derived for M_k as $\epsilon \to 0+$.

Section 7.11, "End Notes", provides a summary of some of the other results of Polymath8b. Their preprint server, web pages and articles cover much more than what has been presented here.

Figure 7.2 provides a graph of the main dependencies between results in this chapter.

7.4 Preliminary Results

We begin with some pivotal preliminary lemmas. The first extracts the essence of the GPY method, using upper and lower bounds rather than asymptotic estimates for the sums.

Recall the definitions. The symbol x represents a real positive variable, certainly $x > 10^7$.

$$w := \log\log\log x,$$

$$W := \prod_{p \le w} p,$$

$$B := \frac{\varphi(W)}{W} \log x.$$

Lemma 7.1 **(Criterion for an infinitude of primes in k-tuples)** *[175, lemma 18]*
Let $k \ge 2$ be an integer and $\rho > 0$ real. Suppose that for each admissible k-tuple

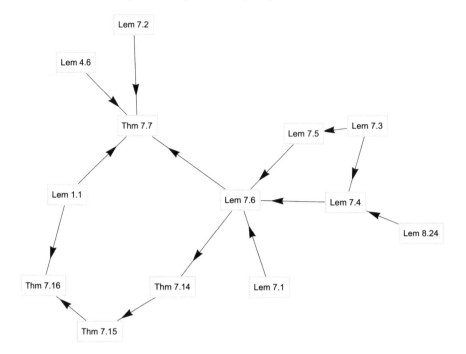

Figure 7.2 Some Chapter 7 dependencies.

$\mathcal{H} = \{h_1, \ldots, h_k\}$ *and n in any residue class of a modulo W such that $(n + h_i, W) = 1$ for all $1 \le i \le k$, there is a nonnegative weight function $v \colon \mathbb{N} \to [0, \infty)$ and constants $\alpha > 0$, $\beta_1, \ldots, b_k \ge 0$ such that for all $x > 0$ we have an upper bound*

$$\sum_{\substack{x \le n < 2x \\ n \equiv a \bmod W}} v(n) \le (\alpha + o(1)) \frac{x}{W B^k},$$

and for $x > 0$ a lower bound, for all i with $1 \le i \le k$,

$$\sum_{\substack{x \le n < 2x \\ n \equiv a \bmod W}} v(n)\theta(n + h_i) \ge (\beta_i + o(1)) \frac{x}{\varphi(W) B^{k-1}},$$

and there exists $m \in \mathbb{N}$ such that

$$\beta_1 + \cdots + \beta_k > \alpha m,$$

then we are able to assert DHL[k, m + 1]. *In other words, in this situation there are infinitely many translates $n + \mathcal{H}$ which contain at least $m + 1$ primes.*

Proof **(1)** Let $\mathcal{H} = \{h_1, \ldots, h_k\}$ be given, admissible and fixed. There is at least one residue class $a \bmod W$ such that $(a + h_i, W) = 1$ for all h_i. Define

$$N(x) := \sum_{\substack{x \le n < 2x \\ n \equiv a \bmod W}} v(n) \left(\left(\sum_{i=1}^k \theta(n + h_i) \right) - m \log(3x) \right).$$

Then, combining the estimates of the lemma statement, we get

$$N(x) \geq (\beta_1 + \cdots + \beta_k - o(1)) \frac{x}{B^{k-1}\varphi(W)} - (m\alpha + o(1)) \frac{x}{B^k W} \log(3x).$$

Then since $B = (\varphi(W)/W) \log x$ and $\beta_1 + \cdots + \beta_k > m\alpha$, if x is sufficiently large we get $N(x) > 0$.

(2) Therefore, since the $v(n)$ are nonnegative, there exists at least one $n \in [x, 2x)$ such that

$$\left(\sum_{i=1}^{k} \theta(n + h_i) \right) - m \log(3x) > 0,$$

which implies the sum in this expression is positive for at least $m + 1$ indices i, so $n + h_i$ must be prime for at least that many indices. Therefore, for all sufficiently large x there is an $n \in [x, 2x)$ such that $n + h_i$ is prime for at least $m + 1$ values of i with $1 \leq i \leq k$. Thus, since $\mathcal{H}$ is an arbitrary tuple admissible of size k, we have DHL[k, m + 1], and this completes the proof. □

Maynard chooses symmetric polynomials for his computation of M_k. Polymath proves this is a good choice when the supremum is taken over all functions in a Hilbert space. First recall some definitions:

$$\mathcal{R}_k := \{t \in [0, \infty)^k : t_1 + \cdots + t_k \leq 1\},$$
$$\mathcal{L}_k := \{F : \mathcal{R}_k \to \mathbb{R} : F \text{ is square integrable and not zero a.e.}\},$$
$$I(F) := \int_{[0,\infty)^k} F(t_1, \ldots, t_k)^2 \, dt_1 \ldots dt_k,$$
$$J_i(F) := \int_{[0,\infty)^{k-1}} \left(\int_0^\infty F(t_1, \ldots, t_k) dt_i \right)^2 dt_1 \ldots dt_{i-1} dt_{i+1} \ldots dt_k, \quad 1 \leq i \leq k,$$
$$M_k := \sup_{F \in \mathcal{L}_k} \frac{\sum_{i=1}^k J_i(F)}{I(F)}.$$

Lemma 7.2 (Optimization over symmetric functions) *[175, lemma 41] Let $\mathcal{L}_k$ be the set of all symmetric square integrable functions on $\mathcal{R}_k$ which are not equivalent in Lebesgue measure to the zero function, extended with value zero on $[0, \infty)^k \setminus \mathcal{R}_k$. Then*

$$M_k = \sup_{F \in \mathcal{L}_k} \frac{k J_1(F)}{I(F)}.$$

Proof **(1)** Note that $I(|F|) = I(F)$, $J_i(|F|) \geq J_i(F)$ and that $\mathcal{L}_k$ is closed when taking absolute values, so we can take the supremum in the definition of M_k over the smaller class of nonnegative functions. We can also assume $0 < M_k < \infty$: for the lower bound, consider $F(t) = 1$ on $\mathcal{R}_k$, and for the upper bound, assuming

$M_k = \infty$ implies there is an index i and sequence $(F_n)_{n \in \mathbb{N}}$ with $I(F_n) = 1$ and $J_i(F_n) \to \infty$, but considering the integral of nonnegative step functions on $\mathcal{R}_k$ gives $J_i(F_n) \leq I(F_n)$.

(2) Thus there is a sequence of nonnegative functions $(F_n)_{n \in \mathbb{N}}$ in $\mathcal{L}_k$ such that as $n \to \infty$ we have

$$M_n = \lim_{n \to \infty} \frac{\sum_{i=1}^{k} J_i(F_n)}{I(F_n)}.$$

Multiplying each F_n by a suitable positive real constant, we can assume $I(F_n) = 1$, so

$$M_n = \lim_{n \to \infty} \left(\sum_{i=1}^{k} J_i(F_n) \right).$$

(3) In this step, we symmetrize the F_n. For each permutation $\sigma \in S_k$, the symmetric group and each $t = (t_1, \ldots, t_k) \in \mathbb{R}^k$, let

$$\sigma \cdot t = (t_{\sigma(1)}, \ldots, t_{\sigma(k)})$$

and define

$$\widehat{F_n}(t) := \frac{1}{k!} \sum_{\sigma \in S_k} F_n(\sigma \cdot t).$$

Because the F_n are nonnegative, we get $\widehat{F_n} \geq (1/k!) F_n$ and so, uniformly in n,

$$I(\widehat{F_n}) \geq \frac{1}{(k!)^2} I(F_n) = \frac{1}{(k!)^2} > 0.$$

Since $M_k = \sup_{F \in \mathcal{L}_k} \sum_{i=1}^{k} J_i(F)/I(F)$, the form defined by

$$Q(F) := M_k \cdot I(F) - \sum_{i=1}^{k} J_i(F),$$

which is quadratic for linear combinations of fixed functions F, is positive semidefinite. Using Fubini's theorem and defining $F^\sigma(t) := F(\sigma \cdot t)$, we have $I(F^\sigma) = I(F)$ and $J_i(F^\sigma) = J_{\sigma(i)}$ so $Q(F)$ is invariant under permutations, i.e., $Q(F^\sigma) = Q(F)$, $F \in \mathcal{L}_k$. By the triangle inequality in the space with norm $\|F\| = \sqrt{Q(F)}$, we get for $n \in \mathbb{N}$,

$$\|\widehat{F}\| \leq \frac{1}{(k!)^2} \sum_{\sigma \in S_n} \|F^\sigma\| = \|F\| \implies Q(\widehat{F_n}) \leq Q(F_n),$$

so $Q(F_n) \to 0+$ implies $Q(\widehat{F_n}) \to 0+$, which implies

$$\frac{k J_1(\widehat{F_n})}{I(\widehat{F_n})} = \frac{\sum_{i=1}^{k} J_i(\widehat{F_n})}{I(\widehat{F_n})} \to M_k$$

so

$$M_k = \sup \frac{k J_1(F)}{I(F)}$$

where the $F \in \mathscr{L}_k$ are symmetric. This completes the proof. □

Recall the definitions $w := \log\log\log x$, $W := \prod_{p \leq w} p$ and $B = \varphi(W) \log x / W$. We also extend the usual definition of the LCM so $[n_1, n_2]$ is the usual least common multiple of the positive integers n_1 and n_2 and $[n_1, \ldots, n_m] = [[n_1, \ldots, n_{m-1}], n_m]$ for $m > 2$.

Lemma 7.3 (Multidimensional Selberg sieve approximation) *[175, lemma 30]*
Let $k \geq 1$ and $N \in \mathbb{N}$ satisfy $(N, W) = 1$ and such that $\log N \ll (\log x)^{O(1)}$. Let

$$F_1, \ldots F_k, G_1, \ldots, G_k \colon [0, \infty) \to \mathbb{R}$$

be smooth real functions which are compactly supported in $(0, \infty)$. Let

$$c := \prod_{j=1}^{k} \int_0^{\infty} F_j'(t) G_j'(t) \, dt.$$

Then as $x \to \infty$, we have

$$E(N, x) := \sum_{\substack{d, e \in \mathbb{N}^k \\ [d_1, e_1], \ldots, [d_k, e_k], W, N \text{ coprime}}} \prod_{j=1}^{k} \frac{\mu(d_j)\mu(e_j) F_j(\log_x d_j) G_j(\log_x e_j)}{[d_j, e_j]}$$

$$= (c + o(1)) \frac{N^k}{B^k \varphi(N)^k}. \tag{7.1}$$

The same is true if we replace each denominator $[d_j, e_j]$ by $\varphi([d_j, e_j])$.

Proof (1) First extend the real functions $t \to e^t F_j(t), e^t G_j(t)$ to smooth compactly supported functions on all of $\mathbb{R}$. As such, we can take the inverse Fourier transforms, with $t, \xi \in \mathbb{R}$,

$$e^t F_j(t) = \int_{\mathbb{R}} f_j(\xi) e^{-it\xi} \, d\xi \quad \text{and} \quad e^t G_j(t) = \int_{\mathbb{R}} g_j(\xi) e^{-it\xi} \, d\xi,$$

where the functions $f_j, g_j \colon \mathbb{R} \to \mathbb{C}$ will be smooth and rapidly decreasing in the sense that for any fixed $A > 0$ and all $\xi \in \mathbb{R}$ and all j, we have

$$f_j(\xi), g_j(\xi) \ll_A \frac{1}{(1 + |\xi|)^A}.$$

To see this, we apply theorems of the Paley–Weiner type based on integration by parts. See for example [199, theorems 7.2.1, 7.2.2]. For background on the Fourier transform, see for example [21, appendix E].

Therefore, for all $d_j \geq 1$ we have

$$F_j(\log_x d_j) = \int_{\mathbb{R}} \frac{f_j(\xi)}{d_j^{\frac{1+i\xi}{\log x}}} \, d\xi \text{ and } G_j(\log_x d_j) = \int_{\mathbb{R}} \frac{g_j(\xi)}{d_j^{\frac{1+i\xi}{\log x}}} \, d\xi.$$

(2) Next, we **claim** that

$$\sum_{d_j, e_j \in \mathbb{N}} \frac{\mu(d_j)^2 \mu(e_j)^2}{[d_j, e_j](d_j e_j)^{1/\log x}} = \prod_p \left(1 + \frac{2}{p^{1+1/\log x}} + \frac{1}{p^{1+2/\log x}} \right) \leq e^{O(\log\log x)},$$

where the product is over all rational primes. To see this, consider the cases $(d_j, e_j) = 1$ and $(d_j, e_j) = g_j > 1$ with d_j, e_j, g_j and $[d_j, e_j]$ squarefree, and note how the terms appear in the Euler product for a given $p \in \mathbb{P}$. The upper bound comes from taking logarithms of the product and using for $x > 1$,

$$\sum_{n \in \mathbb{N}} \frac{1}{n^{1+1/\log x}} \ll \log x.$$

Therefore, if we substitute the expressions for F_j, G_j in terms of the inverse Fourier transforms from Step (1) in the left-hand side of (7.1), the resulting expression will converge absolutely. Thus we can apply Fubini's theorem to the left-hand side of (7.1) using real k vectors ξ, η and the definition

$$K(\xi, \eta) := \sum_{\substack{d, e \in \mathbb{N}^k \\ [d_1, e_1], \ldots, [d_k, e_k], W, N \text{ coprime}}} \prod_{j=1}^{k} \frac{\mu(d_j)\mu(e_j)}{[d_j, e_j] d_j^{\frac{1+i\xi_j}{\log x}} e_j^{\frac{1+i\eta_j}{\log x}}},$$

to write

$$E(N, x) = \int_{\mathbb{R}^{2k}} K(\xi, \eta) \prod_{j=1}^{k} f_j(\xi_j) g_j(\eta_j) \, d\xi_j d\eta_j. \qquad (7.2)$$

(3) Next we factor $K(\xi, \eta)$ as $K = \prod_{p \nmid WN} K_p(\xi, \eta)$, where

$$K_p(\xi, \eta) := 1 + \frac{1}{p} \left(\sum_{\substack{d_1, \ldots, d_k, e_1, \ldots, e_k \\ p=[d_1, \ldots, d_k, e_1, \ldots, e_k] \\ [d_1, e_1], \ldots, [d_k, e_k] \text{ coprime}}} \prod_{j=1}^{k} \frac{\mu(d_j)\mu(e_j)}{d_j^{\frac{1+i\xi_j}{\log x}} e_j^{\frac{1+i\eta_j}{\log x}}} \right).$$

(4) We now **claim** that the Euler factor K_p has an estimate

$$K_p(\xi, \eta) = \left(1 + O\left(\frac{1}{p^2}\right)\right) \prod_{j=1}^{k} \frac{\left(1 - \frac{1}{p^{1+\frac{1+i\xi_j}{\log x}}}\right)\left(1 - \frac{1}{p^{1+\frac{1+i\eta_j}{\log x}}}\right)}{1 - \frac{1}{p^{1+\frac{2+i\xi_j+i\eta_j}{\log x}}}}. \tag{7.3}$$

To see this, expand the sum in parentheses in the definition of $K_p(\xi, \eta)$, considering the cases where one $e_j = p$ and each of the other e_j are 1, one $d_j = p$ and each of the others are 1 and where $d_j = e_j = p$, noting that all of the terms other than the leading term are $O(1/p^2)$ and are finite in number. Then expand the denominator of the expression in parentheses in (7.3) as a geometric series with remainder $O(1/p^2)$ and assemble terms.

Now we note that because $w = \log\log\log x$, taking the exponential we have

$$\prod_{p>w} (1 + O(p^{-2})) = \exp(O(1/w)) = 1 + o(1).$$

Thus defining

$$\zeta_n(s) := \prod_{p \nmid n} \left(1 - \frac{1}{p^s}\right)^{-1},$$

since we have defined $K = \prod_{p \nmid WN} K_p$, for $\Re s > 1$ we have

$$K(\xi, \eta) = (1 + o(1)) \prod_{j=1}^{k} \frac{\zeta_{WN}\left(1 + \frac{2+i\xi_j+i\eta_j}{\log x}\right)}{\zeta_{WN}\left(1 + \frac{1+i\xi_j}{\log x}\right)\zeta_{WN}\left(1 + \frac{1+i\eta_j}{\log x}\right)}.$$

(5) Now for $\sigma = \Re s \geq 1 + 1/(\log x)$, we can derive a bound independent of N and W, taking logarithms to get the second line

$$|\zeta_{NW}| \leq |\zeta(s)| \leq \prod_p \left(1 + \frac{1}{p^{1+1/(\log x)}} + O\left(\frac{1}{p^2}\right)\right)$$
$$\ll \exp\left(\frac{1}{p^{1+1/(\log x)}}\right)$$
$$\leq \exp(\log\log x + O(1))$$
$$\ll \log x.$$

Therefore, using the expression derived in Step (4), we get $K(\xi, \eta) \ll (\log x)^{3k}$.

(6) Combining this estimate with the rapid decrease of the f_j and g_j, outside of the cube $\xi \in C_x := [-\sqrt{\log x}, \sqrt{\log x}]^{2k}$, the contribution to the integral in the rewritten form, (7.2), can be made arbitrarily small. Thus to prove the first claim of the theorem statement, we need only demonstrate that

$$E(N,x) = \int_{C_x} K(\xi,\eta) \prod_{j=1}^{k} f_j(\xi_j) g_j(\eta_j) \, d\xi_j d\eta_j = (c + o(1)) \frac{N^k}{B^k \varphi(N)^k}.$$

When $|\xi_j| \leq \sqrt{\log x}$, because $\zeta(s)$ has a simple pole at $s = 1$, we can write as $x \to \infty$

$$\zeta\left(1 + \frac{1 + i\xi_j}{\log x}\right) = (1 + o(1)) \frac{\log x}{1 + i\xi_j}. \tag{7.4}$$

In addition, for ξ_j with this bound, we have

$$1 - \frac{1}{p^{1 + \frac{1 + i\xi_j}{\log x}}} = 1 - \frac{1}{p} + O\left(\frac{\log p}{p\sqrt{\log x}}\right).$$

Next, recall we are assuming $\log N \ll (\log x)^{O(1)}$ and note from the definition of W and this bound on N that $\log WN \ll (\log x)^{O(1)}$ to get

$$\prod_{p|WN} \left(1 - \frac{1}{p^{1 + \frac{1 + i\xi_j}{\log x}}}\right) = \frac{\varphi(WN)}{WN} \exp\left(O\left(\sum_{p|WN} \frac{\log p}{p\sqrt{\log x}}\right)\right) = (1 + o(1)) \frac{\varphi(WN)}{WN}.$$

To verify the second equality, note that because WN has at most $M := O(\log N)$ prime divisors, the sum will be bounded above by the same sum over the first M primes, which in turn, by [4, theorem 4.10], as $x \to \infty$ is bounded by

$$\sum_{p \leq p_M} \frac{\log p}{p\sqrt{\log x}} \ll \frac{\log p_M}{\sqrt{\log x}} \ll \frac{\log\log N}{\sqrt{\log x}} = o(1).$$

Therefore, recalling the definition $B = (\varphi(W)/W) \log x$, $(W, N) = 1$ and using (7.4), we get

$$\zeta_{WN}\left(1 + \frac{1 + i\xi_j}{\log x}\right) = \frac{(1 + o(1))B\varphi(N)}{(1 + i\xi_j)N},$$

with corresponding simplifications when we replace $1 + i\xi_j$ by $1 + i\eta_j$ or $2 + i\xi_j + i\eta_j$. Hence from the result of Step (4), we obtain the form

$$K(\xi,\eta) = (1 + o(1)) \frac{N^k}{B^k \varphi(N)^k} \prod_{j=1}^{k} \frac{(1 + i\xi_j)(1 + i\eta_j)}{2 + i\xi_j + i\eta_j}. \tag{7.5}$$

(7) Dividing the inverse Fourier transform expressions from Step (1) by e^t and differentiating under the integral sign we get, for real ξ, η,

$$F_j'(t) = -\int_{\mathbb{R}} (1+i\xi)e^{-t(1+i\xi)} f_j(\xi)\, d\xi \text{ and } G_j'(t) = -\int_{\mathbb{R}} (1+i\eta)e^{-t(1+i\eta)} g_j(\eta)\, d\eta.$$

Therefore, using Fubini's theorem, and

$$\int_0^\infty e^{-t(2+i\xi_j+i\eta_j)}\, dt = \frac{1}{2+i\xi_j+i\eta_j},$$

we have

$$\int_{\mathbb{R}}\int_{\mathbb{R}} \frac{(1+i\xi_j)(1+i\eta_j)}{2+i\xi_j+i\eta_j} f_j(\xi_j)g_j(\eta_j)\, d\xi_j d\eta_j = \int_0^\infty F_j'(t)G_j'(t)\, dt,\ 1 \le j \le k.$$

Thus, again by Fubini's theorem and the result of Step (6) we get

$$\int_{\mathbb{R}^{2k}} \prod_{j=1}^k \frac{(1+i\xi_j)(1+i\eta_j)}{2+i\xi_j+i\eta_j} f_j(\xi_j)g_f(\eta_j)\, d\xi_j d\eta_j = \prod_{j=1}^k \int_0^\infty F_j'(t)G_j'(t)\, dt = c.$$

(8) With these preliminaries, we can now complete the first part of the proof. By Step (2), we have

$$E(N,x) = \int_{\mathbb{R}^{2k}} K(\xi,\eta) \prod_{j=1}^k f_j(\xi_j)g_j(\eta_j)\, d\xi_j d\eta_j.$$

By Step (6), we can at the cost of a negligible quantity replace the domain of integration by a bounded box C_x. We also showed

$$K(\xi,\eta) = (1+o(1)) \frac{N^k}{B^k\varphi(N)^k} \prod_{j=1}^k \frac{(1+i\xi_j)(1+i\eta_j)}{2+i\xi_j+i\eta_j}.$$

Inserting this in the expression derived for $E(N,x)$ gives

$$E(N,x) = (1+o(1)) \frac{N^k}{B^k\varphi(N)^k}$$

$$\times \int_{\mathbb{R}^{2k}} K(\xi,\eta) \prod_{j=1}^k \frac{(1+i\xi_j)(1+i\eta_j)}{2+i\xi_j+i\eta_j} f_j(\xi_j)g_j(\eta_j)\, d\xi_j d\eta_j.$$

In Step (7), the integral was evaluated to get finally

$$E(N,x) = (1+o(1)) \frac{N^k}{B^k\varphi(N)^k} c = (c+o(1)) \frac{N^k}{B^k\varphi(N)^k}.$$

(9) To obtain the second part, if we replace the terms $[d_j, e_j]$ with $\varphi([d_j, e_j])$ in $E(N,x)$, then the only change that occurs in the expression for K_p in Step (3) is

$1/p$ would need to be replaced by $1/(p-1) = 1/p + O(1/p^2)$. This completes the proof.   □

The lemma we have proved, Lemma 7.3, is the workhorse which plays an essential role in deriving asymptotic expressions for prime and nonprime sums in Lemmas 7.4 and 7.5 respectively. Note that each constant which is derived takes the form

$$\prod_{j=1}^{k} \int_{0}^{\infty} F_j'(t) G_j'(t) \, dt.$$

First, recall the definitions

$$S(F) := \sup\{x \geq 0 : F(x) \neq 0\},$$

$$\lambda_F(n) := \sum_{d|n} \mu(d) F(\log_x d),$$

$$W := \prod_{p \leq \log\log\log x} p,$$

$$B := \frac{\varphi(W)}{W} (\log x).$$

We also define the **signed discrepancy** for $x > 0$ and $f : \mathbb{N} \to \mathbb{C}$ an arithmetic function with bounded support and $(a, q) = 1$ let

$$\Delta(f, q, a) := \sum_{n \equiv a \bmod q} f(n) - \frac{1}{\varphi(q)} \sum_{(n,q)=1} f(n).$$

For example, using this notation, the expression EH$[\theta]$ (EH is for Elliott–Halberstam) means that if for all $\epsilon > 0$, given $A > 0$, and $Q \leq x^{\theta - \epsilon}$, we have as $x \to \infty$

$$\sum_{q \leq Q} \sup_{\substack{1 \leq a \leq q \\ (a,q)=1}} \left| \Delta(\Lambda \cdot \chi_{[1,x]}, q, a) \right| \ll_{A,\epsilon} \frac{x}{(\log x)^A}.$$

Remarks: Polymath use the notation $X \prec\prec Y$, which we write as $X \prec^o Y$, which means $X \leq x^{o(1)} Y$ as $x \to \infty$. Note that it is a stronger condition than $X \ll x^\epsilon Y$ for some and indeed all $\epsilon > 0$. For example, $\log x \leq x^\epsilon \cdot 1$ for all x sufficiently large but $\log x \leq x^{1/\log x} \cdot 1$ is false for $x > e^e$. The notation needs to be used with some care. For example, for $B > 0$,

$$\frac{\sqrt{x}}{(\log x)^B} = x^{\frac{1}{2} - B \frac{\log\log x}{\log x}} = x^{\frac{1}{2} + o(1)} \prec^o x^{\frac{1}{2}},$$

but we also have $\sqrt{x} (\log x)^B \prec^o x^{\frac{1}{2}}$, which is not a suitable upper bound for the moduli q in the Bombieri–Vinogradov estimate.

Note that in Step (5) of the approximating prime sums lemma we use a number of divisors summation formula from Chapter 8. There it is proved as part of Lemma 8.24, along with some other useful summation formulas.

Lemma 7.4 (**Approximating prime sums**) *[175, lemma 19(i)] Let $\mathcal{H} = \{h_1, \ldots, h_k\}$ be a fixed admissible k-tuple with $k \geq 2$ and let a mod W be such that $(a + h_i, W) = 1$ for all $1 \leq i \leq k$. Let $1 \leq i_0 \leq k$ be a fixed index and suppose that for $1 \leq i \leq k$, $i \neq i_0$, we are given*

$$F_i, G_i : [0, \infty) \to \mathbb{R},$$

which are smooth and compactly supported functions on $(0, \infty)$. Assume also that there is a ϑ in $(0, 1)$ such that EH$[\vartheta]$ is true, and

$$\sum_{\substack{1 \leq i \leq k \\ i \neq i_0}} (S(F_i) + S(G_i)) < \vartheta.$$

If we define, in analogy with Lemma 7.3,

$$\beta_{i_0} := \prod_{\substack{1 \leq i \leq k \\ i \neq i_0}} \int_0^1 F_i'(t) G_i'(t) \, dt,$$

then we have

$$\sum_{\substack{x \leq n < 2x \\ n \equiv a \bmod W}} \theta(n + h_{i_0}) \prod_{\substack{1 \leq i \leq k \\ i \neq i_0}} \lambda_{F_i}(n + h_i) \lambda_{G_i}(n + h_i) = (\beta_{i_0} + o(1)) \frac{x}{B^{k-1} \varphi(W)}.$$

Proof Assume for ease of expression throughout the proof that $i_0 = k$ and $d_1, \ldots, e_{k-1}$ is short for $d_1, \ldots, d_{k-1}, e_1, \ldots, e_{k-1}$.

(1) Let

$$\widetilde{S}(d_1, \ldots, d_{k-1}, e_1, \ldots, e_{k-1}) := \sum_{\substack{x \leq n < 2x \\ n \equiv a \bmod W \\ \forall i < k \ n + h_i \equiv 0 \bmod [d_i, e_i]}} \theta(n + h_k).$$

Then, expanding the definitions of λ_{F_i} and λ_{G_i} we get

$$\sum_{\substack{x \leq n < 2x \\ n \equiv a \bmod W}} \theta(n + h_k) \prod_{1 \leq i < k} \lambda_{F_i}(n + h_i) \lambda_{G_i}(n + h_i)$$

$$= \sum_{d_1, \ldots, e_{k-1}} \left(\prod_{i=1}^{k-1} \mu(d_i) \mu(e_i) F_i(\log_x d_i) G_i(\log_x e_i) \right) \widetilde{S}(d_1, \ldots, e_{k-1}). \quad (7.6)$$

As we have seen before in Chapter 6, if a is chosen so that all of the primes dividing any $n + h_i$ are larger than w, and we can assume x is sufficiently large so that for all $i \neq j$ we have $|h_i - h_j| < w$, we get $\widetilde{S}(d_1, \ldots, d_{k-1}, e_1, \ldots, e_{k-1})$ vanishes unless each of the $[d_i, e_i]$ which appear is coprime to each of the other LCMs and to W. Thus the summand in (7.6) on the right vanishes unless the modulus

$$q := W[d_1, e_1] \cdots [d_{k-1}, e_{k-1}]$$

is squarefree. If that is so, using the Chinese Remainder Theorem we can rewrite the congruences in the definition of $\tilde{S}$ as a single congruence modulo an integer q, i.e.,

$$n + h_k \equiv b \bmod q \text{ with } q = q(W, d_1, \ldots, e_{k-1}),$$

where b depends on W and all of the d_i and e_i with $1 \leq i \leq k - 1$, i.e., $b = b(W, d_1, \ldots, e_{k-1})$ also. Using the discrepancy of an arithmetic function and $(a, q) = 1$, we can then write

$$\tilde{S}(d_1, \ldots, e_{k-1}) = \frac{1}{\varphi(q)} \left(\sum_{x+h_k \leq n < 2x+h_k} \theta(n) \right) + \Delta(\theta \cdot \chi_{[x+h_k, 2x+h_k]}, q, b).$$

(2) Now the prime number theorem gives as $x \to \infty$,

$$\sum_{x+h_k \leq n < 2x+h_k} \theta(n) = (1 + o(1))x.$$

By Lemma 7.3 with $N = 1$, the contribution of this sum to the right-hand side of (7.6) is

$$(\beta_k + o(1)) \frac{x}{B^{k-1} \varphi(W)}.$$

Note that the leading term is independent of the d_i, e_i and thus of any b.

(3) Since $W \ll (\log\log x)^{O(1)}$ and $B := (\varphi(W)/W) \log x$, to complete the proof it will be sufficient to show that for any fixed $A > 0$, which can be chosen sufficiently large, we have for $b = b(W, d_1, \ldots, e_{k-1})$ and $q = q(W, d_1, \ldots, e_{k-1})$, the contribution to the error in (7.6) arising from the second term of the expression for $(\tilde{S})$ derived in Step (1) is

$$\sum_{d_1, \ldots, e_{k-1}} \left(\prod_{i=1}^{k-1} |F_i(\log_x d_i)| |G_i(\log_x e_i)| \right) \left| \Delta(\theta \cdot \chi_{[x+h_k, 2x+h_k]}, q, b) \right| \ll \frac{x}{(\log x)^A},$$

$$(7.7)$$

where we can assume the only terms which appear in the sum are those for which the q's are squarefree. We will derive this in Steps (4) and (5).

(4) First, note we can replace each instance of the product in (7.7) by 1 since the F_i, G_i are fixed with compact support and thus uniformly bounded. Next, by the hypothesis

$$\sum_{\substack{1 \leq i \leq k \\ i \neq i_0}} (S(F_i) + S(G_i)) < \vartheta,$$

we can write, when the summand in (7.6) is nonzero, for some $\epsilon > 0$,

$$q(W, d_1, \ldots, e_{k-1}) \leq x^{\vartheta - \epsilon}.$$

Each choice of q is associated with $O(\tau(q)^{O(1)})$ choices of the $d_1, \ldots, e_{k-1}$ (see for example [153, theorem 7.6]). It follows that the overall contribution of the $\Delta(\cdot)$ terms to the error in (7.6) is

$$O \left(\sum_{q \leq x^{\vartheta - \epsilon}} \tau(q)^{O(1)} \sup_{\substack{b: 1 \leq b \leq q \\ (b, q) = 1}} \left| \Delta(\theta \cdot \chi_{[x + h_k, 2x + h_k)}, q, b) \right| \right).$$

(5) Next we use for $C > 0$, an estimate from Lemma 8.24 (iv), namely

$$\sum_{n \ll x} \frac{\tau(n)^C}{n} \ll_C (\log x)^{O(1)}.$$

We also use the bound

$$\left| \Delta(\theta \cdot \chi_{[x + h_k, 2x + h_k)}, q, b) \right| \ll \frac{x (\log x)^{O(1)}}{q},$$

to derive

$$\sum_{q \leq x^{\vartheta - \epsilon}} \tau(q)^{O(1)} \sup_{\substack{b: 1 \leq b \leq q \\ (b, q) = 1}} \left| \Delta(\theta \cdot \chi_{[x + h_k, 2x + h_k)}, q, b) \right| \ll x (\log x)^{O(1)}.$$

In addition, we have

$$\left| \Delta(\theta \cdot \chi_{[x + h_k, 2x + h_k)}, q, b) - \Delta(\Lambda \cdot \chi_{[x + h_k, 2x + h_k)}, q, b) \right| \leq \left(\frac{x}{q} \right)^{\frac{1}{2} + \epsilon/2}$$

and EH[ϑ] gives for $q \leq x^{\vartheta - \epsilon}$,

$$\sum_{q \leq x^{\vartheta - \epsilon}} \sup_{\substack{b: 1 \leq b \leq q \\ (b, q) = 1}} \left| \Delta(\Lambda \cdot \chi_{[x + h_k, 2x + h_k)}, q, b) \right| \ll \frac{x}{(\log x)^A}.$$

Therefore,

$$\sum_{q \leq x^{\vartheta - \epsilon}} \sup_{\substack{b: 1 \leq b \leq q \\ (b, q) = 1}} \left| \Delta(\theta \cdot \chi_{[x + h_k, 2x + h_k)}, q, b) \right| \ll \frac{x}{(\log x)^A}.$$

Combining these estimates, using Cauchy–Schwarz, for a_q, b_q nonnegative and in a finite set $Q = [1, x^{\vartheta - \epsilon}] \cap \mathbb{N}$, in the form

$$\sum_{q \in Q} \sqrt{a_q} \cdot b_q = \sum_{q \in Q} \sqrt{a_q b_q} \sqrt{b_q} \leq \sqrt{\sum_{q \in Q} a_q b_q} \cdot \sqrt{\sum_{q \in Q} b_q}$$

with $a_q = \tau(q)^{O(1)}$ and

$$b_q = \sup_{\substack{b:1 \leq b \leq q \\ (b,q)=1}} \left| \Delta(\theta \cdot \chi_{[x+h_k, 2x+h_k]}, q, b) \right|,$$

we get

$$\sum_{q \leq x^{\vartheta - \epsilon}} \tau(q)^{O(1)} \sup_{\substack{b:1 \leq b \leq q \\ (b,q)=1}} \left| \Delta(\theta \cdot \chi_{[x+h_k, 2x+h_k]}, q, b) \right| \ll \frac{x}{(\log x)^A},$$

which completes the proof. $\qquad \square$

Recall again the definitions

$$S(F) := \sup\{x \geq 0 : F(x) \neq 0\},$$
$$w := \log\log\log x,$$
$$W := \prod_{p \leq w} p,$$
$$B := (\varphi(W)/W) \log x,$$
$$\lambda_F(n) := \sum_{d \mid n} \mu(d) F(\log_x d).$$

Lemma 7.5 (Approximating nonprime sums) *[175, theorem 20(i)] Let $\mathcal{H} = \{h_1, \ldots, h_k\}$ be a fixed admissible k-tuple with $k \geq 2$ and let $a \bmod W$ be such that $(a + h_i, W) = 1$ for all $1 \leq i \leq k$. Suppose that for $1 \leq i \leq k$, we have*

$$F_i, G_i : [0, \infty) \to \mathbb{R},$$

are smooth functions which are compactly supported in $(0,1)$. Assume uniformly bounded supports satisfying

$$\sum_{i=1}^{k} (S(F_i) + S(G_i)) < 1,$$

and let, as in Lemma 7.3,

$$c := \prod_{i=1}^{k} \int_0^1 F_i'(t) G_i'(t) \, dt.$$

Then as $x \to \infty$, we have

$$\sum_{\substack{x \le n < 2x \\ n \equiv a \bmod W}} \prod_{i=1}^{k} \lambda_{F_i}(n + h_i)\lambda_{G_i}(n + h_i) = (c + o(1))\frac{x}{B^k W}.$$

Proof **(1)** If we define for $d, e \in \mathbb{R}^k$,

$$S(d, e) := \sum_{\substack{x \le n < 2x \\ n \equiv a \bmod W \\ \forall i \; n + h_i \equiv 0 \bmod [d_i, e_i]}} 1,$$

the left-hand side of the estimate of the lemma can be expanded as before to get

$$\sum_{\substack{x \le n < 2x \\ n \equiv a \bmod W}} \prod_{i=1}^{k} \lambda_{F_i}(n + h_i)\lambda_{G_i}(n + h_i)$$

$$= \sum_{d,e} \left(\prod_{i=1}^{k} \mu(d_i)\mu(e_i)F_i(\log_x d_i)G_i(\log_x e_i) \right) S(d, e).$$

As in Lemma 7.4, since for all i, $a + h_i$ is coprime to W and for distinct i, j we have $|h_i - h_j| < w$, then a particular $S(d, e)$ will be zero unless all of the $[d_i, e_i]$ are coprime to each other and to W. Thus the sum in the expression for $S(d, e)$ will be over an arithmetic progression in $[x, 2x)$ with common difference $W[d_1, e_1] \cdots [d_k, e_k]$, giving

$$S(d, e) = \frac{x}{W[d_1, e_1] \cdots [d_k, e_k]} + O(1).$$

(2) Note that $S(F_i), S(G_i) < 1$, so we can restrict integrals to $[0, 1]$. Using Lemma 7.3 with $N = 1$, we see the contribution of the terms of the form $x / W[d_1, e_1] \cdots [d_k, e_k]$ to the estimate of the lemma is

$$(c + o(1))\frac{x}{B^k W}.$$

The contribution to the error term is

$$O\left(\sum_{d,e} \left(\prod_{i=1}^{k} |F_i(\log_x d_i)||G_i(\log_x e_i)| \right) \right).$$

The hypotheses of this lemma imply for some fixed $\epsilon > 0$, we have

$$\sum_{j=1}^{k} S(F_j) + S(G_j) \le 1 - \epsilon \implies \sum_{j=1}^{k} \frac{\log d_j}{\log x} + \frac{\log e_j}{\log x} \le 1 - \epsilon$$

and therefore we have a nonzero term in the sum only when

$$[d_1, e_1] \cdots [d_k, e_k] \le \prod_{j=1}^{k} d_j e_j \le x^{1-\epsilon}.$$

Since $\tau(n) \le n^{o(1)}$, each choice of $[d_1, e_1] \cdots [d_k, e_k]$ arises from $x^{o(1)}$ choices of d, e. It follows that the contribution of the $O(1)$ terms to the error in the lemma estimate is $O(x^{1-\epsilon+o(1)})$, which completes the proof. $\square$

Next we give the proof of Lemma 7.6. This is Maynard's Lemma 6.9. However, Polymath gave a different proof, introducing a scaling technique that is used in Theorem 7.14, and which underpins the result $p_{n+1} - p_n \le 246$ for an infinite number of $n \in \mathbb{N}$. Like Polymath's Selberg sieve Lemma 7.3, which uses the Fourier transform and Euler products in essential ways, this result also uses analytical tools. These are smooth convolutions and the Stone–Weierstrass theorem.

Recall the definitions: the expression DHL[k, j] means that for a given admissible k-tuple $\mathcal{H}$ for an infinite number of positive integers n each translate $n + \mathcal{H}$ contains at least j prime values.

The expression EH[θ] means that given $A > 0$, then for all $\epsilon > 0$ and $Q \le x^{\theta-\epsilon}$, we have

$$\sum_{q \le Q} \sup_{\substack{1 \le a \le q \\ (a,q)=1}} \left| \sum_{\substack{x \le n < 2x \\ n \equiv a \bmod q}} \Lambda(n) - \frac{1}{\varphi(q)} \sum_{\substack{x \le n < 2x \\ (n,q)=1}} \Lambda(n) \right| \ll_{A,\epsilon} \frac{x}{(\log x)^A}.$$

Lemma 7.6 (Maynard's lemma) *[138, lemma 6.3] [175, lemma 22] Let $k \ge 2$ and $m \in \mathbb{N}$ be integers and $F \colon [0,\infty)^k \to \mathbb{R}$ a compactly supported square integrable function. Define the functionals*

$$I(F) := \int_{[0,\infty)^k} F(t_1, \dots, t_k)^2 \, dt_1 \dots dt_k,$$

$$J_i(F) := \int_{[0,\infty)^{k-1}} \left(\int_0^\infty F(t_1, \dots, t_k) dt_i \right)^2 dt_1 \dots dt_{i-1} dt_{i+1} \dots dt_k, \ 1 \le i \le k.$$

Let $\mathcal{L}_k$ be all square integrable functions supported on the simplex

$$\mathcal{R}_k := \left\{ (t_1, \dots, t_k) \in [0,1]^k : t_1 + \cdots + t_k \le 1 \right\}$$

and not equivalent in Lebesgue measure to the zero function, and

$$M_k := \sup_{F \in \mathcal{L}_k} \frac{\sum_{m=1}^k J_m(F)}{I(F)}.$$

Assume there is a fixed θ with $0 < \theta < 1$ such that EH[θ] is true and such that $\theta M_k > 2m$. Then DHL[k, m + 1] is true.

Proof **(1)** From the given hypotheses, there is a square integrable function $F_0 \colon [0,\infty)^k \to \mathbb{R}$ supported on $\mathcal{R}_k$, and not equivalent to the zero function, such that

$$\sum_{i=1}^{k} \frac{J_i(F_0)}{I(F_0)} > \frac{2m}{\theta}.$$

(2) We now show that we can replace F_0 with a smooth function f_4, with a larger support, which satisfies a corresponding revised inequality.

To this end, let $\delta_1 > 0$ be sufficiently small so that $\alpha := \theta/2 - \delta_1 > 0$ and define $F_1 \colon [0,\infty)^k \to \mathbb{R}$ to be

$$F_1(t) = F_0(t/\alpha) \implies F_1(\alpha t) = F_0(t),$$

where division of a vector by a scalar is component by component. This gives a function which is square integrable and supported on the reduced simplex

$$\alpha \cdot \mathcal{R}_k = \{t \in [0,\infty)^k : \sum_{i=1}^{k} t_i \le \alpha\}.$$

By making δ_1 sufficiently small, changing variables in the integrals, we can assume

$$\sum_{i=1}^{k} \frac{J_i(F_1)}{I(F_1)} > m.$$

(3) Next, we let $\hat{o}_2 > 0$ be such that $\delta_2 < \delta_1$ is sufficiently small, and define $F_2 \colon [0,\infty)^k \to \mathbb{R}$ by

$$F_2(t) := F_1(t - \delta_2 \mathbf{1}) \text{ where } \mathbf{1} := (1, 1, \ldots, 1) \text{ has dimension } k,$$

in case each $t_i \ge \delta_2$ and zero on $\mathbb{R}^k$ otherwise. Then F_2 is square integrable, compactly supported and not zero and, because the integrals are continuous with respect to translations of the integrand variables, for δ_2 sufficiently small,

$$\sum_{i=1}^{k} \frac{J_i(F_2)}{I(F_2)} > m.$$

Because $\theta/2 - \delta_1 < \theta/2 - \delta_2$ and the support of F_1 from Step (2), considering the limits of the integrals, we can assume F_2 is supported on

$$\{t \in \mathbb{R}^k : t_1 + \cdots + t_k \le \theta/2 - \delta_2, \ t_1, \ldots, t_k \ge \delta_2\},$$

so F_2 is zero in a neighbourhood which includes each of the boundary faces of $\mathcal{R}_k$.

(4) Now let

$$F_3(t) := (\rho_\epsilon * F_2(t)) = \int_{[0,\infty)^k} \rho_\epsilon(y) F_2(t-y) \, dy_1 \cdots dy_k,$$

using the standard convolution of a smooth ρ_ϵ with integral 1, and support in a k-dimensional Euclidean ball with centre 0 and radius $\epsilon > 0$, and with ϵ sufficiently small so the support of F_3 is a subset of

$$\{t \in \mathbb{R}^k : t_1 + \cdots + t_k \leq \theta/2 - \delta_2/2, \ t_1, \ldots, t_k \geq \delta_2/2\}.$$

Then $F_3(t)$ is not identically zero and

$$\sum_{i=1}^k \frac{J_i(F_3)}{I(F_3)} > m.$$

Note that $\theta/2 - \delta_2 < \theta/2 - \delta_2/2$, so we have enlarged the simplex slightly in this step.

(5) Next, define

$$f_3(t) := \int_{s_1 \geq t_1, \ldots, s_k \geq t_k} F_3(s) \, ds_1 \cdots ds_k \implies F_3(t) = (-1)^k \frac{\partial^k}{\partial t_1 \cdots \partial t_k} f_3(t),$$

so f_3 is smooth, not identically zero and, because when $t_i < \delta/2$, we have

$$f_3(t) = \int_{s_i \geq t_i} F_3(s) \, ds_1 \cdots ds_k = \int_{s_i \geq \delta/2} F_3(s) \, ds_1 \cdots ds_k,$$

$f_3(t)$ has support contained in

$$\mathscr{R}_1 := \{t \in \mathbb{R}^k : \sum_{i=1}^k \max(t_i, \delta_2/2) \leq \theta/2 - \delta_2/2\}.$$

Let

$$\tilde{I}(f_3) := \int_{[0,\infty)^k} \left(\frac{\partial^k}{\partial t_1 \cdots \partial t_k} f_3(t) \right)^2 dt_1 \cdots dt_k = I(F_3),$$

$$\tilde{J}_i(f_3) := \int_{[0,\infty)^{k-1}} \left(\frac{\partial^{k-1}}{\partial t_1 \ldots \partial t_{i-1} \partial t_{i+1} \ldots \partial t_k} f_3(t_1, \ldots, t_{i-1}, 0, t_{i+1}, \ldots, t_k) \right)^2$$
$$dt_1 \ldots dt_{i-1} dt_{i+1} \ldots dt_k = J_i(F_3), \ 1 \leq i \leq k,$$

so by the result of Step (4),

$$\sum_{i=1}^k \frac{\tilde{J}_i(f_3)}{\tilde{I}(f_3)} > m.$$

(6) Next, we use the Stone–Weierstrass theorem to uniformly approximate f_3 by the algebra $\mathscr{A}$ of functions of the form

$$f(t) = \sum_{j=1}^{J} c_j f_{1,j}(t_1) \cdots f_{k,j}(t_k),$$

where the c_j are real and the $f_{i,j} : \mathbb{R} \to \mathbb{R}$ are smooth, compactly supported real functions such that the support of each product is in $\mathscr{R}_2$

$$\mathscr{R}_2 := \{t \in \mathbb{R}^k : \sum_{i=1}^{k} \max(t_i, \delta_2/4) \leq \theta/2 - \delta_2/4\},$$

so $\mathscr{R}_1 \subset \mathscr{R}_2$, where $\mathscr{R}_2$ was defined in Step (5). This is an algebra under pointwise function multiplication because if $f, g \in \mathscr{A}$, and

$$f(t) = \sum_{i=1}^{I} c_i f_{1,i}(t) \cdots f_{k,i}(t) \in \mathscr{A} \text{ and}$$

$$g(t) = \sum_{j=1}^{J} d_j g_{1,j}(t) \cdots g_{k,j}(t) \in \mathscr{A}, \text{ then we can write}$$

$$(f \cdot g)(t) = \sum_{i=1}^{I} \sum_{j=1}^{J} c_i d_j (f_{1,i} \cdot g_{1,j})(t) \cdots (f_{k,i} \cdot g_{k,j})(t),$$

so $f \cdot g$ is also in $\mathscr{A}$.

Provided the ρ_ϵ is symmetric, at the cost of making a small enlargement of the simplex, the convolution of any $f \in \mathscr{A}$ by ρ_ϵ is also in $\mathscr{A}$, and we can therefore assume that f_3 can be expressed as the uniform limit of smooth functions from $\mathscr{A}$, with each product in the sum supported in

$$\mathscr{R}_2 := \left\{ t \in \mathbb{R}^k : \sum_{i=1}^{k} \max(t_i, \delta_2/4) \leq \theta/2 - \delta_2/8 \right\}.$$

Hence there is an

$$f_4(t) = \sum_{j=1}^{J} c_j f_{1,j}(t_1) \cdots f_{k,j}(t_k),$$

not identically zero, with

$$\sum_{i=1}^{k} \frac{\tilde{J}_i(f_4)}{\tilde{I}(f_4)} > m,$$

and for all j with $1 \leq j \leq J$, $\mathrm{Supp}(f_{1,j}) + \cdots + \mathrm{Supp}(f_{k,j}) < \theta/2 \leq \frac{1}{2}$. From now on, we fix f_4 and the related $f_{i,j}$ with these properties for use later in the proof.

(7) In this step, we treat nonprime sums, defining

$$F_i(t) = G_i(t) := \sum_{j=1}^{J} c_j f_{i,j}(t),$$

and below using the $f_{i,j}$ to define the sieve weights. Note that because the support of a sum is contained in the union of the supports of the summands, by the final inequality of Step (6) we get

$$\sum_{i=1}^{k} S(F_i) + S(G_i) < 1.$$

Next, recall the definition $\lambda_f(n) := \sum_{d|n} \mu(d) f(\log_x d)$, where $f : [0, \infty) \to \mathbb{R}$. We now define sieve weights $\upsilon : \mathbb{N} \to \mathbb{R}$ in terms of the $\lambda_{f_{i,j}}$. Since

$$\lambda_{F_i}(n) = \sum_{d|n} \mu(d) F_i(\log_x d) = \sum_{j=1}^{J} c_j \sum_{d|n} \mu(d) f_{i,j}(\log_x d) = \sum_{j=1}^{J} c_j \lambda_{f_{i,j}}(n),$$

we set the sieve weights

$$\upsilon(n) := \left(\sum_{j=1}^{J} c_j \lambda_{f_{1,j}}(n + h_1) \cdots \lambda_{f_{k,j}}(n + h_k) \right)^2.$$

Thus, if we define

$$\alpha := \prod_{i=1}^{k} \int_0^1 F_i'(t) G_i'(t) \, dt$$

$$= \sum_{j=1}^{J} \sum_{j'=1}^{J} c_j c_{j'} \prod_{i=1}^{k} \int_0^\infty f_{i,j}'(t_i) f_{i,j'}'(t_i) \, dt_i$$

$$= \prod_{i=1}^{k} \int_0^1 \left(\sum_{j=1}^{J} c_j f_{i,j}'(t) \right) \left(\sum_{j'=1}^{J} c_{j'} f_{i,j'}'(t) \right)$$

$$= \int_{[0,\infty)^k} \left(\frac{\partial^k}{\partial t_1 \cdots \partial t_k} f_4(t_1, \ldots, t_k) \right)^2 \, dt_1 \ldots dt_k$$

$$= \tilde{I}(f_4),$$

interchanging sums and products

$$\prod_{i=1}^{k} \lambda_{F_i}(n+h_i)\lambda_{G_i}(n+h_i) = \prod_{i=1}^{k}\left(\sum_{j=1}^{J} c_j \lambda_{f_{i,j}}(n+h_i)\right)^2$$

$$= \left(\sum_{j=1}^{J} c_j \prod_{i=1}^{k} \lambda_{f_{i,j}}(n+h_i)\right)^2$$

$$= v(n).$$

Therefore, by Lemma 7.5 we get

$$\sum_{\substack{x \le n < 2x \\ n \equiv a \bmod W}} v(n) = (\alpha + o(1))\frac{x}{B^k W}.$$

(8) We now treat prime sums using the same sieve weights. For the sake of simplicity of indexing, we consider, for $h_k \in \mathcal{H}$ the sum

$$\sum_{\substack{x \le n < 2x \\ n \equiv a \bmod W}} v(n)\theta(n+h_k),$$

but the same derivation would apply to any fixed h_i with $1 \le i \le k$. If n in this sum is such that $v(n)\theta(n+h_k) \ne 0$, then $n \ge x$ and $n + h_k$ must be prime. Therefore, because $S(f_{k,j}) \subset [0,1)$, for such an n,

$$\lambda_{f_{k,j}}(n+h_k) = \sum_{d|n+h_k} \mu(d) f_{k,j}(\log_x(d)) = f_{k,j}(0).$$

Thus since by Step (6), we have

$$f_4(t_1,\ldots,t_{k-1},0) = \sum_{j=1}^{J} c_j f_{k,j}(0) \prod_{i=1}^{k} f_{i,j}(t_j)$$

if we define β_k and use Step (5) to get the last line we get

$$\beta_k := \sum_{j=1}^{J}\sum_{j'=1}^{J} c_j c_{j'} f_{k,j}(0) f_{k,j'}(0) \prod_{i=1}^{k-1} \int_0^\infty f'_{i,j}(t_i) f'_{i,j'}(t_i)\, dt_i$$

$$= \int_{[0,\infty)^{k-1}} \left(\frac{\partial^k}{\partial t_1 \cdots \partial t_{k-1}} f_4(t_1,\ldots,t_{k-1},0)\right)^2 dt_1 \ldots dt_{k-1}$$

$$= \tilde{J}_k(f_4).$$

Then by Lemma 7.4 and EH[θ], we get

$$\sum_{\substack{x \leq n < 2x \\ n \equiv a \bmod W}} v(n)\theta(n + h_k) = (\beta_k + o(1))\frac{x}{B^{k-1}\varphi(W)}$$

and generally for $1 \leq i \leq k$,

$$\sum_{\substack{x \leq n < 2x \\ n \equiv a \bmod W}} v(n)\theta(n + h_i) = (\beta_i + o(1))\frac{x}{B^{k-1}\varphi(W)}.$$

(9) Assembling the results of Steps (6) through (8), and using Lemma 7.1, we get DHL[k, m + 1], thus completing the proof. $\qquad\square$

7.5 Polymath's Algorithm for M_k

In this section, we give an overview of Polymath's algorithm used to compute a lower bound for M_k. This goes along the same path as Maynard's algorithm, but uses a different set of basis functions and some refinements in the final part of the calculation.

Define the **signature** α of a symmetric polynomial $P(t)$ in k variables as a nonincreasing sequence of natural numbers of length not greater than k. We pad the tail of the signature with zeros so that its length is precisely k. Let the **degree** of α, $\deg \alpha := \alpha_1 + \cdots + \alpha_k$.

We then define the symmetric polynomial

$$P_\alpha(t) := \sum_{a : a = \sigma(\alpha)} t_1^{a_1} \cdots t_k^{a_k},$$

so we include all of the permutations $\sigma \in S_k$ of the signature in the monomials of $P_\alpha(t)$, which are distinct. Some examples: if $\alpha = (3)$ and $k = 4$, then

$$P_\alpha(t) = t_1^3 + t_2^3 + t_3^3 + t_4^3.$$

Similarly, we have $P_{(2)}(t) = \sum_{1 \leq i \leq k} t_i^2$, and if $\alpha = (3,2)$ and $k \geq 2$, then

$$P_\alpha(t) = \sum_{1 \leq i < j \leq k} (t_i^2 t_j^3 + t_i^3 t_j^2).$$

Polymath, after some experimentation, used a basis, called $\mathscr{B}_d$, parameterized by a natural number d, defined by

$$\mathscr{B}_d := \{P = (1 - P_{(1)})^a P_\alpha : a \geq 0, \ \alpha \text{ has all even entries}\},$$

where

$$0 \leq a + \alpha_1 + \cdots + \alpha_k \leq d.$$

The algorithm can be described briefly as follows. It is included in PGpack as `RayleighQuotientPolymath`, which is a lexically scoped version of Polymath's MFinder.

(1) Compute a look-up table of integer constants $c(\alpha, \beta, \gamma)$ derived from all of the products

$$P_\alpha P_\beta = \sum_\gamma c(\alpha, \beta, \gamma) P_\gamma, \quad \deg(\alpha) + \deg(\beta) \leq d.$$

(2) Then, using (1), rewrite all of the integrands of the entries of the matrices A and B, which appear in Lemma 6.10, as integer linear combinations of monomials of the form $(1 - P_{(1)})^a t_1^{a_1} \cdots t_k^{a_k}$.
(3) Next, use Lemma 6.13 to compute the integrals as exact rational numbers.
(4) Then find an eigenvector x corresponding to the maximum eigenvalue, which according to Lemma 6.10, is an approximation to the maximum value of C in the form

$$x^T Bx - C x^T Ax \geq 0.$$

(5) Take y, a rational approximation to x, and verify, using exact arithmetic, that the corresponding value of C, namely

$$C' = \frac{y^T By}{y^T Ay} > 4.$$

Alternatively, use a bisection method and a positive definite test to find an approximate optimum for C, given some knowledge of its range.

Using Polymath's algorithm, the lower bound $M_{54} > 4$ was verified computationally.

We now give an application of this algorithm to prime gaps. Recall the notations $H(k)$ represents the minimum width of any admissible k-tuple (Section 1.6), and

$$H_m := \liminf_{n \to \infty} (p_{n+m} - p_n).$$

Theorem 7.7 *We have* $H_1 \leq 270$.

Proof By Lemma 7.2 and Polymath's algorithm for M_k, we have $M_{54} > 4$, and by EH[θ] with $\theta = \frac{1}{2}$ by Bombieri–Vinogradov's theorem Lemma 4.6, using Lemma 7.6 we can write for $m = 1$,

$$M_{54} > \frac{2m}{\frac{1}{2}} \implies DHL[54, 2]$$

is true. By Lemma 1.1, we have the width $H(54) = 270 \leq 270$. Therefore, there exist an infinite number of translates of $\mathcal{H}_{54}$ which contain at least two primes, so $H_1 \leq 270$. This completes the proof. □

7.6 Limits to These Techniques: Upper Bound for M_k

In this section, we derive an upper bound for M_k which is valid for all $k \geq 2$. We have seen how $M_k > 4$ enables bounded gaps to be derived unconditionally. An upper bound $B \geq M_k$ for a given k with $B \leq 4$ shows that such a derivation would fail.

We need the following concept: the **essential supremum** of a real-valued Lebesgue measurable function on a set A is the supremum of f on $A \setminus B$, where B ranges over all subsets of measure zero. For background reading on Lebesgue measure, see for example [22, appendix D].

Lemma 7.8   *[175, lemma 34] Let $k \geq 2$ and $G_i : \mathcal{R}_k \to (0,\infty)$, $1 \leq i \leq k$ be measurable functions which we set to zero outside of $\mathcal{R}_k$, and which are such that for all $t_1, \ldots, t_{i-1}, t_{i+1} \ldots, t_k \geq 0$ and all i, we have*

$$\int_0^\infty G_i(t_1, \ldots, t_k) \, dt_i \leq 1.$$

Then

$$M_k \leq \text{ess sup}_{t \in \mathcal{R}_k} \sum_{i=1}^k \frac{1}{G_i(t)}.$$

Proof   For any integrable F on $[0,\infty)^k$ which is supported on $\mathcal{R}_k$, using the Cauchy–Schwarz inequality we have for each fixed i and $t_1, \ldots, t_{i-1}, t_{i+1}, \ldots, t_k \geq 0$,

$$\left(\int_0^\infty F(t) \, dt_i \right)^2 = \left(\int_0^\infty \frac{F(t)\sqrt{G_i(t)}}{\sqrt{G_i(t)}} \, dt_i \right)^2 \leq \int_0^\infty \frac{F(t)^2}{G_i(t)} \, dt_i.$$

Thus

$$J_i(F) = \int_{\mathcal{R}_{k-1}} \left(\int_0^\infty F(t) \, dt_i \right)^2 dt_1 \cdots dt_{i-1} dt_{i+1} \cdots dt_k \leq \int_{\mathcal{R}_k} \frac{F(t)^2}{G_i(t)} \, dt_1 \cdots dt_k.$$

Summing over i with $d\mu_k := dt_1 \cdots dt_k$, we get

$$M_k = \sup_{F \in \mathcal{L}_k} \frac{\sum_{i=1}^k J_i(F)}{I(F)} \leq \frac{\int_{\mathcal{R}_k} F(t)^2 \left(\sum_{i=1}^k \frac{1}{G_i(t)} \right) d\mu_k}{\int_{\mathcal{R}_k} F(t)^2 \, d\mu_k} \leq \text{ess sup}_{t \in \mathcal{R}_k} \sum_{i=1}^k \frac{1}{G_i(t)},$$

which completes the proof. □

The following corollary has considerable promise in that it gives an alternative characterization of M_k. It requires having an eigenfunction $F(t)$, or at least knowing its general form.

Corollary 7.9 *[175, corollary 35] Let $k \geq 2$, and suppose we have a positive measurable function $F: \mathcal{R}_k \to (0, \infty)$ which satisfies for some $\lambda > 0$ the equation*

$$\lambda F(t) = \sum_{i=1}^{k} \int_0^\infty F(t_1, \ldots, t_{i-1}, s, t_{i+1}, \ldots, t_k) \, ds, \ \forall t \in [0, \infty)^k,$$

where we have extended F to all of $[0, \infty)^k$ by giving it the value zero outside of $\mathcal{R}_k$, then $\lambda = M_k$.

Proof First, using the given equation, the definitions of $I(F)$ and $J_i(F)$ and Fubini's theorem, we have

$$\lambda I(F) = \int_{\mathcal{R}_k} \lambda F(t) \cdot F(t) \, dt_1 \cdots dt_k$$

$$= \sum_{i=1}^{k} \int_{\mathcal{R}_{k-1}} \left(\left(\int_0^\infty F(t_1, \ldots, t_k) dt_i \right) F(t_1, \ldots, t_k) dt_i \right) dt_1 \cdots dt_{i-1} dt_{i+1} \cdots dt_k$$

$$= \sum_{i=1}^{k} \int_{\mathcal{R}_{k-1}} \left(\int_0^\infty F(t) \, dt_i \right)^2 dt_1 \cdots dt_{i-1} dt_{i+1} \cdots dt_k = \sum_{i=1}^{k} J_i(F).$$

Therefore, $M_k \geq \lambda$.

Next, using Lemma 7.8 with the choice

$$G_i(t) := \frac{F(t)}{\int_0^\infty F(t_1, \ldots, t_{i-1}, s, t_{i+1}, \ldots, t_k) \, ds}$$

so $\int_0^\infty G_i(t) \, dt_i \leq 1$ for $1 \leq i \leq k$, and therefore we get

$$M_k \leq \text{esssup}_{t \in \mathcal{R}_k} \sum_{i=1}^{k} \frac{1}{G_i(t)}$$

$$= \text{esssup}_{t \in \mathcal{R}_k} \frac{\sum_{i=1}^{k} \int_0^\infty F(t_1, \ldots, t_{i-1}, s, t_{i+1}, \ldots, t_k) \, ds}{F(t)}$$

$$= \lambda.$$

This completes the proof. ☐

The following rather miraculous application of Corollary 7.9 delivers an exact value for M_2, and raises the possibility of extensions to higher values of k. It uses the simple form $F(x, y) = f(x) + f(y)$. The author has experimented with the special form for $k = 3$

$$F(x, y, z) = f(x + y) + f(y + z) + f(z + x)$$

without success.

Corollary 7.10 *[175, corollary 36] Here we let W be Lambert's function, i.e., the unique positive solution to the functional equation $x = W(x) \exp(W(x))$. Then*

$$M_2 = \frac{1}{1 - W(1/e)} = 1.3859332759981942539\dots.$$

Proof Let $\lambda := 1/(1 - W(1/e)) > 1$ and define $f : [0, 1] \to [0, \infty)$ by

$$f(x) := \frac{1}{\lambda - 1 + x} + \frac{1}{2\lambda - 1} \log\left(\frac{\lambda - x}{\lambda - 1 + x}\right).$$

Using the given functional equation of $W(x)$, we get

$$2\lambda - 1 = \lambda \log \lambda - \lambda \log(\lambda - 1),$$

so therefore

$$\int_0^{1-x} f(y)\, dy = \frac{\lambda - 1 + x}{2\lambda - 1} \log\left(\frac{\lambda - x}{\lambda - 1 + x}\right) + \frac{\lambda \log \lambda - \lambda \log(\lambda - 1)}{2\lambda - 1}$$
$$= (\lambda - 1 + x) f(x).$$

Next, define $F : \mathscr{R}_2 \to (0, \infty)$ by $F(x, y) := f(x) + f(y)$. This implies

$$\int_0^{1-x} F(s, y)\, ds + \int_0^{1-y} F(x, s)\, ds = \lambda F(x, y),$$

so the corollary follows using Corollary 7.9. $\qquad \square$

We can now give, in Corollary 7.11, the simple derivation of Polymath's useful upper bound for M_k. Of course, we want k such that $M_k > 4$ and the upper bound shows that we must have $k \geq 51$ for any method based on M_k to give results on prime gaps. For $k = 50$, the bound has value 3.991860209620557 and for $k = 51$ the value 4.01046214537881.

Corollary 7.11 *[175, corollary 37] For all $k \geq 2$, we have*

$$M_k \leq \frac{k \log k}{k - 1}.$$

Proof   Define $G_i : \mathscr{R}_k \to (0, \infty)$ by

$$G_i(t) := \frac{k-1}{\log k} \frac{1}{1 - \left(\sum_{i=1}^{k} t_i\right) + k t_i},$$

and extend G_i to all of $[0, \infty)^k$ by setting it to zero outside of $\mathscr{R}_k$. Integrating, for all $t_1, \dots, t_{i-1}, t_{i+1}, \dots, t_k \geq 0$, we get

$$\int_0^\infty G_i(t)\, dt_i = 1 \leq 1.$$

In addition, inverting the definition and summing the right-hand side gives

$$\sum_{i=1}^{k} \frac{1}{G_i(t)} = \frac{k \log k}{k-1}.$$

Therefore, by Lemma 7.8 we have

$$M_k \leq \operatorname{ess\,sup}_{t \in \mathscr{R}_k} \sum_{i=1}^{k} \frac{1}{G_i(t)} = \operatorname{ess\,sup}_{t \in \mathscr{R}_k} \frac{k \log k}{k-1} = \frac{k \log k}{k-1},$$

which completes the proof.   □

Remark: This is an excellent bound. As indicated before, it shows that with Polymath's method, so-called "sieving on the standard simplex" described in Section 7.5, we can do no better than $k = 51$. As seen in Theorem 7.7 we achieve $k = 54$ with `RayleighQuotientPolymath`. In Section 7.9, Polymath obtained $k = 50$ by allowing the simplex to vary – see Section 7.9. But that small improvement comes at a considerable cost in analysis and complexity of the computational code. And the corresponding bound for $M_{k,\epsilon}$, defined later in this chapter, is not particularly close. See Section 7.10.

7.7 Bogaert's Krylov Basis Method

For small values of k, Polymath described the so-called **Krylov basis method**, implemented in PGpack with the function RayleighQuotientKrylov, and based on the approach taken by Ignace Bogaert [8]. First we state and prove the beta function identity used to compute the Krylov operator, then describe the operator and finally derive the lower bounds. The quoted results of Bogaert (see Table 7.1) demonstrate, using the upper bound of Corollary 7.11, namely $(k \log k)/(k-1)$, that the method could give excellent results. We were not able to reach anything close to Bogaert's matrix size used in the final optimization step – see the remark at the end of this section.

Recall the definition $\Gamma(s) = \int_0^\infty t^{s-1} e^{-t} \, dt$, where $s \in \mathbb{C} \setminus \{0, -1, -2, \ldots\}$. Then $\Gamma(n+1) = n!$ for all integers $n \geq 0$. Also recall the beta function improper integral

$$B(r,s) := \int_0^1 x^{r-1}(1-x)^{s-1} \, dx = \frac{\Gamma(r)\Gamma(s)}{\Gamma(r+s)}, \quad r > 0, s > 0.$$

Lemma 7.12 is a multidimensional version of this integral.

Lemma 7.12 *Let a and all the a_j for $1 \leq j \leq k$ be real and nonnegative. Then*

$$\int_{\mathscr{R}_k} \left(1 - \sum_{j=1}^k t_j\right)^a t_1^{a_1} \cdots t_k^{a_k} \, dt_1 \cdots dt_k = \frac{\Gamma(1+a)\Gamma(1+a_1) \cdots \Gamma(1+a_k)}{\Gamma(1+k+a+a_1+\cdot+a_k)}.$$

Proof **(1)** Integrating the right-hand side with respect to t_{k+1}, we get

$$\int_{\mathscr{R}_k} \left(1 - \sum_{j=1}^k t_j\right)^a t_1^{a_1} \cdots t_k^{a_k} \, dt_1 \cdots dt_k = a \int_{\mathscr{R}_{k+1}} t_1^{a_1} \cdots t_k^{a_k} t_{k+1}^{a-1} \, dt_1 \cdots dt_{k+1}.$$

Therefore, we need only consider the case where the first factor of the integrand does not occur, i.e., $a = 0$ in the lemma statement for all $k \geq 2$, and then use the expression we have derived on the right for $a > 0$.

(2) Define

$$X := \int_{t_1 + \cdots t_k = 1} t_1^{a_1} \cdots t_k^{a_k} \, dt_1 \cdots dt_k.$$

Multiplying by the given power of r and changing variables $rt_j \to t_j$, for each $r > 0$ we get

$$X r^{a_1 + \cdots + a_k + k - 1} = \int_{t_1 + \cdots t_k = r} t_1^{a_1} \cdots t_k^{a_k} \, dt_1 \cdots dt_k. \tag{7.8}$$

Then, integrate both sides with respect to r over $[0, 1]$ to get

$$\frac{X}{a_1 + \cdots + a_k + k} = \int_{\mathscr{R}_k} t_1^{a_1} \cdots t_k^{a_k} \, dt_1 \cdots dt_k.$$

Next, multiply both sides of (7.8) by $\exp(-r)$ and integrate with respect to r over $[0, \infty)$ to get

$$X \int_0^\infty r^{a_1 + \cdots + a_k + k - 1} e^{-r} \, dr = \int_{[0,\infty)^k} t_1^{a_1} \cdots t_k^{a_k} e^{-t_1 - \cdots - t_k} \, dt_1 \cdots dt_k$$

$$= \Gamma(1+a_1) \cdots \Gamma(a_k + 1).$$

Therefore,

$$X = \frac{\Gamma(1+a_1)\cdots\Gamma(1+a_k)}{\Gamma(k+a_1+\cdot+a_k)},$$

which, using Step (1), completes the proof. ☐

Remark: The reader might wish to compare Maynard's Lemmas 6.12 and 6.13, in which cases the powers are integers.

Now we describe the Krylov subspace method for bounding M_k below. Let $b_1,\ldots,b_n$ be symmetric real polynomials in k real variables $t_1,\ldots,t_k$. Truncate to $\mathscr{R}_k$ and then define a self-adjoint positive semidefinite operator $\mathscr{L}$ on the real Hilbert space $H = L^2(\mathscr{R}_k)$ by

$$\mathscr{L}(F)(t) := \sum_{j=1}^{k} \int_0^{1-\left(\sum_{i=1}^{k} t_i\right)+t_j} F(t_1,\ldots,t_{j-1},s,t_{j+1},\ldots,t_k)ds.$$

Then, if A and B are the matrices defined in Section 6.6, we can write

$$A = \left(\langle b_i,b_j\rangle\right)_{1\leq i,j\leq n} \text{ and } B = \left(\langle \mathscr{L}b_i,b_j\rangle\right)_{1\leq i,j\leq n}.$$

We then choose $b_1 = 1$, representing the constant function with value 1 on $\mathscr{R}_k$ and for $i \geq 2$ set $b_i := \mathscr{L}^{i-1}1 = \mathscr{L}b_{i-1}$. Then we get with F_0, representing b_1 and using the notation employed by Bogaert, namely

$$P[k,n] := x_1^n + \cdots + x_k^n, \ k,n \in \mathbb{N},$$

and the notation that follows, which is a list of terms $\{\{\alpha_1,\ldots,\alpha_L\} \to coef\}$, represents each of the given $F_j = \mathscr{L}^j F_{j-1}$ for $j \geq 1$ in the form of a coefficient function of k times a list of powers corresponding to a monomial

$$P[k,1]^{\alpha_1} P[k,2]^{\alpha_2} \cdots P[k,L]^{\alpha_L}.$$

Using the PGpack function FnewMon starting with $F_0 = \{\{0\} \to 1\}$, we obtained the following (where we have explained this Mathematica data type giving F_1 as an example):

F_1: $\{\{0\} \to k, \{1\} \to 1 - k\}$ representing $k + (1 - k)P[k,1]^1$

F_2: $\Big\{\{0,0\} \to \dfrac{k^2}{2} + \dfrac{k}{2}, \{0,1\} \to \dfrac{k}{2} - \dfrac{1}{2},$
$\{1,0\} \to k - k^2, \{2,0\} \to \dfrac{k^2}{2} - \dfrac{3k}{2} + 1\Big\}$

$$F_3: \quad \left\{ \{0,0,0\} \to \frac{k^3}{6} + \frac{2k^2}{3} + \frac{k}{6}, \right.$$

$$\{0,0,1\} \to \frac{k^2}{6} - \frac{5k}{6} + \frac{2}{3}, \{0,1,0\} \to k^2 - k,$$

$$\{1,0,0\} \to -\frac{k^3}{2} - \frac{k^2}{2} + \frac{3k}{2} - \frac{1}{2}, \{1,1,0\} \to -k^2 + \frac{5k}{2} - \frac{3}{2},$$

$$\left. \{2,0,0\} \to \frac{k^3}{2} - k^2 - \frac{k}{2} + 1, \{3,0,0\} \to -\frac{k^3}{6} + \frac{5k^2}{6} - \frac{7k}{6} + \frac{1}{2} \right\}$$

$$F_4: \quad \left\{ \{0,0,0,0\} \to \frac{k^4}{24} + \frac{5k^3}{12} + \frac{13k^2}{24}, \right.$$

$$\{0,0,0,1\} \to \frac{k^3}{24} - \frac{3k^2}{4} + \frac{47k}{24} - \frac{5}{4},$$

$$\{0,0,1,0\} \to \frac{k^3}{3} - \frac{11k^2}{6} + \frac{7k}{6} + \frac{1}{3},$$

$$\{0,1,0,0\} \to \frac{3k^3}{4} + \frac{3k^2}{4} - \frac{11k}{4} + \frac{5}{4},$$

$$\{0,2,0,0\} \to \frac{k^2}{2} - \frac{5k}{4} + \frac{3}{4},$$

$$\{1,0,0,0\} \to -\frac{k^4}{6} - k^3 + \frac{7k^2}{6} + \frac{k}{3} - \frac{1}{3},$$

$$\{1,0,1,0\} \to -\frac{k^3}{3} + \frac{5k^2}{2} - \frac{13k}{3} + \frac{13}{6},$$

$$\{1,1,0,0\} \to -\frac{3k^3}{2} + \frac{5k^2}{2} + 2k - 3,$$

$$\{2,0,0,0\} \to \frac{k^4}{4} + \frac{k^3}{2} - \frac{15k^2}{4} + 4k - 1,$$

$$\{2,1,0,0\} \to \frac{3k^3}{4} - \frac{13k^2}{4} + \frac{15k}{4} - \frac{5}{4},$$

$$\{3,0,0,0\} \to -\frac{k^4}{6} + \frac{k^3}{3} + \frac{11k^2}{6} - 5k + 3,$$

$$\left. \{4,0,0,0\} \to \frac{k^4}{24} - \frac{k^3}{4} + \frac{5k^2}{24} + \frac{2k}{3} - \frac{2}{3} \right\}$$

PGpack also includes three functions for manipulating these polynomial representations, namely `MonToPoly`, `PolyToMon` and `PolyMonDegrees`. We then
form the matrices with $n = 3$:

$$A = \left(\langle \mathcal{L}^{i+j-2} 1, 1 \rangle \right)_{1 \le i, j \le n} \text{ and } B = \left(\langle \mathcal{L}^{i+j-1} 1, 1 \rangle \right)_{1 \le i, j \le n}.$$

We are able, using the PGpack functions OperatorL and IntegrateRk, to obtain explicit expressions for matrix entries as functions for $0 \leq i \leq 6$, verifying Ignace Bogaert's published formulas [175, page 61] in those cases. See for example Table 7.2.

Table 7.1 *Bogaert's lower bounds for M_k for $49 \leq k \leq 54$.*

k	Largest eigenvalue
49	3.9182810686921876017389790449597522271308
50	3.9358660345688540805957816293688861369686
51	3.9530417305030088203431426334330003565530
52	3.9698202731187838542379390159650898810231
53	3.9862131121051335673523749444316473685571
54	4.0022311023539785113505189887355557690242

We computed additional values for these inner products which the reader might find useful:

$$\langle \mathcal{L}^7 1, 1 \rangle = \frac{2k}{(k+7)!}(715k^6 + 5562k^5 + 10604k^4 + 4036k^3 - 827k^2$$
$$+ 62k + 8),$$

$$\langle \mathcal{L}^8 1, 1 \rangle = \frac{k}{(k+8)!}(4862k^7 + 54445k^6 + 167092k^5 + 143156k^4 - 1064k^3$$
$$- 7909k^2 + 2558k - 260),$$

$$\langle \mathcal{L}^9 1, 1 \rangle = \frac{2k}{(k+9)!}(8398k^8 + 129503k^7 + 592974k^6 + 908256k^5$$
$$+ 272646k^4 - 118197k^3 + 19826k^2 + 1846k - 852),$$

$$\langle \mathcal{L}^{10} 1, 1 \rangle = \frac{2k}{(k+10)!}(29393k^9 + 602895k^8 + 3899983k^7 + 9467924k^6$$
$$+ 7063858k^5 - 847193k^4 - 526463k^3 + 337004k^2 - 73139k$$
$$+ 4138),$$

$$\langle \mathcal{L}^{11} 1, 1 \rangle = \frac{4k}{(k+11)!}(52003k^{10} + 1379755k^9 + 12100286k^8 + 43146780k^7$$
$$+ 58213838k^6 + 12987985k^5 - 11142804k^4 + 3082388k^3$$
$$+ 191145k^2 - 311644k + 50668).$$

In Table 7.5, we give weak lower bounds using the Krylov method and just 3×3 and 6×6 matrices A and B, and upper bounds given by Corollary 7.11 and Theorem 7.17 for a variety of values of k up to 54. Table 7.5 at $k = 50$ with an upper bound less than 4 for M_k shows that it is impossible to do better with this method than perhaps $k = 51$, using larger values of r and optimizing over $\mathcal{R}_k$.

In Table 7.2, we give Bogaert's stated values computed using 25×25 matrices, which would have taken serious amounts of computer resources for symbolic computation. In addition to computing the action of the operator $\mathcal{L}$, integrating products of these expressions over $\mathcal{R}_k$ was needed to derive as functions of k the matrix entries. The integrals were built up by deriving and solving many recurrence relations in k. A simple example:

$$\int_{\mathcal{R}_k} P[k,n] = \frac{k(n!)}{(k+n)!}.$$

Alternatively, on the face of it this might be completed by expanding the products of the $P[k,n]$ and then using Lemma 7.12.

Table 7.2 *Verifying Bogaert's values for $\langle \mathcal{L}^i 1, 1 \rangle$, $0 \le i \le 6$.*

i	Form	Bogaert	$k=2$	$k=3$	$k=4$
0	$\langle 1,1 \rangle$	$\frac{1}{k!}$	$\frac{1}{2}$	$\frac{1}{6}$	$\frac{1}{24}$
1	$\langle \mathcal{L}1,1 \rangle$	$\frac{2k}{(k+1)!}$	$\frac{2}{3}$	$\frac{1}{4}$	$\frac{1}{15}$
2	$\langle \mathcal{L}^2 1,1 \rangle$	$\frac{k(5k+1)}{(k+2)!}$	$\frac{11}{12}$	$\frac{2}{5}$	$\frac{7}{60}$
3	$\langle \mathcal{L}^3 1,1 \rangle$	$\frac{2k^2(7k+5)}{(k+3)!}$	$\frac{19}{15}$	$\frac{13}{20}$	$\frac{22}{105}$
4	$\langle \mathcal{L}^4 1,1 \rangle$	$\frac{k(42k^3+69k^2+10k-1)}{(k+4)!}$	$\frac{19}{15}$	$\frac{13}{20}$	$\frac{22}{105}$
5	$\langle \mathcal{L}^5 1,1 \rangle$	$\frac{k^2(132k^3+406k^2+196k-14)}{(k+5)!}$	$\frac{631}{360}$	$\frac{223}{210}$	$\frac{1277}{3360}$
6	$\langle \mathcal{L}^6 1,1 \rangle$	$\frac{k(429k^5+2186k^4+2310k^3+184k^2-79k+10)}{(k+6)!}$	$\frac{1529}{630}$	$\frac{487}{280}$	$\frac{97}{140}$

We implemented Bogaert's approach in Mathematica – see the PGpack function `RayleighQuotientBogaert`. At the heart of this algorithm is the construction of a large database of symbolic expressions being the integrals extending Table 7.2. These integrals require integrals of monomials expressions in the $P[k,n]$ which are computed recursively based on Lemma 7.13:

Lemma 7.13 *Let $k,n \in \mathbb{N}$ and define*

$$I[k,\alpha] = I[k,\{\alpha_1,\dots,\alpha_N\}] := \int_{\mathcal{R}_k} P[k,1]^{\alpha_1} \cdots P[k,N]^{\alpha_N} \, dx_1 \cdots dx_N.$$

Then for variable k and explicit $\{\alpha_1,\dots,\alpha_N\}$ and nonnegative integers α_j, we have

$$I[k+1,\{\alpha_1,\dots,\alpha_N\}] = \frac{1}{1+k+\alpha_1+2\alpha_2+\cdots+N\alpha_N} I[k,\{\alpha_1,\dots,\alpha_N\}]$$
$$+ \sum_{\beta} c(\beta,k) E[k,\beta_1,\dots,\beta_N],$$

where the sum is over an explicit set of β's with $0 \le \beta_i \le \alpha_i$ such that at least one $\beta_j < \alpha_j$, and where E is an explicit function of k and the β_j.

Proof First, let $x^k := (x_1, \ldots, x_k)$ and $dx^k := dx_1 \cdots dx_k$. Using the binomial theorem, we can expand the monomials

$$\prod_{n=1}^{N} P[k+1,n]^{\alpha_n} = \sum_{\beta_1=0}^{\alpha_1} \cdots \sum_{\beta_N=0}^{\alpha_n} \prod_{n=1}^{N} \binom{\alpha_n}{\beta_n} P[k,n]^{\alpha_n - \beta_n} x_{k+1}^{n\beta_n}.$$

For a polynomial $f(x^k, x_{k+1})$ following a change of variables

$$x_1 = (1 - x_{k+1})y_1, \ldots x_k = (1 - x_{k+1})y_k,$$

we get the integral formula

$$\int_{\mathcal{R}_{k+1}} f(x^k, x_{k+1}) \, dx^{k+1} = \int_0^1 (1 - x_{k+1})^k \left(\int_{\mathcal{R}_k} f((1 - x_{k+1})x^k, x_{k+1}) \, dx^k \right) dx_{k+1}.$$

Insert the expansion of the monomial and then use the integral formula to get

$$I[k+1, \alpha] = \sum_{\beta_1=0}^{\alpha_1} \cdots \sum_{\beta_N=0}^{\alpha_n} \int_0^1 (1 - x_{k+1})^k$$

$$\times \left(\int_{\mathcal{R}_k} \prod_{n=1}^{N} \binom{\alpha_n}{\beta_n} (1 - x_{k+1})^{n(\alpha_n - \beta_n)} P[k,n]^{\alpha_n - \beta_n} x_{k+1}^{n\beta_n} \, dx^k \right) dx_{k+1}$$

$$= \sum_{\beta_1=0}^{\alpha_1} \cdots \sum_{\beta_N=0}^{\alpha_n} \left(\prod_{n=1}^{N} \binom{\alpha_n}{\beta_n} \right)$$

$$\times \int_0^1 (1 - x_{k+1})^k \prod_{n=1}^{N} (1 - x_{k+1})^{n(\alpha_n - \beta_n)} x_{k+1}^{n\beta_n} \, dx_{k+1} I[k, \alpha - \beta]$$

$$= \sum_{\beta_1=0}^{\alpha_1} \cdots \sum_{\beta_N=0}^{\alpha_n} \left(\prod_{n=1}^{N} \binom{\alpha_n}{\beta_n} \right)$$

$$\times \int_0^1 (1 - x_{k+1})^{k + \sum_{n=1}^{N} n(\alpha_n - \beta_n)} x_{k+1}^{\sum_{n=1}^{N} n\beta_n} \, dx_{k+1} I[k, \alpha - \beta]$$

$$= \sum_{\beta_1=0}^{\alpha_1} \cdots \sum_{\beta_N=0}^{\alpha_n} \left(\prod_{n=1}^{N} \binom{\alpha_n}{\beta_n} \right)$$

$$\times B \left[1 + k + \sum_{n=1}^{N} n(\alpha_n - \beta_n), 1 + \sum_{n=1}^{N} n\beta_n \right] dx_{k+1} I[k, \alpha - \beta]$$

$$= \frac{1}{1 + k + \alpha_1 + 2\alpha_2 + \cdots + N\alpha_N} I[k, \{\alpha_1, \ldots, \alpha_N\}]$$

$$+ \sum_{\beta} c(\beta, k) E[k, \{\beta_1, \ldots, \beta_N\}],$$

where B is the beta function and the leading term comes from $\beta_j = 0$ for $1 \le j \le N$. This completes the derivation. $\qquad\square$

7.8 Bogaert's Algorithm

Here we give an overview of the main steps in Bogaert's method for finding a lower bound for M_k. Recall the definition

$$\mathcal{L}(F)(t) := \sum_{j=1}^{k} \int_0^{1-\left(\sum_{i=1}^{k} t_i\right)+t_j} F(t_1, \ldots, t_{j-1}, s, t_{j+1}, \ldots, t_k)ds.$$

(1) Set $F[0] = 1$ and ensure k is symbolic. Let $[a,b]$ be the target range of k values. Set r the rank of the target matrices A and B. Set $DB = \{\}$, i.e., clear the database of integrals of monomials in the $P[k,n]$.

(2) Compute $F[n + 1] = \mathcal{L}F[n]$ for $0 \le n \le 2r - 1$ as monomials in the $P[k,n]$ with coefficients being polynomials in k with rational coefficients.

(3) Analyze the expressions for the $F[n]$ to find the maximum monomial degree.

(4) Compute the integrals of all of the monomials up to the maximum monomial degree and store in DB.

(5) Use the $F[n]$ from Step (3) and the integrals from Step (4) to evaluate the matrix entries $\langle \mathcal{L}^i 1, 1 \rangle$, $0 \le i \le 2r$.

(6) Form the matrices A and B. They will have entries which are functions of k.

(7) For each explicit k with $a \le k \le b$, solve numerically to find the largest eigenvalue of $Inverse[A].B$, which is then returned as an approximate lower bound for M_k.

Remark: The critical variable is the rank r. Bogaert managed to reach $r = 25$ with 64 GBytes of RAM. We implemented this algorithm in PGpack as `Rayleigh QuotientKrylov` and found the algorithm ran in exponential time, with space not the limiting factor. We extrapolated from the results for $r \in \{2, \ldots, 6\}$ and found an extrapolated value for the objective being greater than 4 was $r = 11$. See Figure 7.3, where the dots represent values obtained using the PGpack function and the smooth curve the extrapolating function which was obtained using an objective form $a + br/\log(r) + cr/\log(r)^2$ with the computed value using the Mathematica function NonlinearModelFit,

$$2.72813 + \frac{0.392826r}{\log(r)} - \frac{0.262308r}{\log^2(r)},$$

with value at $r = 11$ being 4.02835.

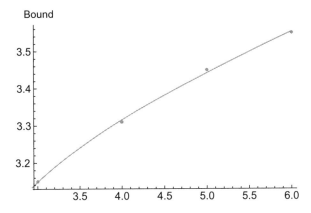

Figure 7.3 Interpolating Krylov bounds for $2 \le r \le 6$.

However, for $r \ge 9$ the compute time on a reasonably current workstation extrapolated to more than 20 years. We used the Mathematica function AbsoluteTime and NonlinearModelFit with objective $ar^2 + br + c$ to model the logarithm of the time for small values of r and extrapolated to find $r = 7$ was the upper limit for reasonable computation times using our particular implementation. See Figure 7.4, where the smooth approximation is $0.238504r^2 + 0.290605r - 1.63937$. The number of integrals of monomials in the $P[k, j]$ is the limiting factor and this would be

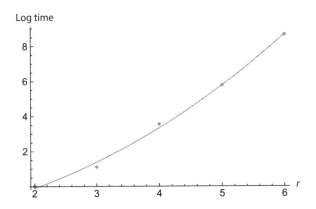

Figure 7.4 Interpolating Krylov absolute time for $2 \le r \le 6$.

more than $r^2 2^{r-3}$. For $r = 25$, this would be more than 2.5×10^9 integrals. For the reader's possible amusement here, the following is a representation of one of the integrals. Note the first sublist represents the powers of the $P[k, j]$, and the rational function of k is the value of the integral of their product over $\mathscr{R}_k$:

$\{\{13, 5, 3, 2, 2, 1\} \leftarrow (1/(56 + k)!)41278242816000$

$(131483491526015462805187383519976059069150069612625133568000000 \, k +$

$35863200028153779153851160635488610613575054880266987765760000 \, k^2 +$

$4528479778710268403543142111950582704207501693559736864476800 \, k^3 +$

$350824923140989102763121342864281109484442366027397666201601 \, k^4 +$

$18627571755642554052967321857948353168868106904201937530881 \, k^5 +$

$7171035330403381300814822414731312560621420655638711014411 \, k^6 +$

$20645153935873610415281310617038503886478083744758556800 \, k^7 +$

$451799940257442534146700986518092788798930603632572800 \, k^8 +$

$7561854548474015972137963140631106714480722455551201 \, k^9 +$

$9658636937368523455291556795453924087413441123200 \, k^{10} +$

$93208463161867129891859174613488314952482027180 \, k^{11} +$

$6673753790170615161197838252082179674468503801 \, k^{12} +$

$345302611914652920996017724344177911156556511 \, k^{13} +$

$125028490595515560541900827945901551482851 \, k^{14} +$

$308712708233535939574239632518436161301 \, k^{15} +$

$52169973370018986237638006989590330 \, k^{16} +$

$605917736507613535016876494554551 \, k^{17} +$

$47812156862182175520321933255 \, k^{18} +$

$24893276865288993884269480 \, k^{19} + 837442330642530546828011 \, k^{20} +$

$1981625235535566475 \, k^{21} + 363143912009115 \, k^{22} +$

$62655054730 \, k^{23} + 14313570 \, k^{24} + 4297 \, k^{25} + k^{26})\}.$

7.9 How the Gap Bound $p_{n+1} - p_n \le 246$ Is Derived

This is one of the most important sections of the book. It is the culmination of Chapter 6 and this chapter, and calls upon PGpack's `RayleighQuotientPolymath`. It is based on perturbing the standard simplex in a simple symmetric manner to introduce an additional degree of freedom, denoted by a real variable ϵ, and making some corresponding changes to the basis functions used in the optimization step. There is a reference back to much of the sieving lemma (Maynard's lemma) Lemma 7.6.

The following sieve result is interesting. It is one place where the flexibility of Lemma 7.1, namely using upper and lower bounds for the nonprime and prime sums, is used, and then only for the latter case.

Theorem 7.14 (**Sieving on an enlarged simplex**) *[175, theorem 26] Let* $k \geq 2$ *and* $m \in \mathbb{N}$ *be integers,* $F: [0, \infty)^k \to \mathbb{R}$ *a compactly supported square integrable function and let* ϵ *be given with* $0 < \epsilon < 1$. *Define the functionals for* $1 \leq i \leq k$,

$$I(F) := \int_{[0,\infty)^k} F(t_1, \ldots, t_k)^2 \, dt_1 \ldots dt_k,$$

$$J_{i,1-\epsilon}(F) := \int_{(1-\epsilon)\mathcal{R}_{k-1}} \left(\int_0^\infty F(t_1, \ldots, t_k) dt_i \right)^2 dt_1 \ldots dt_{i-1} dt_{i+1} \ldots dt_k.$$

Let $\mathcal{L}_{k,\epsilon}$ *be all square integrable functions supported on the simplex*

$$(1+\epsilon)\mathcal{R}_k := \left\{ (t_1, \ldots, t_k) \in [0, \infty)^k : t_1 + \cdots + t_k \leq 1 + \epsilon \right\},$$

and not equivalent in Lebesgue measure to the zero function, and

$$M_{k,\epsilon} := \sup_{F \in \mathcal{L}_{k,\epsilon}} \frac{\sum_{i=1}^k J_{i,1-\epsilon}(F)}{I(F)}.$$

Let the primes have level of distribution θ *with* $0 < \theta < \frac{1}{2}$ *and suppose also that* $\epsilon > 0$ *is sufficiently small so* $1 + \epsilon < 1/\theta$. *If*

$$\frac{1}{2} \theta M_{k,\epsilon} > m \quad then \quad DHL[k, m + 1].$$

Proof **(1)** In Steps (1) through (4), we derive an asymptotic expression for the prime sum related to the problem, and in Steps (5) through (8) a lower bound for the nonprime sum. In Step (9), the proof is completed.

Let $F: [0, \infty)^k \to \mathbb{R}$ be a Lebesgue square integrable function supported on $(1 + \epsilon) \cdot \mathcal{R}_k$, which is not equivalent to the zero function, and which satisfies

$$\frac{\sum_{i=1}^k J_{i,1-\epsilon}(F)}{I(F)} > \frac{2m}{\theta}.$$

(2) Next, apply Steps (1) through (6) of Lemma 7.6 to obtain a smooth function $f_4: \mathbb{R}^k \to \mathbb{R}$ of the form

$$f_4(t) = \sum_{j=1}^J c_j f_{1,j}(t_1) \cdots f_{k,j}(t_k),$$

with $S(f_{i,j}) = \sup\{t : f_{i,j}(t) \neq 0\}$ for $1 \leq j \leq J$ such that

$$S(f_{1,j}) + \cdots + S(f_{k,j}) \leq (1 + \epsilon)\theta/2 - \delta_2/8.$$

(3) Let $\delta_3 < \delta_2 < \delta_1$ be sufficiently small. Using a smooth partition of unity on $\mathbb{R}$, we can assume each of the $f_{i,j}$ is supported on an interval of length at most δ_3, and that the sum

$$\sum_{j=1}^{J} |c_j| |f_{1,j}(t_1) \cdots f_{k,j}(t_k)|$$

is bounded uniformly in all of the t_i and in δ_3.

As before in Lemma 7.6 Step (7), let $v(n)$ be defined by

$$v(n) := \left(\sum_{j=1}^{J} c_j \lambda_{f_{1,j}}(n + h_1) \cdots \lambda_{f_k,j}(n + h_k) \right)^2,$$

where $\lambda_f(n) := \sum_{d|n} \mu(d) f(\log_x d)$. Consider

$$\sum_{\substack{x \leq n < 2x \\ n \equiv a \bmod W}} v(n) = \sum_{\substack{x \leq n < 2x \\ n \equiv a \bmod W}} \sum_{(j,j') \in \mathscr{J}} c_j c'_j \prod_{i=1}^{k} \lambda_{f_{i,j}}(n + h_i) \lambda_{f_{i,j'}}(n + h_i),$$

where $\mathscr{J}$ is the set of elements (j, j') of $\{1, \ldots, J\}^2$ for which each of the $\lambda_{f_{i,j}}(n + h_i)$ and $\lambda_{f_{i,j'}}(n + h_i)$ is nonzero.

(4) Using the hypothesis EH[θ] with $\theta < \frac{1}{2}$, and the sum from Step (2), we can write

$$\sum_{1 \leq i \leq k} S(f_{i,j}) + S(f_{i,j'}) \leq (1 + \epsilon)\theta < 1.$$

This means we can write using Lemma 7.5 for each $(j, j') \in \mathscr{J}$,

$$\prod_{i=1}^{k} \lambda_{f_{i,j}}(n + h_i) \lambda_{f_{i,j'}}(n + h_i) = (c_{j,j'} + o(1)) \frac{x}{B^k W},$$

with

$$c_{j,j'} := \prod_{i=1}^{k} \int_0^1 f'_{i,j}(t) f'_{i,j'}(t) \, dt.$$

Assembling the linear combination required to construct $v(n)$ and defining $c :=$ $\sum_{(j,j') \in \mathscr{J}} c_{j,j'}$, gives

$$\sum_{\substack{x \leq n < 2x \\ n \equiv a \bmod W}} v(n) = (c + o(1)) \frac{x}{B^k W}.$$

(5) Now we consider the prime sum

$$\sum_{\substack{x \leq n < 2x \\ n \equiv a \bmod W}} v(n) \theta(n + h_k).$$

Steps (1) and (2) give a smooth function f_4 on $[0, \infty)^k$ bounding the $M_{k,\epsilon}$ functional from below and having the form

$$f_4(t) = \sum_{j=1}^{J} c_j f_{1,j}(t_1) \cdots f_{k,j}(t_k).$$

Next, we partition the indices j with $1 \leq j \leq J$ in two disjoint sets J_1 and J_2 where each $j \in J_1$ satisfies

$$S(f_{1,j}) + \cdots + S(f_{k-1,j}) < (1 - \epsilon)\theta/2$$

and $J_2 = \{1, \ldots, J\} \setminus J_1$ is the complement.

Because for all real x, y we have $(x + y)^2 = x^2 + 2xy + y^2 \geq (x + 2y)x$, using the expansion for $v(n)$ from Step (3), we get for each $n \in \mathbb{N}$,

$$v(n) \geq \left(\left(\sum_{j \in J_1} + 2 \sum_{j \in J_2} \right) c_j \lambda_{f_{1,j}}(n + h_1) \cdots \lambda_{f_{k,j}}(n + h_k) \right)$$

$$\times \left(\sum_{j' \in J_1} c_{j'} \lambda_{f_{1,j'}}(n + h_1) \cdots \lambda_{f_{k,j'}}(n + h_k) \right).$$

(6) For each $j \in J$ and $j' \in J_1$, using Step (2) and the definition of J_1, we get

$$\sum_{i=1}^{k-1} \left(S(f_{i,j}) + S(i, j') \right) < \theta.$$

Using Step (2) again for $j \in J$ and $1 \leq i \leq k$, we also have for ϵ sufficiently small,

$$S(f_{i,j}) \leq (1 + \epsilon)\theta/2 < \frac{1}{2}.$$

Therefore, for $n \geq x$ with $n + h_k$ prime, for all nonzero terms in the prime sum we have $\lambda_{f_{k,j'}}(n + h_k) = f_{k,j'}(0)$. Thus multiplying the lower estimate for $v(n)$ by $\theta(n + h_k)$, we get

$$v(n)\theta(n + h_k) \geq \left(\left(\sum_{j \in J_1} + 2 \sum_{j \in J_2} \right) c_j \lambda_{f_{1,j}}(n + h_1) \cdots \lambda_{f_{k-1,j}}(n + h_{k-1}) f_{k,j}(0) \right)$$

$$\times \left(\sum_{j' \in J_1} c_{j'} \lambda_{f_{1,j'}}(n + h_1) \cdots \lambda_{f_{k-1,j'}}(n + h_{k-1}) f_{k,j'}(0) \right) \theta(n + h_k).$$

(7) Now let

$$f_{4,1}(t) := \sum_{j \in J_1} c_j f_{1,j}(t_1) \cdots f_{k,j}(t_k) \text{ and } f_{4,2}(t) := \sum_{j \in J_2} c_j f_{1,j}(t_1) \cdots f_{k,j}(t_k),$$

so $f_4(t) = f_{4,1}(t) + f_{4,2}(t)$, and we can then define

$$\beta_k := \left(\sum_{j \in J_1} + 2 \sum_{j \in J_2} \right) \sum_{j' \in J_1} c_j c_{j'} f_{k,j}(0) f_{k,j'}(0) \prod_{i=1}^{k-1} \int_0^\infty f'_{i,j}(t_i) f'_{i,j'}(t_i) \, dt_i$$

$$= \int_{[0,\infty)^{k-1}} \left(\frac{\partial^{k-1}}{\partial t_1 \cdots \partial t_{k-1}} f_{4,1}(t_1, \ldots, t_{k-1}, 0) \right.$$

$$\left. + 2 \frac{\partial^{k-1}}{\partial t_1 \cdots \partial t_{k-1}} f_{4,2}(t_1, \ldots, t_{k-1}, 0) \right)$$

$$\times \frac{\partial^{k-1}}{\partial t_1 \cdots \partial t_{k-1}} f_{4,1}(t_1, \ldots, t_{k-1}, 0) \, dt_1 \cdots dt_{k-1}.$$

By Lemma 7.4, using the linear combination of weights approach of Step (7) of Lemma 7.6 and EH[θ], we then obtain

$$\sum_{\substack{x \leq n < 2x \\ n \equiv a \bmod W}} v(n)\theta(n + h_k) \geq (\beta_k + o(1)) \frac{x}{B^{k-1}\varphi(W)}.$$

(8) In Step (3), we showed that the functions $f_{4,1}$ and $f_{4,2}$ were both bounded for each n and that their supports overlapped on a set of measures at most some fixed constant times δ_3, which we have chosen to be as small as needed to complete this step. In Step (5), we partitioned the indices $J = J_1 \cup J_2$ so that for $j \in J_1$ we have

$$S(f_{1,j}) + \cdots + S(f_{k-1,j}) < (1 - \epsilon)\theta/2.$$

This implies we can write

$$\beta_k = \tilde{J}_k(f_{4,1}) + O(\delta_3) = \tilde{J}_{k,(1-\epsilon)\theta/2}(f_4) + O(\delta_3).$$

Replacing k by an arbitrary index i and redoing the proof gives for $1 \leq i \leq k$,

$$\beta_i = \tilde{J}_{i,(1-\epsilon)\theta/2}(f_4) + O(\delta_3),$$

so using Step (7) we get for each i,

$$\sum_{\substack{x \leq n < 2x \\ n \equiv a \bmod W}} v(n)\theta(n + h_i) \geq (\beta_i + o(1)) \frac{x}{B^{k-1}\varphi(W)}.$$

(9) Combining the results from Steps (5) and (8) and using Lemma 7.1 completes the proof. □

In Table 7.3, for a particular value of ϵ we give the values of `Rayleigh QuotientPolymath` for three values of k. In Table 7.4, we fix the feasible value of k at 50 and vary ϵ.

Table 7.3 *Lower bounds for* $M_{k,\epsilon}$ *for* $49 \leq k \leq 51$.

k	Degree	ϵ	$M_{k,\epsilon}$ lower bound
49	25	1/27	3.9950000000000000
50	25	1/27	4.0001812744140625
51	25	1/27	4.0049993896484375

Table 7.4 *Varying the choice of* ϵ *for* $M_{50,\epsilon}$ *computations.*

k	Degree	ϵ	$M_{k,\epsilon}$ lower bound
50	25	1/26	4.0000860595703125
50	25	1/27	4.0001812744140625
50	25	1/28	4.0001593017578125

Recall the definitions: $\mathscr{L}_{k,\epsilon}$ is the set of all square integrable functions supported on the simplex

$$(1+\epsilon)\mathscr{R}_k := \left\{ (t_1, \ldots, t_k) \in [0,\infty)^k : t_1 + \cdots + t_k \leq 1+\epsilon \right\}$$

and not equivalent in Lebesgue measure to the zero function, and

$$M_{k,\epsilon} := \sup_{F \in \mathscr{L}_{k,\epsilon}} \frac{\sum_{m=1}^{k} J_{m,1-\epsilon}(F)}{I(F)}.$$

Recall also the definition DHL[k, m]: for $k \geq 2$ and any admissible k-tuple $\mathscr{H}$, there exist infinitely many translates $n + \mathscr{H}$ which contain at least m primes.

Theorem 7.15   (Weak DHL conjecture for two primes) *We have* $M_{50,1/27} > 4.0018$ *and thus* DHL[50, 2].

Proof   Replacement for Step (2) of Section 7.5: This result is obtained by an algorithm which is the same as that described in the original Step (2), except in this case we vary the basis polynomials. Indeed, to compute the matrix entries for A, Polymath use the form

$$(1 + \epsilon - P_{(1)})^a P_\alpha,$$

where the signature α is without any 1 entries, over the region $(1 + \epsilon) \cdot R_k$. For the numerator matrix B, and for the region $(1 - \epsilon) \cdot R_k$, they use the polynomials

$$(1 - \epsilon - P_{(1)})^a P_\alpha = ((1 + \epsilon - P_{(1)}) - 2\epsilon)^a P_\alpha.$$

This showed that $M_{50, 1/27} > 4.00018$. So by Theorem 7.14, and EH[$1/2 -$ $1/10^6$], we have for $m = 1$,

$$M_{50, 1/27} > \frac{2 \cdot 1}{\frac{1}{2}} \text{ and therefore DHL}[50, 2].$$

This completes the proof.   □

Recall the definition: for $m \geq 1$, we have

$$H_m := \liminf_{n \to \infty} (p_{n+m} - p_n),$$

so the twin primes conjecture is the identity $H_1 = 2$.

Theorem 7.16   (An improved upper bound for H₁) *[173, theorem 4]*

$$H_1 \leq 246.$$

Proof   This follows directly from Lemma 1.1 and Theorem 7.15.   □

Remark: This result of Polymath8b is the best prime gap reported in the book. To some, because it relies on a computer algorithm for computing a lower bound for $M_{k, \epsilon}$, it could be a disappointment. However, the program is described with pseudocode in Section 7.5 and Theorem 7.15, and there is Mathematica source code in the PGpack package, with a manual entry `RayleighQuotientPolymath` in Appendix I. The reader might observe how the program processes its data inputs by turning on the verbose flag, and so gain a better understanding of how the method works. As examples, this program was used to derive Tables 7.3 and 7.4.

7.10 Limits to This Approach for $\mathbf{M}_{k, \epsilon}$

Similar to Corollary 7.11, we can derive an upper bound for $M_{k, \epsilon}$, which we do in this section. First recall the definitions:

$$J_{i, 1-\epsilon}(F) := \int_{(1-\epsilon) \cdot \mathcal{R}_{k-1}} \left(\int_0^\infty F(t) \, dt_i \right)^2 dt_1 \ldots dt_{i-1} dt_{i+1} \ldots dt_k, \ 1 \leq i \leq k,$$

$$I(F) := \int_{[0, \infty)^k} F(t)^2 \, dt_1 \cdots dt_k,$$

$$M_{k, \epsilon} := \sup_{F \in \mathcal{L}_{k, \epsilon}} \frac{\sum_{i=1}^k J_{i, 1-\epsilon}(F)}{I(F)},$$

where the supremum is taken over all square integrable real-valued functions supported on the enlarged simplex $(1 + \epsilon) \cdot \mathcal{R}_k$.

Theorem 7.17 (upper bound for $M_{k,\epsilon}$**)** *[175, proposition 38] For* $k \geq 2$ *and* $0 < \epsilon < 1$, *we have*

$$M_{k,\epsilon} \leq \frac{k \log(2k-1)}{k-1}.$$

Proof **(1)** Let $F: [0,\infty)^k \rightarrow \mathbb{R}$ be square integrable on and supported on the enlarged simplex $(1+\epsilon) \cdot \mathscr{R}_k$. Let

$$s_i := 1 + t_i - \sum_{j=1}^{k} t_j,$$

where $1 \leq i \leq k$ is fixed and s_i is constant if each t_j other than t_i is fixed. Then if

$$(t_1, \ldots, t_{i-1}, 0, t_{i+1}, \ldots, t_k) \in (1-\epsilon) \cdot \mathscr{R}_{k-1},$$

we have necessarily $s_i \geq \epsilon$, and because

$$\sum_{j=1}^{k} t_j \leq 1 + \epsilon,$$

we can assume $0 \leq t_i \leq s_i + \epsilon$. Therefore, since s_i does not depend on t_i,

$$\int_0^{s_i+\epsilon} \frac{1}{1 - \sum_{j=1}^{k} t_j + kt_i} \, dt_i = \int_0^{s_i+\epsilon} \frac{1}{s_i + (k-1)t_i} \, dt_i$$

$$= \frac{1}{k-1} \log\left(\frac{ks_i + (k-1)\epsilon}{s_i}\right)$$

$$\leq \frac{\log(2k-1)}{k-1}.$$

(2) By the Cauchy–Schwarz inequality, using

$$F(t) = F(t)\sqrt{s_i + (k-1)t_i} \times \frac{1}{\sqrt{s_i + (k-1)t_i}},$$

we then get

$$\left(\int_0^\infty F(t) \, dt_i\right)^2 \leq \frac{\log(2k-1)}{k-1} \int_0^\infty (s_i + (k-1)t_i) F(t)^2 \, dt_i.$$

Thus, integrating and using $\sum_{i=1}^{k}(s_i + (k-1)t_i) = k$, we get

$$\sum_{i=1}^{k} \int_0^\infty \cdots \int_0^\infty \left(\int_0^\infty F(t) \, dt_i\right)^2 dt_1 \ldots dt_{i-1} dt_{i+1} \ldots dt_k$$

$$\leq \frac{k \log(2k-1)}{k-1} \int_{[0,\infty)^k} F(t)^2 \, dt_1 \cdots dt_k$$

and so

$$M_{k,\epsilon} \leq \frac{k \log(2k - 1)}{k - 1},$$

which completes the proof. □

Remark: This is a clever proof. However, it might be of some use to have an upper bound for $M_{k,\epsilon}$ which is a function of ϵ as well as k, with its value tending to the bound of Corollary 7.11 as $\epsilon \to 0+$. This can be achieved by finessing the proof of Theorem 7.17. In Step (1), we derive

$$
\begin{aligned}
\int_0^{s_i+\epsilon} \frac{1}{1 - \sum_{j=1}^k t_j + kt_i} \, dt_i &= \int_0^{s_i+\epsilon} \frac{1}{s_i + (k-1)t_i} \, dt_i \\
&= \frac{1}{k-1} \log\left(\frac{ks_i + (k-1)\epsilon}{s_i} \right) \\
&\leq \frac{1}{k-1} \log\left(k\left(1 + \frac{\epsilon}{s_i}\right) \right) \\
&\leq \frac{\log k}{k-1} + \epsilon \frac{1}{s_i(k-1)}.
\end{aligned}
$$

Then in Step (2), using the same factorization for Cauchy–Schwarz, and noticing that since $0 \leq t_i \leq s_i + \epsilon$ and by Step (1) $s_i \geq \epsilon$, we have $t_i/s_i \leq 2$ and can write

$$
\begin{aligned}
\left(\int_0^\infty F(t) \, dt_i \right)^2 &\leq \frac{\log k}{k-1} \int_0^\infty (s_i + (k-1)t_i) F(t)^2 \, dt_i \\
&\quad + \frac{\epsilon}{k-1} \int_0^\infty (1 + (k-1)t_i/s_i) F(t)^2 \, dt_i \\
&\leq \frac{\log k}{k-1} \int_0^\infty (s_i + (k-1)t_i) F(t)^2 \, dt_i + \frac{\epsilon(2k-1)}{k-1} \int_0^\infty F(t)^2 \, dt_i.
\end{aligned}
$$

Integrating and summing, we get

$$
\begin{aligned}
\sum_{i=1}^k \int_0^\infty \cdots \int_0^\infty &\left(\int_0^\infty F(t) \, dt_i \right)^2 dt_1 \ldots dt_{i-1} dt_{i+1} \ldots dt_k \\
&\leq \left(\frac{k \log(k)}{k-1} + \epsilon \left(2k + 1 + \frac{1}{k-1} \right) \right) \int_{[0,\infty)^k} F(t)^2 \, dt_1 \cdots dt_k.
\end{aligned}
$$

The bound

$$M_{k,\epsilon} \leq \frac{k \log(k)}{k-1} + \epsilon \left(2k + 1 + \frac{1}{k-1} \right)$$

is superior to the bound of Theorem 7.17 only for small values of ϵ, say of the order of 0.01 in the range of interest $k \in [2,50]$. More precisely, the revised bound is better than that of Polymath for

$$\epsilon < \frac{\log(2 - 1/k)}{2k - 1}.$$

To illustrate, for $\epsilon = 1/100$ when $k = 27$, the bound has value 3.9729844377737263463, and when $k = 28$ it has value 4.0259898624039151804. Hence the bound excludes (i.e., shows that the method gives no information on primes in the corresponding k-tuples), at that value of ϵ, values of k which are 27 or less – values well out of reach with `RayleighQuotientPolymath`. Further refinements of the Polymath bound would be very useful in guiding computational approaches.

We carried out a number of computer experiments to try and improve the Polymath $k = 50$. For example, with $k = 49$ and including all signatures, not only those which have even entries and using Maynard's eigenvector method in the final optimization step, we reached the bound 3.94448755624583370949 with the parameter $d = 20$. Extrapolating, we would have needed more than seven years of compute time to reach $d = 22$, and that would fall well short of the goal of 4 for the $M_{k,\epsilon}$ lower bound. Projecting the bound value, using all signatures we would need about $d = 30$ to reach the bound of 4. Hence it appears a new approach is certainly needed to go lower than the gap size of 246.

In Table 7.5, we have an overview of the variation in values of M_k and $M_{k,\epsilon}$ together with the Polymath and Krylov bounds which were able to be calculated. The first column is obtained using 6×6 matrices, compared with Bogaert, who was able to go to 25×25 and thus obtain better (larger) lower bounds – see Table 7.1.

7.11 End Notes

Here we give a summary of seven other main results of Polymath8b given in [175].

(1) The generalized Elliott–Halberstam conjecture is like EH[θ], but is for a more general class of arithmetic functions than EH or Bombieri–Vinogradov's theorem. It has the statement: given two fixed arithmetic functions α, β with supports in $[N, 2N]$ and $[M, 2M]$ respectively (which in general depend on the variable $x \to \infty$), and fixed real quantities $\epsilon > 0$, $A \geq 1$, where the positive integers M, N satisfy $x^\epsilon \ll N$, $M \ll x^{1-\epsilon}$, $NM \asymp x$ with

$$|\alpha(n)|, |\beta(n)| \ll \tau(n)^{O(1)} \quad \text{and} \quad |\Delta(\beta \cdot \chi_{(\cdot,r)}, q, a)| \ll_M \frac{\tau(qr)^{O(1)}}{(\log x)^A}.$$

Table 7.5 *Lower and upper bounds for M_k, $M_{k,\epsilon}$.*

k	Krylov 3×3	Krylov 6×6	$M_k \leq \frac{k \log k}{k-1}$	$M_{k,\epsilon} \leq \frac{k \log(2k-1)}{k-1}$	$H(k)$
2	1.385653465476342	1.3859330644	1.386294361	2.197224577	2
3	1.643611561968082	1.64642355294	1.647918433	2.414156869	6
4	1.8359573765001243	1.84527776476	1.848392481	2.594546865	8
5	1.9871225877658369	2.00666305549	2.011797391	2.746530722	12
6	2.1100485077021487	2.14252653150	2.150111363	2.877474327	16
7	2.2124804211913840	2.25970996124	2.270228507	2.992440917	20
8	2.2994197352279831	2.36249880095	2.376504619	3.094914516	26
9	2.3742840003678202	2.45374931712	2.471877650	3.187365012	30
10	2.4395116835826685	2.53546608570	2.558427881	3.271598866	32
11	2.4969022324187126	2.60912558072	2.637684800	3.348974681	36
12	2.5478196022526849	2.67586634907	2.710807254	3.420539145	42
13	2.5933204586302768	2.73660177099	2.778695137	3.487115477	48
14	2.6342383917833828	2.79208638982	2.842061740	3.549362779	50
15	2.6712411872791738	2.84295525863	2.901482358	3.607816961	56
16	2.7048709175332324	2.88974850686	2.957427970	3.662919685	60
17	2.7355727027784807	2.93292816460	3.010289178	3.715039284	66
18	2.7637157803396478	2.97289087508	3.060393626	3.764486183	70
19	2.7896092214996854	3.00997818321	3.108018922	3.811524463	76
20	2.8135138429654361	3.04448513700	3.153402393	3.856380680	80
21	2.8356513615220155	3.07666754125	3.196748560	3.899250670	84
22	2.8562115181628150	3.10674806245	3.238234951	3.940304883	90
23	2.8753576844456153	3.13492134130	3.278016680	3.979692603	94
24	2.8932313192741978	3.16135825608	3.316230084	4.017545324	100
25	2.9099555445580937	3.18620946905	3.352995651	4.053979477	110
26	2.9256380381799022	3.20960837511	3.388420400	4.089098658	114
27	2.9403733927748918	3.23167355643	3.422599822	4.122995449	120
28	2.9542450527292089	3.25251083103	3.455619492	4.155752933	126
29	2.9673269153658959	3.27221496828	3.487556395	4.187445956	130
30	2.9796846626984849	3.29087113136	3.518480050	4.218142183	136
31	2.9913768754595906	3.30855609545	3.548453445	4.247902993	140
32	3.0024559700128895	3.32533928122	3.577533835	4.276784234	146
33	3.0129689902834099	3.34128363569	3.605773423	4.304836872	152
34	3.0229582803165865	3.35644638636	3.633219934	4.332107547	156
35	3.0324620580126248	3.37087968974	3.659917122	4.358639049	158
36	3.0415149066230505	3.38463119143	3.685905194	4.384470731	162
37	3.0501481974784062	3.39774451176	3.711221188	4.409638870	168
38	3.0583904549448173	3.41025966857	3.735899299	4.434176981	176
39	3.0662676726362994	3.42221344664	3.759971163	4.458116091	182
40	3.0738035883287155	3.43363972149	3.783466107	4.481484977	186
41	3.0810199237460567	3.44456974437	3.806411368	4.504310384	188
42	3.0879369435569556	3.45503239367	3.828832292	4.526617208	196
43	3.0945718934666578	3.46505439745	3.850752499	4.548428667	200
44	3.1009426542350738	3.47466053102	3.872194044	4.569766447	210
45	3.1070643926135970	3.48387379276	3.893177546	4.590650833	212

Table 7.5 *Cont.*

k	Krylov 3 × 3	Krylov 6 × 6	$M_k \leq \frac{k \log k}{k-1}$	$M_{k,\epsilon} \leq \frac{k \log(2k-1)}{k-1}$	H(k)
46	3.1129514338129335	3.49271556098	3.913722316	4.611100829	216
47	3.1186170244549645	3.50120573427	3.933846463	4.631134265	226
48	3.1240734322733267	3.50936285729	3.953566990	4.650767889	236
49	3.1293320349468696	3.51720423385	3.972899888	4.670017457	240
50	3.1344033994245584	3.52474602867	3.991860210	4.688897806	246
51	3.1392973529094784	3.53200335923	4.010462145	4.707422927	252
52	3.1440230465083742	3.53899037880	4.028719086	4.725606027	254
53	3.1485890124165864	3.54572035160	4.046643681	4.743459588	264
54	3.1530032153922031	3.55220572103	4.064247896	4.760995416	270

then for all $0 < Q \leq x^{\theta-\epsilon}$, we have the estimate

$$\sum_{q \leq Q} \sup_{\substack{1 \leq a \leq q \\ (a,q)=1}} \left| \sum_{n \equiv a \bmod q} (\alpha \star \beta)(n) - \frac{1}{\varphi(q)} \sum_{(n,q)=1} (\alpha \star \beta)(n) \right| \ll_{\epsilon,A} \frac{x}{(\log x)^A}.$$

This is denoted GEH[θ] if we assume the estimate holds for a given θ. For fixed θ with $0 < \theta < 1$, we have GEH[θ] $\implies$ EH[θ] [175, proposition 13], and GEH[θ] is true for all $0 < \theta < \frac{1}{2}$ [175, proposition 14].

Polymath, assuming GEH[1], were able to derive $H_1 \leq 6$.

(2) They also showed, assuming GEH[1], if $\{h_1, h_2, h_3\}$ is any admissible tuple, there exist infinitely many $n \in \mathbb{N}$ for which at least two of

$$\{n + h_1, n + h_2, n + h_3\}$$

are prime, which on the face of it is stronger than $H_1 \leq 6$.

(3) Again under GEH[1], either there are an infinite number of twin primes ($H_1 = 2$), or there is an $N_0 \in \mathbb{N}$ such that for all even $N \geq N_0$, there are primes p, q such that $N = p + q$, i.e., the even Goldbach conjecture, is true asymptotically, or both of these statements are true.

(4) In Section 2.10, we described Selberg's parity obstruction argument to obtaining ultimately precise results from sieve-based arguments. Polymath modified this argument to show that $H_1 \leq 6$ was the best possible result using any type of sieve. Note that the most common prime gap size up to some large integer is 6, but that this is misleading – see Section 1.7 on jumping champions.

(5) Polymath improved Maynard's asymptotic bound $H_m \ll m^3 e^{4m}$ (see Theorem 6.17) to obtain unconditionally $H_m \ll \exp((4-28/157)m + \log m)$. Assuming EH[1], they obtained $H_m \ll \exp(2m + \log m)$.

(6) They also obtained the explicit unconditional values (i.e., assuming EH[$\frac{1}{2}$]):

$$H_2 \le 398130,$$
$$H_3 \le 24797814,$$
$$H_4 \le 1431556072,$$
$$H_5 \le 80550202480.$$

They obtained smaller values for these bounds using EH[1] and also GEH[1]. For example, assuming EH[1] they obtained $H_2 \le 270$.

(7) Polymath improved on the asymptotic lower bound for M_k of Maynard, namely $M_k \ge \log k - 2\log\log k - 2$, using inequalities based on probability and a complicated set of parametric conditions – see [175, theorems 23(xi) and 40, page 54]. They show there is an absolute effective constant $C > 0$ such that for all $k \ge C$ we have $M_k \ge \log k - C$, which, other than finding an explicit value for the smallest possible value of C, is expected to be the last word on this subject.

Finally, here is a list of the PGpack functions which relate to the work of this chapter: BesselZeroUpperBound, ExpandSignature, GetKrylovWs, KrylovF, KrylovLowerBound, KrylovNewMon, MonToPoly, OperatorL, PolyToMon, Poly-MonDegrees, RayleighQuotientBogaert, RayleighQuotientPolymath, SignatureQ, SymmetricPolynomialQ. Manual entries may be found in Appendix I.

8

Variations on Bombieri–Vinogradov

8.1 Introduction

The previous chapters represented a linear time-based progression of ideas. This chapter departs from that sequence in several ways. Here we report on the enhancements of Zhang's work made by Polymath8a [173, 174]. These enhancements show that Zhang's original prime gap bound of 70,000,000 [215] could be very significantly reduced with more work and new ideas, but fell far short of Maynard's 600 [138]. But the ideas are valuable, and not just their historical record. Variations and/or improvements to Bombieri–Vinogradov will find their way to assist with the solutions to many other problems in number theory, as well as prime gaps.

In their paper [72], Goldston, Pintz and Yildirim (GPY) had already shown that (conjecturally) increasing the so-called level of distribution of primes in Bombieri–Vinogradov would produce reductions in the prime gap bound. This sowed the seed for Zhang's variation, which enabled him to establish his result. Then Polymath's refinement of Zhang's variation enabled them to go further.

Work on variations of Bombieri–Vinogradov had been under way for several decades, although at the time of writing no complete extension of the range of moduli $q \leq Q = x^{\frac{1}{2}}/(\log x)^B$ has been found. This work was undertaken especially by Fouvry and Iwaniec [54, 55], and by Bombieri, Friedlander and Iwaniec [15, 16, 18], but the restrictions on residue classes $a \bmod q$ needed to obtain their results were too severe for them to be used in prime gap studies. However, their methods have been the source of many good ideas.

Zhang, as we saw in Chapter 5, Theorem 5.8, varied Bombieri–Vinogradov by restricting to smooth moduli, enabling a small increase in the value of the distribution of primes beyond $1/2$. The degree of smoothness and the size of the extension were interrelated, although through his using fixed numerical constants the relationship was hidden. Polymath8a's variation generalized the smoothness requirement by imposing a weaker parametric condition, so-called multiple dense divisibility, thus extracting the essence of smoothness for the application. It is expected that this new concept, when fully developed, will find many applications in number theory. In addition, Polymath8a derived explicit relationships for all of the important parameters.

The variations of Bombieri–Vinogradov in their full generality, either those of Zhang or Polymath8a, require some very sophisticated techniques, some of which are described in this chapter. However, others are based on the use of multivariable exponential sums, which come from Deligne's solution of the Riemann Hypothesis for varieties over function fields, and are out of scope for this book – see [22, section 9.14] for an overview and references. However, using the one-dimensional form (the Weil conjectures for "curves" – Appendix E), we do include a description of Polymath's significant improvement of Zhang's bound, but not their best improvement, which is summarized in Section 8.12. Indeed, that section has quite extensive information and references for readers on how they might approach background material from algebraic geometry and advanced analytic number theory, including a recent correction to a lemma which underpins some basic well-known results.

Fundamental references for the main part of the chapter are the two Polymath8 papers "New Equidistribution Estimates of Zhang Type, and Bounded Gaps Between Primes" [174] and [173]. In addition to this chapter, the reader is invited to consult the appendices, which contain essential supplementary material for these developments. In particular, Appendix A gives properties of Bessel functions of the first kind, Appendix B develops the spectral theorem for compact operators and then in Appendix C this is used to show why the particular function chosen by Polymath is optimal, Appendix D gives the derivation of a Brun–Titchmarsh style of inequality, due to Shiu, which applies to a range of multiplicative functions, Appendix E gives a derivation of Weil's famous inequality for additive character sums for polynomials over a finite field of prime order, and Appendix G gives a description of the simplest form of the dispersion method of Linnik. It is of interest that each appendix contains a completely different type of mathematics, underlying the wide scope of Zhang and Polymath's work.

Finally in this introduction we have in Figure 8.1 an overview of the work discussed in this chapter and in many of the appendices. It has been adapted from that in [150], which was originally presented by Polymath.

8.2 Special Notations and Definitions

The chapter proper begins with some notations and definitions. These are repeated at appropriate points in the text, but are needed for the summary which follows in Section 8.3.

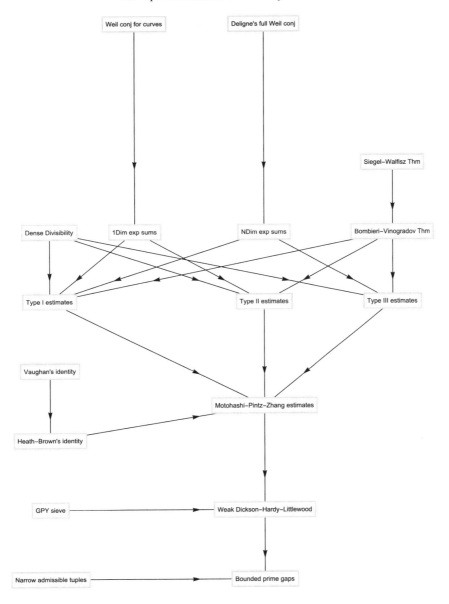

Figure 8.1 Some mathematical dependencies for Chapter 8.

The normal Dirichlet product of two arithmetic functions α and β is represented by $\alpha \star \beta$. We use the Landau–Vinogradov symbols in the normal way. We also adopt a variant of Polymath's notation $f(x) \prec^o g(x)$, which means $f(x) \leq x^{o(1)} g(x)$ (while hoping that neither Polymath's or our notations become widely used!). Sometimes we use the expression $L_X(Y) := (\log X)^Y$ for $X > 0$ and real and Y real. This is because in this type of study it is frequently the power of the logarithm which is the most "active" variable.

Now we define the concepts which are central to the development. For the first, the reader might refer back to Section 1.6 in Chapter 1.

Definition 8.1 (Dickson–Hardy–Littlewood property) Let $k \geq 2$ be an integer. We say DHL$[k, 2]$ is true if for any admissible k-tuple $\mathcal{H} = (h_1, \ldots, h_k)$ there exist infinitely many $n \in \mathbb{N}$ such that the translates $n + \mathcal{H}$ contain at least two primes.

Definition 8.2 (Motohashi–Pintz–Zhang property) Let $\varpi, \delta > 0$ be real numbers and $P_\Omega := \prod_{p \in \Omega} p$. We say MPZ$[\varpi, \delta]$ is true if $\Omega \subset [1, x^\delta]$ is a subinterval and $Q \prec^o x^{\frac{1}{2} + 2\varpi}$, $a \bmod P_\Omega$ is a primitive residue class and $A \geq 1$ is a given real number, then with an implied constant depending only on A, ϖ and δ, we have

$$\sum_{\substack{q \leq Q \\ q \in S_\Omega}} \left| \sum_{\substack{x \leq n < 2x \\ n \equiv a \bmod q}} \Lambda(n) - \frac{1}{\varphi(q)} \sum_{\substack{x \leq n < 2x \\ (n,q)=1}} \Lambda(n) \right| \ll \frac{x}{L_x(A)}, \tag{8.1}$$

where S_Ω is the set of all squarefree numbers being products of primes in Ω.

Definition 8.3 (coefficient sequence) We say a sequence of complex numbers $(c_n)_{n \in \mathbb{N}}$ is a coefficient sequence if it is finitely supported for each $x > 0$ (with support generally depending on x) and satisfies the bound for all $n \in \mathbb{N}$

$$|c(n)| \ll \tau(n)^{O(1)} (\log x)^{O(1)}.$$

For example, $\mu(n) \cdot \chi_{[x, 2x)}(n)$ and for $\alpha, \beta > 0$, $\tau(n)^\alpha \cdot (\log n)^\beta \cdot \chi_{[x, 2x)}(n)$ are coefficient sequences, but $n^\alpha \cdot \chi_{[x, 2x)}(n)$ is not.

Recall the definition of the **signed discrepancy** of an arithmetic function $\alpha(n)$ having finite support: if $(a, q) = 1$, let

$$\Delta(\alpha, q, a) := \sum_{\substack{n \in \mathbb{N} \\ n \equiv a \bmod q}} \alpha(n) - \frac{1}{\varphi(q)} \sum_{\substack{n \in \mathbb{N} \\ (n,q)=1}} \alpha(n).$$

Definition 8.4 (located at scale) We say a coefficient sequence $(c_n)_{n \in \mathbb{N}}$ is located at scale N if its support is contained in an interval $[uN, vN]$ where $1 \ll u < v \ll 1$ are fixed. This implies

$$\Delta(c, q, a) \ll \frac{N (\log x)^{O(1)}}{\varphi(q)}$$

for each integer a with $(a, q) = 1$. For example, if $f : \mathbb{N} \to \mathbb{C}$ is an arithmetic function, then $c(n) = f(n) \cdot \chi_{[N, 2N)}(n)$, where N could be a function of x, is located at scale N.

Definition 8.5   (Siegel–Walfisz property) A complex sequence $(\alpha_n)_{n \in \mathbb{N}}$ is said to obey the Siegel–Walfisz property if for any fixed $A > 0$ and any natural numbers q, r and primitive residue class represented by a modulo q, we have as $x \to \infty$,

$$\left| \sum_{\substack{n \le x \\ n \equiv a \bmod q \\ (n,r)=1}} \alpha(n) - \frac{1}{\varphi(q)} \sum_{\substack{n \le x \\ (n,q)=1 \\ (n,r)=1}} \alpha(n) \right| \ll_A \tau(qr)^{O(1)} \frac{x}{L_x(A)}. \tag{8.2}$$

Definition 8.6   (multiple dense divisibility) Let $y > 1$ and $i \ge 0$. If $i = 0$, every natural number is said to be 0-tuply y-densely divisible. If $i \ge 1$, we define a natural number n to be i-tuply densely divisible recursively as follows: for every pair of nonnegative integers j, k with $j + k = i - 1$ and real numbers R with $1 \le R \le yn$, we have a factorization

$$n = qr \text{ such that } \frac{R}{y} \le r \le R$$

such that q is j-tuply y-densely divisible and r is k-tuply y-densely divisible. Let $\mathcal{D}(i, y)$ be the set of all i-tuply y-densely divisible positive integers, and $\mathcal{D}_\Omega(i, y)$ those which are squarefree and have all prime factors in a given interval Ω. Examples are given in Section 8.5.

Definition 8.7   (densely divisible MPZ property) Let $i \ge 0$ be an integer. We say $\mathrm{MPZ}^{(i)}[\varpi, \delta]$ is true if for any bounded subset $\Omega \subset \mathbb{R}$ which may depend on x, and $Q \prec^o x^{\frac{1}{2}+2\varpi}$, if $(a, P_\Omega) = 1$ and $A \ge 1$ is a fixed real number, then

$$\sum_{\substack{q \le Q \\ q \in \mathcal{D}_\Omega(i, x^\delta)}} \left| \sum_{\substack{x \le n < 2x \\ n \equiv a \bmod q}} \Lambda(n) - \frac{1}{\varphi(q)} \sum_{\substack{x \le n < 2x \\ (n,q)=1}} \Lambda(n) \right| \ll \frac{x}{L_x(A)}.$$

Definition 8.8   (Type I estimate) Let $i \ge 1$, ϖ with $0 < \varpi < \frac{1}{4}$, δ with $0 < \delta < \frac{1}{4} + \varpi$, σ with $\frac{1}{6} < \sigma < \frac{1}{2}$ and $\sigma > 2\varpi$ be such that ϖ, δ and σ are fixed real numbers. Let Ω be a fixed bounded subset of $(0, \infty)$ and define as before the squarefree number

$$P_\Omega = \prod_{p \in \Omega} p.$$

Let a be an integer with $(a, P_\Omega) = 1$ and $a \bmod P_\Omega$ the corresponding congruence class. We say the estimate $\mathrm{Type}_I^{(i)}[\varpi, \delta, \sigma]$ holds if for all Ω and class $a \bmod P_\Omega$ and any integers $M, N \gg 1$ such that $x \ll MN \ll x$ and for some fixed $c > 0$

$$x^{\frac{1}{2}-\sigma} \prec^o N \prec^o x^{\frac{1}{2}-2\varpi-c}.$$

and for any $Q \prec^o x^{\frac{1}{2}+2\varpi}$ and any coefficient sequence α at scale M and β at scale N, with β having the Siegel–Walfiz property, we have for any fixed $A > 0$,

$$\sum_{\substack{q \le Q \\ q \in \mathcal{D}_\Omega(i, x^\delta)}} \left| \sum_{n \equiv a \bmod q} (\alpha \star \beta)(n) - \frac{1}{\varphi(q)} \sum_{(n,q)=1} (\alpha \star \beta)(n) \right| \ll_A \frac{x}{L_x(A)}. \qquad (8.3)$$

Definition 8.9 (Type II estimate) We say the estimate $\mathrm{Type}_{\mathrm{II}}^{(i)}[\varpi, \delta]$ holds if for all Ω and class a mod P_Ω and any integers $M, N \gg 1$ such that $x \ll MN \ll x$ and for some sufficiently small but fixed $c > 0$,

$$x^{\frac{1}{2}-2\varpi-c} \prec^o N \prec^o x,$$

and for any $Q \prec^o x^{\frac{1}{2}+2\varpi}$ and any coefficient sequence α at scale M and β at scale N, with β having the Siegel–Walfiz property, we have for any fixed $A > 0$ the estimate (8.3).

8.3 Chapter Summary

Following some general remarks, the body of the chapter began with some special notations and definitions, Section 8.2. This is followed by this chapter summary, which gives the principal lemmas and theorems. The statements of some preliminary results are given in Section 8.4 with mostly references and no proof. These are the Siegel–Walfisz theorem, Vaughan's identity and a form of the large sieve inequality. We do however give the proof of the generalized Bombieri–Vinogradov theorem. Then in Section 8.5, on multiple dense divisibility, there is the fundamental Lemma 8.5 setting out and demonstrating the main properties of this type of natural number, which is a generalization of smooth integer, and underpinning Polymath's main improvements.

The chapter splits roughly into two parts, linked by Theorem 8.19 (in summary, $\mathrm{MPZ}^{(i)} \implies \mathrm{DHL}$), which is a generalization of Theorem 8.17 ($\mathrm{MPZ} \implies \mathrm{DHL}$). The first part provides a significant improvement of Zhang's bound, without using dense divisibility, and the second part exploits this concept to advantage.

In Section 8.6, "Improving Zhang", we begin with a simple criteria which gives DHL:

Lemma 8.7 (Sufficient condition for DHL) *Given fixed $k \ge 2$, $\mathcal{H}$ an admissible k-tuple, $W = \prod_{p \le \mathrm{logloglog}\, x} p$ a product of very small primes, b mod W a congruence class with $(b + h, W) = 1$ for all $h \in \mathcal{H}$, a function $v \colon \mathbb{N} \to [0, \infty)$, fixed parameters $\alpha, \beta > 0$ and $R = R(x) > 0$, $B = B(x) > 0$ such that as $x \to \infty$, we have an upper bound*

$$\sum_{\substack{x \le n < 2x \\ n \equiv b \bmod W}} v(n) \le (\alpha + o(1)) \frac{Bx}{W},$$

and for all $h_j \in \mathcal{H}$, a lower bound

$$\sum_{\substack{x \le n < 2x \\ n \equiv b \bmod W}} v(n)\theta(n + h_j) \ge (\beta + o(1)) \frac{Bx}{W} \log R,$$

and, in addition, we have the key inequality

$$\frac{\log R}{\log x} > \frac{\alpha}{k\beta}$$

for a sequence of values x with limit value infinity, then $\mathrm{DHL}[\mathrm{k}, 2]$ is true.

Many times in the development there is a need to replace a sum with an integral. The workhorse for this is Lemma 8.8:

Lemma 8.8 (**Fundamental sieve estimate**) *Let* $k \in \mathbb{N}$ *and* $\beta \colon \mathbb{N} \to \mathbb{R}$ *be multiplicative with*

$$p \le w \text{ and } j \ge 1 \implies \beta(p^j) = 0,$$
$$p > w \text{ and } j > 1 \implies \beta(p^j) \ll \exp(O(j)),$$
$$p > w \implies \beta(p) = k + O\left(\frac{1}{p}\right).$$

Given any $g \colon \mathbb{R} \to \mathbb{R}$ *which is Riemann integrable and has compact support, a fixed* $c > 0$ *with* $R > x^c$, *and a function* $w = w(x)$ *which grows sufficiently slowly with* x *as* $x \to \infty$, *then we have*

$$\sum_{n \in \mathbb{N}} \frac{\beta(n)}{n} g\left(\frac{\log n}{\log R}\right) = \left(\frac{\varphi(W)}{W \log R}\right)^k \left(\int_0^\infty g(t) \frac{t^{k-1}}{(k-1)!} \, dt + o(1)\right).$$

The same estimate holds also where the summation on the left is restricted to the set of squarefree n *with all prime factors greater than* w, *which is denoted* $\mathcal{S}_{(w,\infty)}$.

The development is made in terms of a fixed continuous function $f \colon [0, \infty) \to [0, \infty)$, which has compact support in $[0, 1]$, and has all derivatives existing and continuous on $[0, 1]$, where we need only consider left and right derivatives at 1 and 0 respectively. Particular functions are chosen so that the bounds derived using them are optimal in given circumstances. The background to this optimization is not simple, but is set out in Appendices B and C. We also need a set of definitions which tie the weight function $v(n)$ to, in particular, f, B and W. See Section 8.6.

Lemma 8.9   (Generic upper bound) *With the notation set out in Lemma 8.7, and any function $f(t)$ which satisfies the conditions given previously, if we set*

$$\alpha := \int_0^1 f'(t)^2 \frac{t^{k-1}}{(k-1)!}\, dt,$$

then as $x \to \infty$, we have

$$\sum_{\substack{x \le n < 2x \\ n \equiv b \bmod W}} v(n) \le (\alpha + o(1)) \frac{x B^k}{W}.$$

Section 8.7 gives a fundamental technical result which provides a sufficient condition in the form of a single inequality for DHL[k, 2] in terms of k and just two parameters. The particular class of functions $f(t)$ is specified in the statement of Theorem 8.17.

Theorem 8.17   (MPZ implies DHL) *[174, theorem 2.13] Given an integer $k \ge 2$, a real ϖ with $0 < \varpi < \frac{1}{4}$, and a real δ with $0 < \delta < \frac{1}{4} + \varpi$, such that*

$$\kappa_1 := \int_\theta^1 (1-t)^{(k-1)/2} \frac{dt}{t},$$

$$\kappa_2 := (k-1) \int_\theta^1 (1-t)^{k-1} \frac{dt}{t},$$

$$\theta := \frac{\delta}{\frac{1}{4} + \varpi},$$

and where $f(t) := t^{1-k/2} J_{k-2}(j_{k-2}\sqrt{t})$ for $0 < t \le 1$ and 0 elsewhere, where j_k is the smallest positive zero of the Bessel function $J_k(x)$, and such that we have the key inequality

$$(1 + 4\varpi)(1 - 2\kappa_1 - 2\kappa_2) > \frac{j_{k-2}^2}{k(k-1)}.$$

If MPZ[ϖ, δ] is true, then so is DHL[k, 2].

We then find $k = 34429$ satisfies the key inequality with $\varpi = \delta = 1/1168$. Using Zhang's theorem (see Section 5.4) with these values and $k = 34429$ and an admissible k-tuple begin the first k primes greater than k, we get a prime gap upper bound size of 420878.

The following extension of Theorem 8.17 provides a link between the first and second major parts of this chapter. The sufficient condition for DHL now is in terms of the dense divisibility parameter i and the only difference between the two key inequalities is the small parameter κ_3. We have skipped over the full statement, but warn the reader that the proof is not to be taken lightly!

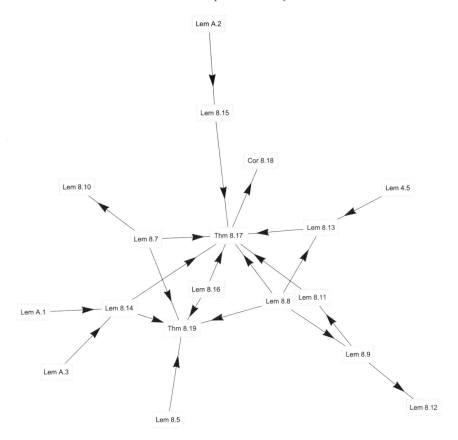

Figure 8.2 Some Chapter 8 dependencies.

Theorem 8.19 (MPZ$^{(i)}$ **implies** DHL) *[174, theorem 2.16] Let* $f(t) := t^{1-k/2}$ $J_{k-2}(j_{k-2}\sqrt{t})$ *for* $0 < t \leq 1$ *and 0 elsewhere. Given* $k \geq 2$ *and* $i \geq 1$, ϖ *with* $0 < \varpi < \frac{1}{4}$, δ, δ' *with* $0 < \delta \leq \delta' < \frac{1}{4} + \varpi$ *and* $A \geq 0$, *where the parameters*

$$\kappa_1, \ \kappa_2, \ \kappa_3, \ \eta, \ \theta, \ \omega, \ \text{and } \xi$$

are defined in the theorem statement, and where for all $k \in \mathbb{Z}$, J_k *is the Bessel function of the first kind of order* k *and* j_k *is its smallest positive zero. Then if*

$$(1 + 4\varpi)(1 - 2\kappa_1 - 2\kappa_2 - 2\kappa_3) > \frac{j_{k-2}^2}{k(k-1)}$$

and MPZ$^{(i)}[\varpi, \delta]$ *is true, so is* DHL$[k, 2]$.

Heath-Brown's generalization of Vaughan's identity is proved in Lemma 8.21. Using this, and some nontrivial combinatorial concepts, we get the lemma which provides a good sufficient condition for MPZ$^{(i)}$.

Lemma 8.22 **(Main combinatorial lemma)** *[174, lemma 2.22] Let $i \geq 1$, ϖ with $0 < \varpi < \frac{1}{4}$, δ with $0 < \delta < \frac{1}{4} + \varpi$, σ with $\frac{1}{6} < \sigma < \frac{1}{2}$ and $\sigma > 2\varpi$, be fixed real numbers such that we have the estimates $\mathrm{Type}_I^{(i)}[\varpi,\delta,\sigma]$ and $\mathrm{Type}_{II}^{(i)}[\varpi,\delta]$. Then we have $\mathrm{MPZ}^{(i)}[\varpi,\delta]$.*

The definitions of Polymath's Type I and II estimates are set out in Section 8.2, and in Section 8.10 we are able to describe the practical estimates which arise from these definitions and the hard work on exponential sums which precedes them in Section 8.9. This is the crucial Theorem 8.34:

Theorem 8.34 **(Type I and II estimates)** *[174, theorem 7.1] Let ϖ, δ, σ be fixed positive real numbers, $i \geq 1$ a fixed integer, $\Omega \subset \mathbb{R}$ bounded and a mod P_Ω a congruence class with $(a, P_\Omega) = 1$. Furthermore, let M, N be positive integers which are sufficiently large, with $MN \asymp x$ and $x^{\frac{1}{2}-\sigma} \prec^o N \prec^o x^{\frac{1}{2}}$.*
Furthermore, let α, β be complex coefficient sequences at scales M, N respectively and such that β has the Siegel–Walfisz property. Assume also that one of the following situations regarding the parameters applies:
(i) Type I: $i = 1, j = 0, k = 0$, $54\varpi + 15\delta + 5\sigma < 1$ and $N \prec^o x^{\frac{1}{2}-2\varpi-c}$ for some fixed $c > 0$, or
(ii) Type I: $i = 2, j = 1, k = 0$, $56\varpi + 16\delta + 4\sigma < 1$ and $N \prec^o x^{\frac{1}{2}-2\varpi-c}$ for some fixed $c > 0$, or
(iv) Type II: $i = 1, j - 0, k = 0$, $68\varpi + 14\delta < 1$ and $x^{\frac{1}{2}-2\varpi-c} \prec^o N$ for some sufficiently small fixed $c > 0$.
Then we have the estimate

$$\sum_{\substack{q \prec^o x^{\frac{1}{2}+2\varpi} \\ q \in \mathscr{D}_\Omega(i,x^\delta)}} |\Delta(\alpha \star \beta, q, a)| \ll \frac{x}{(\log x)^A}.$$

Note the omitted Part (iii).

In Section 8.9 there is an extensive development of what is needed by way of particular exponential sums, incorporating Weil's bound and Linnik's dispersion method. This includes Lemmas 8.23, 8.26, 8.27, 8.28 8.29, 8.30, 8.31 and 8.32 and culminates in Theorem 8.33.

With all of this preparation completed, we are then able to make an application to prime gaps, which is done in Section 8.11.

Theorem 8.37 **(A Motohashi–Pintz–Zhang style estimate)** *[174, theorem 2.17(ii)] For any fixed pair $\varpi, \delta > 0$ with $168\varpi + 48\delta < 1$, we have $\mathrm{MPZ}^{(2)}[\varpi,\delta]$.*

Combining results, we get Theorem 8.38, namely for $k = 1783$ we have DHL[k, 2]. This is followed by what is really an example, stated as Theorem 8.39. This is an explicit construction of an admissible k-tuple with $k = 1783$ and width 14950, which immediately translates to Theorem 8.40, and which is such that for an infinite set of natural numbers n, we have $p_{n+1} - p_n \leq 14950$. Note that the inductive tuple of size 1783 has width 15792, and the corresponding prime tuple has width over 16000.

In Section 8.12, "End Notes", we summarize the results Polymath8a obtained using multivariable exponential sums based on the work of Deligne, combined with dense divisibility.

This chapter is quite complex, and to describe the main relationships we have three graphs: Figures 8.2, 8.3 and 8.7.

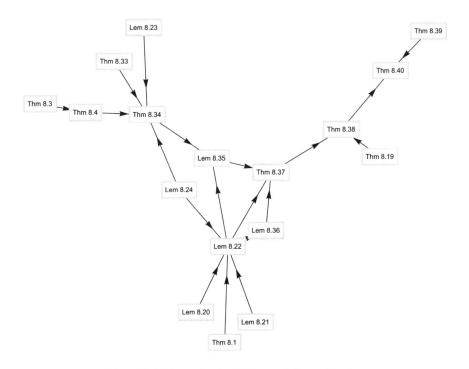

Figure 8.3 Some further Chapter 8 dependencies.

8.4 Preliminary Results

In this section we give four important backgound theorems. References are given for the first three. The last result is due essentially to Motohashi and we state his "inductive" form at the end. We prove the version of Bombieri–Friedlander–Iwaniec, since that is the form which is applied by Polymath.

We begin with the well-known Sigel–Walfisz theorem. It is the prime number theorem for primes in an arithmetic progression.

Theorem 8.1 (**Siegel–Walfisz theorem**) *[117, corollary 5.29] Let $A > 0$ be fixed. If $(a,q) = 1$ and $q \leq (\log x)^A$, then there is an absolute constant $c_1 > 0$ such that*

$$\psi(x,q,a) = \frac{x}{\varphi(q)} + O_A\left(x \exp(-c_1\sqrt{\log x})\right).$$

Let $\mathbf{1}(n) := 1$ for all $n \in \mathbb{N}$ and $\star$ represent the Dirichlet convolution product.

The following famous identity of Vaughan is used by Polymath, so it is included for completeness. It is proved in [22, lemma 12.28]. Here we use the more general form Heath-Brown's identity Lemma 8.21 which is proved in Section 8.8.

Theorem 8.2 (**Vaughan's identity**) *Let $\Lambda(n)$ be the von Mangoldt arithmetic function, $L(n) = \log n$ and $\mu(n)$ the Möbius function. Let $u,v > 1$ be arbitrary real parameters and define the restricted functions*

$$\Lambda_{\geq}(n) := \Lambda(n) \cdot \chi_{[v,\infty)}, \quad \Lambda_{<}(n) := \Lambda(n) \cdot \chi_{[1,v)},$$
$$\mu_{\geq}(n) := \mu(n) \cdot \chi_{[u,\infty)}, \quad \mu_{<}(n) := \mu(n) \cdot \chi_{[1,u)}. \tag{8.4}$$

With these definitions, we have

$$\Lambda_{\geq} = \mu_{<} \star L - \mu_{<} \star \Lambda_{<} \star \mathbf{1} + \mu_{\geq} \star \Lambda_{\geq} \star \mathbf{1}.$$

Another useful family of results goes by the name "large sieve inequality". Here we give a form due originally to Gallagher.

Theorem 8.3 (**A large sieve inequality**) *[22, theorems 12.19, 12.21] Let $(a_n)_{n\in\mathbb{N}}$ be a complex sequence and let χ be a Dirichlet character. Then there is an absolute constant such that for all $N, Q \in \mathbb{N}$, we have*

$$\sum_{q \leq Q} \sum_{\substack{\chi \bmod q \\ primitive}} \left| \sum_{n \leq N} a_n \chi(n) \right|^2 \ll (Q^2 + N)\|a\|^2,$$

where $\|a\|^2 := |a_1|^2 + \cdots + |a_N|^2$.

The general Bombieri–Vinogradov estimate given next is an essential part of this work. The proof is relatively short but hides a lot of complexity. The original idea came from Motohashi [146], but we give the form derived by Bombieri–Friedlander–Iwaniec [15, theorem 0] and follow with the statement of Motohashi's theorem.

Remark: Another preliminary set of results which are used frequently in this chapter is the multipart Lemma 8.24. This gives bounds for a variety of divisor sums built around $\sum_{n \leq x} \tau(n)^C$. The lemma will be used here in Theorem 8.4, but it really comes into its own later in this chapter, and is proved in full just before its main applications.

Theorem 8.4 (**The general Bombieri–Vinogradov estimate of Motohashi–Bombieri–Friedlander–Iwaniec**) *Let x be real and positive with $x \to \infty$ and let $\epsilon > 0$ be a fixed parameter. Let $M, N \gg 1$ be functions of x such that $x \ll MN \ll x$ and $N \le x^\epsilon$. Let α, β be complex sequences at scale M and N respectively, and let β have the Siegel–Walfiz property. Then for any fixed $A > 0$ there exists a $B_1 > 0$ depending only on A and ϵ and the scale parameters, such that for $Q \le \sqrt{x}/(\log x)^{B_1}$, we have*

$$\sum_{q \le Q} \max_{\substack{a \bmod q \\ (a,q)=1}} \left| \sum_{\substack{n \in \mathbb{N} \\ n \equiv a \bmod q}} (\alpha \star \beta)(n) - \frac{1}{\varphi(q)} \sum_{\substack{n \in \mathbb{N} \\ (n,q)=1}} (\alpha \star \beta)(n) \right|$$

$$= \sum_{q \le \sqrt{x}/L_x(B)} \max_{\substack{a \bmod q \\ (a,q)=1}} |\Delta(\alpha \star \beta, q, a)| \ll_{A,\epsilon} \frac{x}{L_x(A)}. \tag{8.5}$$

Proof (**1**) First, using the definition of the signed discrepancy and the orthogonality of Dirichlet characters modulo q, we derive

$$\frac{1}{\varphi(q)} \sum_{\chi \ne \chi_0} \overline{\chi}(a) \left(\sum_m \alpha_m \chi(m) \right) \left(\sum_n \beta_n \chi(n) \right)$$

$$= \sum_{\substack{n \le x \\ n \equiv a \bmod q}} (\alpha \star \beta)(n) - \frac{1}{\varphi(q)} \sum_{\substack{n \le x \\ (n,q)=1}} (\alpha \star \beta)(n) = \Delta(\alpha \star \beta, q, a).$$

(**2**) Next, we obtain an upper bound by reducing to primitive characters using

$$\sum_n \alpha_n \chi(n) = \sum_n \alpha_n \chi_1(n) \cdot \psi(n) = \sum_{(n,q)=1} \alpha_n \psi(n),$$

where χ_1 is the principal character modulo q and χ is induced by ψ, to get

$$T := \sum_{q \le Q} \max_{(a,q)=1} |\Delta(\alpha \star \beta, q, a)|$$

$$\le \sum_{q \le Q} \frac{1}{\varphi(q)} \left(\sum_{2 \le r \le Q/q} \frac{1}{\varphi(r)} \sum_{\substack{\chi \bmod r \\ primitive}} \left| \sum_{(m,q)=1} \alpha_m \chi(m) \right| \left| \sum_{(n,q)=1} \beta_n \chi(n) \right| \right)$$

$$= \sum_{q \le Q} \frac{1}{\varphi(q)} (T_1 + T_2),$$

where T_1 is the second inner sum in the range $2 \le r \le R$ and T_2 is the same sum but with $R < r \le Q/q$, where R is chosen in Step (5) so $2 \le R \le Q/q$.

(3) Using the Cauchy–Schwarz inequality we get

$$T_1 \le \left(\sum_{2 \le r \le R} \frac{1}{\varphi(r)} \sum_{\substack{\chi \bmod r \\ primitive}} \left| \sum_{(m,q)=1} \alpha_m \chi(m) \right|^2 \right)^{\frac{1}{2}}$$

$$\times \left(\sum_{2 \le r \le R} \frac{1}{\varphi(r)} \sum_{\substack{\chi \bmod r \\ primitive}} \left| \sum_{(n,q)=1} \beta_n \chi(n) \right|^2 \right)^{\frac{1}{2}}.$$

For the α_m term, we use the large sieve inequality, Theorem 8.3, and assume $R < \sqrt{M}$ to obtain the upper estimate $\sqrt{M}\|\alpha\|$. We then split the multiple sum in the second factor into progressions modulo r and use the Siegel–Walfisz property, Definition 8.5, satisfied by (β_n). Let

$$S := \sum_{(n,q)=1} \beta(n)\chi(n) \quad \text{and} \quad W := \|\beta\| \frac{N^{\frac{1}{2}}}{(\log N)^{A'}} \tau(q)^B.$$

Then

$$S = \sum_{1 \le l \le r} \chi(l) \sum_{\substack{n \equiv l \bmod r \\ (n,q)=1}} \beta(n),$$

and by Siegel–Walfisz

$$\left| \sum_{\substack{n \equiv l \bmod r \\ (n,q)=1}} \beta(n) - \frac{1}{\varphi(r)} \sum_{(n,qr)=1} \beta(n) \right| \ll W.$$

Therefore

$$\left| \sum_{l=1}^{r} \chi(l) \sum_{\substack{n \equiv l \bmod r \\ (n,q)=1}} \beta(n) - \frac{1}{\varphi(r)} \left(\sum_{l=1}^{r} \chi(l) \right) \sum_{(n,qr)=1} \beta(n) \right| \ll Wr.$$

But since χ is nonprinciple, we have $\sum_{1 \le l \le r} \chi(l) = 0$, so we get

$$|S| = \left| \sum_{(n,q)=1} \beta(n)\chi(n) \right| = \left| \sum_{1 \le l \le r} \chi(l) \sum_{\substack{n \equiv l \bmod r \\ (n,q)=1}} \beta(n) \right| \lll Wr.$$

Therefore,

$$\sum_{(n,q)=1} \beta(n)\chi(n) \ll \|\beta\| \frac{N^{\frac{1}{2}}}{(\log N)^{A'}} \tau(q)^B r,$$

and then

$$\sum_{2 \leq r \leq R} \frac{1}{\varphi(r)} \sum_{\substack{\chi \bmod r \\ \chi \text{ primitive}}} \left| \sum_{(n,q)=1} \beta(n)\chi(n) \right|^2 \ll \tau(q)^{2B} R^3 \|\beta\|^2 \frac{N}{(\log N)^{2A'}}.$$

Assuming as before $R < \sqrt{M}$, we then get

$$T_1 \ll \|\alpha\| \times \|\beta\| x^{\frac{1}{2}} \tau(q)^B \frac{R^{3/2}}{(\log N)^{A'}}.$$

(4) For T_2, we split the sum into intervals with $R \leq V < Q/q$, applying Cauchy–Schwarz to each such sum. We then apply the large sieve inequality Theorem 8.3 to each sum to get

$$T_2 \ll (\log Q)^2 \|\alpha\| \times \|\beta\| \sup_{R \leq V \leq Q} (V^2 + M)^{\frac{1}{2}} (V^2 + N)^{\frac{1}{2}} / V$$

$$\ll (\log Q)^2 \|\alpha\| \times \|\beta\| \sup_{R \leq V \leq Q} \left(\sqrt{V} + \frac{\sqrt{M}}{\sqrt{V}} \right) \left(\sqrt{V} + \frac{\sqrt{N}}{\sqrt{V}} \right)$$

$$\ll (\log Q)^2 \|\alpha\| \times \|\beta\| \sup_{R \leq V \leq Q} \left(V + \sqrt{M} + \sqrt{N} + \sqrt{MN}/V \right)$$

$$\ll (\log Q)^2 \|\alpha\| \times \|\beta\| \left(Q + M^{\frac{1}{2}} + N^{\frac{1}{2}} + M^{\frac{1}{2}} N^{\frac{1}{2}} / R \right).$$

(5) Using a divisor bound from Lemma 8.24(iv), we choose $B_2 > 0$, depending only on B (say $B_2 = 2^B + 1$), such that

$$\sum_{q \leq Q} \frac{\tau(q)^B}{\varphi(q)} \ll (\log x)^{B_2},$$

use the estimate [196]

$$\sum_{q \leq Q} \frac{1}{\varphi(q)} \ll \log Q$$

and make the choices

$$B_1 = A + 2,$$
$$A' = 5A/2 + B_2 + 3,$$
$$Q = \frac{\sqrt{x}}{(\log x)^{B_1}},$$
$$R = (\log x)^{B_1} < M^{\frac{1}{2}}.$$

Finally we combine the estimates from Steps (2) through (4) to get

$$
\begin{aligned}
T &\ll \sum_{q \leq Q} \left(\sqrt{x} \frac{\tau(q)^B}{\varphi(q)} \frac{(\log x)^{3B_1/2}}{(\log N)^{A'}} \right. \\
&\quad + \left. \frac{(\log Q)^2}{\varphi(q)} \left(Q + M^{\frac{1}{2}} + N^{\frac{1}{2}} + M^{\frac{1}{2}}N^{\frac{1}{2}}/(\log x)^{B_1} \right) \right) \|\alpha\| \|\beta\| \\
&\ll \sqrt{x} \left(\frac{1}{(\log x)^{A'-B_2-3B_1/2}} + \frac{1}{(\log x)^{B_1}} \right) \|\alpha\| \|\beta\| \\
&\ll \frac{\sqrt{x}}{(\log x)^A} \|\alpha\| \|\beta\|,
\end{aligned}
$$

to complete the proof. $\qquad\square$

Motohashi's theorem [146, theorem 1] provides a more symmetric version of Theorem 8.4, so it offers some advantages. Consider first three properties of arithmetic functions f:

(a) We have $f(n) \ll \tau(n)^{C_f}$.
(b) If X, a nonprinciple character, has conductor which is $O((\log x)^D)$, then

$$
\sum_{n \leq x} f(n) X(n) \ll \frac{x}{(\log x)^{3D}}.
$$

(c) If we define the deficiency for $(q,l) = 1$,

$$
E(y,q,l,f) := \sum_{\substack{n \leq x \\ n \equiv l \bmod q}} f(n) - \frac{1}{\varphi(q)} \sum_{\substack{n \leq x \\ (n,q)=1}} f(n),
$$

and for all $A > 0$, if $Q \leq x^{\frac{1}{2}}/(\log x)^B$, with B depending on A, then

$$
\sum_{q \leq Q} \max_{y \leq x} \max_{(q,l)=1} |E(y,q,l,f)| \ll \frac{x}{(\log x)^A}.
$$

In this, C_f is fixed, but A, B and D can be arbitrarily large.

Motohashi's Theorem 1 states that if f and g satisfy (a) through (c), then so does $f \star g$. Thus in this context, the Bombieri–Vinogradov variation iterates naturally.

The proof, which we don't give here since further developments in the book are based on Theorem 8.4, involves a nonsymmetric subdivision of a subset of the first quadrant of positive integers under a hyperbola, so it could afford some variations.

8.5 Multiple Dense Divisibility

We begin with the definition and then derive the properties of dense divisibility which are needed subsequently and totally central to Polymath's approach. Of particular interest is Lemma 8.5(iv), which shows we have a genuine generalization of smoothness and (v), which gives a sufficient condition. The lemma is followed by an example which was produced using the PGpack function `DenselyDivisibleQ` and a brief indication regarding the asymptotics of these numbers.

> **Definition 8.10** Let $y \geq 2$ and $i \geq 0$ be integers. If $i = 0$, every natural number is said to be 0-tuply y-densely divisible. If $i \geq 1$, we define a natural number n to be i-tuply **densely divisible** recursively as follows: for every pair of nonnegative integers j, k with $j + k = i - 1$ and real numbers R with $1 \leq R \leq yn$, we have a factorization
>
> $$n = rq \text{ such that } \frac{R}{y} \leq r \leq R$$
>
> such that r is j-tuply y-densely divisible and q is k-tuply y-densely divisible.

Note that in this definition we can replace the constraint on R by $y \leq R \leq n$ since when $n < R \leq yn$ we always have a factorization $n = n \cdot 1$ and when $R < y$ we have $n = 1 \cdot n$.

Recall we defined $\mathcal{D}(i, y)$ to be the set of all i-tuply y-densely divisible positive integers, and $\mathcal{D}_\Omega(i, y)$ those which are squarefree and have prime factors in a given interval Ω.

> **Lemma 8.5 (Dense divisibility)** *[174, lemma 4.12] Let the integer $i \geq 0$ and real $y > 1$. Then*
>
> (i) *if $y \leq z$, then $\mathcal{D}(i, y) \subset \mathcal{D}(i, z)$,*
> (ii) *if $i \leq j$, then $\mathcal{D}(j, y) \subset \mathcal{D}(i, y)$,*
> (iii) *if $m, n \in \mathscr{D}(1, y)$, then the LCM of n and m, $[n, m] \in \mathscr{D}(1, y)$ also,*
> (iv) *if n is y-smooth, it is in $\mathcal{D}(i, y)$ for all $i \geq 0$,*
> (v) *if for some $z \geq y$, the natural number n is z-smooth and squarefree and*
>
> $$\prod_{\substack{p \mid n \\ p \leq y}} p \geq \frac{z^i}{y};$$
>
> *then n is i-tuply y-densely divisible.*

Proof **(i)** This follows since $R/y \leq r$ implies $R/z \leq r$.

(ii) For $j \geq 1$, $j - 1 = (j - 1) + 0$ and $n = n \cdot 1$, so $\mathcal{D}(j, y) \subset \mathcal{D}(j - 1, y)$.

(iii) Assume $m \le n$ and let $a = [m,n]/n \le m \le n$. Let R satisfy $1 \le R \le [m,n]$. In case $R \le n$, then we have a factorization $n = qr$ with $R/y \le r \le R$. This gives $[m,n] = aqr$ for some integer a, so the integer r suffices to deduce $[m,n] \in \mathscr{D}(1,y)$. If however $n < R \le [m,n] = an$, we have

$$1 \le \frac{n}{a} \le \frac{R}{a} \implies 1 \le \frac{R}{a}.$$

Therefore, since $n \in \mathscr{D}(1,y)$ there exists a factorization $n = rq$ such that

$$\frac{R}{ay} \le r \le \frac{R}{a} \implies \frac{R}{y} \le ar \le R.$$

But $[m,n] = (ar)q$, so $[m,n] \in \mathscr{D}(1,y)$.

(iv) Let n be y-smooth. Then $n \in \mathscr{D}(0,y)$, so let $i > 0$ and assume $n \in \mathscr{D}(j,y)$ for all $0 \le j < i$. Let $j,k \ge 0$ have $j + k = i - 1$ and suppose $y \le R \le n$. Let $r \ge 1$ be the largest divisor of n with $r \le R$ and set $q = n/r$ so q and r are y-smooth. Hence by the inductive hypothesis, we have $r \in \mathscr{D}(j,y)$ and $q \in \mathscr{D}(k,y)$. In addition, assuming as we may $q \ne 1$, because all prime divisors p of q have $p \le y$, we have $R/y \le R/p \le n/p \le r$. Also by the choice we have made for r, $r \le R$. Hence

$$R/y \le r \le R.$$

Therefore, $n \in \mathscr{D}(i,y)$ and we have completed the inductive step, so $n \in \mathscr{D}(i,y)$ for all $i \ge 0$.

(v) This part is more difficult than (i) through (iv) and requires both induction on i and case analysis. If $i = 0$, the result is immediate since $\mathscr{D}(0,y) = \mathbb{N}$. If $i = 1$, let $1 \le R \le n$ and define

$$s_1 = \prod_{\substack{p \le y \\ p \mid n}} p, \quad r_1 = \prod_{\substack{p > y \\ p \mid n}} p.$$

In case $r_1 \le R$, because $n/r_1 = s_1$ is y-smooth we have $s_1 \in \mathscr{D}(1,y)$. Because $n = r_1 s_1$ and $R \le n$, $1 \le R/r_1 \le s_1$, and we can find a factorization $s_1 = q_2 r_2$ with $R/(r_1 y) \le r_2 \le R/r_1$. Then $n = q_2(r_1 r_2)$ and $R/y \le r_1 r_2 \le R$ so $n \in \mathscr{D}(1,y)$.

If, however, we had $r_1 > R$, because n and r_1 are z-smooth we can write $r_1 = r_2 q_2$ with $R/z \le r_2 \le R$. Let r_3 be the smallest divisor of s_1 with $r_2 r_3 \ge R/y$. This divisor exists since $R/y \le z r_2/y \le s_1 r_2$, where the second inequality comes from the estimate of the hypothesis of (v) with $i = 1$. Because s_1 is y-smooth, we **claim** $r_2 r_3 \le R$. If not, we have $r_3 \ne 1$, so r_3 would be divisible by a prime $p \le y$, and r_3/p would be a smaller divisor of s_1 obeying the constraint $r_3 r_2/p > R/y$, a contradiction. Therefore, $n = q(r_2 r_3)$ with $R/y \le r_2 r_3 \le R$, completing the case $i = 1$.

Next consider $i > 1$ and assume the result (v) is true for all $j < i$. Let $j, k \ge 0$ be such that $j + k = i - 1$, Using the same definition for the integers r_1 and s_1 as before, we have

$$z^j \cdot z^k \cdot \frac{z}{y} = \frac{z^i}{y} \le s_1.$$

Thus we can **claim** we can write $s_1 = n_1 n_2 n_3$ with

$$z^j/y \le n_1 \le z^j \text{ and } z^k/y \le n_2 \le z^k \implies n_3 \ge \frac{z}{y}. \tag{8.6}$$

To see this, first, there is an $n_1 \mid s_1$ so $z^j/y \le n_1 \le z^j$ and then an $n_2 \mid s_1/n_1$ with

$$\frac{z^k}{n_1 y} \le n_2 \le \frac{z^k}{n_1}.$$

Finally, n_3 is the complementary factor so that $s_1 = n_1 n_2 n_3$.

Now we consider three cases, depending on the value of R, so that we might find a factorization of n.

First, assume $n_1 \le R \le n/n_2$: then

$$1 \le \frac{R}{n_1} \le \frac{n}{n_1 n_2},$$

and by the final inequality of (8.6), $n/(n_1 n_2) = r_1 n_3$ satisfies the assumptions of (v) for $i = 1$. Thus, by that case we can find a factorization $r_1 n_3 = q'r'$ with $R/(n_1 y) \le r' \le R/n_1$. Set $r = n_1 r'$ and $q = n_2 q'$, and note that by the inequalities (8.6), r satisfies the assumptions of (v) for $i = j$ and q satisfies the assumptions of (v) for $i = k$. By induction, $n = qr \in \mathscr{D}(i, y)$.

Next, assume $R < n_1$: because $n_1 \mid s_1$ is y-smooth, by (iv) we can find a divisor r of n_1 such that $R/y \le r \le R$. Then $q = n/r$ is a multiple of n_2 so, since all divisors of n are squarefree, they must satisfy

$$\prod_{\substack{p \le y \\ p \mid q}} p \ge n_2 \ge z^k/y.$$

It follows by induction that $q \in \mathscr{D}(k, y)$. Because r is y-smooth, r is in $\mathscr{D}(j, y)$ using property (iv), and thus we have an appropriate factorization $n = qr$ to get $n \in \mathscr{D}(i, y)$.

Finally, assume $n \geq R > n/n_2$: this can be written $n/R < n_2$. Since n_2 is y-smooth, we can find a divisor $q \mid n_2$ such that $n/(Ry) \leq q \leq n/R$. Then $r = n/q$ is a multiple of n_1 and thus satisfies

$$\prod_{\substack{p \leq y \\ p \mid r}} p \geq n_1 \geq z^j/y.$$

Thus by induction $r \in \mathscr{D}(j, y)$. Since q is in $\mathscr{D}(k, y)$ by (iv) we have the desired factorization to get $n \in \mathscr{D}(i, y)$, and this completes the proof. $\qquad\square$

Examples: (1) By (v) we get $6 \in \mathscr{D}(1, 2)$ and then by (i) or (iv) $6 \in \mathscr{D}(1, 3)$.

(2) The following are all the i-tuply 5-densely divisible numbers, other than those which are 5-smooth, in the range $[1, 200]$, i.e., $\mathscr{D}(i, 5) \cap [1, 200]$:

$i = 1$: $\{14, 21, 28, 33, 35, 39, 42, 44, 52, 55, 56, 63, 65, 66, 68, 70, 76,$
$\qquad\qquad 78, 84, 85, 88, 95, 98, 99, 102, 104, 105, 110, 112, 114, 115, 117,$
$\qquad\qquad 126, 130, 132, 136, 138, 140, 147, 152, 153, 154, 156, 165, 168, 170,$
$\qquad\qquad 171, 174, 175, 176, 182, 184, 189, 190, 195, 196, 198\}$

$i = 2$: $\{28, 42, 56, 63, 70, 84, 99, 105, 112, 117, 126, 132, 140, 156, 165,$
$\qquad\qquad 168, 175, 176, 189, 195, 196, 198\}$

$i = 3$: $\{56, 84, 112, 126, 140, 168, 189\}$

$i = 4$: $\{112, 168\}$

There are no nonsquarefree elements of $\mathscr{D}(i, 5)$ in the given range for all $i \geq 5$.

(3) It would be desirable to have a more intuitive equivalent to the concept of this section. For example, all elements of $\mathscr{D}(1, y)$ have a y-smooth divisor, but this is not sufficient since for instance $38 = 2 \times 19$ has no divisor d with

$$\left\lceil \frac{R}{5} \right\rceil \leq d \leq \lfloor R \rfloor$$

for any R with $10 < R < 19$, so $38 \notin \mathscr{D}(1, 5)$, i.e., its divisors are not "sufficiently dense" on the 5 scale.

(4) In Figure 8.4, we give a numerical count of the number of elements of $\mathscr{D}(1, 5)$ divided by the number of 5-smooths up to x for x up to 10^4. Recalling that the number of 5-smooth numbers up to x is asymptotic to $(\log x)^3$ (see for example [22, section 13.1]), a cursory inspection might indicate that the number of densely divisible numbers goes like x times $\log\log x$ over some power of $\log x$, i.e., for fixed y these densely divisible numbers are much more plentiful than smooth numbers and the same applies to the subset $\mathscr{D}(2, y)$.

We can now state the definition of Polymath's variation on Bombieri–Vinogradov. Recall $\mathscr{D}_\Omega(i, x^\delta)$ represents the set of densely divisible numbers with all prime

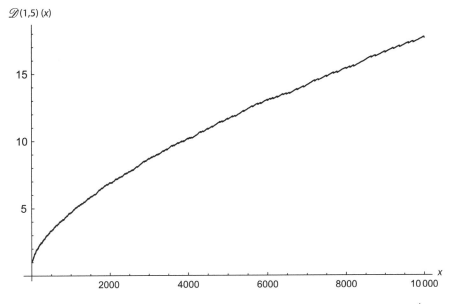

Figure 8.4 A plot of no. of $\mathscr{D}(1,5)$ elements up to x for $2 \leq x \leq 10^4$ vs 5-smooths.

factors in a given subset $\Omega \subset \mathbb{P}$, and that MPZ is the acronym for Motohashi–Pintz–Zhang. This definition represents Polymath's variation on Bombieri–Vinogradov. The reader will note that the moduli are restricted to being squarefree, i-tuply x^δ-densely divisible, not greater than $x^{\frac{1}{2}+2\varpi}$ and with all prime factors in a given subset Ω. The symbols δ, ϖ represent real positive parameters.

Definition 8.11 (**Densely divisible MPZ property**) Let $i \geq 0$ be an integer. We say $\mathrm{MPZ}^{(i)}[\varpi, \delta]$ is true if for any bounded subset $\Omega \subset \mathbb{R}$ which may depend on x, and $Q \prec^o x^{\frac{1}{2}+2\varpi}$, if $(a, P_\Omega) = 1$ and $A \geq 1$ is a fixed real number, then

$$\sum_{\substack{q \leq Q \\ q \in \mathcal{D}_\Omega(i, x^\delta)}} \left| \sum_{\substack{x \leq n < 2x \\ n \equiv a \bmod q}} \Lambda(n) - \frac{1}{\varphi(q)} \sum_{\substack{x \leq n < 2x \\ (n,q)=1}} \Lambda(n) \right| \ll \frac{x}{(\log x)^A}. \tag{8.7}$$

8.6 Improving Zhang

First we show how Polymath improved Zhang's bound $p_{n+1} - p_n \leq 7 \times 10^7$ using Zhang's variation on Bombieri–Vinogradov's estimate. No use is made in this section of dense divisibility, and the improvement is not small but modest compared with later developments.

Recall the definitions $W(x) = \prod_{p \leq w(x)} p$, where $w(x) \ll \log\log\log x$ tends to infinity with x sufficiently slowly, and $\theta(n) := \log n$ for prime n and is 0 otherwise. Also recall that we have defined an admissible k-tuple to be a set of integers distinct $\mathscr{H} = \{h_1, \ldots, h_k\}$ for which

$$v_p(\mathscr{H}) := \#\{h_j \bmod p : 1 \leq j \leq k\} < p.$$

Finally, recall

$$P(n) := (n + h_1) \cdots (n + h_k).$$

The following lemma shows that by restricting to integers n in the congruence class of a particular integer b (depending on W) we can ensure that elements $P(n)$ all have no small prime divisors. This of course improves the chances that some of the polynomial factors will be prime.

Lemma 8.6 *Given $k \geq 2$ and $\mathscr{H}$ an admissible k-tuple, then there exists a congruence class $b \bmod W$ with $(b + h, W) = 1$ for all $h \in \mathscr{H}$.*

Proof We **claim** that, since $\mathscr{H}$ is admissible, for all $p \mid W$ there is an integer b_p and an r_p with $p \nmid r_p$ such that

$$P(b_p) \equiv r_p \bmod p.$$

To see this, since $\mathscr{H}$ is admissible, we have

$$\#\{-h_i \bmod p : 1 \leq i \leq k\} < p,$$

so choose $b_p \not\equiv -h_i \bmod p$ for all $1 \leq i \leq k$. Then $p \nmid P(b_p) =: r_p$. By the Chinese Remainder Theorem, there is a b such that $b \equiv b_p \bmod p$ for all $p \mid W$, so $b \equiv b_p \bmod W$ and $P(b) \equiv r_p \bmod p$. Thus $p \nmid P(b)$ so $(P(b), W) = 1$ and so $(p + h_i, W) = 1$ for all i. This completes the proof. □

The next lemma gives a general criterion for an infinite number of translates of an admissible k-tuple to each contain at least two prime members. The weight function $v(n)$ and supplementary functions $B(x)$ and $R(x)$ are not fixed, and the criterion is given in terms of a key inequality. This is at the heart of the method of GMPY and Zhang, and the proof is a great place to start!

Lemma 8.7 **(Sufficient condition for DHL)** *[174, lemma 4.1] Given fixed $k \geq 2$, $\mathscr{H}$ an admissible k-tuple, $b \bmod W$ a congruence class with $(b + h, W) = 1$ for all $h \in \mathscr{H}$, a function $v \colon \mathbb{N} \to [0, \infty)$, fixed parameters $\alpha, \beta > 0$ and $R = R(x) > 0$, $B = B(x) > 0$ such that as $x \to \infty$, we have an upper bound*

$$\sum_{\substack{x \leq n < 2x \\ n \equiv b \bmod W}} v(n) \leq (\alpha + o(1)) \frac{B^k x}{W}, \tag{8.8}$$

and for all $h_j \in \mathcal{H}$ a lower bound

$$\sum_{\substack{x \le n < 2x \\ n \equiv b \bmod W}} v(n)\theta(n + h_j) \ge (\beta + o(1))\frac{B^k x}{W} \log R, \tag{8.9}$$

and, in addition, we have the key inequality

$$\frac{\log R}{\log x} > \frac{\alpha}{k\beta} \tag{8.10}$$

for all x sufficiently large, then DHL[k, 2] *is true.*

Proof **(1)** Because $\mathcal{H}$ is admissible, by Lemma 8.6 there is a class $b \bmod W$ such that $(b + h_j, W) = 1$ for all $1 \le j \le k$. Let

$$N(x) := \sum_{\substack{x \le n < 2x \\ n \equiv b \bmod W}} v(n)\left(\left(\sum_{j=1}^{k} \theta(n + h_j)\right) - \log(3x)\right).$$

Using the inequalities (8.8) and (8.9), we get

$$N(x) \ge ((\beta + o(1))k \log R - (\alpha + o(1)) \log(3x))\frac{B^k x}{W}.$$

Using the bound (8.10), we have for all x sufficiently large, $N(x) > 0$.

(2) The result of Step (1) implies that for $x > h_k$ and sufficiently large, for at least one $n \in [x, 2x)$, the expression

$$\left(\sum_{j=1}^{k} \theta(n + h_j)\right) - \log(3x)$$

must be positive for at least two indices j. Therefore, for all x sufficiently large, there must be an integer $n \in [x, 2x)$ such that $n + h_j$ is prime for at least two $j \in [1, k]$. Because $\mathcal{H}$ is an arbitrary admissible k-tuple, we have DHL[k, 2]. This completes the proof. □

Important Remark: From now on in this chapter, we fix $0 < \varpi < \frac{1}{4}$, $R(x) = x^{1/4 + \varpi}$ and $B(x) = \frac{\varphi(W)}{W} \log R$, where as before

$$W = W(x) := \prod_{p \le w(x)} p, \quad w(x) \ll \log\log\log x, \quad w(x) \to \infty \text{ as } x \to \infty.$$

These choices imply $\log R / \log x = \frac{1}{4} + \varpi$, so the key sufficient inequality of Lemma 8.7 becomes $k(\frac{1}{4} + \varpi) > \alpha/\beta$.

The next lemma enables us to exchange a sum of values of a multiplicative function times an integrable function with compact support with an integral. It is a fundamental device for this approach to the prime gaps problem and is used subsequently in an essential way in many lemmas or theorems. Note that there are some differences from the original proof and that the "W-trick" plays an essential role. To see this, the reader might wish to experiment with its application to the sum

$$\sum_{n \leq x} \frac{2^{\Omega(n)}}{n}$$

with $W = 1$ and $g = \chi_{[0,1]}$.

First, we define an equicontinuous family enabling a variation of the function on both sides of the expression of the lemma.

Definition 8.12   We say a family $\mathscr{F}$ of functions $g \colon \mathbb{R} \to \mathbb{R}$ is **equicontinuous** if for all $\epsilon > 0$ there is a $\delta > 0$ such that $|x - y| < \delta$ implies $|f(x) - f(y)| < \epsilon$ for all $x, y \in \mathbb{R}$ and all $f \in \mathscr{F}$. So in particular, members of $\mathscr{F}$ are uniformly continuous on $\mathbb{R}$.

Lemma 8.8   (Fundamental sieve estimate) *[174, lemma 4.2] Given $k \in \mathbb{N}$ and $\beta \colon \mathbb{N} \to \mathbb{R}$ a multiplicative function such that for $p \in \mathbb{P}$, we have*

$$p \leq w \text{ and } j \geq 1 \implies \beta(p^j) = 0,$$
$$p > w \text{ and } j > 1 \implies \beta(p^j) \ll \exp(O(j)),$$
$$p > w \implies \beta(p) = k + O\left(\frac{1}{p}\right).$$

Given any $g \colon \mathbb{R} \to \mathbb{R}$ which is Riemann integrable and has compact support, a fixed $c > 0$ with $R > x^c$, and a function $w = w(x)$ which grows sufficiently slowly with x as $x \to \infty$, then we have

$$\sum_{n \in \mathbb{N}} \frac{\beta(n)}{n} g\left(\frac{\log n}{\log R}\right) = B^k \left(\int_0^\infty g(t) \frac{t^{k-1}}{(k-1)!} \, dt + o(1) \right). \tag{8.11}$$

The same estimate holds also where the summation on the left is restricted to the set of squarefree n with all prime factors greater than w. This set is denoted $S_{(w,\infty)}$.

In addition, both of these estimates hold uniformly in g for g in any equicontinuous family of real functions having a fixed compact support in $\mathbb{R}$.

Proof   The proof is by induction on k and has five steps.
(1) Let $k = 1$ and $W = 1$: first evaluate the left-hand side when $g(t)$ is the characteristic function of a bounded interval and then a step function, to see that as $x \to \infty$, we have

$$\sum_{n \in \mathbb{N}} \frac{1}{n} g\left(\frac{\log n}{\log R}\right) = (\log R)\left(\int_0^\infty g(t)\, dt + o(1)\right).$$

We refine this sum to exclude small prime divisors from the n's, i.e., those with $p \le w = \log\log\log x$. First, fix p:

$$\sum_{\substack{n \in \mathbb{N} \\ p \mid n}} \frac{1}{n} g\left(\frac{\log n}{\log R}\right) = \sum_{n \in \mathbb{N}} \frac{1}{pn} g\left(\frac{\log(pn)}{\log R}\right)$$

$$= \frac{1}{p} \sum_{n \in \mathbb{N}} \frac{1}{n} g\left(\frac{\log(n)}{\log R} + \frac{\log(p)}{\log R}\right)$$

$$= \frac{1}{p}(\log R)\left(\int_{\frac{\log p}{\log R}}^\infty g(t)\, dt + o(1)\right)$$

$$= \frac{1}{p}(\log R)\left(\int_0^\infty g(t)\, dt + o(1)\right).$$

Thus

$$\sum_{\substack{n \in \mathbb{N} \\ (p,n)=1}} \frac{1}{n} g\left(\frac{\log n}{\log R}\right) = \left(1 - \frac{1}{p}\right)(\log R)\left(\int_0^\infty g(t)\, dt + o(1)\right).$$

Applying this to a finite product of a small number (i.e., all these up to w) then gives

$$\sum_{\substack{n \in \mathbb{N} \\ (W,n)=1}} \frac{1}{n} g\left(\frac{\log n}{\log R}\right) = \frac{\varphi(W)}{W}(\log R)\left(\int_0^\infty g(t)\, dt + o(1)\right).$$

(2) Next, we define a multiplicative function h by $h(p^j) := 0$ for $j \ge 1$ and $p \le w$, and $h(p^j) := \beta(p^j) - \beta(p^{j-1})$ for $p > w$ and all $j \ge 1$. Then

$$p \le w \text{ and } j \ge 1 \implies h(p^j) = 0,$$
$$p > w \text{ and } j > 1 \implies h(p^j) \ll \exp(O(j)),$$
$$p > w \implies h(p) = k - 1 + O\left(\frac{1}{p}\right),$$

and, in addition,

$$\beta(n) = \sum_{\substack{a|n \\ (a,W)=1}} h(a).$$

In particular, h satisfies the same set of conditions as β but with k replaced by $k-1$.

(3) Assuming (8.11) is true for k replaced by $k - 1$ and $k \geq 2$, and all suitable functions $\beta(n)$, we can apply this inductive hypothesis with $\beta(n)$ replaced by $h(n)$. This implies

$$\sum_{\substack{a\in\mathbb{N} \\ (a,W)=1}} \frac{\beta(a)}{a} g\left(\frac{\log a}{\log R}\right) = \sum_{\substack{a\in\mathbb{N} \\ (a,W)=1}} \frac{h(a)}{a} \sum_{\substack{d\in\mathbb{N} \\ (d,W)=1}} \frac{1}{d} g\left(\frac{\log a}{\log R} + \frac{\log d}{\log R}\right).$$

If we rewrite the inner sum, we get

$$\sum_{\substack{d\in\mathbb{N} \\ (d,W)=1}} \frac{1}{d} g\left(\frac{\log a}{\log R} + \frac{\log d}{\log R}\right) = \sum_{b|W} \frac{\mu(b)}{b} \sum_{d\in\mathbb{N}} \frac{1}{d} g\left(\frac{\log a}{\log R} + \frac{\log b}{\log R} + \frac{\log d}{\log R}\right).$$

Then using the result of Step (1) and the uniform integrability of translates of $g(x)$, using the prime number theorem (so $W(x) \ll \log\log x$) to mop up terms of order $o(\log R)$, we get

$$\sum_{\substack{d\in\mathbb{N} \\ (d,W)=1}} \frac{1}{d} g\left(\frac{\log a}{\log R} + \frac{\log d}{\log R}\right) = \left(B \int_{\log a/\log R}^{\infty} g(t)\, dt + o(1)\right).$$

Thus, for $w(x)$ growing sufficiently slowly we have

$$\sum_{n\in\mathbb{N}} \frac{\beta(n)}{n} g\left(\frac{\log n}{\log R}\right)$$

$$= \frac{\varphi(W)(\log R)}{W} \left(\sum_{\substack{a\in\mathbb{N} \\ (a,W)=1}} \frac{h(a)}{a} \int_{\log a/\log R}^{\infty} g(t)\, dt + o\left(\sum_{\substack{a\in\mathbb{N} \\ (a,W)=1}} \frac{|h(a)|}{a}\right)\right).$$

$$\text{(8.12)}$$

(4) In this step, we consider $k = 1$ with $W(x) \to \infty$. In this case, we get

$$\sum_{\substack{a \in \mathbb{N} \\ (a, W)=1}} \frac{|h(a)|}{a} = \prod_{p>w} \left(1 + \sum_{j=1}^{\infty} \frac{|h(p^j)|}{p^j} \right) = 1 + o(1).$$

Therefore, since $h(1)/1 = 1$, we get

$$\sum_{\substack{a>1 \\ (a, W)=1}} \frac{|h(a)|}{a} = o(1),$$

and the equation of the lemma, (8.11), in case $k = 1$, follows directly from this.

(5) Next, assume $k \geq 2$ and that the equation of the lemma is true replacing k with $k - 1$. Applying this to $h(n)$, replacing $g(t)$ with the function $y \to \int_y^\infty g(t)\, dt$ and using integration by parts gives for the first term in the second factor of the result of Step (2):

$$\sum_{\substack{a \in \mathbb{N} \\ (a, W)=1}} \frac{h(a)}{a} \int_{\log a/ \log R}^{\infty} g(t)\, dt = B^{k-1} \left(\int_0^\infty g(t) \frac{t^{k-1}}{(k-1)!}\, dt + o(1) \right).$$

In addition, this expression also implies

$$\sum_{\substack{a \leq R \\ (a, \overline{W})=1}} \frac{|h(a)|}{a} \ll B^{k-1},$$

which, using (8.12), gives (8.11) for the function $g(n)$ and k, thus completing the inductive step.

(6) To obtain the restricted sum over $n \in \mathcal{S}_{(w, \infty)}$, we simply replace $\beta(n)$ by

$$\widetilde{\beta}(n) = \begin{cases} \beta(n), & \text{if } n \text{ is squarefree,} \\ 0, & \text{otherwise.} \end{cases}$$

This completes the proof. $\qquad\qquad\qquad\qquad\qquad\qquad\qquad\qquad\qquad\quad \square$

To apply the program outlined in Lemma 8.7, we must determine α, which is done in Lemma 8.9, and β, which is the subject of Lemma 8.13. The development is made in terms of a fixed continuous function $f : [0, \infty) \to [0, \infty)$, which has compact support in $[0, 1]$, and has all derivatives existing and continuous on $[0, 1]$, where we need only consider left and right derivatives at 1 and 0 respectively, which do not necessarily vanish. Particular functions are chosen later, parameterized by k, so that the bounds derived using them are not trivial.

Recall the definitions $W = W(x) := \prod_{p \leq w} p$, $\mathcal{X} \subset (w, \infty)$, $\mathcal{S}_{(w,\infty)}$ are the set of squarefree positive integers with all prime factors greater than w, and

$$B = B(x) := \frac{\varphi(W)}{W} \log R,$$

$$P(n) := \prod_{i=1}^{k} (n + h_i),$$

$$\Phi(d) := \prod_{p|d} \frac{p - k}{k},$$

$$\rho(d) := \prod_{p|d} \frac{k}{p},$$

$$a_d := \frac{1}{\rho(d)\Phi(d)} \sum_{\substack{q \in \mathcal{S}_{(w,\infty)} \\ (q,d)=1 \\ qd \in \mathcal{X}}} \frac{1}{\Phi(q)} f'\left(\frac{\log dq}{\log R}\right),$$

$$v(n) := \left(\sum_{\substack{d \in \mathcal{S}_{(w,\infty)} \\ d | P(n)}} \mu(d) a_d \right)^2.$$

In particular, we have now specified the weights $v(n)$ up to making f explicit with the restrictions given previously. In Lemma 8.9, the parameter α, which appears in Lemma 8.7, is defined in terms of the undefined $f(t)$ and the upper bound required by Lemma 8.7 verified. Note that the sum-to-integral technique of Lemma 8.8 is used twice and the Selberg sieve method (Chapter 2) is applied. Note that $\Omega(d)$ is the total number of primes in the unique factorization of $d \in \mathbb{N}$, including multiplicity.

Lemma 8.9 (**generic upper bound**) *[174, proposition 4.3] With the notation set out in Lemma 8.7, and any function $f(t)$ which satisfies the conditions given previously, if we set*

$$\alpha := \int_0^1 f'(t)^2 \frac{t^{k-1}}{(k-1)!} dt; \tag{8.13}$$

then as $x \to \infty$, we have

$$\sum_{\substack{x \leq n < 2x \\ n \equiv b \bmod W}} v(n) \leq (\alpha + o(1)) \frac{x B^k}{W}.$$

Proof **(1)** We begin by expanding the definition of $v(n)$ and summing to get

$$\sum_{\substack{x \leq n < 2x \\ n \equiv b \bmod W}} v(n) = \sum_{d,e \in \mathcal{S}_{(w,\infty)}} \mu(d)a_d\mu(e)a_e \sum_{\substack{x \leq n < 2x \\ [d,e] \mid P(n) \\ n \equiv b \bmod W}} 1.$$

From the definition, we see the product $a_d a_e$ is supported on squarefree positive integers $d, e \leq R$. If x is sufficiently large that all of the $h_j \in \mathcal{H}$ are pairwise noncongruent modulo any prime $p > w$ (e.g., if $w \geq h_k$), then there are precisely k classes n modulo p such that $P(n) \equiv 0 \bmod p$. By the Chinese Remainder Theorem and the definition of $\rho(d)$, because d begin squarefree implies $\rho(d)d$ is the number of roots of $P(n)$ modulo d for $(d, W) = 1$, we get

$$\sum_{\substack{x \leq n < 2x \\ [d,e] \mid P(n) \\ n \equiv b \bmod W}} 1 = \rho([d,e])\frac{x}{W} + O([d,e]\rho([d,e])).$$

Therefore,

$$\sum_{\substack{x \leq n < 2x \\ n \equiv b \bmod W}} v(n) = \sum_{d,e \in \mathcal{S}_{(w,\infty)}} \mu(d)a_d\mu(e)a_e\rho([d,e])\frac{x}{W}$$

$$+ O\left(\sum_{d,e \in \mathcal{S}_{(w,\infty)}} |a_d||a_e|k^{\Omega([d,e])}\right).$$

(2) Now consider the error term in this expression. Since for all $d, e \in \mathbb{N}$,

$$\Omega(d) + \Omega(e) \geq \Omega([d,e]),$$

we have

$$\sum_{\substack{d,e \in \mathcal{S}_{(w,\infty)} \\ d,e \leq R}} |a_d||a_e|k^{\Omega([d,e])} \leq \left(\sum_{\substack{d \in \mathcal{S}_{(w,\infty)} \\ d \leq R}} |a_d|k^{\Omega(d)}\right)^2.$$

The definition of a_d and Lemma 8.8 with $\beta(p) = kp/(p-k) = k+O(1/p), p > w$, give for $d \leq R$ using

$$\sum_{\substack{q \in \mathcal{S}_{(w,\infty)} \\ (q,d)=1 \\ qd \in \mathcal{X}}} \frac{1}{\Phi(q)} = \sum_{\substack{q \in \mathcal{S}_{(w,\infty)} \\ (q,d)=1 \\ qd \in \mathcal{X}}} \frac{\beta(q)}{q} \ll B^k,$$

the estimate

$$|a_d| \ll \frac{1}{\Phi(d)\rho(d)}B^k.$$

Thus, using Lemma 8.8 again with the same definition of $\beta(p)$ and noting that since $w < d$

$$\sum_{\substack{d \in \mathcal{S}_{(w,\infty)} \\ d \le R}} |a_d| k^{\Omega(d)} = \sum_{\substack{d \in \mathcal{S}_{(w,\infty)} \\ d \le R}} \beta(d)$$

$$\le \sum_{\substack{d \in \mathcal{S}_{(w,\infty)} \\ d \le R}} \frac{\beta(d)}{d} R \ll B^k R,$$

we get the error estimate

$$O\left(R^2 \left(\frac{\varphi(W) \log R}{W} \right)^{O(1)} \right) = o\left(\frac{x B^k}{W} \right).$$

(3) We now consider the coefficient of the main term

$$\sum_{d,e \in \mathcal{S}_{(w,\infty)}} \mu(d) a_d \mu(e) a_e \rho([d,e]).$$

We use the Selberg sieve method, which is to consider this as a quadratic form in the variables (a_d), which we proceed to diagonalize. To this end, we let

$$\xi(d) := \rho(d) \prod_{p|d} (1 - \rho(p)) \quad \text{and} \quad y_d := \sum_{de \in \mathcal{S}_{(w,\infty)}} \rho(e) \mu(de) a_{de}.$$

Then we **claim**

$$\rho([d,e]) = \sum_{c|(d,e)} \xi(c) \rho\left(\frac{d}{c}\right) \rho\left(\frac{e}{c}\right).$$

To see this, recall d, e are squarefree and ρ is multiplicative, so we can write for the left-hand side,

$$\rho([d,e]) = \rho\left(\frac{de}{(d,e)} \right) = \rho\left(\frac{d}{(d,e)} \right) \rho(e) = \frac{\rho(d)\rho(e)}{\rho((d,e))}.$$

The right-hand side is a little more demanding. We use $\mathbf{1} \star \mu = \delta$, where $\mathbf{1}(n) = 1$ for all $n \in \mathbb{N}$ and δ is the identity for Dirichlet multiplication $\star$.

$$\sum_{c|(d,e)} \xi(c) \rho\left(\frac{d}{c}\right) \rho\left(\frac{e}{c}\right) = \sum_{c|(d,e)} \rho(c) \prod_{p|c} (1 - \rho(p)) \rho\left(\frac{d}{c}\right) \rho\left(\frac{e}{c}\right)$$

$$= \sum_{c|(d,e)} \rho(c) \sum_{c'|c} \mu(c')\rho(c') \rho\left(\frac{d}{c}\right) \rho\left(\frac{e}{c}\right)$$

$$= \sum_{c|(d,e)} \sum_{c'|c} \mu(c')\rho(c)\rho(c') \rho\left(\frac{d}{c}\right) \rho\left(\frac{e}{c}\right)$$

$$= \rho(d)\rho(e) \sum_{c|(d,e)} \sum_{c'|c} \mu(c') \frac{1}{\rho(c/c')}$$

$$= \rho(d)\rho(e) \sum_{c|(d,e)} \left(\mu \star \frac{1}{\rho} \right)(c)$$

$$= \rho(d)\rho(e) \left(1 \star \mu \star \frac{1}{\rho} \right)((d,e)) = \frac{\rho(d)\rho(e)}{\rho((d,e))},$$

which completes the proof of the claim.

Substituting the expression we have derived for $\rho([d,e])$ then implies

$$\sum_{d,e \in S_{(w,\infty)}} \mu(d)a_d\mu(e)a_e\rho([d,e]) = \sum_{d \in S_{(w,\infty)}} \xi(d)y_d^2,$$

which is in diagonal form.

(4) Since in any summand in the definition of y_d given in Step (3), necessarily the variables d and e are squarefree and coprime, using the definition of the a_d, and noting that $(e,q) = 1$ and $(d,e) = 1$ can be assumed, we get

$$y_d = \sum_{de \in \mathcal{S}_{(w,\infty)}} \rho(e)\mu(de)a_{de}$$

$$= \sum_{de \in S_{(w,\infty)}} \frac{\rho(e)\mu(d)\mu(e)}{\Phi(de)\rho(de)} \sum_{\substack{q \in S_{(w,\infty)} \\ deq \in \mathcal{X}}} \frac{1}{\Phi(q)} f'\left(\frac{\log(deq)}{\log R} \right)$$

$$= \frac{\mu(d)}{\Phi(d)\rho(d)} \sum_{\substack{m \in S_{(w,\infty)} \\ dm \in \mathcal{X}}} \frac{1}{\Phi(m)} f'\left(\frac{\log(dm)}{\log R} \right) \sum_{eq=m} \mu(e)$$

$$= \frac{\mu(d)}{\rho(d)\Phi(d)} f'\left(\frac{\log d}{\log R} \right) \quad \text{or 0 if } d \notin \mathcal{X}.$$

(5) In this final step, first note that

$$\rho(d)^2 \Phi(d) = \rho(d) \prod_{p|d} \left(\frac{p-k}{k} \right) \frac{k}{p} = \xi(d),$$

so by the result of Step (4), we get

$$\sum_{d \in S_{(w,\infty)}} \xi(d)y_d^2 = \sum_{d \in \mathcal{X}} \frac{1}{\Phi(d)} f'\left(\frac{\log d}{\log R} \right)^2.$$

Each term is positive, so we get an upper bound by including all $d \in S_{(w,\infty)}$ in the sum on the right. We use Lemma 8.8 as before with for $d \in S_{(w,\infty)}$, $\beta(d) :=$

$\prod_{p\mid d} pk/(p-k)$, and the results of Steps (1) through (3) to get using (1) to get the first equality, (3) for the second, (4) for the third and Lemma 8.8 with

$$\beta(p) = \frac{k}{1 - \frac{k}{p}}, \quad p > w, \text{ and } g(t) = f'(t)^2 \chi_{[0,1]}(t)$$

for the final upper bound:

$$\sum_{\substack{x \le n < 2x \\ n \equiv b \bmod W}} v(n) = \sum_{d,e \in \mathscr{S}_{(w,\infty)}} \mu(d)\mu(e) a_d a_e \rho([d,e]) \frac{x}{W} + o\left(\frac{x B^k}{W}\right)$$

$$= \left(\sum_{d \in S_{(w,\infty)}} \xi(d) y_d^2\right) \frac{x}{W} + o\left(\frac{x B^k}{W}\right)$$

$$= \sum_{d \in \mathscr{X}} \frac{1}{\Phi(d)} \left(f'\left(\frac{\log d}{\log R}\right)\right)^2 \frac{x}{W} + o\left(\frac{x B^k}{W}\right)$$

$$\le (\alpha + o(1)) \frac{x B^k}{W}.$$

This completes the proof.　　　　　　　　　　　　　　　　　　　　□

Note that we have now derived the first part of what we need to apply Lemma 8.7. Now recall that $0 < \varpi < \frac{1}{4}$ is fixed and $R := x^{\frac{1}{4}+\varpi}$ (so $\log R / \log x = \frac{1}{4} + \varpi$ in Lemma 8.7). Let $\mathscr{H}$ be a fixed k-tuple and $\mathscr{X}$ a fixed subset of $S_{(w,\infty)}$. Recall the definitions:

$$P(n) := \prod_{1 \le j \le k} (n + h_j),$$

$$P_i(n) := \prod_{\substack{1 \le j \le k \\ j \ne i}} (n + h_j - h_i), \quad \deg P_i = k - 1,$$

$$C_i(q) := \{n : 1 \le n \le q, \ (n,q) = 1, \ P_i(n) \equiv 0 \bmod q\},$$

$$E_i(q) = \sum_{a \in C_i(q)} \left| \left(\sum_{\substack{x \le n < 2x \\ n \equiv b + h_i \bmod W \\ n \equiv a \bmod q}} \Lambda(n) \right) - \frac{x}{\varphi(Wq)} \right|,$$

$$\hat{\rho}(q) := \frac{|C_i(q)|}{\varphi(q)} = \prod_{p \mid q} \frac{k - 1}{p - 1}, \quad q \in S_{(w,\infty)}.$$

We now derive an expression for a particular sum which will be used later. Note that the main term is independent of the index i, but the error depends on E_i.

Lemma 8.10 **(Generic lower bound)** *[174, page 45]*

$$\sum_{\substack{x \leq n < 2x \\ n \equiv b \bmod W}} \Lambda(n)\theta(n + h_i) = \frac{x}{\varphi(W)} \sum_{\substack{d,e > w \\ squarefree}} \mu(d)a_d\mu(e)a_e \hat{\rho}([d,e])$$

$$+ O\left(\sum_{\substack{d,e > w \\ squarefree}} |a_d||a_e|E_i([d,e]) \right) + O\left(R^2 x^{o(1)} \right). \quad (8.14)$$

Proof **(1)** First, we note that ([121, section 4.1]) for $1 \leq i \leq k$,

$$\sum_{x \leq n < 2x} |\Lambda(n + h_i) - \theta(n + h_i)| \prec^o x^{\frac{1}{2}}.$$

This implies, using the divisor bound $\tau(n) \ll n^{o(1)}$, the definition of $v(n)$ (given before Lemma 8.9) and the bound for $|a_d|$ given in the estimate derived in Lemma 8.7 Step (2), so $|a_d| \leq B^k$,

$$\left| \sum_{\substack{x \leq n < 2x \\ n \equiv b \bmod W}} v(n)\Lambda(n + h_i) - \sum_{\substack{x \leq n < 2x \\ n \equiv b \bmod W}} v(n)\theta(n + h_i) \right| \prec^o \sqrt{x} \sup_{x \leq n < 2x} v(n)$$

$$\prec^o \sqrt{x} = o\left(\frac{xB \log R}{W} \right).$$

Thus we can obtain a lower bound for the sum with $\theta(n + h_i)$ by first obtaining one for that with $\Lambda(n + h_i)$, which turns out to be easier.

Note that

$$\sum_{\substack{x \leq n < 2x \\ n \equiv b \bmod W}} v(n)\Lambda(n + h_i) = \sum_{d,e \in \mathcal{S}_{(w,\infty)}} \mu(d)a_d\mu(e)a_e \left(\sum_{\substack{x \leq n < 2x \\ [d,e] \mid P(n) \\ n \equiv b \bmod W}} \Lambda(n + h_i) \right).$$

$$(8.15)$$

In this expression, the inner sum is a logarithmic weighted sum of primes in arithmetic progression modulo $W[d,e]$, which we then "average" over d,e. Step (2) shows how this can be done.

(2) We have $d,e \leq R$ so $[d,e] \leq R^2 = x^{\frac{1}{2}+2\varpi}$. Thus for x sufficiently large, we get

$$\sum_{\substack{x \leq n < 2x \\ [d,e] \mid P(n) \\ n \equiv b \bmod W}} \Lambda(n + h_i) = \sum_{a \in \mathscr{C}_i([d,e])} \sum_{\substack{x \leq n < 2x \\ n \equiv b \bmod W \\ n \equiv a - h_i \bmod [d,e]}} \Lambda(n + h_i) + O(x^{o(1)}), \quad (8.16)$$

where the error comes from integers n such that $n + h_i$ is a power greater than 1 of a prime which divides $[d, e]$, $\tau(n) \le n^{o(1)}$ and $\sqrt{[d, e]} \le R$.

(3) From the definition of $\hat{\rho}(n)$ for all $q \in S_{(w, \infty)}$, we have

$$\frac{|\mathscr{C}_i(q)|}{\varphi(q)} =: \hat{\rho}(q) = \prod_{p|q} \frac{k-1}{p-1}.$$

Using $\hat{\rho}$, (8.15) and E_i, we can now rewrite (8.16), for each i with $1 \le i \le k$, in the form

$$\sum_{\substack{x \le n < 2x \\ n \equiv b \bmod W}} v(n)\Lambda(n + h_i) = \frac{x}{\varphi(W)} \sum_{\substack{d, e > w \\ \text{squarefree}}} \mu(d) a_d \mu(e) a_e \hat{\rho}([d, e])$$

$$+ O\left(\sum_{\substack{d, e > w \\ \text{squarefree}}} |a_d| |a_e| E_i([d, e]) \right) + O\left(R^2 x^{o(1)} \right).$$

(8.17)

This completes the proof. ∎

Next, we show that the first error term in the sum of Lemma 8.10 is sufficiently small. Recall the definitions $P_\Omega := \prod_{p \in \Omega} p$, $\rho(n) = \prod_{p|n} k/p$, and $\Phi(n) := \prod_{p|n} (p - k)/k$.

Lemma 8.11 (Distribution of primes) *[174, lemma 4.4] Let $\Omega \subset \mathbb{R}$ be bounded and $\mathcal{Y} \subset S_\Omega$ such that $W[d, e] \in \mathcal{Y}$ when $d, e \in S_{(w, \infty)}$ are nonzero positive integers with a_d, a_e nonzero. If for all integers a with $(a, P_\Omega) = 1$ and all fixed $A > 0$, we have the estimate as $x \to \infty$*

$$\sum_{q \in \mathcal{Y}} \left| \sum_{\substack{x \le n < 2x \\ n \equiv a \bmod q}} \Lambda(n) - \frac{1}{\varphi(q)} \sum_{\substack{x \le n < 2x \\ (n, q) = 1}} \Lambda(n) \right| \ll \frac{x}{(\log x)^A},$$ (8.18)

then, if as defined previously,

$$E_i(q) := \sum_{a \in C_i(q)} \left| \left(\sum_{\substack{x \le n < 2x \\ n \equiv b + h_i \bmod W \\ n \equiv a \bmod q}} \Lambda(n) \right) - \frac{x}{\varphi(Wq)} \right|;$$

we also have

$$E := \sum_{d, e \in S_{(w, \infty)}} |a_d| |a_e| E_i([d, e]) \ll \frac{x}{(\log x)^A} = o((xB/W) \log R).$$

Proof (**1**) Because $a_d = 0$ for $d > R$, we can assume that

$$\mathcal{Y} \subset W\mathcal{S}_{(w,\infty)} \cap [1, WR^2].$$

(**2**) Using the estimate from Step (2) of Lemma 8.9 to bound $|a_d|, |a_e|$, $W \ll$ $(\log\log x)^{O(1)}$, $R = x^{\frac{1}{4}+\varpi}$, and Cauchy–Schwarz in the final step, if we define a multiplicative function

$$h(q) := \sum_{\substack{d,e \in \mathcal{S}_{(w,\infty)} \\ q=[d,e]}} \frac{1}{\Phi(d)\rho(d)\Phi(e)\rho(e)},$$

then we can derive

$$E \ll B^{2k} \sum_{\substack{d,e \in \mathcal{S}_{(w,\infty)} \\ W[d,e] \in \mathcal{Y}}} \frac{E_i([d,e])}{\Phi(d)\rho(d)\Phi(e)\rho(e)}$$

$$\ll (\log x)^{2k} \sum_{Wq \in \mathcal{Y}} h(q) E_i(q)$$

$$\ll (\log x)^{2k} \left(\sum_{Wq \in \mathcal{Y}} h(q)^2 |C_i(q)| E_i(q) \right)^{\frac{1}{2}} \times \left(\sum_{Wq \in \mathcal{Y}} \frac{E_i(q)}{|C_i(q)|} \right)^{\frac{1}{2}}.$$

(**3**) From the definition of E_i, we have the bound

$$E_i(q) \ll x(\log x)|C_i(q)|/\varphi(q).$$

If we replace k by $9k^4$, the function

$$p \to h(p)^2 |C_i(p)|^2/\varphi(p) \ll \beta(p := \left(\frac{3k^2 p^2}{(1-k/p)^2} \right)^2 = 9k^4 + O\left(\frac{1}{p} \right)$$

satisfies the assumptions of Lemma 8.8, so using that lemma, the first two factors of the right-hand side of the result of Step (2) are bounded by

$$(\log x)^{2k} \left(\sum_{Wq \in \mathcal{Y}} h(q)^2 |C_i(q)| E_i(q) \right)^{\frac{1}{2}} \ll x^{\frac{1}{2}} (\log x)^{O(1)}.$$

(**4**) For the third factor, let $\Omega := (1, WR^2]$ so $\mathcal{Y} \subset \Omega$. Using the assumption of the lemma but restricting it to moduli Wq, which are multiples of a fixed q, and then averaging over all classes $a \bmod P_\Omega$ such that $a \bmod P_\Omega / W \in C_i(P_\Omega / W)$ and $a \equiv b + h_i \bmod W$, we obtain the estimate which is valid for any fixed $A > 0$:

$$\sum_{Wq\in\mathcal{Y}} \frac{1}{|C_i(q)|} \sum_{a\in C_i(q)} \left| \sum_{\substack{x\le n<2x \\ n\equiv a \bmod q \\ n\equiv b+h_i \bmod W}} \Lambda(n) - \frac{1}{\varphi(qW)} \sum_{\substack{x\le n<2x \\ (n,qW)=1}} \Lambda(n) \right| \ll \frac{x}{(\log x)^A}.$$

Comparing this with the definition of E_i, and noting that, because there are at most $o(q)$ p's to remove and $q \le R^2 = x^{\frac{1}{2}+2\varpi}$,

$$x \sim \sum_{\substack{x\le p<2x \\ (p,Wq)=1}} \log p,$$

then gives

$$\sum_{q:Wq\in\mathcal{Y}} \frac{E_i(q)}{|C_i(q)|} \ll \frac{x}{(\log x)^A},$$

which gives a suitable bound for the third factor, and, using the expression derived in Step (2) with this bound together with that derived in Step (3), completes the proof. ∎

We now derive an expression for the main term in the sum of Lemma 8.10, i.e., the lower bound, in terms of the function $f(t)$. Recall the definitions: $S_{(w,\infty)}$ is the set of all squarefree numbers with every prime divisor greater than $w = w(x) > 0$ with $w(x) \to \infty$ sufficiently slowly; $\mathcal{X} \subset S_{(w,\infty)}$ is a given subset; $P(n) = (n+h_1)\cdots(n+h_k)$ is a polynomial with integer coefficients; $f(t)$ is a fixed real function in the family described before Lemma 8.9, namely $f: [0,\infty) \to [0,\infty)$ is continuous with support in $[0,1]$ and nonnegative and smooth on $[0,1]$; and $\hat\rho$ is defined before Lemma 8.10 and satisfies $\hat\rho(q) = \prod_{p|q}(k-1)/(p-1)$ for $q \in S_{(w,\infty)}$. Note that $\hat\rho(p) < \rho(p) = k/p$ whenever $p > k$. Recall also that $\Phi(p) = k/(p-k)$ for $p > w$.

Lemma 8.12 (Main term of the lower bound) *[174, lemma 4.5] Let a multiplicative function $h: \mathbb{N} \to \mathbb{R}$ be defined for suitable n by*

$$h(n) := n \left(\frac{n}{\varphi(n)}\right)^2 \prod_{p|n} \frac{k-1}{p-k}.$$

Then we can rewrite the main term of the lower bound in the form

$$\sum_{d,e \in S_{(w,\infty)}} \mu(d)a_d\mu(e)a_e\hat\rho([d,e]) = \sum_{d \in S_{(w,\infty)}} \frac{h(d)}{d} \left(\sum_{\substack{m \in S_{(w,\infty)} \\ (m,d)=1 \\ md \in \mathcal{X}}} \frac{1}{\varphi(m)} f'\left(\frac{\log(dm)}{\log R}\right) \right)^2.$$

Proof **(1)** Define a multiplicative function $\hat\xi$ and optimization variables z_d by

$$\hat\xi(d) := \hat\rho(d)\prod_{p|d}(1 - \hat\rho(p)) \text{ and } z_d := \sum_{de \in S_{(w,\infty)}} \hat\rho(e)\mu(de)a_{de},$$

respectively. Then diagonalizing as in Lemma 8.9 Step (3), we get

$$\sum_{d,e \in S_{(w,\infty)}} \mu(d)a_d\mu(e)a_e\hat\rho([d,e]) = \sum_{d \in S_{(w,\infty)}} \hat\xi(d)z_d^2.$$

(2) Next, expanding the definition of a_{de}, using the pairwise coprimality of d, e, q and interchanging the order of summation with $m = eq$, we get

$$z_d = \frac{\mu(d)}{\Phi(d)\rho(d)} \sum_{de \in S_{(w,\infty)}} \frac{\hat\rho(e)\mu(e)}{\Phi(e)\rho(e)} \sum_{deq \in \mathcal{X}} \frac{1}{\Phi(q)} f'\left(\frac{\log(deq)}{\log R}\right)$$

$$= \frac{\mu(d)}{\Phi(d)\rho(d)} \sum_{dm \in \mathcal{X}} \frac{1}{\Phi(m)} f'\left(\frac{\log dm}{\log R}\right) \sum_{eq=m} \frac{\hat\rho(e)\mu(e)}{\rho(e)}$$

$$= \frac{\mu(d)}{\Phi(d)\rho(d)} \sum_{dm \in \mathcal{X}} \frac{1}{\Phi(m)} f'\left(\frac{\log(dm)}{\log R}\right) \prod_{p|m}\left(1 - \frac{\hat\rho(p)}{\rho(p)}\right)$$

$$= \frac{\mu(d)}{\Phi(d)\rho(d)} \sum_{dm \in \mathcal{X}} \frac{1}{\Phi(m)} f'\left(\frac{\log(dm)}{\log R}\right) \frac{\Phi(m)}{\varphi(m)}.$$

Therefore, if we define

$$\gamma(d) = \frac{\hat\xi(d)}{\Phi(d)^2\rho(d)^2} = \frac{h(d)}{d},$$

we can write

$$\sum_{d,e \in S_{(w,\infty)}} \mu(d)a_d\mu(e)a_e\hat\rho([d,e]) = \sum_{d \in S_{(w,\infty)}} \gamma(d)\left(\sum_{md \in \mathcal{X}} \frac{1}{\varphi(m)} f'\left(\frac{\log(dm)}{\log R}\right)\right)^2.$$

This completes the proof. $\qquad\qquad\qquad\qquad\qquad\qquad\qquad\qquad\qquad\square$

At last, we are in a position to define the parameter β and show that the sum in the main term from Lemma 8.10 satisfies the inequality needed for Lemma 8.7.

In what follows in this chapter, we use the term **negligible error** to refer to any quantity which is $o(B^{k+1})$.

Recall that in Lemma 8.12 we have defined

$$h(n) := \frac{n^3}{\varphi(n)^2} \prod_{p|n} \frac{k-1}{p-k},$$

and we use this definition also in Lemma 8.13.

Lemma 8.13   *[174, lemma 4.7] Let $\mathcal{X} = \mathcal{S}_{(w,\infty)}$ and $\mathcal{Y} = \mathcal{X} \cap [1, WR^2]$. Then if*

$$\beta := \int_0^1 f(t)^2 \frac{t^{k-2}}{(k-2)!} \, dt,$$

we have, up to negligible errors, i.e., errors of order $o(B^{k+1})$,

$$\sum_{d\in\mathcal{S}_{(w,\infty)}} \frac{h(d)}{d} \left(\sum_{\substack{m\in\mathcal{S}_{(w,\infty)} \\ (m,d)=1 \\ md\in\mathcal{X}}} \frac{1}{\varphi(m)} f'\left(\frac{\log(md)}{\log R}\right) \right)^2 \geq \beta B^{k+1}. \qquad (8.19)$$

Proof   (1) Because $\mathcal{X} = \mathcal{S}_{(w,\infty)}$, the constraint in the inner sum $md \in \mathcal{X}$ means $m \in \mathcal{S}_{(w,\infty)}$ and $(m,d) = 1$. For fixed d, using Lemma 8.8 with $\beta(p) = p/(p-1) = 1 + O(1/p)$, we get a bound for the inner sum in the statement on the left, with an implied constant not dependent on d,

$$\left| \sum_{m\in\mathcal{S}_{(w,\infty)}} \frac{1}{\varphi(m)} f'\left(\frac{\log(dm)}{\log R}\right) \right| \ll B. \qquad (8.20)$$

We also have, uniformly for all d since $y \to f'(y_0 + y)$ is equicontinuous as y_0 varies,

$$\sum_{m\in\mathcal{S}_{(w,\infty)}} \frac{1}{\varphi(m)} f'\left(\frac{\log(md)}{\log R}\right) = B\left(\int_0^\infty f'\left(t + \frac{\log d}{\log R}\right) dt + o(1) \right)$$

$$= -B\left(f\left(\frac{\log d}{\log R}\right) + o(1) \right).$$

Next, note that for any prime $p \mid d$, we have

$$\sum_{\substack{m \in \mathcal{S}_{(w,\infty)} \\ p|m}} \frac{1}{\varphi(m)} f'\left(\frac{\log(md)}{\log R}\right) \leq \frac{1}{\varphi(p)} \sum_{m \in \mathcal{S}_{(w,\infty)}} \frac{1}{\varphi(m)} f'\left(\frac{\log(mdp)}{\log R}\right)$$

$$= -\frac{B}{\varphi(p)} \left(f\left(\frac{\log(pd)}{\log R}\right) + o(1) \right) \ll \frac{B}{p},$$

so uniformly in d

$$\left| \sum_{\substack{m \in \mathcal{S}_{(w,\infty)} \\ (d,m) \neq 1}} \frac{1}{\varphi(m)} f'\left(\frac{\log(md)}{\log R}\right) \right| \ll B \sum_{p \leq R} \frac{1}{p} \ll B \log\log R.$$

(2) Now take the left-hand side of the inequality of the lemma, expand the square of the form we have derived in Step (1) and sum over d to get, since f is bounded,

$$B^2 \left(\sum_{d \in \mathcal{S}_{(w,\infty)}} \frac{h(d)}{d} f\left(\frac{\log d}{\log R}\right)^2 + o(1) \times \sum_{\substack{d \in \mathcal{S}_{(w,\infty)} \\ d \leq R}} \frac{h(d)B^2}{d} \right).$$

Up to negligible errors, because $h(p)/p = k - 1 + O(1/p)$ and f has support in $[0, 1]$, by Lemma 8.12 the main term is

$$B^{k+1} \int_0^1 f(t)^2 \frac{t^{k-2}}{(k-2)!} \, dt.$$

In addition, Lemma 8.8 applied again with $g = \chi_{[0,1]}$ gives

$$\sum_{\substack{n \in \mathcal{S}_{(w,\infty)} \\ n \leq R}} \frac{h(n)}{n} = o\left(B^{k-1}\right), \tag{8.21}$$

so the error is negligible, and the proof is complete. $\qquad\qquad\square$

Remark: A comment on the choice of weight function $f(t)$ as defined in the statement of Lemma 8.14: it is based on the Bessel function $J_k(x)$. That this might be so was first suggested by Brian Conrey in unpublished calculations. A more complete development was presented by Farkas, Pintz and Révész in [48] using the calculus of variations. This rather remarkable derivation is set out in Appendices A and B. The reader may wish to consult these appendices, where complete proofs are given of the Bessel function identities and optimization results using the spectral theory of compact operators on a Hilbert space which are needed to derive the form. There is also a plot of the explicit weight function $f(t)$, which is used in Lemma 8.14, for several values of $k \geq 2$.

First recall the definition of Bessel functions of the first class. If $k \geq 2$ and $k \in \mathbb{Z}$, then

$$J_k(x) := \sum_{n=0}^{\infty} \frac{(-1)^n}{n!\,(n+k)!} \left(\frac{x}{2}\right)^{2n+k}.$$

In Lemma 8.14, the weight function $f(t)$ is defined explicitly in terms of the Bessel function, and the properties of the $f(t)$ set out and proved. Significant use is made of results from the appendices.

Lemma 8.14 **(The weight function)** *[174, lemma 4.10] Let the weight function $f : [0,\infty) \to [0,\infty)$ be defined to be zero outside of $[0,1]$ and on that interval defined by*

$$f(t) := t^{1-k/2} J_{k-2}\left(j_{k-2}\sqrt{t}\right), \tag{8.22}$$

where j_k is the first strictly positive zero of $J_k(x)$.

(i) *The function f is nonnegative and $\log f$ is concave downwards on $[0,1)$.*
(ii) *For all $r,s \geq 0$, we have $f(r+s) \leq f(r)f(s)/f(0)$.*
(iii) *If $\eta := j_{k-2}^2/(4(k-1))$, then $\eta > 0$ and for all $r,s \geq 0$, we have $f(r+s) \leq \exp(-\eta s)f(r)$.*
(iv) *f is nonincreasing.*

Proof (i) By Lemma A.1 and Lemma A.4, we have

$$-f(t)^2(\log f(t))'' = f'(t)^2 - f(t)f''(t)$$

$$= \frac{j_{k-2}^2 t^{1-k}}{4}\left(J_{k-1}(j_{k-2}\sqrt{t})^2 - J_{k-2}(j_{k-2}\sqrt{t})J_k(j_{k-2}\sqrt{t})\right)$$

$$= \frac{j_{k-2}^2 t^{-k}}{4}\int_0^t J_{k-1}(j_{k-2}\sqrt{u})^2\,du \geq 0.$$

Therefore, $(\log f(t))'$ is nonincreasing on $[0,1)$ and (i) follows.

(ii) Let $g(x)$ be a suitable function with $g''(x) \leq 0$ on (a,b). Consider the two-term Taylor expansion for $g(x)$ about y. We have for a point ξ which lies between x and x_0 and use $g''(x) \leq 0$ on (a,b) to get

$$g(x) = g(y) + g'(y)(x-y) + \frac{1}{2}g''(\xi)(x-y)^2 \leq g(y) + g'(y)(x-y).$$

Let $y = \alpha u + (1-\alpha)v$ and $x = u$ so

$$g(u) \leq g(y) + g'(y)(u - \alpha u - (1-\alpha)v) = g(y) + (1-\alpha)g'(y)(u-v).$$

Similarly, if we let $x = v$, with as before $y = \alpha u + (1-\alpha)v$, we get

$$g(v) \leq g(y) + \alpha g'(y)(v-u).$$

Multiplying the first inequality by α and the second by $(1 - \alpha)$ gives

$$\alpha g(u) + (1 - \alpha)g(v) \le g(y) = g(\alpha u + (1 - \alpha)v).$$

Now let $g(x) := \log f(x)$ so $g''(x) \le 0$ and by (i) we have

$$\alpha g(u) + (1 - \alpha)g(v) \le g(\alpha u + (1 - \alpha)v), \quad u, v \in [0, b).$$

Let $u = 0$, $v = x + y$, $\alpha = x/(x + y)$ so

$$\frac{x}{x + y}g(0) + \frac{y}{x + y}g(x + y) \le g\left(\frac{x}{x + y} \times 0 + \frac{y}{x + y}(x + y)\right) = g(y).$$

Similarly,

$$\frac{y}{x + y}g(0) + \frac{x}{x + y}g(x + y) \le g\left(\frac{y}{x + y} \times 0 + \frac{x}{x + y}(x + y)\right) = g(x).$$

Adding gives $g(0) + g(x + y) \le g(x) + g(y)$, so

$$\log f(x + y) \le \log(f(x) \cdot f(y)) - \log f(0) \implies f(x + y) \le \frac{f(x)f(y)}{f(0)},$$

which gives (ii).

(iii) The Taylor expansion of $J_k(x)$ about $x = 0$ is

$$J_k(x) = \frac{1}{k!}\frac{x^k}{2^k} + O(x^{k+1}) \implies \frac{f'(0)}{f(0)} = -\frac{j_{k-2}^2}{4(k - 1)} =: -\eta.$$

By (i), we have $f'(t)^2 - f(t)f''(t) \ge 0$ so $(f'(t)/f(t))' \le 0$. Therefore, for $0 \le t < 1$ we get

$$\frac{f'(t)}{f(t)} \le \frac{f'(0)}{f(0)} = -\eta.$$

Hence

$$\log f(u) - \log f(0) = \int_0^u \frac{f'(t)}{f(t)} \, dt \le -\eta(u - 0) = -\eta u.$$

Using (ii),

$$\log\left(\frac{f(t + u)}{f(t)}\right) \le \log\left(\frac{f(u)}{f(0)}\right) \le -\eta u \implies f(t + u) \le f(t)\exp(-\eta u),$$

which gives (iii).

(iv) This follows from (iii). $\qquad\qquad\qquad\qquad\qquad\qquad\qquad\qquad\qquad\square$

Now that $f(t)$ has been specified, we can consider the definitions of both α and β from Lemmas 8.9 and 8.13 respectively, and derive a useful expression for α/β in terms of k and the first positive Bessel function zero for J_{k-2}. Note that we have assumed $k \geq 2$ (and will be quite a lot larger than 2).

Lemma 8.15   *[174, page 52] If $f(t) := t^{1-k/2} J_{k-2}\left(j_{k-2}\sqrt{t}\right)$ and*

$$\alpha := \int_0^1 f'(t)^2 \frac{t^{k-1}}{(k-1)!} \, dt,$$

$$\beta := \int_0^1 f(t)^2 \frac{t^{k-2}}{(k-2)!} \, dt,$$

then we have

$$\frac{\alpha}{\beta} = \frac{j_{k-2}^2}{4(k-1)}.$$

Proof   By Lemma A.1, we can write

$$f'(t) = -\frac{j_{k-2} t^{\frac{1}{2}-k/2}}{2} J_{k-1}\left(j_{k-2}\sqrt{t}\right). \tag{8.23}$$

We also have $J_{k-2}(j_{k-2}) = 0$ and, using Lemma A.2, $J_k(x) = (x/(2k))(J_{k-1}(x) + J_{k+1}(x))$, we get $J_{k-3}(j_{k-2}) = -J_{k-1}(j_{k-2})$. These give, using Lemma A.3,

$$\alpha = \frac{j_{k-2}^2}{4(k-1)!} J_{k-1}(j_{k-2})^2,$$

and

$$\beta = \frac{1}{(k-2)!} J_{k-1}(j_{k-2})^2.$$

Therefore, for this choice of $f(t)$, it follows that

$$\frac{\alpha}{\beta} = \frac{j_{k-2}^2}{4(k-1)},$$

which completes the proof.   □

Note that the following lemma has, on the face of it, stronger hypotheses than the corresponding Lemma 4.11 in [174].

Lemma 8.16   *[174, lemma 4.11] Let $g: \mathbb{R} \to \mathbb{R}$ have compact support in $(0,1)$ and be smooth on $\mathbb{R}$. Let $R = x^c$ for some fixed c with $0 < c < 1$. Then*

$$\sum_p \left(\frac{1}{p}\right) g\left(\frac{\log p}{\log R}\right) = \sum_p \left(\frac{1}{p}\right) g\left(\frac{\log p}{c \log x}\right) = \int_0^\infty \frac{g(t)}{t} + o(1).$$

Proof **(1)** This follows from the general form of Abel's summation theorem [4, theorem 4.2] with Mertens' theorem [4, theorem 4.10]. Let as $x \to \infty$,

$$A(x) := \sum_{p \leq x} \frac{\log p}{p} = \log x + O(1) \text{ and } f(t) := g\left(\frac{\log t}{c \log x}\right) / (\log t), \ t \geq 1.$$

We have

$$\sum_{p \in \mathbb{P}} \frac{1}{p} g\left(\frac{\log p}{c \log x}\right) = \sum_{1 < p \leq x^c} \frac{1}{p} g\left(\frac{\log p}{c \log x}\right)$$

$$= A(x^c) f(x^c) - A(1) f(1) - \int_1^{x^c} A(t) f'(t) \, dt$$

$$= 0 + \int_1^{x^c} (\log t + O(1)) \left(\frac{g\left(\frac{\log t}{c \log x}\right)}{t \log^2 t} - \frac{g'\left(\frac{(\log t)}{c \log x}\right)}{c(\log x) t (\log t)} \right) dt.$$

(2) To estimate this expression, let

$$s = \frac{\log t}{c \log x} \implies \frac{ds}{s} = \frac{dt}{t \log t} \text{ and } \frac{ds}{(c \log x) s^2} = \frac{dt}{t \log^2 t}$$

and let $t = 1, x^c$ correspond to $s = 0, 1$. We **claim** using $g(s) = g'(s) = O(1)$ and $g(0) = g'(0) = g(1) = g'(1) = 0$ the sum can be written

$$\int_0^1 \frac{g(s)}{s} \, ds + O(1/\log x) = \int_0^\infty \frac{g(s)}{s} \, ds + o(1).$$

To see this, expand the expression we derived in Step (1):

$$\int_1^{x^c} \frac{g\left(\frac{\log t}{c \log x}\right)}{t (\log t)} \, dt = \int_0^1 \frac{g(s)}{s} \, ds,$$

$$O(1) \int_1^{x^c} f'(t) dt = O(1)(f(x^c) - f(1)) = 0,$$

$$\int_1^{x^c} \frac{g'\left(\frac{\log t}{c \log x}\right)}{t (c \log x)} \, dt \leq \frac{1}{c(\log x)} \int_1^{x^c} \frac{g'\left(\frac{\log t}{c \log x}\right)}{} dt \ll \frac{1}{\log x}.$$

This verifies the claim and completes the proof. □

Recall some definitions:

$$B := \frac{\varphi(W)}{W} \log R,$$

$$h(n) := \frac{n^3}{\varphi(n)^2} \prod_{p|n} \frac{k-1}{p-k}.$$

Also recall the so-called Motohashi–Pintz–Zhang property for positive real numbers ϖ and δ, which means the following. Let $P_\Omega := \prod_{p \in \Omega} p$. We say MPZ$[\varpi, \delta]$ is true if $\Omega \subset [1, x^\delta]$ is a subinterval, S_Ω is the set of all squarefree numbers which are products of primes in Ω, $Q \prec^\circ x^{\frac{1}{2} + 2\varpi}$, $a \bmod P_\Omega$ is a primitive residue class and $A \geq 1$ is a given real number, then with an implied constant depending only on A, ϖ and δ, we have

$$
\sum_{\substack{q \leq Q \\ q \in S_\Omega}} \left| \sum_{\substack{x \leq n < 2x \\ n \equiv a \bmod q}} \Lambda(n) - \frac{1}{\varphi(q)} \sum_{\substack{x \leq n < 2x \\ (n, q) = 1}} \Lambda(n) \right| \ll \frac{x}{L_x(A)}. \tag{8.24}
$$

Finally, DHL$[k, 2]$, the so-called Dickson–Hardy–Littlewood property for $k \geq 2$, is true if for any admissible k-tuple $\mathcal{H} = (h_1, \ldots, h_k)$, there exist infinitely many $n \in \mathbb{N}$ such that the translates $n + \mathcal{H}$ contain at least two primes.

With these central concepts at hand, we can now state and prove a parameterized form of the key inequality of Lemma 8.7.

Theorem 8.17   (MPZ implies DHL) *[174, theorem 2.13] Given an integer $k \geq 2$, a real ϖ with $0 < \varpi < \frac{1}{4}$ and a real δ with $0 < \delta < \frac{1}{4} + \varpi$, such that*

$$
\kappa_1 := \int_\theta^1 (1 - t)^{(k-1)/2} \frac{dt}{t},
$$

$$
\kappa_2 := (k - 1) \int_\theta^1 (1 - t)^{k-1} \frac{dt}{t}, \text{ where}
$$

$$
\theta := \frac{\delta}{\frac{1}{4} + \varpi},
$$

and where, as in Lemma 8.14, $f(t) := t^{1-k/2} J_{k-2}(j_{k-2}\sqrt{t})$ for $0 \leq t \leq 1$ and 0 elsewhere, and such that we have

$$
(1 + 4\varpi)(1 - 2\kappa_1 - 2\kappa_2) > \frac{j_{k-2}^2}{k(k - 1)}, \tag{8.25}
$$

and MPZ$[\varpi, \delta]$ is true, so is DHL$[k, 2]$.

Proof   First, we give an outline of the proof, and then the details follow in eight steps. Good use is made of Lemma 8.11 (three times) and Lemma 8.16 (twice). Let $\mathcal{X} = S_{(w, x^\delta]}$ be the set of squarefree integers that are x^δ-smooth and have no prime factor less than or equal to w. Define the subset $\mathcal{Y}$ by

$$
\mathcal{Y} = S_{[1, x^\delta]} \cap [1, WR^2],
$$

which is the set of all x^δ-smooth squarefree numbers in the range $[1, WR^2]$. Then $\mathcal{Y}$ satisfies the preliminary conditions for Lemma 8.11. The given assumption

MPZ$[\varpi, \delta]$, the setting $R = x^{\frac{1}{4}+\varpi}$ and the fact that $a_d = 0$ for $d > R$ will enable us to apply Lemma 8.11.

Using that lemma, for this choice of sets $\mathcal{X}$ and hence $\mathcal{Y}$, we are able to derive the lower bound

$$\sum_{n \in \mathcal{S}_{(w,\infty)}} \frac{h(n)}{n} \left(\sum_{\substack{m \in \mathcal{S}_{(w,\infty)} \\ (m,n)=1 \\ mn \in \mathcal{X}}} \frac{1}{\varphi(m)} f' \left(\frac{\log mn}{\log R} \right) \right)^2 \geq \beta B^{k+1} \qquad (8.26)$$

up to terms of order

$$o(B^{k+1}) = o \left(\frac{\varphi(W) \log R}{W} \right)^{k+1} = O \left((\log x)^{O(1)} \right).$$

Then, we show β satisfies the key condition of Lemma 8.7, namely

$$\beta k \log R > \alpha \log x,$$

with the explicit $f(t)$ from Lemma 8.14 and

$$\alpha = \int_0^1 f'(t)^2 \frac{t^{k-1}}{(k-1)!} \, dt,$$

the final implication is then verified using the assumption

$$(1 + 4\varpi)(1 - 2\kappa_1 - 2\kappa_2) > \frac{j_{k-2}^2}{k(k-1)}.$$

We now commence the proof proper.

(1) First we define for $n \in \mathcal{S}_{(w,\infty)}$

$$F(n) := - \sum_{\substack{m \in \mathcal{S}_{(w,\infty)} \\ (m,n)=1}} \frac{1}{\varphi(m)} f' \left(\frac{\log(mn)}{\log R} \right) \qquad (8.27)$$

and

$$\widetilde{F}(n) := - \sum_{\substack{m \in \mathcal{S}_{(w,\infty)} \\ (m,n)=1 \\ mn \in \mathcal{X}}} \frac{1}{\varphi(m)} f' \left(\frac{\log mn}{\log R} \right). \qquad (8.28)$$

Then since $f'(t) \leq 0$, we have $0 \leq \widetilde{F}(n) \leq F(n)$. Thus the left-hand side of the inequality (8.26) can be written

$$\sum_{n \in \mathcal{S}_{(w, \infty)}} \frac{h(n)\widetilde{F}(n)^2}{n}.$$

Rearranging $(\widetilde{F}(n) - F(n))^2 \geq 0$, we get

$$\widetilde{F}(n)^2 \geq F(n)^2 - 2F(n)(F(n) - \widetilde{F}(n)).$$

Therefore,

$$\sum_{n \in \mathcal{S}_{(w, \infty)}} \frac{h(n)\widetilde{F}(n)^2}{n} \geq \sum_{n \in \mathcal{S}_{(w, \infty)}} \frac{h(n)F(n)^2}{n} - 2\sum_{n \in \mathcal{S}_{(w, \infty)}} \frac{h(n)F(n)}{n}(F(n) - \widetilde{F}(n)).$$

$$(8.29)$$

In Lemma 8.13, we saw that if

$$\beta_0 := \int_0^1 f(t)^2 \frac{t^{k-2}}{(k-2)!}\,dt,$$ $$(8.30)$$

then up to negligible errors we have

$$\sum_{n \in \mathcal{S}_{(w, \infty)}} \frac{h(n)F(n)^2}{n} \geq \beta_0 B^{k+1},$$ $$(8.31)$$

which is a lower bound for the leading term in the estimate (8.29).

(2) Next consider the second term on the right in (8.29). Let

$$T := 2 \sum_{d \in \mathcal{S}_{(w, \infty)}} \frac{h(n)F(n)}{n}(F(n) - \widetilde{F}(n)).$$

The proof of Lemma 8.13 gives the representation

$$F(n) = B\left(f\left(\frac{\log n}{\log R}\right) + o(1)\right)$$

with as before recalling $f(t)$ is fixed

$$0 \leq F(n) \ll B.$$

Therefore, if we set

$$T_1 := 2B \sum_{n \in \mathcal{S}_{(w, \infty)}} \frac{h(n)}{n}(F(n) - \widetilde{F}(n))f\left(\frac{\log n}{\log R}\right),$$

$$T_2 := 2B \sum_{n \in \mathcal{S}_{(w, \infty)}} \frac{h(n)}{n}(F(n) - \widetilde{F}(n))\epsilon(n),$$

we get $T = T_1 + T_2$. We now proceed to derive upper bounds for each of these quantities in Steps (3) through (7).

(3) For T_2, because $0 \leq F(n) - \widetilde{F}(n) \leq F(n)$ using the estimate (8.21), we get

$$|T_2| \leq 2B \sum_{n \in \mathcal{S}_{(w,\infty)}} \frac{h(n)}{n} F(n) \epsilon(n),$$

$$\ll B^2 \sum_{n \in \mathcal{S}_{(w,\infty)}} \frac{h(n)}{n} o(1),$$

$$= o\left(B^{k+1}\right),$$

which is negligible.

(4) For T_1, recalling $\mathcal{X} = \mathcal{S}_{(w,x^\delta]}$, we have

$$F(n) - \widetilde{F}(n) = -\sum_{\substack{m \in \mathcal{S}_{(w,\infty)} \\ (m,n)=1 \\ mn \notin \mathcal{X}}} \frac{1}{\varphi(m)} f'\left(\frac{\log mn}{\log R}\right). \tag{8.32}$$

Because $\mathcal{X}$ is a set of x^δ-smooth squarefree numbers, if $m, n \in \mathcal{S}_{(w,\infty)}$, then $mn \notin \mathcal{X}$ only if at least one of m, n is divisible by a prime $p > x^\delta$. Because also $f(t) \geq 0$ and $-f'(t) \geq 0$ we are able to find an upper bound for $|T_1|$ by considering the different possible cases for m and n which occur in (8.32) and adding the related upper bounds.

(5) First consider the contribution, denoted T_{11}, to T_1 of the (m,n), where m is divisible by a prime $p > x^\delta$. We get

$$T_{11} \leq 2B$$

$$\times \sum_{x^\delta < p \leq R} \left(\sum_{n \in \mathcal{S}_{(w,\infty)}} \frac{h(n)}{n} f\left(\frac{\log n}{\log R}\right) \times \left(-\sum_{m \in \mathcal{S}_{(w,\infty)}} \frac{1}{\varphi(m)\varphi(p)} f'\left(\frac{\log(pmn)}{\log R}\right) \right) \right).$$

Next, define

$$G(r,s) := \int_0^1 f(t+r) f(t+s) \frac{t^{k-2}}{(k-2)!} \, dt. \tag{8.33}$$

Using Lemma 8.8 in the sum over m in this bound, the function G and Lemma 8.16 in the final step gives, apart from negligible quantities,

$$T_{11} \le 2B^2 \sum_{x^\delta < p \le R} \frac{1}{\varphi(p)} \left(\sum_{n \in \mathcal{S}_{(w,\infty)}} \frac{h(n)}{n} f\left(\frac{\log n}{\log R}\right) \times f\left(\frac{\log n}{\log R} + \frac{\log p}{\log R}\right) \right)$$

$$\le 2B^{k+1} \sum_{x^\delta < p \le R} \frac{1}{\varphi(p)} G\left(0, \frac{\log p}{\log R}\right)$$

$$\le 2B^{k+1} \int_\theta^1 \frac{G(0,t)}{t}\, dt,$$

where, as defined before

$$\theta := \frac{\delta}{\frac{1}{4} + \varpi} = \frac{\log(x^\delta)}{\log R}.$$

(6) Now we consider the contribution, denoted T_{12}, to T_1 of the (m,n), where n is divisible by a prime $p > x^\delta$. We get

$$T_{12} \le 2B \sum_{x^\delta < p \le R} \frac{h(p)}{p} \left(\sum_{n \in \mathcal{S}_{(w,\infty)}} \frac{h(n)}{n} f\left(\frac{\log(pn)}{\log R}\right) F(pn) \right).$$

Because $-f'(t) \ge 0$, we can estimate the sum (8.27) by relaxing the constraint $(m,np) = 1$ in the sum over m, and then applying the workhorse Lemma 8.8. This gives

$$F(pn) \le B \left(f\left(\frac{\log pn}{\log R}\right) + o(1) \right).$$

Therefore, using Lemma 8.8 yet again and omitting negligible quantities,

$$T_{12} \le 2B^2 \sum_{x^\delta < p \le R} \frac{h(p)}{p} \left(\sum_{n \in \mathcal{S}_{(w,\infty)}} \frac{h(n)}{n} f\left(\frac{\log(pn)}{\log R}\right)^2 \right)$$

$$\le 2B^{k+1} \sum_{x^\delta < p \le R} \frac{h(p)}{p} G\left(\frac{\log p}{\log R}, \frac{\log p}{\log R}\right).$$

Using $h(p) = k - 1 + O(1/p)$ from Step (2) of Lemma 8.8, and applying Lemma 8.16 to the sum over p, we can write finally in this step

$$T_{12} \le 2(k-1)B^{k+1} \int_\theta^1 \frac{G(t,t)}{t}\, dt.$$

(7) Combining the results of Steps (5) and (6), we get

$$T \leq 2B^{k+1} \left(\int_\theta^1 \frac{G(0,t)}{t} \, dt + (k-1) \int_\theta^1 \frac{G(t,t)}{t} \, dt \right).$$

Applying this to the results of Steps (1) and (2), if we define

$$\beta := \beta_0 - 2 \int_\theta^1 \frac{G(0,t)}{t} \, dt - 2(k-1) \int_\theta^1 \frac{G(t,t)}{t} \, dt,$$

this implies, up to negligible quantities, we have

$$\sum_{n \in S_{(w,\infty)}} \frac{h(d)}{d} \widetilde{F}(n)^2 \geq \beta B^{k+1}.$$

Therefore, considering Lemmas 8.10 and 8.12 we get

$$\sum_{\substack{x \leq n < 2x \\ n \equiv b \bmod W}} v(n) \Lambda(n + h_i) \geq \frac{x}{\varphi(W)} \left(\sum_{n \in S_{(w,\infty)}} \frac{h(d)}{d} \widetilde{F}(n)^2 \right) \geq \frac{\beta x B^{k+1}}{\varphi(W)} = \frac{\beta B^k}{W} x \log R.$$

(8) In this final step, we simplify the expression defining β, given in Step (7), and introduce some key parameters. First note that (8.30) gives

$$\beta_0 = \int_0^1 f(t)^2 \frac{t^{k-2}}{(k-2)!} \, dt \implies \beta_0 = G(0,0),$$

where as indicated before we have chosen $f(t) := t^{1-k/2} J_{k-2}(j_{k-2}\sqrt{t})$ for $0 \leq t \leq 1$ and 0 elsewhere.

Next we **claim** for $r \in [0,1]$ and $s \in [0,1)$, we have $f(r+s) \leq f(r/(1-s))$. To see this, note that

$$\frac{r}{1-s} \leq r+s \iff r+s \leq 1.$$

Hence, because $f(r+s) = 0$ for $r+s > 1$, and by Lemma 8.14, $f(t)$ is decreasing and nonnegative, we must have $f(r+s) \leq f(r/(1-s))$ for r, s in the given ranges. This implies for $0 \leq t \leq 1$,

$$G(t,t) = \int_0^1 f(t+u)^2 \frac{u^{k-2}}{(k-2)!} \, du$$

$$\leq \int_0^1 f(u/(1-t))^2 \frac{u^{k-2}}{(k-2)!} \, du$$

$$= (1-t)^{k-1} \int_0^{1/(1-t)} f(v)^2 \frac{v^{k-2}}{(k-2)!} \, dv$$

$$\leq (1-t)^{k-1} G(0,0).$$

Using the Cauchy–Schwarz inequality applied to the integral definition of G and using this bound, we get

$$G(0,t) \leq \sqrt{G(0,0)}\sqrt{G(t,t)} \leq (1-t)^{(k-1)/2}G(0,0).$$

Applying these bounds to the expression for β we have defined in Step (7), together with the definitions

$$\kappa_1 := \int_\theta^1 \frac{(1-t)^{(k-1)/2}}{t}\,dt \geq \beta_0 \int_\theta^1 \frac{G(0,t)}{t}\,dt,$$

$$\kappa_2 := (k-1)\int_\theta^1 \frac{(1-t)^{k-1}}{t}\,dt \geq \beta_0(k-1)\int_\theta^1 \frac{G(t,t)}{t}\,dt$$

gives

$$\beta \geq \beta_0(1 - 2\kappa_1 - 2\kappa_2). \tag{8.34}$$

This with the assumption inequality (8.25), which is

$$(1+4\varpi)(1 - 2\kappa_1 - 2\kappa_2) > \frac{j_{k-2}^2}{k(k-1)},$$

together with $\beta_0 = G(0,0)$ derived previously, and the result of Lemma 8.15, namely

$$\frac{j_{k-2}^2}{k(k-1)} = \frac{4\alpha}{k\beta_0},$$

gives

$$(1+4\varpi)(1 - 2\kappa_1 - 2\kappa_2) > \frac{4\alpha}{k\beta_0},$$

which in turn, using the inequality (8.34), implies the key inequality

$$\frac{1}{4} + \varpi = \frac{\log R}{\log x} > \frac{\alpha}{k\beta}.$$

Therefore, by Lemma 8.7 we have DHL[k, 2], which completes the proof. □

Finally, in this part we use Zhang's theorem, Theorem 5.8, to derive a prime gap result.

Corollary 8.18 *[174, theorem 2.13 corollary] Zhang's work implies* MPZ[ϖ, δ] *for* $1168 = 1/\varpi = 1/\delta < 0.0008562$. *By Theorem 8.17, we first compute the smallest k for which the inequality of the theorem is satisfied with this value of ϖ, namely $k = 34429$. This gives a prime gap size using the admissible prime k-tuple for that value of k, which has diameter 420878. To check this computation, the reader could use the* **PGpack** *functions* SmallestKPolymath0, PrimeTuple *and* TupleDiameter.

Table 8.1 gives the best possible (i.e., the smallest) value of k, namely k^*, for which the inequality of Theorem 8.17 is satisfied with the small parameters κ_i set to 0. This is for a range of values of ϖ so that we see variations around the value $k* = 1783$. This value was chosen because, looking ahead to Theorem 8.34, $k = 1783$ was the best value for k (corresponding to a prime gap size of 14950) obtained by Polymath8a without using multidimensional exponential sum bounds. The reduction from 420878 to 14950 more than justifies the introduction the new ideas which made it possible.

Table 8.1 *The best value of k setting $\kappa_i = 0$ for $0.00585 \leq \varpi \leq 0.00605$.*

ϖ	$k*$
0.00585	1830
0.00586	1825
0.00587	1820
0.00588	1815
0.00589	1811
0.00590	1806
0.00591	1801
0.00592	1796
0.00593	1792
0.00594	1787
0.00595	1783
0.00596	1778
0.00597	1773
0.00598	1769
0.00599	1764
0.00600	1760
0.00601	1755
0.00602	1751
0.00603	1746
0.00604	1742
0.00605	1737

This completes Part I of the chapter. In Part II, we use the same overall approach as in Part I, but replace Zhang's Bombieri–Vinogradov variation with one based on multiple dense divisibility, but only use the classes $\mathscr{D}(i, y)$ for $i \leq 2$. To do this, we need to strengthen some of the ideas from Part I and some new ideas, in particular MPZ becomes MPZ$^{(i)}$, and there will be a large section on exponential sums.

8.7 A Fundamental Technical Result

We call the following lemma "technical" because of its intense formulaic and analytic nature. For background on the Bessel function J_k, see Appendix A.

Recall the definitions $W = W(x) := \prod_{p \le w} p$, $\mathcal{X} \subset (w, \infty)$, and

$$\Phi(d) := \prod_{p|d} \frac{p - k}{k},$$

$$\rho(d) := \prod_{p|d} \frac{k}{p},$$

$$a_d := \frac{1}{\rho(d)\Phi(d)} \sum_{\substack{q \in \mathcal{S}_{(w,\infty)} \\ (q,d)=1 \\ qd \in \mathcal{X}}} \frac{1}{\Phi(q)} f'\left(\frac{\log dq}{\log R}\right),$$

$$h(n) := \frac{n^3}{\varphi(n)^2} \prod_{p|n} \frac{k - 1}{p - k}.$$

We also use the Dickson–Hardy–Littlewood property and dense divisibility. The latter concept was developed near the start of the chapter in Section 8.5, and goes to the heart of Polymath8a's significant improvements to Zhang.

We say $n \in \mathcal{D}_{\Omega}(i, x^{\delta})$ if n is a positive integer which is i-tuply x^{δ}-densely divisible and all prime divisors of n are in Ω.

Definition 8.11 (Densely divisible MPZ property) *Let $i \ge 0$ be an integer. We say* MPZ$^{(i)}[\varpi, \delta]$ *is true if for any bounded subset $\Omega \subset \mathbb{R}$ which may depend on x, and $Q \prec^{o} x^{\frac{1}{2}+2\varpi}$, if $(a, P_{\Omega}) = 1$ and $A \ge 1$ is a fixed real number, then*

$$\sum_{\substack{q \le Q \\ q \in \mathcal{D}_{\Omega}(i, x^{\delta})}} \left| \sum_{\substack{x \le n < 2x \\ n \equiv a \bmod q}} \Lambda(n) - \frac{1}{\varphi(q)} \sum_{\substack{x \le n < 2x \\ (n,q)=1}} \Lambda(n) \right| \ll \frac{x}{(\log x)^A}.$$

The parameters in Theorem 8.19 are complicated. However, the cost over Theorem 8.17 is mainly the additional small parameter κ_3 and its supports A, θ, ω and δ'. The dense divisibility level i only occurs in the definition of ω.

Terry Tao in his blog post "A Truncated Elementary Selberg Sieve of Pintz", gives some insight into the meaning of the new use of δ and the new parameter δ', acknowledging Janos Pintz as the discoverer of the idea. We paraphrase Tao's comments:

The new sieve decouples Zhang's smoothness parameter δ into two parameters, δ which now measures dense divisibility as used in Elliott–Halberstam (or Bombieri–Vinogradov) estimates, but also δ' which also measures smoothness, but is allowed to be significantly larger than δ. This enables the $\kappa_1, \kappa_2, \kappa_3$ losses (see the left hand side of the bound (8.35)) to be almost completely eliminated (say by having δ' close to $\frac{1}{4} + \varpi$), whereas the Elliott–Halberstam part of the argument is subject only to the value of δ (which thus can be set in the unconditionally true range). With a larger left-hand side for (8.35) we can choose a smaller value of k on the right.

The proof of Theorem 8.19 takes ten steps and is an essential part of the Polymath8a prime gaps result. We still need a sufficient condition for MPZ$^{(i)}$, which is developed later in the chapter. The result of the theorem is an inequality with the same structure as (8.25), but with the additional small parameter κ_3. This ends up making a significant difference on the size of the consecutive primes gap.

The parameter δ appears in Zhang's approach in the bound for the smoothness of the modulus x^δ and satisfies $0 < \delta < \frac{1}{4} + \varpi$. The parameter δ' with $0 < \delta \leq \delta' < \frac{1}{4} + \varpi$ was suggested to Polymath8a by Janos Pintz. The bound x^δ rather than smoothness of the modulus, now relates to its dense divisibility level and can be reasonably small. Smoothness is now constrained by $x^{\delta'}$, which can be significantly larger. By playing with these two parameters, we can reduce errors in the sieve. See the Polymath web pages under the heading "A Truncated Selberg Sieve of Pintz" and the blog posts linked to this write-up for a fuller explanation.

Theorem 8.19 (MPZ$^{(i)}$ implies DHL) *[174, theorem 2.16] Let* $f(t) := t^{1-k/2}$ $J_{k-2}(j_{k-2}\sqrt{t})$ *for* $0 < t \leq 1$ *and* 0 *elsewhere. Given* $k \geq 2$ *and* $i \geq 1$, ϖ *with* $0 < \varpi < \frac{1}{4}$, δ, δ' *with* $0 < \delta \leq \delta' < \frac{1}{4} + \varpi$ *and* $A \geq 0$, *define parameters*

$$\kappa_1 := \int_\theta^1 (1-t)^{(k-1)/2} \frac{dt}{t},$$

$$\kappa_2 := (k-1) \int_\theta^1 (1-t)^{k-1} \frac{dt}{t},$$

$$\kappa_3 := \frac{\omega^2 \left(J_{k-2}(j_{k-2}\omega)^2 - J_{k-3}(j_{k-2}\omega) J_{k-1}(j_{k-2}\omega) \right)}{J_{k-1}(j_{k-2})^2}$$
$$\times \exp\left(A + (k-1) \int_\xi^\theta e^{-(A+2\eta)t} \frac{dt}{t} \right),$$

$$\eta := \frac{j_{k-2}^2}{4(k-1)},$$

$$\theta := \frac{\delta'}{\frac{1}{4} + \varpi},$$

$$\omega := \sqrt{\frac{(i\delta' - \delta)/2 + \varpi}{\frac{1}{4} + \varpi}},$$

$$\xi := \frac{\delta}{\frac{1}{4} + \varpi},$$

where J_k *is the Bessel function of the first kind of order* k *and* j_k *is its smallest positive zero. Then if*

$$(1 + 4\varpi)(1 - 2\kappa_1 - 2\kappa_2 - 2\kappa_3) > \frac{j_{k-2}^2}{k(k-1)}, \tag{8.35}$$

and MPZ$^{(i)}[\varpi, \delta]$ *is true, so is* DHL$[k, 2]$.

Proof We begin by noting as before we use the weight function $f(t) = t^{1-k/2}$ $J_{k-2}(j_{k-2}\sqrt{t})$. To complete the proof, apply Lemma 8.7, which gives a sufficient condition for DHL[$k, 2$], namely $\log R/\log x > \alpha/(k\beta)$. The proof follows along the path taken by Theorem 8.17 and takes ten steps.

(1) Let $\epsilon > 0$ be small and define the subset $\mathcal{X} \subset \mathcal{S}_{(w, x^{\delta'})}$ by (recall $1 \leq i$ is a fixed integer and $\delta \leq \delta' < \frac{1}{4} + \varpi$)

$$\mathcal{X} := \mathcal{S}_{(w, x^{\delta'})} \cap \left\{ n : \prod_{\substack{p|n \\ p < x^{\delta}}} p \geq x^{(i\delta' - \delta)/2 + \varpi + \epsilon/2} \right\}. \qquad (8.36)$$

Recall $\mathscr{D}(i, x^{\delta})$ represents the set of i-tuply x^{δ}-densely divisible integers. Then we also define $\mathcal{Y}$ by

$$\mathcal{Y} := \left(\mathcal{S}_{[1, x^{\delta'})} \cap [1, WR^2] \cap \mathscr{D}(i, x^{\delta}) \right) \cup [1, Wx^{\frac{1}{2} - \epsilon}]$$

and note that $\mathcal{Y} \subset \mathcal{S}_{\Omega}$, where $\Omega := [1, x^{\delta'}) \cup [1, Wx^{\frac{1}{2} - \epsilon}]$.

(2) Having defined these subsets, first, we **claim** that if $m, n \in \mathcal{S}_{(w, \infty)}$ are such that $a_m a_n \neq 0$ and we define $q = W[m, n]$, then $q \in \mathcal{Y}$. To see this, note that because $m \notin \mathcal{S}_{(w, x^{\delta'})}$ or $m > R$ implies $a_m = 0$, we have $m, n \leq R$ and $m, n \in \mathcal{S}_{(w, x^{\delta'})}$. Thus $q \in \mathcal{S}_{[1, x^{\delta'})} \cap [1, WR^2]$. Considering $\mathcal{Y}$, the proof of the claim will thus be complete if we can show $q > Wx^{\frac{1}{2} - \epsilon}$ implies $q \in \mathscr{D}(i, x^{\delta})$.

So, in the given context, let $q > Wx^{\frac{1}{2} - \epsilon}$. From the definition of a_m and a_n, and because the support of f is in $[0, 1]$, there are integers r, s such that

$$mr, ns \in \mathcal{X}, \text{ and } mr, ns \leq R = x^{\frac{1}{4} + \varpi}.$$

Thus

$$\prod_{\substack{p|mr \\ p < x^{\delta}}} p \geq x^{(i\delta' - \delta)/2 + \varpi + \epsilon/2} \implies r \prod_{\substack{p|m \\ p < x^{\delta}}} p \geq x^{(i\delta' - \delta)/2 + \varpi + \epsilon/2}.$$

Similarly,

$$s \prod_{\substack{p|n \\ p < x^{\delta}}} p \geq x^{(i\delta' - \delta)/2 + \varpi + \epsilon/2}.$$

The product of these two inequalities gives

$$rs(m, n) \prod_{\substack{p|[m, n] \\ p < x^{\delta}}} p \geq x^{i\delta' - \delta + 2\varpi + \epsilon}.$$

We also have

$$rsmn \leq R^2 = x^{\frac{1}{2} + 2\varpi},$$

so eliminating rs, we get

$$\frac{1}{[m,n]} \prod_{\substack{p|[m,n] \\ p<x^\delta}} p \geq x^{-\frac{1}{2}+i\delta'-\delta+\epsilon}.$$

But we have assumed $q = W[m,n] > Wx^{\frac{1}{2}-\epsilon}$, which implies $[m,n] > x^{\frac{1}{2}-\epsilon}$, so

$$\prod_{\substack{p|q \\ p\leq x^\delta}} p \geq \prod_{\substack{p|[m,n] \\ p\leq x^\delta}} p \geq x^{'-\delta} = \frac{x^{i\delta'}}{x^\delta}.$$

Because q is $x^{\delta'}$-smooth and squarefree, using the sufficient condition Lemma 8.5(v), we have $q \in \mathscr{D}(i, x^\delta)$, which completes the proof of the claim.

(3) We next check the prime distribution condition. By the Bombieri–Vinogradov theorem, which applies since $W \ll (\log\log x)^{O(1)}$, for all integers a with $(a, P_\Omega) = 1$, where Ω is defined in Step (1), we have

$$\sum_{q\leq Wx^{\frac{1}{2}-\epsilon}} \left| \sum_{\substack{x\leq n<2x \\ n\equiv a \bmod q}} \Lambda(n) - \frac{1}{\varphi(q)} \sum_{\substack{x\leq n<2x \\ (n,q)=1}} \Lambda(n) \right| \ll \frac{x}{(\log x)^A}.$$

We also have the assumption $\mathrm{MPZ}^{(i)}[\varpi, \delta]$. This implies

$$\sum_{\substack{q\leq WR^2 \\ q\in\mathscr{S}_{[1,x^{\delta'})}\cap\mathscr{D}(i,x^\delta)}} \left| \sum_{\substack{x\leq n<2x \\ n\equiv a \bmod q}} \Lambda(n) - \frac{1}{\varphi(q)} \sum_{\substack{x\leq n<2x \\ (n,q)=1}} \Lambda(n) \right| \ll \frac{x}{(\log x)^A}.$$

Combining these two estimates gives the sufficient prime distribution condition (8.18) of Lemma 8.11 for small errors in the prime sum of Lemma 8.10, which is

$$\sum_{q\in\mathcal{Y}} \left| \sum_{\substack{x\leq n<2x \\ n\equiv a \bmod q}} \Lambda(n) - \frac{1}{\varphi(q)} \sum_{\substack{x\leq n<2x \\ (n,q)=1}} \Lambda(n) \right| \ll \frac{x}{(\log x)^A}. \tag{8.37}$$

(4) We now switch our attention to deriving an explicit value for the lower bound parameter β of (8.26). Using the arguments set out in Lemma 8.16 Steps (1) through (4), with the same definitions for β_0, h, F, $\widetilde{F}$ and T, but with the explicit definition for $\mathcal{X}$ given in Step (1), we have

$$T := 2 \sum_{n\in\mathcal{S}_{(w,\infty)}} \frac{h(n)}{n} F(n)(F(n) - \widetilde{F}(n)) \text{ and } \sum_{n\in\mathcal{S}_{(w,\infty)}} \frac{h(n)}{n} F(n)^2 \geq \beta_0 B^{k+1} - T.$$

With the decomposition $T = T_1 + T_2$ as before, T_2 is a negligible quantity, i.e., $T_2 = o\left(B^{k+1}\right)$, so it can be omitted. Deriving an upper bound for T_1 requires more work, which is the subject of Steps (5) through (8).

(5) To this end, first recall

$$T_1 := 2B \sum_{d \in \mathcal{S}_{(w,\infty)}} \frac{h(n)}{n} (F(n) - \widetilde{F}(n)) f\left(\frac{\log n}{\log R}\right).$$

We have as before, in Step (4) of Lemma 8.16,

$$F(n) - \widetilde{F}(n) = - \sum_{\substack{m \in \mathcal{S}_{(w,\infty)} \\ (m,n)=1 \\ mn \notin \mathcal{X}}} \frac{1}{\varphi(m)} f'\left(\frac{\log(mn)}{\log R}\right).$$

If a pair (m,n) appears in this summation, and $mn \notin \mathcal{X}$, then at least one of the following conditions holds:

(i) there is a prime p in $[x^{\delta'}, R]$ with $p \mid m$,

(ii) there is a prime p in $[x^{\delta'}, R]$ with $p \mid n$,

(iii) $n \in \mathcal{S}_{(w, x^{\delta'})}$ and

$$\prod_{\substack{p \mid n \\ p \leq x^{\delta}}} p \leq \prod_{\substack{p \mid mn \\ p \leq x^{\delta}}} p < x^{(i\delta' - \delta)/2 + \varpi + \epsilon/2}.$$

We treat cases (i) and (ii) in the same manner as before in Theorem 8.17, using δ' in place of δ as necessary. Case (iii) is treated in the following three steps, where we denote by T_{13} the corresponding upper bound for T_1.

(6) Let $\mathcal{Z} := \{n : n \text{ satisfies condition (iii)}\}$. By Theorem 8.17 Step (2), we get

$$F(n) \leq B \left(f\left(\frac{\log n}{\log R}\right) + o(1) \right),$$

which implies

$$T_{13} \leq 2B^2 \sum_{n \in \mathcal{Z}} \frac{h(n)}{n} f\left(\frac{\log n}{\log R}\right)^2.$$

(7) Next, recall from the statement that $\theta := \delta'/(\frac{1}{4} + \varpi)$ and $\xi := \delta/(\frac{1}{4} + \varpi)$. Also define

$$\widetilde{\theta}_\epsilon := \frac{((i\delta' - \delta)/2 + \varpi + \epsilon/2) \log x}{\log R} = \frac{(i\delta' - \delta)/2 + \varpi + \epsilon/2}{\frac{1}{4} + \varpi}.$$

Then we can write

$$\mathcal{Z} = \left\{ n \in \mathcal{S}_{(w, R^\theta)} : n \leq R, \prod_{\substack{p \mid n \\ p \leq R^\xi}} p \leq R^{\tilde{\theta}_\epsilon} \right\}.$$

This shows that any $n \in \mathcal{Z}$ can be factored in the form $n = p_1 \cdots p_j m$, where $m \in [1, R^{\tilde{\theta}_\epsilon}]$ has $(p_1 \cdots p_j, m) = 1$, and we can assume

$$R^\xi \leq p_1 < p_2 < \cdots < p_J \leq R^\theta,$$

and $0 \leq J \leq 1/\xi$. Therefore,

$$T_{13} \leq 2B^2 \sum_{0 \leq J \leq 1/\xi} \left(\sum_{R^\xi \leq p_1 < \cdots < p_j \leq R^\theta} \frac{h(p_1 \cdots p_J)}{p_1 \cdots p_J} \right.$$

$$\left. \times \sum_{m \leq R^{\tilde{\theta}_\epsilon}} \frac{h(m)}{m} f \left(\frac{\log(p_1 \cdots p_J m)}{\log R} \right)^2 \right).$$

(8) Now define, using a restricted domain for the functions $G(r,s)$ from Theorem 8.17,

$$G_\epsilon(r,s) := \int_0^{\tilde{\theta}_\epsilon} f(t+r) f(t+s) \frac{t^{k-2}}{(k-2)!} \, dt.$$

For the formula derived in Step (7) we can then, using Lemma 8.8 again, rewrite the inner sum in the closed form

$$G_\epsilon \left(\frac{\log(p_1 \cdots p_J)}{\log R}, \frac{\log(p_1 \cdots p_J)}{\log R} \right).$$

We also have from the properties of $h(p)$,

$$h(p_1 \cdots p_J) = (k-1)^J + O \left(\frac{1}{p_1} \right) = (k-1)^J + O \left(\frac{1}{R^\xi} \right).$$

Therefore, other than negligible terms, we have

$$T_{13} \leq B^{k+1} \sum_{0 \leq J \leq 1/\xi} (k-1)^J$$

$$\times \int_{\xi \leq t_1 < \cdots < t_J \leq \theta} \frac{G_\epsilon(t_1 + \cdots + t_J, t_1 + \cdots + t_J)}{t_1 \cdots t_J} \, dt_1 \cdots dt_J.$$

(9) Having made these preparations, we are now able to complete the proof. Define

$$\beta := \beta_0(1 - 2\kappa_1 - 2\kappa_2 - 2\kappa_3'),$$

where

$$\kappa_3' := \frac{1}{G(0,0)} \sum_{0 \le J \le 1/\xi} (k-1)^J$$

$$\times \int_{\xi \le t_1 < \cdots < t_J \le \theta} \frac{G_\epsilon(t_1 + \cdots + t_J, t_1 + \cdots + t_J)}{t_1 \cdots t_J} \, dt_1 \cdots dt_J. \qquad (8.38)$$

Then with this value of β and the corresponding κ_3' we obtain the key estimate (8.26), namely

$$\sum_{n \in \mathcal{S}_{(w,\infty)}} \frac{h(n)}{n} \left(\sum_{\substack{m \in \mathcal{S}_{(w,\infty)} \\ (m,n)=1 \\ mn \in \mathcal{X}}} \frac{1}{\varphi(m)} f'\left(\frac{\log mn}{\log R} \right) \right)^2 \ge \beta B^{k+1}. \qquad (8.39)$$

(10) As a postscript, we weaken some estimates so that we can replace κ_3' by something more manageable than the multiple integral definition. First, note that $G_\epsilon(t,t) = 0$ for all $t \ge 1$. By Lemma 8.14(iii), we get for $0 \le t \le 1$ the estimate, using the value defined in the theorem statement and $\eta := j_{k-2}^2/(4(k-1))$,

$$G_\epsilon(t,t) \le e^{-2\eta t} G_\epsilon(0,0).$$

Thus, for any fixed $A > 0$, we have for all $t \ge 0$,

$$G_\epsilon(t,t) \le e^{A(1-t)} e^{-2\eta t} G_\epsilon(0,0).$$

Replacing G_ϵ by this estimate in (8.38), recalling the definition of the parameter ω from the theorem statement, and letting $\epsilon \to 0$ so $\tilde{\theta}_\epsilon \to \omega^2$ to get the last line, then gives

$$\kappa_3' \le \frac{G_\epsilon(0,0)}{G(0,0)} e^A \sum_{0 \le J \le 1/\xi} \frac{(k-1)^J}{J!} \left(\int_\xi^\theta \frac{e^{-(A+2\eta)t}}{t} \, dt \right)^J$$

$$\le \frac{G_\epsilon(0,0)}{G(0,0)} \exp\left(A + (k-1) \int_\xi^\theta \frac{e^{-(A+2\eta)t}}{t} \, dt \right)$$

$$\le \frac{1}{G(0,0)} \int_0^{\omega^2} f(t)^2 \frac{t^{k-2}}{(k-2)!} \, dt \times \exp\left(A + (k-1) \int_\xi^\theta \frac{e^{-(A+2\eta)t}}{t} \, dt \right).$$

From the definition of the weight function $f(t)$ and the Bessel function identity of Lemma A.4, we get

$$\frac{1}{G(0,0)} \int_0^{\omega^2} f(t)^2 \frac{t^{k-2}}{(k-2)!} \, dt = \frac{\omega^2 \left(J_{k-2}(j_{k-2}\omega)^2 - J_{k-3}(j_{k-2}\omega) J_{k-1}(j_{k-2}\omega) \right)}{J_{k-1}(j_{k-2})^2}.$$

Also recall, from Lemma 8.15, that

$$G(0,0) = \beta_0 = \int_0^1 f(t)^2 \frac{t^{k-2}}{(k-2)!} \, dt = \frac{J_{k-1}(j_{k-2})^2}{(k-2)!}.$$

Finally, we choose ϵ sufficiently small so that we can have κ_3' replaced by κ_3 defined in the theorem statement, namely

$$\kappa_3 := \frac{\omega^2 \left(J_{k-2}(j_{k-2}\omega)^2 - J_{k-3}(j_{k-2}\omega) J_{k-1}(j_{k-2}\omega) \right)}{J_{k-1}(j_{k-2})^2}$$
$$\times \exp\left(A + (k-1) \int_\xi^\theta e^{-(A+2\eta)t} \frac{dt}{t} \right),$$

and obtain $\beta \geq \beta_0(1 - 2\kappa_1 - 2\kappa_2 - 2\kappa_3)$. The result then follows by Lemma 8.7, in an analogous manner to Theorem 8.17 Step (8). This completes the proof. $\qquad\square$

Remarks: Following Polymath, we recomputed values of the parameters and checked constraints using the PGpack functions `SmallestKStar`, `FeasibleK` `PolymathQ` and `SmallestKPolymath1`. Frequently we found the value returned by `SmallestKStar` to be feasible and that it could not be improved with a smaller feasible value. We needed slightly different parameter settings and obtained values which were close to those of Polymath. See Table 8.2. Polymath's value $k = 1783$ is used in Theorem 8.39 and the constraint

$$c_\varpi \varpi + c_\delta \delta = 168\varpi + 48\delta < 1$$

in Theorem 8.37.

Polymath undertook extensive computations, which are reported in summary in [174, page 71]. The reader should note that this is version 2 of the article with the given title on ArXiv.

8.8 Using Heath-Brown's Identity

It follows from Theorem 8.19 that for an "optimal" $f(t)$ and parameter values δ, ϖ we have an inequality depending on $k \geq 2$ such that when $\mathrm{MPZ}^{(i)}[\varpi, \delta]$ is true so is $\mathrm{DHL}[k, 2]$. So in order to achieve this latter goal, we must find parameter ranges for which we can demonstrate $\mathrm{MPZ}^{(i)}[\varpi, \delta]$. To do this, we will need to tack upwind. The approach taken by Zhang and Polymath is both combinatorial and algebraic,

Table 8.2 *A feasible and nonfeasible value of k and corresponding parameter settings.*

Parameter	Feasible	Nonfeasible
k	1783	1782
ϖ	0.0059495533'40	0.005958'40
δ	9.8965035'40/10^6	9.8965035'40/10^6
δ'	3.7117059'40/10^3	3.7117059'40/10^3
A	757	757
c_ϖ	168	168
c_δ	48	48
i	2	2
$1 - (168\varpi + 48\delta)$	1.343200/10^8	0.0045389678320
$1 - j_{k-2}^2/(k(k-1)(1+4\varpi))$	6.289583/10^6	$-2.1773/10^6$

so dear reader, prepare for a wind shift. We begin with a combinatorial lemma of Roger Heath-Brown. It is hard to motivate its relevance, except to note that the t_i of the lemma will become powers of x, and the division into types is central for both Zhang's and Polymath's approaches.

Note that we have introduced a nonstandard set of terms for subsets consistent with their roles: small, large, stabilizing and enlarging. These are defined in Step (3) of the proof.

Lemma 8.20   *[174, lemma 5.1] Let $1/6 < \sigma < 1/2$, $n \geq 1$, and $t_1, \ldots, t_n \geq 0$ be real numbers with $\sum_{j=1}^n t_j = 1$. Then at least one of the three following situations is true:*

(Type 0) There exists a j such that $t_j \geq \frac{1}{2} + \sigma$, or

(Type I/II) there is a partition $S \sqcup T = \{1, 2, \ldots, n\}$ such that

$$\frac{1}{2} - \sigma < \sum_{j \in S} t_j \leq \sum_{j \in T} t_j < \frac{1}{2} + \sigma, \text{ or}$$

(Type III) for all distinct i, j, k with $2\sigma \leq t_i \leq t_j \leq t_k \leq \frac{1}{2} - \sigma$, then

$$t_i + t_j < \frac{1}{2} + \sigma, \text{ or } t_i + t_k < \frac{1}{2} + \sigma \text{ or } t_j + t_k < \frac{1}{2} + \sigma.$$

In addition, since $1/6 < \sigma$ Type III does not occur.

Proof   (1) If $n = 1$, then $1 = t_1 > \frac{1}{2} + \sigma$, so Type 0 is true. If $n = 2$ and Type I/II is false, then $t_1 \geq \frac{1}{2} + \sigma \iff t_2 \leq \frac{1}{2} - \sigma$ so neither t_1 nor t_2 are in $\left(\frac{1}{2} - \sigma, \frac{1}{2} + \sigma\right)$. Thus, since $t_1 + t_2 = 1$ we have t_1 or t_2 satisfies $t_j \geq \frac{1}{2} + \sigma$, which again is Type 0. Therefore, in the remainder of the proof we can assume $n \geq 3$.

Since $\sigma > 1/6$, we have $2\sigma > \frac{1}{2} - \sigma$, so there are no distinct i, j, k with $\frac{1}{2} - \sigma < 2\sigma \le t_i \le t_j \le t_k \le \frac{1}{2} - \sigma \implies \frac{1}{2} - \sigma < \frac{1}{2} - \sigma$. Thus, with the given constraint on σ, Type III does not occur.

(2) We will now show that at least one of Type 0 and Type I/II must be true. Suppose both are false in order that we might obtain a contradiction, which we will arrive at in Step (5). First, by the Type 0 failure, for all j we have $t_j < \frac{1}{2} + \sigma$. By the Type I/II failure for every subset $S \subset \{1, \dots, n\}$, if T is the complement of S we have

$$\sum_{j \in S} t_j = 1 - \sum_{j \in T} t_j \implies \left(\sum_{j \in S} t_j \notin \left(\tfrac{1}{2} - \sigma, \tfrac{1}{2} + \sigma\right) \iff \text{Type I/II is false} \right).$$

Thus we can assume for all $S \subset \{1, \dots, n\}$, we have $\sum_{j \in S} t_j \notin \left(\frac{1}{2} - \sigma, \frac{1}{2} + \sigma\right)$.

(3) We now give some definitions which are helpful in the proof. If S is such that $\sum_{j \in S} t_j \ge \frac{1}{2} + \sigma$, we say S is **large**. If it is such that $\sum_{j \in S} t_j \le \frac{1}{2} - \sigma$, we say it is **small**. By the result of Step (2), every subset of $\{1, \dots, n\}$ is either large or small and not of course both. Since for all j, $t_j < \frac{1}{2} + \sigma$, singleton subsets are small. The empty set is small since its corresponding sum is zero. Because $t_1 + t_2 + \cdots + t_n = 1$, the complement of a small set is large and vice versa.

We also say an element j with $1 \le j \le n$ is **enlarging** if for at least one small set S in the complement of $\{j\}$, we have $S \cup \{j\}$ large. We say an element j is **stabilizing** if it is not enlarging. Therefore, adding or removing a stabilizing element to or from a subset S cannot change it from small to large or vice versa. Hence the union of a set of stabilizing elements with a small subset gives a small subset.

(4) Next we **claim** that there are exactly three enlarging elements. If S is the set of all stabilizing elements, then from Step (3) it is small, so its complement is large. This complement cannot be a singleton, since for all j we have $t_j < \frac{1}{2} + \sigma$, so it must contain at least two elements. Thus there are at least two enlarging elements. Choose any one of these two, i, say. Then $\{i\}$ being a singleton is small, so adding each element of S successively to this subset, we get $S \cup \{i\}$ is also small, so its complement is large and thus has at least two elements, a, b, say. Each, for example a, is enlarging since $\{a\} \subset \{a, b\}$, which is large and $\{b\}$ is small. Thus there are at least three enlarging elements.

If i is enlarging, then since the gap between any small sum and that when t_i is included differ by at least 2σ we get $t_i \ge 2\sigma$. Thus if i and j are distinct and enlarging, we have, because $\sigma > 1/6$,

$$t_i + t_j \geq 4\sigma > \frac{1}{2} + \sigma.$$

Thus $\{i, j\}$ is large. If however $\{i, j, k, l\}$ was a set of distinct enlarging elements, we would get

$$1 = t_1 + \cdots + t_n \geq t_i + t_j + t_k + t_l \geq 2\left(\frac{1}{2} + \sigma\right) > 1.$$

Hence there are precisely three enlarging elements.

(5) Therefore, we can denote the three distinct enlarging elements as i, j, k and we can assume $t_i \leq t_j \leq t_k$. Then by Step (4) and since Type 0 is false, we have

$$2\sigma \leq t_i \leq t_j \leq t_k < \frac{1}{2} - \sigma.$$

If $\{t_i, t_j\}$ were small, we would have $\frac{1}{2} - \sigma \geq t_i + t_j \geq 4\sigma$, which is false. Therefore, $\{t_i, t_j\}$ is large, and the same proof works for the two other sets of pairs, which corresponds to Type III. But by Step (1), this type does not occur. Thus we must have either Type 0 or Type I/II and the proof is complete.   □

Next we give the derivation of Heath-Brown's generalization of Vaughan's identity Theorem 8.2, enabling the von Mangoldt function to be decomposed in terms of Dirichlet products of the Möbius function and the logarithm.

We use standard definitions of some arithmetic functions on $\mathbb{N}$:

$$\mathbf{1}(n) := 1,$$

$$L(n) := \log n,$$

$$(f \cdot g)(n) := f(n)g(n),$$

$$\mu_{\leq}(n) := \mu(n) \cdot \chi_{[1,(2x)^{1/J}]},$$

$$\mu_{>}(n) := \mu(n) \cdot \chi_{((2x)^{1/J}, \infty)}.$$

In addition, $\alpha_S := \star_{j \in S} \alpha_j$ is the Dirichlet convolution of a finite set of arithmetic functions, and $f^{\star j} = f \star \cdots \star f$ the convolution of f with itself j times, or in other words, for $n \in \mathbb{N}$,

$$f^{\star j}(n) := \sum \cdots \sum_{\substack{(a_1, \ldots, a_j) \in \mathbb{N}^j \\ a_1 \cdots a_j = n}} f(a_1) \cdots f(a_j).$$

In particular, $f^{\star 1}(n) = f(n)$. Finally, we set $f^{\star 0}(n) := \delta(n) = \chi_{\{1\}}(n)$.

Lemma 8.21   (Heath-Brown's identity) *[174, lemma 5.3] For $J \in \mathbb{N}$, we have on* $[x, 2x]$ *the identity*

$$\Lambda = \sum_{j=1}^{J} (-1)^{j-1} \binom{J}{j} \mu_{\leq}^{\star j} \star \mathbf{1}^{\star(j-1)} \star L. \tag{8.40}$$

Vaughan's identity is the case $J = 2$ (see for example [22, lemma 12.28]).

Proof Because $\mu_>^{\star J} = 0$ on $[1, 2x]$, we have $\mu_>^{\star J} \star \mathbf{1}^{\star(J-1)} \star L$ zero on $[0, 2x]$ also. Hence, expanding $\mu_>^{\star J} = (\mu - \mu_\le)^{\star J}$ and using the binomial theorem, we get on $[x, 2x]$ the identity

$$\sum_{j=0}^{J} (-1)^j \binom{J}{j} \mu^{\star(J-j)} \star \mu_\le^{\star j} \star \left(\mathbf{1}^{\star(J-1)} \star L\right) = 0. \tag{8.41}$$

Next, we use the standard properties of the Dirichlet product, such as associativity, commutativity, $\Lambda = \mu \star L$, $\delta = \chi_{\{1\}} = \mu \star \mathbf{1}$ to simplify the terms in this expansion. For the $j = 0$ term of (8.41), we get

$$\mu^{\star J} \star \mathbf{1}^{\star(J-1)} \star L = \mu \star L = \Lambda.$$

For each $j > 0$, we get

$$\mu^{\star(J-j)} \star \mathbf{1}^{\star(J-1)} = \mu^{\star(J-j)} \star \mathbf{1}^{\star J - j} \star \mathbf{1}^{\star(j-1)} = \mathbf{1}^{\star(j-1)}.$$

Therefore,

$$\mu^{\star(J-j)} \star \mu_\le^{\star j} \star \mathbf{1}^{\star(J-1)} \star L = \mu_\le^{\star j} \star \mathbf{1}^{\star(j-1)} \star L.$$

Putting these together completes the proof. □

With these preliminaries, we are nearly ready to state and prove the main combinatorial lemma. Based on Heath-Brown's identity, it gives a sufficient condition in terms of parameter values for $\mathrm{MPZ}^{(i)}[\varpi, \delta]$, and is an essential part of the development.

First, we recall some important definitions. We have defined a sequence of complex numbers $(c_n)_{n \in \mathbb{N}}$ to be a **coefficient sequence** if it is finitely supported for each $x > 0$ (with support possibly depending on x) and satisfies the bound for all $n \in \mathbb{N}$

$$|c(n)| \ll \tau(n)^{O(1)} (\log x)^{O(1)}.$$

The **discrepancy** of an arithmetic function α having finite support is as follows:

$$\Delta(\alpha, q, a) := \sum_{\substack{n \in \mathbb{N} \\ n \equiv a \bmod q}} \alpha(n) - \frac{1}{\varphi(q)} \sum_{\substack{n \in \mathbb{N} \\ (n,q)=1}} \alpha(n).$$

A coefficient sequence $(c)_{n \in \mathbb{N}}$ is said to be **located at scale** N if its support is contained in an interval $[uN, vN]$, where $1 \ll u < v \ll 1$ are fixed. This implies

$$\Delta(c, q, a) \ll \frac{N (\log x)^{O(1)}}{\varphi(q)}$$

for each integer a with $(a, q) = 1$.

A complex sequence $(\alpha_n)_{n \in \mathbb{N}}$ is said to obey the **Siegel–Walfisz property** if for any fixed $A > 0$ and any natural numbers q, r and primitive residue class a modulo q we have as $x \to \infty$

$$\left| \sum_{\substack{n \leq x \\ n \equiv a \bmod q \\ (a,r)=1}} \alpha(n) - \frac{1}{\varphi(q)} \sum_{\substack{n \leq x \\ (n,q)=1 \\ (n,r)=1}} \alpha(n) \right| \ll \tau(qr)^{O(1)} \frac{x}{(\log x)^A}. \tag{8.42}$$

This is equivalent to the sequence $\alpha(n)$ being uniformly distributed in every primitive residue class modulo q for every $q \ll (\log x)^A$ for all $A > 0$.

Let $i \geq 0$ be an integer. Recall also we have defined the predicate $\mathrm{MPZ}^{(i)}[\varpi, \delta]$ to be true if for any bounded subset $\Omega \subset \mathbb{R}$ which may depend on x, and $Q \prec^o x^{\frac{1}{2}+2\varpi}$, if $(a, P_\Omega) = 1$ and $A \geq 1$ is a fixed real number. Then

$$\sum_{\substack{q \leq Q \\ q \in \mathcal{D}_\Omega(i, x^\delta)}} \left| \sum_{\substack{x \leq n < 2x \\ n \equiv a \bmod q}} \Lambda(n) - \frac{1}{\varphi(q)} \sum_{\substack{x \leq n < 2x \\ (n,q)=1}} \Lambda(n) \right| \ll \frac{x}{(\log x)^A}. \tag{8.43}$$

Let $i \geq 1$, ϖ with $0 < \varpi < \frac{1}{4}$, δ with $0 < \delta < \frac{1}{4} + \varpi$, σ with $\frac{1}{6} < \sigma < \frac{1}{2}$ and $\sigma > 2\varpi$ be fixed real numbers, let Ω be a fixed bounded subset of $(0, \infty)$ and define the squarefree number

$$P_\Omega = \prod_{p \in \Omega} p.$$

Let a be an integer with $(a, P_\Omega) = 1$ and $a \bmod P_\Omega$ the corresponding congruence class. We say the estimate $\mathrm{Type}_{\mathrm{I}}^{(i)}[\varpi, \delta, \sigma]$ holds if for all Ω and class $a \bmod P_\Omega$ and any integers $M, N \gg 1$ such that $x \ll MN \ll x$ and for some fixed $c > 0$ sufficiently small,

$$x^{\frac{1}{2}-\sigma} \prec^o N \prec^o x^{\frac{1}{2}-2\varpi-c}$$

and for any $Q \prec^o x^{\frac{1}{2}+2\varpi}$ and any coefficient sequence α at scale M and β at scale N, with β having the Siegel–Walfiz property, we have for any fixed $A > 0$,

$$\sum_{\substack{q \leq Q \\ q \in \mathcal{D}_\Omega(i, x^\delta)}} \left| \sum_{\substack{n \equiv a \bmod q \\ (a,r)=1}} (\alpha \star \beta)(n) - \frac{1}{\varphi(q)} \sum_{(n,q)=1} (\alpha \star \beta)(n) \right| \ll \frac{x}{(\log x)^A}. \tag{8.44}$$

We say the estimate $\mathrm{Type}_{\mathrm{II}}^{(i)}[\varpi, \delta]$ holds if we get the same estimate as for $\mathrm{Type}_{\mathrm{I}}^{(i)}[\varpi, \delta, \sigma]$ but in the range of N values

$$x^{\frac{1}{2}-2\varpi-c} \prec^o N \prec^o x^{\frac{1}{2}}.$$

We also need the following definition.

Definition 8.13 A coefficient sequence α is said to be **smooth at scale** $N \in \mathbb{N}$ if there exists a smooth function $\psi : \mathbb{R} \to \mathbb{C}$ with support in a closed bounded interval $[a, b]$ with $0 < a < b$ such that

$$\alpha(n) = \psi\left(\frac{x}{N}\right) \text{ and } \forall j \geq 0, \ |\psi^{(j)}(x)| \ll_j (\log x)^{O(1)}.$$

If all else is the same but there is an $x_0 \in \mathbb{R}$ with $\alpha(n) = \psi((x - x_0)/N)$, we say the sequence is **shifted smooth at scale** N.

Remark: For background on the Fourier transform, see for example [22, appendix E]. In addition, we need the following useful bound for the Fourier transform of a smooth function $f : \mathbb{R} \to \mathbb{C}$ with compact support. Recall

$$\mathscr{F}f(z) := \int_{\mathbb{R}} f(x)e^{-2\pi i z x} \, dx.$$

In fact, we don't need smoothness to derive the bound. If $f : \mathbb{R} \to \mathbb{C}$ has compact support and at least a continuous second derivative, then for $\alpha \geq 0$ and $z \neq 0$, the Fourier transform of f has a bound

$$|\mathscr{F}f(z)| \ll \frac{1}{|z|^2},$$

where the implied constant depends only on f and α and holds uniformly on the strip $|\Im z| \leq \alpha$. We now derive this estimate.

Proof Let $f(x) = 0$ for $|x| > R > 0$. Then because the series expansion for $\exp(z)$ converges absolutely and uniformly on $[-R, R]$, we can write

$$\mathscr{F}f(z) = \int_{-R}^{R} f(x)e^{-2\pi i z x} \, dx$$

$$= \int_{-R}^{R} f(x) \sum_{n=0}^{\infty} \frac{(-2\pi i z x)^n}{n!} \, dx$$

$$= \sum_{n=0}^{\infty} \frac{(-2\pi i z)^n}{n!} \int_{-R}^{R} x^n f(x) \, dx,$$

so the Fourier transform is an entire function of z.

Let $|f''(x)| \leq N$ on $[-R, R]$. If $|\Im z| \leq \alpha$, we have

$$\left| \int_{-R}^{R} f''(x)e^{-2\pi i z x} \, dx \right| \leq \int_{-R}^{R} |f''(x)|e^{2\pi \alpha |x|} \, dx \leq 2RNe^{2\pi R\alpha}.$$

Integrating by parts twice, we also have

$$\left| \int_{-R}^{R} f(x)e^{-2\pi i z x} \, dx \right| = \left| \frac{1}{(-2\pi i z)^2} \int_{-R}^{R} f''(x)e^{-2\pi i z x} \, dx \right| \leq \frac{2RNe^{2\pi R\alpha}}{4\pi^2 |z|^2}.$$

This implies for $z \neq 0$,

$$\mathscr{F}f(z) \ll \frac{1}{|z|^2}.$$

$\square$

Now we are ready for the proof of the main combinatorial lemma giving a sufficient condition for $\text{MPZ}^{(i)}$. This takes six main steps and quite a few sub-steps. Notice however that to complete the proof we need two very strong assumptions, $\text{Type}_{\text{I}}^{(i)}$ and $\text{Type}_{\text{II}}^{(i)}$. Deriving parameter settings which are sufficient for these will come later and require a study of exponential sums. The proof combines the beautiful combinatorial division into types of Lemma 8.20 with the generalized additive decomposition of the von Mangoldt function of Lemma 8.21.

Lemma 8.22 (**Main combinatorial lemma**) *[174, lemma 2.22] Let $i \geq 1$, ϖ with $0 < \varpi < \frac{1}{4}$, δ with $0 < \delta < \frac{1}{4} + \varpi$, σ with $\frac{1}{6} < \sigma < \frac{1}{2}$ and $\sigma > 2\varpi$, be fixed real numbers such that we have the estimates $\text{Type}_{\text{I}}^{(i)}[\varpi, \delta, \sigma]$ and $\text{Type}_{\text{II}}^{(i)}[\varpi, \delta]$. Then we have $\text{MPZ}^{(i)}[\varpi, \delta]$.*

Proof Let $A > 0$ be fixed, $\Omega \subset (0, \infty)$ any bounded subset and $a \bmod P_\Omega$ any congruence class. Then to prove the lemma, it is sufficient to show that, for $Q \prec^o x^{\frac{1}{2} + 2\varpi}$, we have

$$\sum_{\substack{q \leq Q \\ q \in \mathscr{D}_\Omega(i, x^\delta)}} \left| \sum_{\substack{x \leq n < 2x \\ n \equiv a \bmod q}} \Lambda(n) - \frac{1}{\varphi(q)} \sum_{\substack{x \leq n < 2x \\ (n,q)=1}} \Lambda(n) \right| \ll_{A, \delta, i} \frac{x}{(\log x)^{A + O(1)}}. \qquad (8.45)$$

The proof is in six steps and several substeps.

(1) Let $J \in \mathbb{N}$ be fixed and satisfy $1/J < 2\sigma$. Using Lemma 8.21, it is sufficient to prove for each fixed value of j with $1 \leq j \leq J$,

$$\sum_{\substack{q \leq Q \\ q \in \mathscr{D}_\Omega(i, x^\delta)}} \left| \Delta((\mu_{\leq}^{\star j} \star 1^{\star(j-1)} \star L) \cdot \chi_{[x, 2x]}, q, a) \right| \ll \frac{x}{(\log x)^{A/2 + O(1)}}. \qquad (8.46)$$

(2) Let $\theta := 1 + 1/(\log x)^A$ and $\psi : \mathbb{R} \to \mathbb{R}$ a smooth function supported on $[-\theta, \theta]$ which is 1 on $[-1, 1]$ and which satisfies for each derivative order $m \geq 0$ and each $x > 1$ the bound

$$|\psi^{(m)}(x)| \ll_m (\log x)^{mA}.$$

If we define for $n \geq 0$ and $D := \{n : n \geq 0\}$, the function $\psi_n(x) := \psi(x/\theta^n) - \psi(\theta x/\theta^n)$, which is smooth with support for positive x in $[\theta^{n-1}, \theta^{n+1}]$, we obtain

$$\sum_{n \in D} \psi_n(x) = 1,$$

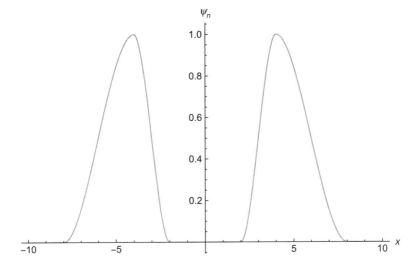

Figure 8.5 A plot of $\psi_2(x)$ for $\theta = 2$.

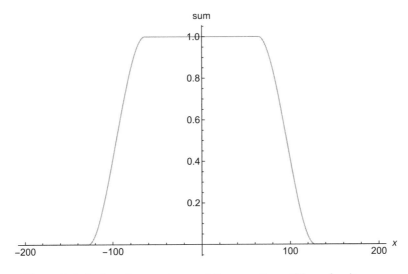

Figure 8.6 A plot of seven terms of the smooth partition of unity sum.

i.e., a smooth partition of unity. See in this regard Figures 8.5 and 8.6.
This gives the decompositions

$$\mu_{\leq} = \sum_{n \in D} \psi_n \cdot \mu_{\leq}, \; 1 = \sum_{n \in D} \psi_n \text{ and } L = \sum_{n \in D} \psi_n \cdot L.$$

Therefore, using the definition $\widehat{\psi_n}(x) := \psi_n(x) \log x / \log(\theta^n)$, for each j with $1 \leq j \leq J$, we get the expansion

$$\left(\mu_{\le}^{\star j} \star \mathbf{1}^{\star(j-1)} \star L\right) \chi_{[x,2x]} =$$

$$\sum \cdots \sum_{n_1,\dots,n_{2j} \in D} \left((\psi_{n_1} \cdot \mu_{\le}) \star \cdots \star (\psi_{n_j} \cdot \mu_{\le}) \star \psi_{n_{j+1}} \star \cdots \star \psi_{n_{2j-1}} \star (\psi_{n_{2j}} \cdot L)\right) \chi_{[x,2x]}$$

$$= \sum \cdots \sum_{n_1,\dots,n_{2j} \in D} \log(\theta^{n_{2j}})$$

$$\times \left((\psi_{n_1} \cdot \mu_{\le}) \star \cdots \star (\psi_{n_j} \cdot \mu_{\le}) \star \psi_{n_{j+1}} \star \cdots \star \psi_{n_{2j-1}} \star \widehat{\psi_{n_{2j}}}\right) \mathbf{1}_{[x,2x]}.$$

(3) Let $N_j = \theta^{n_j}$ for $1 \le j \le J$. Because of the truncation of the Möbius function and the bounded support of each ψ_{n_j}, a summand in this expression, vanishes at $n \in \mathbb{N}$ unless $n = d_1 \cdots d_{2j} \in [x, 2x]$ with $d_i \in [N_i/\theta, N_i\theta]$ for $1 \le i \le 2j$, so necessarily we need

$$N_1, \dots, N_j \ll x^{1/J} \text{ and } \frac{x}{\theta^{2J}} \le N_1 \cdots N_{2j} \le 2x\theta^{2J}.$$

This implies if a summand does not vanish, we must have for some positive constant c, fixed because J is fixed,

$$x\left(1 - \frac{c}{(\log x)^A}\right) \le N_1 \cdots N_{2j} \le 2x\left(1 + \frac{c}{(\log x)^A}\right). \qquad (8.47)$$

In turn, this implies, recalling the N_i are powers of θ and using for $y > 0$,

$$\#\{(n_1, \dots, n_{2j}) : 0 \le n_1 + \cdots + n_{2j} \le y\} \le (y+1)^{2j} \ll y^{2j},$$

there are at most $O((\log x)^{2j(A)})$ tuples $(N_1, \dots, N_{2j})$ for which a summand is nonzero. Let $\mathcal{Z}$ be the set of all of these tuples and define a function on $\mathbb{N}$ by

$$\alpha(n) := \sum_{(N_1,\dots,N_{2j}) \in \mathcal{Z}} \left((\psi_{N_1} \cdot \mu_{\le}) \star \cdots \star (\psi_{N_j} \cdot \mu_{\le}) \star \psi_{N_{j+1}} \star \cdots \star \psi_{N_{2j-1}} \star \widehat{\psi_{N_{2j}}}\right)(n)$$

$$- (\mu_{\le}^{\star j} \star \mathbf{1}^{\star(j-1)} \star L)\chi_{[x,2x]}(n).$$

Note that this sum is related to but not the same as the sum from Step (2).

(4) Since $n \in [x, 2x]$ by Lemma 8.21, the right-hand side of the result of Step (3) is 0, the function α has support in

$$[x(1 - c/(\log x)^A, x] \cup [2x, 2x(1 + c/(\log x)^A].$$

Since J is fixed and if $B_k(n)$ is a bound for $\tau_k(n)$, we have $B_2(n) = \tau(n)$ and $B_k(n) \le \tau(n)B_{k-1}(n)$, and the function α also satisfies, with the implied exponent constants independent of A, the bound

$$\alpha(n) \ll \tau(n)^{O(1)}(\log n)^{O(1)}.$$

Bounding the discrepancy by the number of nonzero $\alpha(n)$ times an upper bound, we also get, for each $q \geq 1$ and some positive constant B,

$$\Delta(\alpha,q,a) \ll \frac{x}{(\log x)^{A-B}}.$$

Hence, in what follows, to prove the estimate of Step (1), we choose α_i as any one of $\psi_{N_i} \cdot \mu_{\leq}$, ψ_{N_i} or $\widehat{\psi}_{N_i}$, where the N_i satisfy the bounds given in Step (3), use the bound $\#\mathcal{Z}$ for the number for which a summand is nonzero and use Heath-Brown's identity Lemma 8.21. It will be (more than) sufficient to show that for each given $A > 0$ we have

$$\sum_{\substack{q \leq Q \\ q \in \mathscr{D}_\Omega(i,x^\delta)}} |\Delta(\alpha_1 \star \cdots \star \alpha_{2j}, a, q)| \ll \frac{x}{(\log x)^A}. \tag{8.48}$$

To do this, we first develop in Step (5) properties of the functions α_i.

(5) In this step and its substeps, we derive some properties of the arithmetic functions α_i and parameters N_i that will be used in the sequel. Let $1 \leq j \leq 2J$ and $S \subset \{1,\ldots,2J\}$ be nonempty. Let $\alpha_S := \star_{j \in S}\alpha_j$ and $N_S := \prod_{j \in S} N_j$. Then we make and demonstrate the following four **claims**. (The reader might want to refer back to the definitions given before this lemma.)

(5.1) Each α_j is a coefficient sequence at scale N_j and the convolution α_S is a function at scale N_S: the first part follows because the support of α_j is in $[N_j/\theta, N_j\theta]$. The second part follows by induction since if α is at scale M and β at N, then $\alpha \star \beta$ is at scale MN.

(5.2) If $N_j \gg x^{2\sigma}$, then α_j is smooth at scale N_j: to see this, by Step (1) $2\sigma > 1/J$ and by Step (3) for $1 \leq j \leq J$, $N_j \ll x^{1/J}$. Thus as $x \to \infty$,

$$N_j \ll x^{1/J} < x^{2\sigma} \ll N_j,$$

which is false. Thus we can have $N_j \gg x^{2\sigma}$ only when $J < j \leq 2J$ so α_j is either ψ_{N_j} or $\widehat{\psi}_{N_j}$, which are both smooth.

(5.3) If $N_j \gg x^\epsilon$ for fixed $\epsilon > 0$ as $x \to \infty$, then α_j satisfies the Siegel–Walfisz property Definition 8.5: if $j \leq J$, then the Siegel–Walfisz property follows from the Siegel–Walfisz Theorem 8.1 [117, corollary 5.29, page 124]. If $j > J$, we use the Poisson summation formula for smooth compactly supported functions, in a similar manner to the Type 0 case set out in Step (6.2).

For general products of coefficient sequences, by induction it is sufficient to check that if α, β are sequences at scales M, N respectively, with $x^\epsilon \ll N \ll x^C$ for fixed $\epsilon, C > 0$, and such that β satisfies the Siegel–Walfisz property, then so does $\alpha \star \beta$. To see this, we need to show, if an arithmetic function X^q is defined by $X^q(n) = 1 \iff (n,q) = 1$, else $X^q(n) = 0$, and any q,r and for fixed $A > 0$ and integer a with $(a,r) = 1$, that

$$|\Delta(\alpha \star \beta) \cdot X^q, r, a)| \ll \tau(qr)^{O(1)} \frac{N}{(\log x)^A}.$$

To show this, we first modify the sequences by setting their values to 0 if $(n, qr) \neq 1$. Also we will assume $r \ll (\log x)^{A+O(1)}$, since otherwise the estimate follows from the trivial bound $\Delta(\alpha \star \beta, a, r) \ll N(\log x)^{O(1)}/\varphi(r)$.

Expanding the convolution with $b\bar{b} \equiv 1 \bmod r$, we get

$$\sum_{\substack{n \in N \\ n \equiv a \bmod r}} (\alpha \star \beta)(n) = \sum_{\substack{b \in \mathbb{Z}/r\mathbb{Z} \\ (b,r)=1}} \left(\sum_{d \equiv b \bmod r} \alpha(d) \right) \left(\sum_{m \equiv \bar{b}a \bmod r} \beta(m) \right),$$

and

$$\sum_{n \in N} (\alpha \star \beta)(n) = \sum_{\substack{b \in \mathbb{Z}/r\mathbb{Z} \\ (b,r)=1}} \left(\sum_{d \equiv b \bmod r} \alpha(d) \right) \left(\sum_{m \in N} \beta(m) \right).$$

These identities imply

$$|\Delta(\alpha \star \beta, r, a)| \le \sum_{\substack{b \in \mathbb{Z}/r\mathbb{Z} \\ (b,r)=1}} \left| \sum_{n \equiv b \bmod r} \alpha(n) \right| \cdot |\Delta(\beta, r, \bar{b}a)|.$$

By the definition of a coefficient sequence and an estimate for powers of the number of divisors summed over an arithmetic progression, for fixed $C > 0$ and given $1 < y \le x^{O(1)}$ and $a \bmod q$ a fixed congruence class with $q \le y^\delta$ with $0 < \delta < 1$, then as $x \to \infty$ we have

$$\sum_{\substack{n \le x \\ n \equiv a \bmod q}} \tau(n)^C \ll \frac{y}{\varphi(q)} \tau(q)^{O(1)} (\log x)^{O(1)}$$

(as proved in Lemma 8.24(iii)), we have

$$\sum_{\substack{n \le N \\ n \equiv b \bmod r}} \alpha(n) \ll \frac{N}{r} \tau(r)^{O(1)} (\log x)^{O(1)} + N^{o(1)}.$$

Also because β has the Siegel–Walfisz property, for any fixed $B > 0$ we have

$$|\Delta(\beta, r, \bar{b}a)| \ll (N/r)\tau(r)^{O(1)} M (\log x)^{-B},$$

with each of these last two estimates holding for any b mod r. Therefore, since $r \ll (\log x)^{A+O(1)}$,

$$|\Delta(\alpha \star \beta, r, a)| \ll \tau(r)^{O(1)} \varphi(r) \left(\frac{N}{r} + N^{o(1)} \right) M (\log x)^{-B+O(1)}$$

$$\ll \tau(r)^{O(1)} M N (\log x)^{-B+O(1)}.$$

Thus $\alpha \star \beta$ satisfies the Siegel–Walfisz property and the proof of Step (5.3) is complete.

(5.4) $N_1 \cdots N_{2J} \asymp x$: this is a direct consequence of the estimates (8.47).

(6) With these preparations, assume we have arithmetic functions α_j as specified in Step (4). By Step (5.4), we can assume there are nonnegative real numbers t_j such that $N_j \asymp x^{t_j}$, $1 \le j \le 2J$ with $t_1 + \cdots + t_{2J} = 1$. By Lemma 8.20, the set of the t_j are of Type 0 or of Type I/II. We consider this latter case first.

(6.1) Suppose the t_j are such that we are in the Type I/II case. Then, by Lemma 8.20 there is a partition such that $\{1, 2, \ldots, 2J\} = S \sqcup T$, and by assembling the related indices we can write

$$\alpha_S \star \alpha_T = \alpha_1 \star \cdots \star \alpha_{2J}.$$

By Step (5.1), α_S is at scale N_S and α_T at scale N_T, and by Step (5.3) both satisfy the Seigel–Walfisz property. By Step (5.4), we have $N_S \star N_T \asymp x$. Directly from Lemma 8.20, we have as $x \to \infty$,

$$x^{\frac{1}{2}-\sigma} \ll N_S \ll N_T \ll x^{\frac{1}{2}+\sigma} \text{ or } x^{\frac{1}{2}-\sigma} \ll N_T \ll N_S \ll x^{\frac{1}{2}+\sigma}.$$

Therefore, if for some sufficiently small fixed $c > 0$ we have either $N_S \le x^{\frac{1}{2}-2\varpi-c}$ from the hypothesis Type$_{\mathrm{I}}^{(i)}[\varpi, \delta]$, or $N_S > x^{\frac{1}{2}-2\varpi-c}$ from the hypothesis Type$_{\mathrm{II}}^{(i)}[\varpi, \delta]$, then we get from the estimate (8.44),

$$\sum_{\substack{q \le Q \\ q \in \mathscr{D}_\Omega(i, x^\delta)}} |\Delta(\alpha_1 \star \cdots \star \alpha_{2j}, a, q)| \ll \frac{x}{(\log x)^A}.$$

(6.2) It remains to consider Type 0 and to derive the estimate (8.48) in this case. Here there is a j with $1 \le j \le 2J$ such that $t_j \ge \frac{1}{2} + \sigma > 2\sigma$. Step (5.2) shows α_j has smooth support, and we will see that it has both large support and is well distributed in arithmetic progressions to large moduli, so much so that the effect of the remaining α_i is small.

To this end, write for $S = \{1, \ldots, 2J\} \setminus \{j\}$,

$$\alpha_j \star \alpha_S = \alpha_1 \star \cdots \star \alpha_{2J}.$$

By Step (5), α_j is smooth at scale $N_j \gg x^{\frac{1}{2}+\sigma}$, and α_S has scale N_S which satisfies $N_S N_j \asymp x$. Then using $n = ml$ we get

$$\Delta(\alpha_j \star \alpha_S, q, a) = \sum_{\substack{m \in \mathbb{Z} \\ (m,q)=1}} \alpha_S(m) \Delta(\alpha_j, q, \bar{m}a).$$

Using the estimate $\sum_{n \leq y} \tau(n)^C \ll y(\log x)^{O(1)}$ from Lemma 8.24(ii), and the property of the coefficient sequence

$$\sum_{n \in \mathbb{N}} |\alpha_S(n)| \prec^o N_S,$$

we have

$$\sum_{\substack{q \leq Q \\ q \in \mathscr{D}_\Omega(i,x^\delta)}} |\Delta(\alpha_1 \star \cdots \star \alpha_{2J}, q, a)| \prec^o N_S \sum_{q \leq Q} \sup_{\substack{b \in \mathbb{Z}/q\mathbb{Z} \\ (b,q)=1}} |\Delta(\alpha_j, q, b)|. \qquad (8.49)$$

Because α_j is smooth at scale N_j, there is a smooth function $\psi : \mathbb{R} \to \mathbb{R}$ supported on a bounded interval with bounded derivatives (since they are smooth and compactly supported), such that $\alpha_j(n) = \psi(n/N_j)$ for all $n \in \mathbb{Z}$. Then the Poission summation formula (for example, [22, theorem E.9]), using the Fourier transform

$$\mathscr{F}\psi(s) := \int_{-\infty}^{\infty} \psi(t) e^{-2\pi i t s} \, dt,$$

since $g(x) := \psi(\lambda x)$ for $\lambda \neq 0$ implies $\mathscr{F}g(s) = \mathscr{F}\psi(s/\lambda)/\lambda$, and we have assumed $\alpha(-n) = \alpha(n)$ since $\psi(-x) = \psi(x)$ (see Step (2)) for $n \in \mathbb{N}$, gives for $q \geq 1$,

$$\sum_{\substack{n \equiv b \bmod q \\ n \in \mathbb{Z}}} \alpha_j(n) = \frac{N_j}{q} \sum_{m \in \mathbb{Z}} e_q(mb) \mathscr{F}\psi\left(\frac{mN_j}{q}\right)$$

$$= \frac{N_j}{q} \mathscr{F}\psi(0) + \frac{N_j}{q} \sum_{m \in \mathbb{Z} \setminus \{0\}} e_q(mb) \mathscr{F}\psi\left(\frac{mN_j}{q}\right).$$

Since ψ is smooth with bounded support, using the bound derived in the remark following the proof of Lemma 8.21, we obtain for $m \neq 0$ and $q \leq Q$,

$$\left| \mathscr{F}\psi\left(\frac{mN_j}{q}\right) \right| \prec^o \left(\frac{q}{mN_j}\right)^2.$$

Therefore,

$$\sum_{\substack{n \in \mathbb{Z} \\ n \equiv b \bmod q}} \alpha_j(n) = \frac{N_j}{q} \mathscr{F}\psi(0) + O\left(\frac{q}{N_j}\right),$$

which from the definition of the discrepancy Δ implies

$$|\Delta(\alpha_j, q, b)| \prec^o \frac{q}{N_j}.$$

Using the estimate (8.49) and recalling $Q \prec^o x^{\frac{1}{2}+2\varpi}$, $N_j \gg x^{\frac{1}{2}+\sigma}$ and $N_S N_j \asymp x$, we then get

$$\sum_{\substack{q \leq Q \\ q \in \mathcal{D}_\Omega(i, x^\delta)}} |\Delta(\alpha_1 \star \cdots \star \alpha_{2J}, q, a)| \prec^o N_S N_j \left(\frac{Q}{N_j}\right)^2 \ll x^{1-2\sigma+4\varpi} \ll \frac{x}{(\log x)^A},$$

where the final estimate on the right comes from the assumption $\sigma > 2\varpi$ of the lemma statement. This completes the proof of Lemma 8.22. $\qquad\square$

8.9 One-Dimensional Exponential Sums

Recall the definition $\mathbf{e}_q(x) := \exp(2\pi i x/q)$ regarded as a mapping $\mathbb{Z}/q\mathbb{Z} \to \mathbb{C}$. We extend this definition by setting for integers a and b, $\mathbf{e}_q(a/b) = \mathbf{e}_q(a\bar{b})$, where $\bar{b}$ is the inverse of b modulo q when $(b, q) = 1$ and 0 otherwise. To be more specific, for prime p we use the convention

$$\mathbf{e}_p\left(\frac{a}{b}\right) = \begin{cases} 0 & p \nmid a, p \mid b, \\ 1 & p \mid a, p \nmid b. \end{cases}$$

Next, we list some properties of the discrete Fourier transform. Let $q > 1$ be an integer and $f \colon \mathbb{Z}/q\mathbb{Z} \to \mathbb{C}$ a function defined on congruences classes modulo q. Define for $x \in \mathbb{Z}/q\mathbb{Z}$,

$$\mathscr{F}_q(f)(x) := \frac{1}{\sqrt{q}} \sum_{h \in \mathbb{Z}/q\mathbb{Z}} f(h) \mathbf{e}_q(hx). \tag{8.50}$$

With this normalization, the transform is unitary with respect to the natural inner product

$$\langle f, g \rangle := \sum_{x \in \mathbb{Z}/q\mathbb{Z}} f(x)\overline{g(x)} \implies \langle f, g \rangle = \langle \mathscr{F}_q(f), \mathscr{F}_q(g) \rangle,$$

which is the Plancherel formula in this setting. We also need the inversion formula

$$\mathscr{F}_q(\mathscr{F}_q(f))(x) = f(-x).$$

Define for $M \geq 1$ the abbreviation $\mathscr{L}_M := (\log M)^{O(1)}$.

The graph of Figure 8.7 shows how the lemmas of this section relate. Note the importance of the Weil bound Theorem E.9, which is proved in Appendix E. Note also how this feeds into the Weil bound for incomplete character sums, which is Lemma 8.29, and then onto the principal result of this section, which is Theorem 8.33 and its estimate (8.67). It is this estimate which is carried forward to the application to prime gaps. The reader who wishes to skip some of the detail might note the use of dense divisibility in Step (3) of Theorem 8.33.

This part of the work requires both conceptual depth and quite complex manipulation of inequalities. This latter aspect is especially error prone, and it is a testament to Polymath8a's work that the results stood up to this writer's scrutiny. However, errors may have been introduced and will be dutifully recorded/corrected in the online "errata and notes" for this book accessible from the author's website, if pointed out.

Lemma 8.23 **(Completing exponential sums)** *[174, lemma 6.9] Let $M \geq 1$ be real, $x_0 \in \mathbb{R}$, ψ a smooth function with support in a bounded interval, with derivative bounds for all $j \geq 0$, $|\psi^{(j)}(x)| \ll (\log M)^{O_j(1)}$. Define a shifted, dilated smooth function*

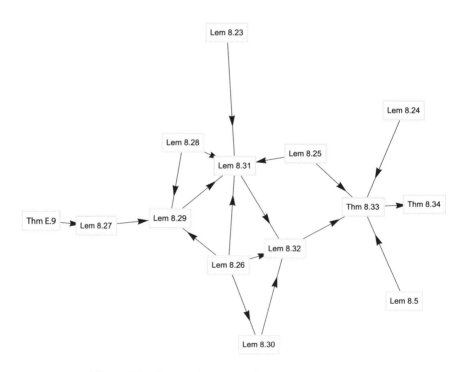

Figure 8.7 Chapter 8 exponential sums dependencies.

$$\psi_M(x) := \psi\left(\frac{x - x_0}{M}\right)$$

and a smooth q-periodic function on $\mathbb{R}$ by

$$\Psi_M(x) := \sum_{n \in \mathbb{Z}} \psi_M(x + qn).$$

Also define $M' := \sum_{m \in \mathbb{N}} \psi_M(m) \ll M\mathscr{L}_M$, and assume $M' \ll M\mathscr{L}_M$. For $q \geq 2$, $f : \mathbb{Z}/q\mathbb{Z} \to \mathbb{C}$ a function. Then if $M \ll q\mathscr{L}_M$, we have for any fixed $A > 0$ and $\epsilon > 0$,

$$\left| \sum_{m \in \mathbb{N}} \psi_M(m) f(m) - \frac{M'}{q} \sum_{m \in \mathbb{Z}/q\mathbb{Z}} f(m) \right| \ll_{A, \epsilon} \mathscr{L}_M \frac{M}{\sqrt{q}} \sum_{0 < |n| \leq qM^{-1+\epsilon}} |\mathscr{F}_q(f)(n)|$$

$$+ \frac{1}{M^A} \sum_{m \in \mathbb{Z}/q\mathbb{Z}} |f(m)|. \qquad (8.51)$$

In addition, we have the following corollary: let I be a finite index set and for $i \in I$, c_i a complex number and a_i mod q a congruence class. Then for fixed $A, \epsilon > 0$, we have

$$\left| \sum_{i \in I} c_i \sum_{m \equiv a_i \bmod q} \psi_M(m) - \frac{M'}{q} \sum_{i \in I} c_i \right|$$

$$\ll \mathscr{L}_M \frac{M}{q} \sum_{0 < |n| \leq qM^{-1+\epsilon}} \left| \sum_{i \in I} c_i e_q(a_i n) \right| + \frac{1}{M^A} \sum_{i \in I} |c_i|. \qquad (8.52)$$

Proof **(1)** Using periodicity, the Plancherel formula, and properties of the discrete Fourier transform, we get

$$\sum_{m \in \mathbb{N}} \psi_M(m) f(m) = \sum_{n \in \mathbb{Z}/q\mathbb{Z}} f(n) \Psi_M(n) = \sum_{n \in \mathbb{Z}/q\mathbb{Z}} \mathscr{F}_q(f)(n) \mathscr{F}_q(\Psi_M)(-n). \quad (8.53)$$

From the term on the right with $n = 0$, we have

$$\mathscr{F}_q(f)(0) \mathscr{F}_q(\Psi_M)(0) = \frac{1}{q} \sum_{m \in \mathbb{Z}/q\mathbb{Z}} f(m) \sum_{m \in \mathbb{Z}/q\mathbb{Z}} \Psi_M(m) = \frac{M'}{q} \sum_{m \in \mathbb{Z}/q\mathbb{Z}} f(m).$$

(2) Next, we consider the terms with $n \neq 0$. By the definition of Ψ_M, if we define also a function Ψ on $\mathbb{R}/\mathbb{Z}$ by

$$\Psi(x) := \sum_{m \in \mathbb{Z}} \psi_M(m) \exp(-2\pi i m x),$$

we get

$$\sqrt{q}\,\mathscr{F}_q(\Psi_M)(-n) = \Psi\left(\frac{n}{q}\right).$$

Summing all terms with $n \neq 0$ then gives

$$\left|\sum_{n\in\mathbb{Z}/q\mathbb{Z}\setminus\{0\}} \mathscr{F}_q(f)(n)\mathscr{F}_q(\Psi_M)(-n)\right| \leq \frac{1}{\sqrt{q}} \sup_{n\in\mathbb{Z}/q\mathbb{Z}\setminus\{0\}} \left|\mathscr{F}_q(f)(n)\right| \sum_{\substack{-q/2<n\leq q/2 \\ n\neq 0}} \left|\Psi\left(\frac{n}{q}\right)\right|.$$

Recalling $\psi_M(x) := \psi((x - x_0)/M)$ and using Poisson summation, we get

$$\Psi(x) = M \sum_{n\in\mathbb{Z}} \hat{\psi}(M(n + x)) \exp(-2\pi i(n + x)x_0).$$

(3) Using applications of integration by parts, and the bounded support of ψ_M, for fixed $A > 0$ and $-\frac{1}{2} < x \leq \frac{1}{2}$, we get the bound

$$|\mathscr{F}\psi(z)| \ll \frac{\mathscr{L}_M}{(1 + |z|)^A} \implies |\Psi(x)| \ll \frac{M\mathscr{L}_M}{(1 + |x|M)^A}.$$

Using $A = 2$ gives

$$\sum_{\substack{-q/2<n\leq q/2 \\ n\neq 0}} \left|\Psi\left(\frac{n}{q}\right)\right| \ll \mathscr{L}_M \sum_{1\leq n\leq q/2} \frac{M}{(1 + |n|M/q)^2} \ll q\mathscr{L}_M.$$

Therefore,

$$\left|\sum_{m\in\mathbb{Z}} \psi_M(m)f(m) - \frac{M'}{q}\sum_{m\in\mathbb{Z}/q\mathbb{Z}} f(m)\right| \ll_{A,\epsilon} \sqrt{q}\,\mathscr{L}_M \sup_{n\in\mathbb{Z}/q\mathbb{Z}\setminus\{0\}} |\mathscr{F}_q(f)(n)|, \tag{8.54}$$

and so for $M \ll q\mathscr{L}_M$ we get

$$\left|\sum_{m\in\mathbb{Z}} \psi_M(m)f(m)\right| \ll \sqrt{q}\,\mathscr{L}_M \|\mathscr{F}_q(f)\|_{\ell^\infty(\mathbb{Z}/q\mathbb{Z})}. \tag{8.55}$$

(4) We are now able to complete the proof of the main estimate of the lemma. Let $A > 0$ and $\epsilon > 0$ be fixed. Then, with the same approach used in Step (2) and the sum from Step (1), we get

$$\left| \sum_{m\in\mathbb{Z}} \psi_M(m) f(m) - \frac{M'}{q} \sum_{m\in\mathbb{Z}/q\mathbb{Z}} f(m) \right| \le \frac{1}{\sqrt{q}} \sum_{\substack{-q/2 < n \le q/2 \\ n \ne 0}} \left| \hat{\Psi}\left(\frac{n}{q}\right) \right| \cdot |\mathscr{F}_q(f)(n)|$$

$$\ll \frac{M\mathscr{L}_M}{\sqrt{q}} \sum_{0 < |n| \le qM^{-1+\epsilon}} |\mathscr{F}_q(f)(n)|$$

$$+ \mathscr{L}_M \left(\sum_{n\in\mathbb{Z}/q\mathbb{Z}} |f(n)| \right) \left(\sum_{|n| > qM^{-1+\epsilon}} \frac{M}{q(1 + |n|M/q)^A} \right).$$

The proof is completed by choosing A sufficiently large.

To derive the corollary, in the estimate (8.51) use

$$f(m) := \sum_{\substack{i\in I \\ m \equiv a_i \bmod q}} c_i \implies \mathscr{F}_q(f)(n) = \frac{1}{\sqrt{q}} \sum_{i\in I} c_i e_q(a_i n).$$

The corollary then follows directly. □

We need, as is often the case in work of this nature, bounds for sums related to the number of divisors of n function $\tau(n)$. These can be found in various places in the literature, but are repeated here for ease of reference, since the proofs are of interest and might give the reader some relief before the mountain climbing which awaits.

Lemma 8.24 (Number of divisors bounds) *[174, lemma 5.1]*
(i) For all $\epsilon > 0$ and n sufficiently large, we have $\tau(n) \le 2^{(1+\epsilon)\log n/\log\log n}$. Thus if $n \le x^{O(1)}$, then $\tau(n) \prec^o 1$.
(ii) If $C > 0$ and $1 < y \le x^{O(1)}$, then

$$\sum_{n \le y} \tau(n)^C \ll y(\log x)^{O(1)}.$$

(iii) If $C > 0$, $1 < y \le x^{O(1)}$ and $a \bmod q$ is a fixed congruence class with $q \le y^\delta$ for some $0 < \delta < 1$, then

$$\sum_{\substack{n \le y \\ n \equiv a \bmod q}} \tau(n)^C \ll \frac{y}{\varphi(q)} \tau(q)^{O(1)} (\log x)^{O(1)}.$$

(iv) For all $C > 0$, there is a $B > 0$, dependent only on C such that as $x \to \infty$,

$$\sum_{n \le x} \frac{\tau(n)^C}{\varphi(n)} \ll (\log x)^B.$$

Proof We prove (i) in Steps (1) through (3), (ii) in (4) through (6), (iii) in (7) and (iv) in (8).

(1) Let $n = p_1^{a_1} \cdots p_m^{a_m}$ so $\tau(n) = \prod_{j=1}^m (1 + a_j)$. Split this product at $f(n)$, to be assigned later in the proof, so $\tau(n) = T_1 \times T_2$ with

$$T_1 = \prod_{\substack{j=1 \\ p_j \leq f(n)}}^m f(n)(1 + a_j) \text{ and } T_2 = \prod_{\substack{j=1 \\ p_j > f(n)}}^m (1 + a_j).$$

To estimate T_1, first write

$$T_1 = \exp\left(\sum_{p_j \leq f(n)} \log(a_j + 1) \right).$$

We **claim** that if n is sufficiently large, then we must have $\log(a_j + 1) < 2 \log\log n$. To see this, we have

$$2^{a_j} \leq p_j^{a_j} \leq n \implies a_j \leq \frac{\log n}{\log 2}.$$

Thus for n sufficiently large, we have

$$1 + a_j \leq 1 + \log n / \log 2 < (\log n)^2 \implies \log(1 + a_j) < 2 \log\log n,$$

proving the claim. Thus using $\pi(x) < 6x / \log x$, which is valid for all $x \geq 2$,

$$T_1 < \exp\left(2 \log\log n \sum_{p_j \leq f(n)} 1 \right) \leq \exp((2 \log\log n)\pi(f(n)))$$

$$< \exp\left(\frac{12 f(n) \log\log n}{\log f(n)} \right) = 2^{12 f(n) \log\log n / (\log 2 \log f(n))}.$$

(2) For T_2, use the inequality $(x + 1) \leq 2^x$ valid for all $x \geq 1$ to get

$$S(n) := \sum_{\substack{j=1 \\ p_j > f(n)}}^m a_j \implies T_2 \leq 2^{S(n)}.$$

We can also write

$$n = \prod_{j=1}^m p_j^{a_j} \geq \prod_{p_j > f(n)} p_j^{a_j} \geq \prod_{p_j > f(n)} f(n)^{a_j} = f(n)^{S(n)}.$$

Therefore,

$$\log n \geq S(n) \log f(n) \implies S(n) \leq \frac{\log n}{\log f(n)} \implies T_2 \leq 2^{\log n / \log f(n)}.$$

(3) In what follows, we let $\eta = 12/\log 2$. Using Steps (1) and (2), we get

$$\tau(n) = T_1 \times T_2 < 2^{g(n)}$$

where

$$g(n) := \frac{\log n + \eta f(n) \log\log n}{\log f(n)} = \frac{\log n}{\log\log n} \times \frac{1 + \frac{\eta f(n) \log\log n}{\log n}}{\frac{\log f(n)}{\log\log n}}.$$

Now make the choice

$$f(n) := \frac{\log n}{(\log\log n)^2},$$

so as $n \to \infty$ we get $f(n) \log\log n / \log n \to 0$ and $\log f(n)/\log\log n \to 1$. With this choice, for n sufficiently large we have

$$g(n) = \frac{\log n(1 + o(1))}{\log\log n(1 + o(1))} = (1 + o(1))\frac{\log n}{\log\log n} < (1 + \epsilon)\frac{\log n}{\log\log n}.$$

Therefore,

$$\tau(n) < 2^{g(n)} < 2^{(1+\epsilon)\frac{\log n}{\log\log n}}.$$

This completes the proof of Part (i).

(4) Now let $f(n)$ be a given nonnegative multiplicative function and assume there is a constant $A > 0$ such that

$$\sum_{p \le x} f(p) \log p \le Ax \quad \text{and} \quad \sum_{p^m, m \ge 2} \frac{f(p^m) \log(p^m)}{p^m} \le A.$$

Then we **claim** as $x \to \infty$ we have

$$\log x \sum_{n \le x} f(n) \ll (A + 1)x \sum_{n \le x} \frac{f(n)}{n}.$$

To see this, first because for $1 \le n \le x$, we have $\log(x/n) \le x/n$, we get

$$\sum_{n \le x} f(n) \log(x/n) \le x \sum_{n \le x} \frac{f(n)}{n}. \tag{8.56}$$

We have for $n \in \mathbb{N}$, $\log n = \sum_{d|n} \Lambda(n)$. If p is prime and $m \in \mathbb{N}$, let $m = p^j r$, $p \nmid r$, $d = p^i$ and $l = i + j$. Then

$$\sum_{n \le x} f(n) \log n = \sum_{d \le x} \Lambda(d) \sum_{m \le x/d} f(md)$$

$$= \sum_{\substack{p,i \ge 1, j \ge 0 \\ p^{i+j} \le x}} (\log p) f(p^{i+j}) \sum_{\substack{r \le x/p^{i+j} \\ p \nmid r}} f(r)$$

$$= \sum_{\substack{p,l \\ p^l \le x}} l(\log p) f(p^l) \sum_{\substack{r \le x/p^l, \, p \nmid r}} f(r).$$

Next, drop the constraint $p \nmid r$ and use

$$\sum_{n \le x} f(n) \le x \sum_{n \le x} \frac{f(n)}{n},$$

replacing x by x/p. By the first assumption, the terms with $l \ge 2$ contribute to the upper bound on the right of $\sum_{n \le x} f(n) \log n$ a quantity

$$x \sum_{p, l \ge 2} \log(p^l) \frac{f(p^l)}{p^l} \sum_{r \le x/p^l} \frac{f(r)}{r} \le Ax \sum_{n \le x} \frac{f(n)}{n}.$$

By the first assumption, the terms with $l = 1$ contribute to the bound

$$\sum_{p \le x} (\log p) f(p) \sum_{r \le x/p} f(r) = \sum_{r \le x} f(r) \sum_{p \le x/r} f(p) \log p \le Ax \sum_{r \le x} \frac{f(r)}{r}.$$

Therefore,

$$\sum_{n \le x} f(n) \log n \ll Ax \sum_{n \le x} \frac{f(n)}{n}.$$

Adding this estimate to that of (8.56), we get

$$\log x \sum_{n \le x} f(n) \ll (A + 1)x \sum_{n \le x} \frac{f(n)}{n},$$

verifying the claim.

(5) We now apply the result of Step (4) to $f(n) = \tau(n)^C$, where $C > 0$ is a fixed real constant. Such an $f(n)$ satisfies all of the conditions in Step (4). In addition, by multiplicativity, we have

$$\sum_{n \le x} \frac{f(n)}{n} \le \prod_{p \le x} \left(1 + \frac{f(p)}{p} + \frac{f(p^2)}{p^2} + \cdots \right),$$

and note also that $f(p) = 2^C$, $f(p^2) = 3^C$, etc. Write

$$\prod_{p \leq x} \left(1 + \frac{2^C}{p} + \frac{3^C}{p^2} + \cdots \right) = \left(\prod_{p \leq x} \left(1 - \frac{1}{p} \right) \right)^{-2^C}$$

$$\times \left(\prod_{p \leq x} \left(1 - \frac{1}{p} \right)^{2^C} \times (1 + \frac{2^C}{p} + \frac{3^C}{p^2} + \cdots) \right)$$

and observe that the second factor tends to a finite positive nonzero limit. The first factor, by Merten's theorem, is equal asymptotically to a positive constant multiple of $(\log x)^{2^C}$. Therefore, using the result of Step (4), we can write

$$\sum_{n \leq x} \tau(n)^C \ll x (\log x)^{2^C - 1} = x (\log x)^{O(1)}.$$

This completes the proof of Part (ii).

(6) Let $g = (a, q)$ be the GCD and set $q = gq'$, $a = ga'$ and $n = gn'$ so in particular $(a', q') = 1$. Then

$$\sum_{\substack{n \leq y \\ n \equiv a \bmod q}} \tau(n)^C = \sum_{\substack{n' \leq y/g \\ n' \equiv a' \bmod q'}} \tau(gn')^C$$

$$\leq \tau(g)^C \sum_{\substack{n' \leq y/g \\ n' \equiv a' \bmod q'}} \tau(n')^C$$

$$\leq \tau(q)^{O(1)} \sum_{\substack{n' \leq y/g \\ n' \equiv a' \bmod q'}} \tau(n')^C.$$

Thus in (iii) we need only consider the case where $(a, q) = 1$ with y replaced by y/g.

(7) Now consider the special case $f(n) = \tau(n)^C$, where $f(n)$ satisfies for some constants A, B and all primes p and $m \in \mathbb{N}$, $f(p^m) \leq A^m$, and by Step (1) $f(n) \leq Bn^\epsilon$ for all $\epsilon > 0$ and $n \in \mathbb{N}$. Therefore, with this definition of $f(n)$ and $(a, q) = 1$, we can use the result of Shiu's Theorem D.1 in Appendix D.
    First we see that since $f(p) = 2^C$, we get

$$\exp \left(\sum_{\substack{p \leq y \\ p \nmid q}} \frac{f(p)}{p} \right) \leq \exp \left(O(1) \sum_{p \leq y} \frac{1}{p} \right) \leq (\log y)^{O(1)} \ll (\log x)^{O(1)}.$$

Identifying x with y in Theorem D.1 and using this bound assuming $(a,q) = 1$, we get

$$\sum_{\substack{n \le y \\ n \equiv a \bmod q}} f(n) \ll \frac{y}{\varphi(q) \log y}(\log x)^{O(1)} \ll \frac{y}{\varphi(q)}(\log x)^{O(1)}.$$

Hence

$$\sum_{\substack{n' \le y/g \\ n' \equiv a' \bmod q'}} f(n') \ll \frac{y/g}{\varphi(q/g)}(\log x)^{O(1)} \ll \frac{y}{\varphi(q)}(\log x)^{O(1)}.$$

Therefore, using the result of Step (6) we get

$$\sum_{\substack{n \le y \\ n \equiv a \bmod q}} \tau(n)^C \ll \frac{y}{\varphi(q)}\tau(q)^{O(1)}(\log x)^{O(1)},$$

which completes the proof of Part (iii).

(8) First, we apply Merten's theorem to get for $n \ge 2$,

$$\frac{\varphi(n)}{n} = \prod_{p|n}\left(1 - \frac{1}{p}\right) \gg \frac{1}{\log n} \implies \frac{1}{\varphi(n)} \ll \frac{\log n}{n}.$$

Using this inequality and the method of Step (5), we derive

$$\sum_{n \le x} \frac{\tau(n)^C}{\varphi(n)} \ll \sum \frac{\tau(n)^C}{n} \log x$$

$$= \log x \prod_{p \le x}\left(1 + \frac{2^C}{p} + \frac{3^C}{p^2} + \cdots\right)$$

$$\ll \log x \prod_{p \le x}\left(1 - \frac{1}{p}\right)^{-2^C}$$

$$\ll (\log x)^{2^C+1},$$

which completes the proof of (iv).   □

It is interesting to see the GCD sum appearing in this context as it has appeared quite a few times in recent years, for example in [20].

Lemma 8.25   (GCD sum bounds) *[174, lemma 1.6] (i) Let $q \ge 2$ and $x \ge 1$, then*

$$\sum_{n \le x}(n,q) \le x\tau(q).$$

(ii) If $q \ll x^{O(1)}$, then

$$\sum_{a \in \mathbb{Z}/q\mathbb{Z}} (a,q) \prec^o q.$$

(iii) If $\epsilon > 0$, then as $x \to \infty$

$$\sum_{n \leq x} (n,q) \ll_{\epsilon} xq^{\epsilon} + q.$$

Proof We prove only (i) – the proofs of (ii) and (iii) are similar.
Since $(n,q) \leq \sum_{d|(n,q)} d$, if $k = \lfloor x \rfloor$ we get

$$\sum_{n \leq x} (n,q) \leq \sum_{\substack{1 \leq n \leq k \\ d|n \\ d|q}} \sum d$$

$$\leq \sum_{d|q} d \sum_{1 \leq e \leq k/d} 1 \leq \sum_{d|q} \frac{kd}{d}$$

$$\leq x\tau(q). \qquad \square$$

Recall the definition of the additive character. For $q \geq 2$, we define

$$\mathbf{e}_q(x) := \exp(2\pi i x/q).$$

Then $\mathbf{e}_q(x + y) = \mathbf{e}_q(x) \cdot \mathbf{e}_q(y)$, $x \equiv y \bmod q$ implies $\mathbf{e}_q(x) = \mathbf{e}_q(y)$, and $q \mid n$ implies $\mathbf{e}_q(n) = 1$.

Lemma 8.26 **(Chinese remainder characters)** *[174, lemma 6.4] Let $q_1, q_2 \in \mathbb{N}$ have $(q_1, q_2) = 1$. Then for any $n \in \mathbb{Z}$, we have*

$$\mathbf{e}_{q_1 q_2}(n) = \mathbf{e}_{q_1}\left(\frac{n}{q_2}\right) \mathbf{e}_{q_2}\left(\frac{n}{q_1}\right). \qquad (8.57)$$

Using induction, we also get for $k \geq 2$ and k pairwize coprime natural numbers q_j

$$\mathbf{e}_{q_1 \cdots q_k}(n) = \prod_{j=1}^{k} \mathbf{e}_{q_j}\left(\frac{n}{\prod_{i \neq j} q_i}\right).$$

Proof Let integers $\overline{q_1}$ and $\overline{q_2}$ be such that $q_1 \overline{q_1} \equiv 1 \bmod q_2$ and $q_2 \overline{q_2} \equiv 1 \bmod q_1$. Then $q_1 q_2 \mid q_1 \overline{q_1} + q_2 \overline{q_2} - 1$, so we can write

$$\mathbf{e}_{q_1 q_2}(n) = \mathbf{e}_{q_1 q_2}(n(q_1 \overline{q_1} + q_2 \overline{q_2}))$$

$$= \mathbf{e}_{q_1 q_2}(nq_1 \overline{q_1}) \mathbf{e}_{q_1 q_2}(nq_2 \overline{q_2})$$

$$= \mathbf{e}_{q_1}(n\overline{q_2}) \mathbf{e}_{q_2}(n\overline{q_1}),$$

which completes the proof. $\qquad \square$

Example: Consider the following sum and use Lemma 8.26, noting that $2 \times 2 \equiv$ 1 mod 3 and $3 \times 1 \equiv 1$ mod 2, to get

$$\sum_{n \in \mathbb{Z}/6\mathbb{Z}} \mathbf{e}_6(n) = \mathbf{e}_6(0) + \mathbf{e}_6(1) + \mathbf{e}_6(2) + \mathbf{e}_6(3) + \mathbf{e}_6(4) + \mathbf{e}_6(5)$$

$$= \mathbf{e}_2(0 \times 1)\mathbf{e}_3(0 \times 2) + \mathbf{e}_2(1 \times 1)\mathbf{e}_3(1 \times 2) + \mathbf{e}_2(2 \times 1)\mathbf{e}_3(2 \times 2)$$

$$+ \mathbf{e}_2(3 \times 1)\mathbf{e}_3(3 \times 2) + \mathbf{e}_2(4 \times 1)\mathbf{e}_3(4 \times 2) + \mathbf{e}_2(5 \times 1)\mathbf{e}_3(5 \times 2)$$

$$= (\mathbf{e}_2(0) + \mathbf{e}_2(1))(\mathbf{e}_3(0) + \mathbf{e}_3(1) + \mathbf{e}_3(2)).$$

The following lemma is a key result for the development. It is a consequence of the Riemann Hypothesis for curves over finite fields, which is proved in Theorem E.9 in Appendix E. A full history of the problem and other proofs are given in [32].

Lemma 8.27 **(Weil bound)** *[174, lemma 6.2] Let $P, Q \in \mathbb{Z}[X]$ be given polynomials and p a prime such that $Q(X)$ mod p is nonzero and such that there is **no** rational function $g(X) \in \mathbb{F}_p(X)$ such that in $\mathbb{F}_p$ we have*

$$\frac{P}{Q} \equiv (g^p - g + c) \bmod p.$$

Then we have the estimate

$$\left| \sum_{x \in \mathbb{F}_p} \mathbf{e}_p \left(\frac{P(x)}{Q(x)} \right) \right| \ll \sqrt{p},$$

where the implied constant depends only on $\max(\deg(P), \deg(Q))$ in a linear manner. If a rational function $g(X)$ with the stated property exists, then we can do no better than the trivial bound p.

Proof   Let

$$\frac{P(x)}{Q(x)} = \frac{P_1(x)}{Q_1(x)},$$

where $P_1(x), Q_1(x)$ are coprime polynomials over $\mathbb{F}_p$. Then if $\overline{Q_1(x)}$ is the multiplicative inverse of $Q_1(x)$ in $\mathbb{F}_p[x]$, we can write, recalling $\mathbf{e}_p(a/b)$ is defined to be 0 if $p \mid b$ and using Theorem E.9, with p sufficiently large so $p > \deg P_1(x)\overline{Q_1(x)})$ to get the last line,

$$\left| \sum_{x\in\mathbb{F}_p} \mathbf{e}_p\left(\frac{P(x)}{Q(x)}\right)\right| = \left| \sum_{\substack{x\in\mathbb{F}_p \\ Q_1(x)\neq 0}} \mathbf{e}_p\left(P_1(x)\overline{Q_1(x)}\right)\right|$$

$$\leq \left| \sum_{x\in\mathbb{F}_p} \mathbf{e}_p\left(P_1(x)\overline{Q_1(x)}\right)\right|$$

$$\ll \sqrt{p}.$$

This completes the proof. $\qquad\square$

If $f \in \mathbb{Z}(X)$ and $q \mid \mathbb{N}$, then we say $q \mid f$ or $f(X) \equiv 0 \bmod q$ provided q divides each coefficient of the numerator of $f(X)$ when written in lowest terms $f = P/Q$ with $P, Q \in \mathbb{Z}[X]$ with $(P, Q) = 1$. In addition, if $q \in \mathbb{N}$ and $f \in \mathbb{Z}[X]$, let (f, q) be the largest factor of q which divides $f(x)$ in $\mathbb{Z}[x]$ and extend this definition to $f(x)$ in $\mathbb{Z}(x)$, the quotient ring, as follows: if $f = P/Q$ with $(P, Q) = 1$ in $\mathbb{Z}[X]$, then $(f, q) := (P, q)$.

> **Lemma 8.28** *[174, lemma 6.5] Let the rational function $f = P/Q$ be in $\mathbb{Q}(X)$ with $P, Q \in \mathbb{Z}[X]$ coprime. Let $p \in \mathbb{P}$ such that $Q(X) \bmod p$ is nonzero.*
> *(i) If $p \mid f'$, and p is sufficiently large depending on $\deg(P)$ and $\deg(Q)$, then there is an integer c such that $p \mid f - c$.*
> *(ii) If q is squarefree, $Q(X) \bmod p$ has degree $\deg Q$ for all $p \mid q$, $\deg P < \deg Q$, and all primes dividing q are sufficiently large in terms of $\deg P$ and $\deg Q$, then $(f', q) \mid (f, q)$. Thus if $(f, q) = 1$, we must have $(f', q) = 1$ also.*

Proof (i) First, we note that

$$f = \frac{P}{Q} \implies f' = \frac{P_1}{Q_1} \text{ with } (P_1, Q_1) = 1 \text{ in } \mathbb{Z}[X].$$

The condition $p \mid f'$ implies $P_1(x) \equiv 0 \bmod p$ for all $x \in \mathbb{Z}/p\mathbb{Z}$.

If $Q_1(X) \bmod p \neq 0$ in $(\mathbb{Z}/p\mathbb{Z})[X]$, then $f' \bmod p$ is well defined except at the at most $\deg Q$ zeros of Q_1 in $\mathbb{Z}/p\mathbb{Z}$. Since $p \mid f'$, f' takes the value zero at not less than $p - \deg Q$ values.

Therefore, if p is sufficiently large with respect to $\deg P$ and $\deg Q$, we must have $f'(X) \equiv 0 \bmod p$ and we can write in $\mathbb{Z}[X]$, $f'(X) = p\widehat{P_1}(X)/Q_1(X)$. This implies $p \mid f^{(n)}$ for all $n \geq 1$, so using the Taylor expansion for $f(X) \bmod p$, we get $f(X) \equiv c \bmod p$. This implies $f \equiv c \bmod p$, so $p \mid f - c$. This completes the proof.

(ii) If $p \mid (f', q)$, by Part (i), there is a $c \in \mathbb{Z}$ such that $p \mid f - c$. If $p \nmid (f, q)$, then $c \neq 0$. In that case, we get

$$p \mid f - c = \frac{P - cQ}{Q} \implies p \mid P - cQ,$$

where $\deg(P - cQ) = \deg Q \geq 1$ since $\deg Q > \deg P$. For $p > \deg Q$, considering $P - cQ$ as a polynomial over $\mathbb{Z}/p\mathbb{Z}$, the degree is 0, which is a contradiction. Therefore, $p \mid (f,q)$, which completes the proof of the lemma. $\quad\square$

Lemma 8.29 (**Weil exponential sum bounds**) *[174, proposition 6.6] Let $q \in \mathbb{N}$ be squarefree and $P, Q \in \mathbb{Z}[X]$ coprime polynomials with Q nonzero modulo p for all primes $p \mid q$. Then if we define a rational function of real x by $f(x) := P(x)/Q(x)$, we have, for a constant $C > 0$, which depends at most on $\deg(P)$ and $\deg(Q)$, the bound*

$$\left| \sum_{n \in \mathbb{Z}/q\mathbb{Z}} \exp(2\pi i f(n)/q) \right| \leq C^{\Omega(q)} \sqrt{q} \, \frac{(f',q)}{\sqrt{(f'',q)}}.$$

Proof (**1**) Using the Chinese Remainder formula of Lemma 8.26 and interchanging the summation and products (see the example following Lemma 8.11), we get

$$\sum_{n \in \mathbb{Z}/q\mathbb{Z}} \mathbf{e}_q(f(n)) = \prod_{p \mid q} \sum_{n \in \mathbb{Z}/p\mathbb{Z}} \mathbf{e}_p\left(\frac{f(n)}{q/p} \right).$$

Note that for each divisor $p \mid q$, the integer q/p is invertible in $\mathbb{Z}/p\mathbb{Z}$. Thus we need only verify the estimates

$$(a) \quad \sum_{n \in \mathbb{Z}/p\mathbb{Z}} \mathbf{e}_p(f(n)) \ll p, \quad \text{if } p \mid f', \, p \mid f'',$$

$$(b) \quad \sum_{n \in \mathbb{Z}/p\mathbb{Z}} \mathbf{e}_p(f(n)) \ll 1, \quad \text{if } p \mid f'', \, p \nmid f',$$

$$(c) \quad \sum_{n \in \mathbb{Z}/p\mathbb{Z}} \mathbf{e}_p(f(n)) \ll \sqrt{p}, \quad \text{if } p \nmid f'', \, p \nmid f', \tag{8.58}$$

where the implied constants depend at most on $\deg(P)$, $\deg(Q)$.

(**2**) If $p \mid q$ is small, the estimates (a) through (c) follow immediately since the left-hand side in each case is bounded above by p. The estimate (a) follows for arbitrary p for this same reason.

(**3**) Consider the estimate (b). By Lemma 8.28, since p can be taken sufficiently large, there is a $c \in \mathbb{Z}/p\mathbb{Z}$ such that $p \mid f' - c$. Then since $p \nmid f'$, c is not zero. But $f'(X) - c = (f(X) - cX)'$. By Lemma 8.28, again there is a $d \in \mathbb{Z}/p\mathbb{Z}$ such that $p \mid f(X) - cX - d$. Therefore, provided n is not a zero of the denominator of $f(X)$, we have $f(n) \equiv cn + d \bmod p$. Now the denominator $Q(X)$ has at most $\deg(Q)$ zeros in $\mathbb{Z}/p\mathbb{Z}$, so we must have $\mathbf{e}_p(f(n)) = \mathbf{e}_p(cn + d)$ for all but at most that number of values $n \in \mathbb{Z}/p\mathbb{Z}$. Therefore,

$$\left| \sum_{n \in \mathbb{Z}/p\mathbb{Z}} \mathbf{e}_p(f(n)) \right| = \left| \sum_{n \in \mathbb{Z}/p\mathbb{Z}} \mathbf{e}_p(f(n)) - \sum_{n \in \mathbb{Z}/p\mathbb{Z}} \mathbf{e}_p(cn + d) \right| \le \deg(Q),$$

which, since $p \nmid c$ implies

$$\sum_{n \bmod p} \mathbf{e}_p(cn + d) = (\sum_{n \bmod p} \mathbf{e}_p(n)) \mathbf{e}_p(\bar{c}d) = 0,$$

gives the estimate (b).

(4) Finally consider the estimate (c). This is follows directly from the Weil bound, Lemma 8.27, provided that the reduction of f modulo p does *not* satisfy an identity for some $g \in \mathbb{F}_p(X)$, which has the form

$$f \equiv g^p - g + c \bmod p.$$

Assume we have such an equivalence. Then any pole of g of order $n \ge 1$ would be a pole of f of order np, so for p larger than the order of any pole of f, $g \bmod p$, and so $f \bmod p$ must be a polynomial. In particular, $g^p \equiv g \bmod p$. Thus $f - c$ is zero modulo p and thus $f = c + ph$ for some rational function h. Therefore, we must have $p \mid f'$, which is not allowed. This completes the proof. $\square$

> **Lemma 8.30** *[174, lemma 6.8] Let q_1, q_2 be squarefree integers (so the LCM $[q_1, q_2]$ is squarefree), and let c_1, c_2, l_1, l_2 be integers. Then if each $p \mid [q_1, q_2]$ is sufficiently large in terms of $|l_1|$ and $|l_2|$, there is a constant $C \ge 1$ such that if $\delta_1 := q_1/(q_1, q_2)$ and $\delta_2 := q_2/(q_1, q_2)$, then*

$$\left| \sum_{n \in \mathbb{Z}/[q_1, q_2]\mathbb{Z}} \mathbf{e}_{q_1}\left(\frac{c_1}{n + l_1} \right) \mathbf{e}_{q_2}\left(\frac{c_2}{n + l_2} \right) \right| \ll_\epsilon C^{\Omega([q_1, q_2])} (c_1, \delta_1)(c_2, \delta_2)(q_1, q_2).$$

Proof First use Lemma 8.26 to reduce to where $[q_1, q_2] = p$, a prime. In case one of the three GCDs (c_i, δ_i) or (q_1, q_2) is p, then the trivial bound for the sum on the left is p, so the result is immediate. Thus we can assume $q_1 = p$, $q_2 = 1$ and $(c_1, \delta_1) = (c_1, p) = 1$.

Then, using Ramanujan's sum Lemma 3.4 (or for example, [4, theorem 8.6]),

$$c_k(n) := \sum_{\substack{m=1 \\ (m,k)=1}}^{k} \mathbf{e}_k(mn) = \sum_{d \mid (n,k)} d\mu(k/d),$$

and the substitution $m = c_1/(n + l_1)$, since p is sufficiently large (e.g., $p > 2|l_i|$) so $p \nmid n + l_i$, and also using $p \nmid c_1$, we get

$$
\left| \sum_{n \in \mathbb{Z}/p\mathbb{Z}} \mathbf{e}_p \left(\frac{c_1}{n+l} \right) \mathbf{e}_1 \left(\frac{c_2}{n+l} \right) \right| = \left| \sum_{n \in \mathbb{Z}/p\mathbb{Z}} \mathbf{e}_p \left(\frac{c_1}{n+l_1} + \frac{pc_2}{n+l_2} \right) \right|
$$

$$
= \left| \sum_{n \in \mathbb{Z}/p\mathbb{Z}} \mathbf{e}_p \left(\frac{c_1}{n+l_1} \right) \right| = \left| \sum_{m \in \mathbb{Z}/p\mathbb{Z}} \mathbf{e}_p(m) \right| = |c_p(1)| = 1.
$$

This completes the proof. $\qquad\qquad\qquad\qquad\qquad\qquad\qquad\qquad\qquad\qquad\square$

Lemma 8.31 (A Weil bound for incomplete character sums) *[174, proposition 6.12] Let $q \in \mathbb{N}$ be squarefree, $P, Q \in \mathbb{Z}[X]$ such that the leading coefficient of Q is coprime with q and $\deg(P) < \deg(Q)$, and $f = P/Q$ be a rational function. Let $q_1 := q/(f,q) = q/(P,q)$. Let a positive real N be given with $N \ll q^{O(1)}$, $x_0 \in \mathbb{R}$, and ψ a smooth function with support in a bounded interval, with derivative bounds for all $j \geq 0$, $|\psi^{(j)}(x)| \ll_j (\log N)^{O(1)}$. Define a shifted, dilated smooth function*

$$
\psi_N(x) := \psi \left(\frac{x - x_0}{N} \right).
$$

(i) We have the so-called Polyá–Vinogradov–Ramanujan–Weil estimate, for every $\epsilon > 0$ with an implied constant depending only on ϵ and degrees of P, Q

$$
\sum_{n \in \mathbb{Z}} \psi_N(n) \mathbf{e}_q(f(n)) \ll q^\epsilon \left(\sqrt{q} + \frac{N}{q} \left| \sum_{n \in \mathbb{Z}/q\mathbb{Z}} \mathbf{e}_q(f(n)) \right| \right). \tag{8.59}
$$

(ii) If $q = rs$ then for any given $\epsilon > 0$, we have the so-called van der Corput–Ramanujan–Weil estimate

$$
\sum_{n \in \mathbb{Z}} \psi_N(n) \mathbf{e}_q(f(n)) \ll q^\epsilon \left(N^{\frac{1}{2}} \left(r^{\frac{1}{2}} + s^{\frac{1}{4}} \right) + \frac{N}{q} \left| \sum_{n \in \mathbb{Z}/q\mathbb{Z}} \mathbf{e}_q(f(n)) \right| \right). \tag{8.60}
$$

Proof **(1)** First, we make some simplifications. We may assume q has no prime factor smaller than a B which depends only on $\deg(P)$, $\deg(Q)$. If this is so, the general case follows by writing $q = q_1 q_2$, where all prime factors p of q_1 have $p \leq B$, and because ψ has compact support the sum over $n \in \mathbb{N}$ can be divided up into a finite number of pieces. We write for $n \equiv b \bmod q_1$,

$$
\sum_{n \in \mathbb{N}} \psi_N(n) \mathbf{e}_q(f(n)) = \sum_{b \bmod q_1} \mathbf{e}_{q_1} \left(\frac{f(b)}{q_2} \right) \sum_{n \in \mathbb{N}} \psi_N(n) \mathbf{e}_{q_2} \left(\frac{f(n)}{q_1} \right).
$$

We may also assume we can replace f by $f/(f,q)$, q by q_1, r by r_1 and s by s_1. This is because if we set $q = q_1 q_2$, then

$$e_q(f(n)) = e_{q_1}\left(\frac{P(n)}{q_2 Q(n)}\right).$$

In other words, we can assume $(f,q) = 1$, $q = q_1$, $r = r_1$, $s = s_1$. In this situation, because $\deg(P) < \deg(Q)$ and so $\deg P'Q - Q'P < \deg Q^2$, we have also by Lemma 8.28 (ii) $(f',q) = (f'',q) = 1$, since we can assume all primes dividing q are sufficiently large using the first simplification.

(2) In this first main step, we apply Lemma 8.23. Taking $A > 0$ sufficiently large and using $N \ll q^{O(1)}$, we get

$$\sum_{n\in\mathbb{Z}} \psi_N(n)e_q(f(n)) \ll \frac{N^{1+\epsilon}}{\sqrt{q}} \sum_{|m|\leq qN^{-1+\epsilon}} \left|\sum_{n\in\mathbb{Z}/q\mathbb{Z}} e_q(f(n)+mn)\right| + 1.$$

In case $N < q$, using Lemma 8.29 for all m, and because

$$(f'',q) = 1 \implies (f'+m,q) = 1,$$

we get

$$\sum_{n\in\mathbb{Z}} \psi_N(n)e_q(f(n)) \ll \frac{N^{1+\epsilon}}{\sqrt{q}} \sum_{|m|\leq qN^{-1+\epsilon}} (f'+m,q)$$

$$\ll \sqrt{q}N^{2\epsilon}.$$

This implies, recalling $q_1 := q/(f,q)$,

$$\sum_{n\in\mathbb{Z}} \psi_N(n)e_q(f(n)) \ll q^\epsilon\left(\sqrt{q_1} + \frac{N}{q_1}\left|\sum_{n\in\mathbb{Z}/q_1\mathbb{Z}} e_{q_1}\left(\frac{f(n)}{(f,q)}\right)\right|\right). \qquad (8.61)$$

Lifting the $\mathbb{Z}/q_1\mathbb{Z}$ sum to a $\mathbb{Z}/q\mathbb{Z}$ sum completes the proof of Part (i). In case $q \leq N$, using the Ramanujan sum estimate

$$\left|\sum_{\substack{n\in\mathbb{Z}/q\mathbb{Z}\\(n,q)=1}} e_q(bn)\right| \ll (b,q),$$

with $n \neq 0$ gives

$$\sum_{n\in\mathbb{Z}} \psi_N(n)e_q(f(n)) \ll \frac{N^{1+\epsilon}}{q}\left|\sum_{n\in\mathbb{Z}/q\mathbb{Z}} e_q(f(n))\right| + \sqrt{q}N^{2\epsilon},$$

which is (8.61) in this case.

(3) In the final three steps, we complete the proof of the lemma. In case $s < N$, this follows from the estimate (8.61). In case $N < r$, we can use the bound

$$\sum_{n \in \mathbb{Z}} \psi_N(n) \mathbf{e}_q(f(n)) \ll N(\log N)^{O(1)} \ll \sqrt{rN}(\log N)^{O(1)}.$$

Thus we can assume $r \le N \le s$.

Let $J = \lfloor N/r \rfloor$. By the translation invariance of the sum on the left, we have

$$\sum_{n \in \mathbb{Z}} \psi_N(n) \mathbf{e}_q(f(n)) = \frac{1}{J} \sum_{n \in \mathbb{Z}} \sum_{j=1}^{J} \psi_N(n + jr) \mathbf{e}_q(f(n + jr)).$$

Also $q = rs$ implies, by Lemma 8.26 and the periodicity in q of $\mathbf{e}_q$,

$$\mathbf{e}_q(f(n + jr)) = r_r(\bar{s} f(n)) \mathbf{e}_s(\bar{r} f(n + jr)).$$

Therefore, because the summand is supported on an interval of length $O(N)$, using the Cauchy–Schwarz inequality,

$$\left| \sum_{n \in \mathbb{Z}} \psi_N(n) \mathbf{e}_q(f(n)) \right| \le \frac{1}{J} \sum_{n \in \mathbb{Z}} \left| \sum_{j=1}^{J} \psi_N(n + jr) \mathbf{e}_s(\bar{r} f(n + jr)) \right|$$

$$\ll \frac{\sqrt{N}}{J} \sqrt{ \sum_{n \in \mathbb{Z}} \left| \sum_{j=1}^{J} \psi_N(n + jr) \mathbf{e}_s(\bar{r} f(n + jr)) \right|^2 }.$$

To simplify the sum on the right, we define

$$A(j,l) := \sum_{n \in \mathbb{Z}} \psi_N(n + jr) \overline{\psi_N(n + lr)} \mathbf{e}_s(\bar{r}(f(n + jr) - f(n + lr))).$$

Expanding the square of the sum gives, in terms of this double sequence,

$$\left| \sum_{n \in \mathbb{Z}} \psi_N(n) \mathbf{e}_q(f(n)) \right|^2 \ll \frac{N}{J^2} \sum_{1 \le j, l \le J} A(j,l). \tag{8.62}$$

For the diagonal values, we have the bound

$$A(j,j) = \sum_{n \in \mathbb{Z}} |\psi_N(n + jr)|^2 \ll N(\log N)^{O(1)} \implies \sum_{j=1}^{J} |A(j,j)| \ll JN(\log N)^{O(1)}.$$

(4) In this step, we derive a bound for each off diagonal term of $A(j,l)$, for fixed j and l with $j \ne l$. Let $f(X) = P(X)/Q(x)$ in lowest terms and define

$$g(n) := \frac{f(n + jr) - f(n + lr)}{r} = \frac{P_1(n)}{Q_1(n)},$$

where $P_1, Q_1 \in \mathbb{Z}[X]$ may be defined by

$$P_1(X) := P(X + jr)Q(X + lr) - Q(X + jr)P(X + lr),$$
$$Q_1(X) := rQ(X + jr)Q(X + lr).$$

Next we **claim** that provided q has all prime factors sufficiently large, we get

$$(g', s) \mid (s, j - l) \text{ and } (g, s) \mid (s, j - l).$$

To see this, by Lemma 8.28(ii), since $\deg(P) < \deg(Q)$ and the degree of Q does not change when it is reduced modulo any prime $p \mid q$, it is sufficient to show $(g, s) \mid (s, j - l)$. So let $p \mid (g, s)$, then changing variables we have using induction,

$$p \mid (f(X + (j - l)r) - f(X), s) \implies (\forall i \in \mathbb{Z}) \; p \mid (f(X + i(j - l)r) - f(X), s).$$

If $p \nmid j - l$ then $(j - l)r$ generates $\mathbb{Z}/p\mathbb{Z}$ so $p \mid (s, f(X + a) - f(X))$ for all $a \in \mathbb{Z}/p\mathbb{Z}$. This implies $f \bmod p$ is constant wherever it is defined. But taking p sufficiently large in terms of $\deg(Q)$, using the same argument as in the proof of Lemma 8.28(ii) we would have $p \mid f$, a contradiction since $(f, s) = 1$. Therefore, $p \mid j - l$ so $(g, s) \mid (s, j - l)$, which implies $(g', s) \mid (s, j - l)$. This completes the proof of the claim.

Using the estimate (8.61) and Lemma 8.29, we then get the estimate

$$A(j, l) \ll q^\epsilon \left(\sqrt{s} + \theta \frac{N}{s} \left| \sum_{n \in \mathbb{Z}/s\mathbb{Z}} \mathbf{e}_s(g(n)) \right| \right)$$

$$\ll q^\epsilon \left(\sqrt{s} + \theta \frac{N}{\sqrt{s}} \sqrt{(s, j - l)} \left| \sum_{n \in \mathbb{Z}/s\mathbb{Z}} \mathbf{e}_s(g(n)) \right| \right),$$

where $\theta = 1$ if $N \geq s/(s, j - l)$, and otherwise $\theta = 0$.

(5) Summing over all $j \neq l$, and using Lemma 8.25, $N < s$, and the bound

$$\theta \leq (s, j - l)^{\frac{1}{2}} (N/s)^{\frac{1}{2}},$$

we derive

$$\sum_{1\le j\ne l\le J} |A(j,l)| \ll q^\epsilon J^2 s^{\frac{1}{2}} + q^\epsilon \left(\frac{N}{s^{\frac{1}{2}}} \sum_{1\le j\ne l\le J} (s, j-l)^{\frac{1}{2}} \theta \right)$$

$$\ll q^\epsilon J^2 s^{\frac{1}{2}} + q^\epsilon \left(\frac{N^{3/2}}{s} \sum_{1\le j\ne l\le J} (s, j-l) \right)$$

$$\ll q^\epsilon J^2 s^{\frac{1}{2}} + q^\epsilon \left(\frac{N^{3/2}}{s} J^2 q^\epsilon \right)$$

$$\ll q^\epsilon J^2 s^{\frac{1}{2}} + q^\epsilon \left(J^2 s^{\frac{1}{2}} q^\epsilon \right) \ll q^{2\epsilon} J^2 \sqrt{s}.$$

Combining this with the bounds from Step (3) gives

$$\left| \sum_{n\in\mathbb{Z}} \psi_N(n) \mathbf{e}_q(f(n)) \right|^2 \ll \frac{q^{2\epsilon} N}{J^2} \left(JN(\log N)^{O(1)} + J^2 s^{\frac{1}{2}} \right) \ll q^\epsilon (Nr + Ns^{\frac{1}{2}}).$$

From this, setting $q_1 = q/(f,q)$, $r_1 = (r, q_1)$ and $(s_1 = (s, q_1)$, we get

$$\sum_{n\in\mathbb{N}} \psi_N(n) \mathbf{e}_q(f(n)) \ll q^\epsilon \left(N^{\frac{1}{2}} (r_1^{\frac{1}{2}} + s_1^{\frac{1}{4}}) + \frac{N}{q_1} \left| \sum_{n\in\mathbb{Z}/q_1\mathbb{Z}} \mathbf{e}_{q_1} \left(\frac{f(n)}{(f,q)} \right) \right| \right), \quad (8.63)$$

where the second term is replaced by zero in case $N < q_1$. To complete the proof of (ii), we lift the $\mathbb{Z}/q_1\mathbb{Z}$ sum to a $\mathbb{Z}/q_1\mathbb{Z}$ sum. This completes the proof of the lemma. □

Lemma 8.32 *[174, corollary 6.16] Let $N \in \mathbb{N}$. Define a smooth function with compact support on $\mathbb{R}$ by $\psi_N(x)$ in the same manner as in Lemma 8.31. Let q_1, q_2 be squarefree integers and $c_1, c_2, l_1, l_2 \in \mathbb{Z}$. Let $y \ge 1$ be real and $[q_1, q_2]$ be y-densely divisible, i.e., in $\mathscr{D}(1, y)$. Let $d \mid [q_1, q_2]$ and $a \bmod d$ a fixed congruence class. Assume also that $N \le [q_1, q_2]^{O(1)}$, that $\epsilon > 0$ is any positive small quantity, and for $1 \le i \le 2$, we have defined the four quantities*

$$\delta_i := \frac{q_i}{(q_1, q_2)} \quad \text{and} \quad \delta_i' := \frac{\delta_i}{(d, \delta_i)}.$$

(i) Then we get

$$\left| \sum_{n\equiv a \bmod d} \psi_N(n) \mathbf{e}_{q_1} \left(\frac{c_1}{n+l_1} \right) \mathbf{e}_{q_2} \left(\frac{c_2}{n+l_2} \right) \right|$$

$$\ll_\epsilon [q_1, q_2]^\epsilon \left(\frac{N^{\frac{1}{2}} [q_1, q_2]^{1/6} y^{1/6}}{d^{\frac{1}{2}}} + \frac{(c_1, \delta_1')(c_2\delta_2')}{d\delta_1'\delta_2'} N \right).$$

(ii) We also have the simpler estimate

$$\left| \sum_{n \equiv a \bmod d} \psi_N(n) e_{q_1} \left(\frac{c_1}{n+l_1} \right) e_{q_2} \left(\frac{c_2}{n+l_2} \right) \right|$$

$$\ll_\epsilon [q_1, q_2]^\epsilon \left(\frac{[q_1, q_2]^{\frac{1}{2}}}{d^{\frac{1}{2}}} + \frac{(c_1, \delta_1')(c_2 \delta_2')}{\delta_1' \delta_2'} N \right).$$

Proof **(1)** Let $q := [q_1, q_2]$ so $d \mid q$. Consider first $d = 1$ so all $n \in \mathbb{Z}$ satisfy $n \equiv a \bmod d$. If we chose $R := y^{1/3} q^{1/3} \leq yq$, y-dense divisibility of q implies there is a factorization $q = rs$ such that

$$y^{-2/3} q^{1/3} \leq r \leq y^{1/3} q^{1/3} \text{ and } y^{-1/3} q^{2/3} \leq s \leq y^{2/3} q^{2/3}.$$

Using Lemma 8.26, there is a rational function $f(n)$ of the form $f = P/Q \in \mathbb{Q}(x)$ with $\deg(P) < \deg(Q)$ such that

$$e_{q_1} \left(\frac{c_1}{n+l_1} \right) e_{q_2} \left(\frac{c_2}{n+l_2} \right) = e_q(f(n)).$$

The bound (i) of the lemma then follows from Lemma 8.31(ii) combined with the estimate from Lemma 8.30, i.e.,

$$\left| \sum_{n \in \mathbb{Z}/q\mathbb{Z}} e_{q_1} \left(\frac{c_1}{n+l_1} \right) e_{q_2} \left(\frac{c_2}{n+l_2} \right) \right| \ll q^\epsilon (c_1, \delta_1)(c_2, \delta_2)(q_1, q_2).$$

The bound (ii) follows from Lemma 8.31(i).

(2) Now let $d > 1$. Substitute $n = n'd + a$ and replace N by N/d with other modifications in Step (1) to reduce this situation to showing that

$$\left| \sum_{\substack{n \in \mathbb{Z}/q\mathbb{Z} \\ n \equiv a \bmod d}} e_{q_1} \left(\frac{c_1}{n+l_1} \right) e_{q_2} \left(\frac{c_2}{n+l_2} \right) \right| \ll q^\epsilon (c_1, \delta_1')(c_2, \delta_2')(q_1', q_2'),$$

where for $1 \leq i \leq 2$, we define $q_i' := q_i/(d, q_i)$, and **claim** that

$$\frac{d(q_1', q_2')}{[q_1, q_2]} = \frac{1}{\delta_1' \delta_2'}.$$

To see this, let $g = (q_1, q_2)$ and note that the given identity is equivalent to

$$d \left(\frac{q_1}{(d, q_1)}, \frac{q_2}{(d, q_2)} \right) = g \left(d, \frac{q_1}{g} \right) \left(d, \frac{q_2}{g} \right). \tag{8.64}$$

Write

$$d = (d, q_1/g)(d, q_2/g)(d, g) =: d_1 * d_2 * d_g,$$

giving a coprime factorization. We next verify (8.64) for each d_1, d_2, d_g separately and then use the identity

$$\left(\frac{n}{(a,n)}, \frac{m}{(a,m)} \right) \times \left(\frac{n}{(b,n)}, \frac{m}{(b,m)} \right) = \frac{(n,m)^2}{(ab,(n,m))},$$

which is valid for all squarefree $m, n, a, b \in \mathbb{N}$ with $(a,b) = 1$. Finally note that

$$\left(\frac{q_1}{(ab,q_1)}, \frac{q_2}{(ab,q_2)} \right) = \frac{(q_1,q_2)}{(ab,(q_1,q_2))} = \frac{g}{(ab,g)},$$

to complete the proof of the claim by combining the factors of d.

Using the given change of variables $n = n'd + a$, the estimate follows again from Lemma 8.30 and Lemma 8.31. This completes the proof.   □

Recall $n = qr$ where

$$q \in \mathscr{D}_{\Omega \cap (D_0, \infty)}(i_Q, x^{\delta + o(1)}) \cap [Q, 2Q),$$

$$r \in \mathscr{D}_{\Omega}(i_R, x^{\delta + o(1)}) \cap [R, 2R).$$

Theorem 8.33   (Application of exponential sum estimates) *[174, theorem 7.8]*
Let ϖ, δ, σ be given positive real numbers, $\Omega \subset \mathbb{R}$ a bounded subset, $j, k \geq 0$ fixed integers and $a, b_1, b_2 \bmod P_\Omega$ fixed congruence classes with $(a, P_\Omega) = (b_1, P_\Omega) = (b_2, P_\Omega) = 1$. Let M, N be sufficiently large integers which satisfy $MN \asymp x$ and $x^{\frac{1}{2} - \sigma} \prec^o N \prec^o x^{\frac{1}{2}}$. Let Q, R be positive real numbers which satisfy

$$x^{-3\epsilon - \delta} N \prec^o R \prec^o x^{-3\epsilon} N, \text{ and } x^{\frac{1}{2}} \prec^o QR \prec^o x^{\frac{1}{2} + 2\varpi}.$$

Let $l \in \mathbb{Z}$ satisfy $1 \leq |l| \ll N/R$ and β a given complex coefficient sequence at scale N.

Let $\mathbf{p} = (m, n, r, q_0, q_1, q_2)$ be a parameter vector, with $q_0 := (q_1, q_2)$, and $\Phi_l(\mathbf{p})$ a phase function defined by

$$\Phi_l(\mathbf{p}) := \mathbf{e}_r \left(\frac{am}{nq_0 q_1 q_2} \right) \times \mathbf{e}_{q_0 q_1} \left(\frac{b_1 m}{nr q_2} \right) \times \mathbf{e}_{q_2} \left(\frac{b_2 m}{(n + lr) r q_0 q_1} \right), \qquad (8.65)$$

where $\mathbf{e}_n(x) := \exp(2\pi i x/n)$.
Let χ^n be the characteristic function of solutions n to the congruence $b_1(n + lr) \equiv b_2 n \bmod q_0$. Define a sum

$$\Upsilon_{l,r}(b_1, b_2, q_0) :=$$

$$\sum_{\substack{q_1, q_2 \asymp Q/q_0 \\ (q_1, q_2) = 1 \\ q_0, q_1, q_0 q_1 \in \mathscr{D}_\Omega(j, x^{\delta + o(1)}) \\ q_0 q_1 r, q_0 q_2 r \in \mathscr{D}_\Omega(x^\delta)}} \quad \sum_{1 \leq |m| \ll x^\epsilon \frac{RQ^2}{q_0 M}} \left| \sum_{n \in \mathbb{N}} \chi^n \beta(n) \overline{\beta(n + lr)} \Phi_l(m, n, r, q_0, q_1, q_2) \right|. \qquad (8.66)$$

If we have at least one of the following situations:

(i) Type I: $i = 1$, $(j,k) = (0,0)$, $54\varpi + 15\delta + 5\sigma < 1$, and $N \prec^o x^{\frac{1}{2} - 2\varpi - c}$ for some fixed $c > 0$, or

(ii) Type I: $i = 2$, $(j,k) = (1,0)$, $56\varpi + 16\delta + 4\sigma < 1$, and $N \prec^o x^{\frac{1}{2} - 2\varpi - c}$ for some fixed $c > 0$, or

(iv) Type II: $i = 1$, $(j,k) = (0,0)$, $68\varpi + 14\delta < 1$, and $x^{\frac{1}{2} - 2\varpi - c} \prec^o N$ for some sufficiently small fixed $c > 0$,

then for all $q_0 \in S_\Omega$ we have the estimate

$$\sum_{r \in \mathscr{D}_\Omega(k, x^{\delta + o(1)}) \cap [R, 2R]} \Upsilon_{l,r}(b_1, b_2, q_0) \prec^o x^{-\epsilon} \frac{Q^2 R N(q_0, l)}{q_0^2}. \tag{8.67}$$

Proof **(1)** In this first step, we demonstrate Part (iv). This is more straightforward than the other parts. For example, we do not need to average over r but **claim**, that for $q_0 \geq 1$, $r \asymp R$ and $l \leq N/R$, that

$$(j,k) = (0,0),\ 68\varpi + 14\delta < 1,\ x^{\frac{1}{2} - 2\varpi - c} \prec^o N$$

$$\implies \Upsilon_{lr}(u, v, q_0) \prec^o x^{-\epsilon} \frac{Q^2 N(q_0, l)}{q_0^2}, \tag{8.68}$$

which is sufficient for our purposes.

(1.1) To this end, let $h := x^\epsilon R Q^2 / (M q_0)$. Using the definition (8.66) and coefficients c_{m, q_1, q_2} with modulus at most 1, we get

$$\Upsilon_{l,r}(u, v, q_0) :=$$

$$\sum_{\substack{q_1, q_2 \asymp Q/q_0 \\ (q_1, q_2) = 1}} \sum_{1 \leq |m| \ll h} c_{m, q_1, q_2} \sum_{n \in \mathbb{N}} \chi^n \beta(n) \overline{\beta(n + lr)} \Phi_l(m, n, r, q_0, q_1, q_2). \tag{8.69}$$

Next we interchange the order of summation, first summing over n. Because $Q \leq N$, χ^n is the characteristic function of at most (q_0, l) congruence classes modulo q_0, we get using divisor bounds (Lemma 8.24) and the Cauchy–Schwarz than inequality

$$\sum_{n \in \mathbb{N}} \chi^n |\beta(n)|^2 |\beta(n + lr)|^2 \prec^o \frac{N(q_0, l)}{q_0}.$$

Using the Cauchy–Schwarz inequality again and multiplying by a coefficient sequence ψ_N which is positive, smooth at scale N and not less 1 on the support of $\beta(n)\overline{\beta(n + lr)}$, defining

$$S_{l,r} = S_{l,r}(m_1, m_2, q_1, q_2, s_1, s_2)$$

$$:= \sum_{n \in \mathbb{Z}} \chi^n \psi_N(n) \Phi(m_1, n, r, q_0, q_1, q_2) \overline{\Phi(m_2, n, r, q_0, s_1, s_2)},$$

we get

$$|\Upsilon_{l,r}(u,v,q_0)|^2 \prec^o N\frac{(q_0,l)}{q_0}\sum_{n\in\mathbb{Z}}\chi^n\psi_N(n)$$

$$\times \left|\sum_{\substack{q_1,q_2\asymp Q/q_0 \\ (q_1,q_2)=1}}\sum_{1\le|m|\le h}c_{m,q_1,q_2}\Phi_l(m,n,r,q_0,q_1,q_2)\right|^2$$

$$\prec^o N\frac{(q_0,l)}{q_0}\sum_{\substack{q_1,q_2,s_1,s_2\asymp Q/q_0 \\ (q_1,q_2)=(s_1,s_2)=1}}\cdots\sum\sum_{1\le m_1,m_2\le h}|S_{l,r}(m_1,m_2,q_1,q_2,s_1,s_2)|.$$

(1.2) To simplify this estimate further, we need to find a suitable bound for $S_{l,r} = S_{l,r}(\mathbf{p})$, where as before $\mathbf{p} := (m_1,m_2,q_1,q_2,s_1,s_2)$ with $(q_0q_1q_2s_1s_2,r) = 1, l \ne 0$ and l,r restricted as before. In addition, we can assume $q_0q_1, q_0q_2, q_0s_1, q_0s_2 \ll Q$ and $r \ll R$. We **claim**

$$|S_{l,r}(\mathbf{p})| \prec^o (q_0,l)\left(\frac{Q^2\sqrt{R}}{q_0^2} + \frac{N}{q_0R}(m_1s_1s_2 - m_2q_1q_2,r)\right).$$

To show this, fix the parameters r,l,q_0,a,u,v and then expand Φ_l in terms of the additive characters $\mathbf{e}_d$ using (8.65). For convenience, we index with a superscript the different additive characters which may depend on those parameters to write

$$\Phi_l(m,n,r,q_0,q_1,q_2) = \mathbf{e}_r^1\left(\frac{m}{q_1q_2n}\right)\mathbf{e}_{q_0q_1}^2\left(\frac{m}{nq_2}\right)\mathbf{e}_{q_2}^3\left(\frac{m}{(n+lr)q_0q_1}\right).$$

We use the abbreviations

$$\Phi_1(n) := \Phi_l(m_1,n,r,q_0,q_1,q_2) \text{ and}$$

$$\Phi_2(n) := \Phi_l(m_2,n,r,q_0,s_1,s_2),$$

and define $d_1 := rq_0[q_1,s_1]$ and $d_2 := [q_2,s_2]$. Then there are integers c_1,c_2 and corresponding additive characters $\mathbf{e}_{d_1}^4, \mathbf{e}_{d_2}^5$ such that

$$\Phi_1(n)\overline{\Phi_2(n)} = \mathbf{e}_r^1\left(\frac{m_1}{q_1q_2n} - \frac{m_2}{s_1s_2n}\right)\mathbf{e}_{q_0q_1}^2\left(\frac{m_1}{nq_2}\right)$$

$$\times \mathbf{e}_{q_0s_1}^2\left(-\frac{m_2}{ns_2}\right)\mathbf{e}_{q_2}^3\left(\frac{m_1}{(n+lr)q_0q_1}\right)\mathbf{e}_{s_2}^3\left(-\frac{m_2}{(n+lr)q_0s_1}\right)$$

$$= \mathbf{e}_{d_1}^4\left(\frac{c_1}{n}\right)\mathbf{e}_{d_2}^5\left(\frac{c_2}{n+lr}\right). \tag{8.70}$$

Again using the fact that χ^n is the characteristic function of not greater than (q_0, l) residue classes modulo q_0, we get

$$|S_{l,r}(\mathbf{p})| \leq (q_0, l) \max_{t \in \mathbb{Z}/q_0\mathbb{Z}} \left| \sum_{n \equiv t \bmod q_0} \psi_N(n) \Phi_1(n) \overline{\Phi_2(n)} \right|.$$

Using Lemma 8.32 with this bound, denoting $\delta_i := d_i/(d_1, d_2)$ and $\delta_i' := \delta_i/(q_0, \delta_i)$, $i = 1, 2$ and using

$$[d_1, d_2] \leq r q_0 q_1 q_2 s_1 s_2 \ll q_0 R \frac{Q^4}{q_0^4} \quad \text{and} \quad \frac{(c_2, \delta_2')}{\delta_2'} \leq 1,$$

we can derive

$$|S_{l,r}| \prec^o (q_0, l) \left(\sqrt{\frac{[d_1, d_2]}{q_0}} + \frac{N(c_1, \delta_1')(c_2, \delta_2')}{q_0 \delta_1' \delta_2'} \right) \prec^o (q_0, l) \left(\sqrt{R} \frac{Q^2}{q_0^2} + \frac{N(c_1, \delta_1')}{q_0 \delta_1'} \right).$$

In addition, since $r \mid \delta_1$ and $(r, q_0) = 1$, we have

$$\frac{(c_1, \delta_1')}{\delta_1'} = \prod_{\substack{p \mid \delta_1 \\ p \nmid c_1, q_0}} p \leq \frac{(c_1, r)}{r}.$$

Finally in this substep, note that a prime p which divides r divides c_1 when and only when the $\mathbf{e}_r$ term of the character identity (8.70) is constant, when and only when $p \mid m_1 s_1 s_2 - m_2 q_1 q_2$. Therefore,

$$S_{l,r} \prec^o (q_0, l) \sqrt{R} \frac{Q^2}{q_0^2} + \frac{(q_0, l) N}{q_0 R} (r, m_1 s_1 s_2 - m_2 q_1 q_2),$$

which completes the proof of the claim.

(1.3) Using the bound from Step (1.2) in the estimate we have derived for Υ in Step (1.1), noting by the definitions of Φ_1, (8.65) and $\mathbf{e}_q$, that $S_{l,r} = 0$ unless $(q_0 q_1 q_2 s_1 s_2, r) = 1$, where h is defined in Step (1.1) and discarding some powers of q_0, gives

$$|\Upsilon|^2 \prec^o N \left(\frac{(q_0, l)}{q_0} \right)^2$$

$$\times \sum_{\substack{q_1, q_2, s_1, s_2 \asymp Q/q_0 \\ (q_1, q_2) = (s_1, s_2) = 1}} \cdots \sum \sum_{1 \leq m_1, m_2 \leq h} \left(\frac{Q^2 \sqrt{R}}{q_0} + \frac{N}{R} (r, m_1 s_1 s_2 - m_2 q_1 q_2) \right).$$

Next, make a change of variables $\Delta := m_1 s_1 s_2 - m_2 q_1 q_2$, where each number Δ has at most $\tau_3(\Delta)$ representations (m_2, q_1, q_2), for fixed m_1, s_1, s_2. Therefore, using

$$\tau_3(n) \leq \sum_{d|n} \sum_{e|n/d} 1 \leq \sum_{d|n} \sum_{e|n} 1 = \tau(n)^2,$$

and Lemmas 8.24 and 8.25, we get

$$\sum_{\substack{q_1,q_2,s_1,s_2 \asymp Q/q_0 \\ (q_1,q_2)=(s_1,s_2)=1}} \cdots \sum \sum_{1 \leq m_1,m_2 \leq h} (r, m_1 s_1 s_2 - m_2 q_1 q_2)$$

$$\leq \sum_{|\Delta| \ll hQ^2/q_0^2} (\Delta, r) \sum_{r_1,s_1,s_2} \tau_3(r_1 s_1 s_2 - \Delta)$$

$$\prec^o \frac{hQ^2}{q_0^2} \sum_{0 \leq |\Delta| \ll hQ^2/q_0^2} (\Delta, r)$$

$$\prec^o \frac{hQ^2}{q_0^2} \left(\frac{hQ^2}{q_0^2} + R \right).$$

Next, simplifying by reducing some of the powers of q_0,

$$|\Upsilon|^2 \prec^o N \frac{(q_0, l)^2}{q_0^2} \left(\frac{h^2 Q^2 \sqrt{R}}{q_0} \left(\frac{Q}{q_0}\right)^4 + \frac{h^2 N}{R} \left(\frac{Q}{q_0}\right)^4 + Nh \left(\frac{Q}{q_0}\right)^2 \right)$$

$$\prec^o \frac{N^2 Q^4 (q_0, l)^2}{q_0^4} \left(\frac{h^2 Q^2 \sqrt{R}}{N} + \frac{h^2}{R} + \frac{h}{Q^2} \right)$$

$$\prec^o \frac{N^2 Q^4 (q_0, l)^2}{q_0^4} \left(x^{2\epsilon} \frac{Q^6 \sqrt{R^5}}{M^2 N} + x^{2\epsilon} \frac{RQ^4}{M^2} + x^\epsilon \frac{R}{M} \right).$$

To deal with this, we derive the following asymptotic bounds in terms of x and N using $N \prec^o M$, the Type II estimate $x^{\frac{1}{2} - 2\varpi - c} \prec^o N$,

$$x^{-3\epsilon - \delta} N \prec^o R \prec^o x^{-3\epsilon} N \text{ and } x^{\frac{1}{2}} \prec^o QR \prec^o x^{\frac{1}{2} + 2\varpi} \implies NQ \prec^o x^{\frac{1}{2} + 2\varpi + \delta + 3\epsilon},$$

to get (c/f the deduction in Step (4) of the proof of Theorem 8.34)

$$x^{\frac{1}{2}} \prec^o QR \prec^o x^{\frac{1}{2} + 2\varpi} \text{ and } NQ \prec^o x^{\frac{1}{2} + 2\varpi + \delta + 3\epsilon},$$

and the estimates from the theorem statement $N \prec^o \sqrt{x}$, $R \prec^o x^{-3\epsilon} N$, and $NM \asymp x$ together with the definition $h := x^\epsilon Q^2 R/(Mq_0)$,

$$
\begin{aligned}
\frac{Q^6 R^{5/2}}{M^2 N} &\asymp \frac{(NQ)(QR)^5}{x^2 R^{5/2}} \prec^o \frac{x^{1+12\varpi+\delta+3\epsilon}}{R^{5/2}} \prec^o \frac{x^{1+12\varpi+7\delta/2+21\epsilon/2}}{N^{5/2}}, \\
\frac{Q^4 R}{M^2} &\asymp \frac{N^2 R Q^4}{x^2} \prec^o \frac{(QR)(NQ)^3}{x^2 N} \prec^o \frac{x^{8\varpi+3\delta+9\epsilon}}{N}, \\
\frac{R}{M} &\asymp \frac{NR}{x} \prec^o \frac{N^2}{x^{1+3\epsilon}} \prec^o x^{-3\epsilon}.
\end{aligned}
\tag{8.71}
$$

We also have for sufficiently small $c > 0$ and $\epsilon > 0$, if ϖ and δ satisfy

$$
1 + 12\varpi + 7\delta/2 < (5/2)\left(\frac{1}{2} - 2\varpi\right) \text{ and } 8\varpi + 2\delta < \frac{1}{2} - 2\varpi,
$$

which hold if and only if

$$
68\varpi + 14\delta < 1 \text{ and } 20\varpi + 6\delta < 1,
$$

which are implied by $68\varpi + 14\delta < 1$, an assumption of Part (iv) of this theorem, Theorem 8.33. For the first line of (8.71), using $x^{\frac{1}{2}-2\varpi-c} \prec^o N$, choosing $c, \epsilon > 0$ sufficiently small and using the estimates from (8.71), we get

$$
\sqrt{x^{2\epsilon} \frac{Q^6 R^{5/2}}{M^2 N}} \prec^o \sqrt{\frac{x^{1+12\varpi+7\delta/2+25\epsilon/2}}{x^{(5/2)(1/2-2\varpi-c)}}} \prec^o x^{-\epsilon}.
$$

A similar (but easier) calculation gives the same upper bound from lines two and three of (8.71). Then apply $\sqrt{a+b+c} \le \sqrt{a} + \sqrt{b} + \sqrt{c}$, to get

$$
|\Upsilon| \prec^o x^{-\epsilon} \frac{N Q^2 (q_0, l)}{q_0^2},
$$

completing the proof of Part (iv).

(2) In this step and its substeps, we demonstrate Part (i). This follows a similar pattern of proof for Step (1). For example, we do not need to average over r but show, for $q_0 \ge 1$, $r \asymp R$ and $l \le N/R$, that

$$
(j,k) = (0,0), \ 54\varpi + 15\delta + 5\sigma < 1, \ N \prec^o x^{\frac{1}{2} - 2\varpi - c}
$$

$$
\implies \Upsilon_{l,r}(u, v, q_0) \prec^o x^{-\epsilon} \frac{Q^2 N (q_0, l)}{q_0^2}, \tag{8.72}
$$

which is easily seen to be sufficient for our purposes.

(2.1) To this end, let as before $h := x^\epsilon R Q^2/(M q_0) > 0$. Using the definition of $\Upsilon_{l,r}$ (8.66) and using coefficients c_{m,q_1,q_2} with modulus at most 1, we get

$$\Upsilon_{l,r}(u,v,q_0) := \sum_{\substack{q_1,q_2 \asymp Q/q_0 \\ (q_1,q_2)=1}} \sum_{1 \leq |m| \ll h} c_{m,q_1,q_2} \sum_n \chi^n \beta(n)\overline{\beta(n+lr)} \Phi_l(m,n,r,q_0,q_1,q_2).$$

$$(8.73)$$

Define $S_{l,r}$ as before in Step (1.1) and assume (q_1,q_2,s_2) satisfy $q_0q_1r, q_0q_2r, q_0s_2r \in \mathscr{D}_\Omega(1,x^\delta)$. Also define

$$\Upsilon_1 := \sum_{q_1 \asymp Q/q_0} \sum_{n \in \mathbb{Z}} \chi^n |\beta(n)|^2 |\beta(n+lr)|^2 \prec^o \frac{N Q(q_0,l)}{q_0^2}, \text{ and}$$

$$\Upsilon_2 := \sum_{n \in \mathbb{Z}} \psi_N(n)\chi^n \sum_{q_1 \asymp Q/q_0} \left| \sum_{\substack{q_2 \asymp Q/q_0 \\ (q_1,q_2)=1}} \sum_{1 \leq |m| \leq h} c_{m,q_1,q_2} \Phi_l(m,n,r,q_0,q_1,q_2) \right|^2$$

$$= \sum_{q_1 \asymp Q/q_0} \sum_{\substack{q_2,s_2 \asymp Q/q_0 \\ (q_1,q_2)=(q_1,s_2)=1}} \sum_{1 \leq m_1,m_2 \leq h} c_{m_1,q_1,q_2}\overline{c_{m_2,q_2,q_2}} S_{l,r}(m_1,m_2,q_1,q_2,q_1,s_2),$$

where $S_{l,r}$ is defined in Step (1).

(2.2) To simplify this estimate further, we seek a suitable bound for $S_{l,r}(\mathbf{p})$, where $\mathbf{p} := (m_1,m_2,q_1,q_2,q_1,s_2)$ with $(q_0q_1q_2s_2,r) = 1, l \neq 0$ and l,r restricted as before so $r \ll R$ and $l \ll N/R$. In addition, we can assume $q_0q_1, q_0q_2, q_0s_2 \ll Q$, $r \ll R$ and

$$q_0q_1r, \ q_0s_2r, \ q_0q_2r \in \mathscr{D}_\Omega(1,x^\delta).$$

We **claim**

$$|S_{l,r}(\mathbf{p})| \prec^o Q^{\frac{1}{2}} N^{\frac{1}{2}} R^{1/6} q_0^{1/6} x^{\delta/6} + \frac{N}{R}(m_1s_2 - m_2q_2, r).$$

To show this, fix the parameters r,l,q_0,a,u,v and then expand Φ_l in terms of the additive characters e_d using (8.65). For convenience, we index the different additive characters which may depend on those parameters to write

$$\Phi_l(m,n,r,q_0,q_1,q_2) = e_r^1 \left(\frac{m}{q_1q_2n}\right) e_{q_0q_1}^2 \left(\frac{m}{nq_2}\right) e_{q_2}^3 \left(\frac{m}{(n+lr)q_0q_1}\right).$$

We now use the abbreviations

$$\Phi_1(n) := \Phi_l(m_1,n,r,q_0,q_1,q_2),$$
$$\Phi_2(n) := \Phi_l(m_2,n,r,q_0,q_1,s_2),$$

and define $d_1 := rq_0q_1$ and $d_2 := [q_2, s_2]$. Then there are integers c_1, c_2 and corresponding additive characters $e_{d_1}^4$, $e_{d_2}^5$ such that, as in Step (1.2),

$$\Phi_1(n)\overline{\Phi_2(n)} = e_{d_1}^4\left(\frac{c_1}{n}\right)e_{d_2}^5\left(\frac{c_2}{n+lr}\right).$$

Because rq_0q_1, rq_0q_2 and rq_0s_2 are each x^δ-densely divisible, by Lemma 8.5(ii), so is the LCM, $[d_1, d_2] = [rq_0q_1, rq_0q_2, rq_0s_2]$.

Again using the fact that χ^n is the characteristic function of not greater than (q_0, l) residue classes modulo q_0, so splitting the sum into residue classes modulo q_0 and applying Lemma 8.32(i) to each class, defining $\delta_i := d_i/(d_1, d_2)$ and $\delta_i' := \delta_i/(q_0, \delta_i)$ for $1 \leq i \leq 2$, we get

$$|S_{l,r}(\mathbf{p})| \leq (q_0, l)\left(\frac{N^{\frac{1}{2}}[d_1, d_2]^{1/6}x^{\delta/6}}{q_0^{\frac{1}{2}}} + \frac{N(c_1, \delta_1')(c_2, \delta_2')}{q_0\delta_1'\delta_2'}\right).$$

Then we use

$$[d_1, d_2] \leq R\frac{Q^3}{q_0}, \quad \frac{(c_2, \delta_2')}{\delta_2'} \leq 1, \text{ and } \frac{(c_1, \delta_1')}{\delta_1'} \leq \frac{(c_1, r)}{r},$$

to simplify the bound. Finally note that a prime p which divides r divides c_1 when and only when the e_r term of the character identity (8.65) is constant, which is the case when and only when $p \mid m_2s_2 - m_2q_2$. Therefore,

$$|S_{l,r}(\mathbf{p})| \prec^o \left(Q^{\frac{1}{2}}N^{\frac{1}{2}}R^{1/6}q_0^{1/6}x^{\delta/6} + \frac{N}{R}(m_1s_2 - m_2q_2, r)\right),$$

which completes the proof of the claim.

(2.3) Using the approach of Step (1.3) to sum the GCDs $(m_1s_2 - m_2q_2, r)$, we get

$$\Upsilon_2 \prec^o \left(\frac{Q}{q_0}\right)^3 h^2\left(q_0^{1/6}N^{\frac{1}{2}}(Q^3R)^{1/6}x^{\delta/6} + \frac{N}{R}\right) + hN\left(\frac{Q}{q_0}\right)^2.$$

Substituting in the bounds derived in Step (2.1), and simplifying by discarding some powers of q_0 as before, gives

$$|\Upsilon|^2 \prec^o \frac{NQ(q_0, l)}{q_0^2}\left(\left(\frac{Q}{q_0}\right)^3 q_0^{1/6}h^2N^{\frac{1}{2}}(Q^3R)^{1/6}x^{\delta/6} + \left(\frac{Q}{q_0}\right)^3\frac{h^2N}{R} + \left(\frac{Q}{q_0}\right)^2 hN\right)$$

$$\prec^o \frac{N^2Q^4(q_0, l)^2}{q_0^4}\left(\frac{h^2Q^{\frac{1}{2}}R^{1/6}x^{\delta/6}}{N^{\frac{1}{2}}} + \frac{h^2}{R} + \frac{h}{Q}\right).$$

Now again as in Step (1.3) we calculate, using $N \prec^o M$, and from Step (4) of the proof of Theorem 8.34 and $N \prec^o x^{\frac{1}{2}-2\varpi-c}$ from the assumptions for Part (i),

$$x^{\frac{1}{2}} \prec^o QR \prec^o x^{\frac{1}{2}+2\varpi} \text{ and } NQ \prec^o x^{\frac{1}{2}+2\varpi+\delta+3\epsilon},$$

to obtain the following asymptotic bounds in terms of x and N:

$$\frac{h^2 Q^{\frac{1}{2}} R^{1/6} x^{\delta/6}}{N^{\frac{1}{2}}} \prec^o \frac{R^{13/6} Q^{9/2}}{M^2 N^{\frac{1}{2}}} x^{\delta/6+2\epsilon} \prec^o \frac{N^{3/2}(QR)^{9/2}}{R^{7/3}} x^{-2+\delta/6+2\epsilon}$$

$$\prec^o \frac{x^{\frac{1}{4}+9\varpi+5\delta/2+9\epsilon}}{N^{5/6}},$$

$$\frac{h^2}{R} \prec^o \frac{x^{8\varpi+3\delta+11\epsilon}}{N},$$

$$\frac{h}{Q} \le \frac{RQ}{M} x^\epsilon \prec^o \frac{x^{\frac{1}{2}+2\varpi+\epsilon}}{M} \prec^o x^{-c+\epsilon}.$$

Using $x^{\frac{1}{2}-\sigma} \prec^o N$, together with $c = 3\epsilon$ with ϵ sufficiently small, and the inequalities

$$\frac{1}{4} + 9\varpi + 5\delta/2 < (5/6)\left(\frac{1}{2} - \sigma\right), \text{ and } 8\varpi + 3\delta < \frac{1}{2} - \sigma,$$

it follows that these inequalities hold if and only if

$$54\varpi + 15\delta + 5\sigma < 1 \text{ and } 16\varpi + 6\delta + 2\sigma < 1.$$

These two constraints are implied by $54\varpi + 15\delta + 5\sigma < 1$, an assumption of Part (i) of this theorem, completing the proof of (i) in that situation, using the same approach as in Step (1.3). This completes the proof of Part (i).

(3) In this final step and its substeps, we prove Part (ii) of the theorem, i.e., when $i = 2$, $i - 1 = j + k$ with $j = 1$, $k = 0$, we will prove

$$56\varpi + 16\delta + 4\sigma < 1, \ N \prec^o x^{\frac{1}{2}-2\varpi-c} \implies \Upsilon_{l,r} \prec^o x^{-\epsilon} \frac{Q^2 N(q_0, l)}{q_0}.$$

In other words, we are able, as in the previous parts, to establish the given estimate for $\Upsilon_{l,r}$ for each individual r, rather than on average. As before, we set $h := x^\epsilon RQ^2/(Mq_0)$, and note that we can assume $h \ge 1$ because the estimate follows directly if $h < 1$ – note that in that case the sum in the estimate from Step (1.1) is empty.

First, observe that in the definition of $\Upsilon_{l,r}$ (8.66), we are given q_1 is $x^{\delta+o(1)}$-densely divisible. Using Lemma 8.5(i), we get q_1 is $q_0 x^{\delta+o(1)}$-densely divisible. In addition, since

$$x^{\frac{1}{2}} \prec^o QR \prec^o x^{\frac{1}{2}+2\varpi}$$

and $MN \asymp x \implies x^{\frac{1}{2}+2\varpi+c} \prec^o M$, we can assume

$$x^{c-3\epsilon} \prec^o \frac{Q}{h} x^{-2\epsilon}$$

Also, from the sum in (8.66), because $h \geq 1$ and $q_1 q_0 \asymp Q$, we have

$$\frac{Q}{h} x^{-2\epsilon} \prec^o q_1 q_0 x^{\delta+o(1)}.$$

Therefore, with $\epsilon > 0$ so small that $c > 3\epsilon$, by dense divisibility, with the R in that definition replaced by $(Q/h)x^{-2\epsilon}$ and the y with $q_0 x^\delta$, for $(Q/h)x^{-2\epsilon} \prec^o q_1$ or using $u_1 = q_1, v_1 = 1$ otherwise, we can factor $q_1 = u_1 v_1$ with each factor squarefree and such that

$$\frac{Q}{q_0 h} x^{-\delta-2\epsilon} \prec^o u_1 \prec^o \frac{Q}{h} x^{-2\epsilon} \text{ and}$$

$$\frac{h}{q_0} x^{2\epsilon} \prec^o v_1 \prec^o h x^{\delta+2\epsilon}.$$

(3.1) Now define, recalling u_1, v_1 are taken to be squarefree,

$$\Upsilon_{U,V} := \sum_{\substack{1 \leq |m| \leq h \\ u_1 \asymp U \\ v_1 \asymp V}} \sum_{\substack{q_2 \asymp Q/q_0 \\ (u_1 v_1, q_0 q_2)=1}} \left| \sum_n \chi^n \beta(n) \overline{\beta(n+lr)} \Phi_l(m,n,r,q_0,u_1 v_1,q_2) \right|.$$

Decomposing the sum dyadically, and recalling the constraints derived in Step (3), it is sufficient to prove that whenever

$$\frac{Q}{q_0 h} x^{-\delta-2\epsilon} \prec^o U \prec^o \frac{Q}{H} x^{-2\epsilon},$$

$$\frac{h}{q_0} x^{2\epsilon} \prec^o V \prec^o h x^{\delta+2\epsilon}, \text{ and}$$

$$UV \asymp \frac{Q}{q_0},$$

we have for each such (U, V) the estimate

$$\Upsilon_{U,V} \prec^o (q_0, l) \frac{Q^2 N}{q_0^2} x^{-\epsilon}. \tag{8.74}$$

(3.2) Now let

$$\Upsilon_1 := \sum_{\substack{u_1 \asymp U \\ q_2 \asymp Q/q_0}} \sum_{n \in \mathbb{Z}} \chi^n |\beta(n)|^2 |\beta(n+lr)|^2 \prec^o (q_0, l) \frac{NQU}{q_0^2},$$

and define

$$T_{l,r} := \sum_{n\in\mathbb{Z}} \chi^n \psi_N(n) \Phi_l(m_1,n,r,q_0,u_1v_1,q_2) \overline{\Phi_l(m_2,n,r,q_0,u_1v_2,q_2)},$$

so we can define

$$\Upsilon_2 := \sum_{\substack{u_1\asymp U \\ q_2\asymp Q/q_0}} \sum_n \chi^n \psi_N(n) \left| \sum_{\substack{v_1\asymp V \\ (u_1v_1,q_0q_2)=1}} \sum_{1\le|m|\le h} c_{m,u_1,v_1,q_2} \Phi_l(m,n,r,q_0,u_1v_1,q_2) \right|^2$$

$$= \sum_{\substack{u_1\asymp U \\ q_2\asymp Q/q_0}} \sum_{\substack{v_1,v_2\asymp V \\ (u_1v_1v_2,q_0q_1)=1}} \sum_{1\le|m_1|,|m_2|\le h} c_{m_1,u_1,v_1,q_2} \overline{c_{m_2,u_1,v_2,q_2}}$$

$$\times T_{l,r}(m_1,m_2,u_1,v_1,v_2,q_2,q_0),$$

whereas before we have simplified the sum for Υ_2 introducing complex number coefficients c_{m,u_1,v_1,q_2}. Using Cauchy–Schwarz, we obtain

$$\left|\Upsilon_{U,V}\right|^2 \le \Upsilon_1 \cdot \Upsilon_2.$$

(3.3) We next **claim** that for any $\mathbf{p}=(m_1,m_2,u_1,v_1,v_2,q_2,q_0)$ satisfying $(u_1v_1v_2,q_0q_2)=(q_0,q_2)=1, l\ne 0$ and $r\ll N$, we have the estimate

$$|T_{l,r}(\mathbf{p})| \prec^o (q_0,l) \left(\frac{N^{\frac{1}{2}}(RhQ^2)^{1/6}}{q_0^{\frac{1}{2}}} x^{\delta/3+\epsilon/3} + \frac{N}{q_0 R}(m_1v_2-m_2v_1,r) \right).$$

To see this as in Steps (1) and (2), first let

$$\Phi_1(n) := \Phi_l(m_1,n,r,q_0,u_1v_1,q_2) \text{ and } \Phi_2(n) := \Phi_l(m_2,n,r,q_0,u_1v_2,q_2),$$

and write using (8.65) with $d_1 := rq_0u_1[v_1,v_2]$ and $d_2 := q_2$ and some integers c_1, c_2,

$$\Phi_1(n)\overline{\Phi_2(n)} = e_{d_1}^4\left(\frac{c_1}{n}\right) e_{d_2}^5\left(\frac{c_2}{n+lr}\right).$$

Lemma 8.5(ii) implies that because $rq_0u_1v_1$, $rq_0u_1v_2$ and rq_0q_2 are all x^δ-densely divisible (see the definition of $\Upsilon_{l,r}$ with $j=1$), their LCM $[d_1,d_2]$ is also x^δ densely divisible. Splitting the factor of χ^n into congruence classes modulo q_0 and applying Lemma 8.32(i) to each class, and defining δ_i and δ_i' as in Step (2), we get

$$|T_{l,r}| \prec^o (q_0,l) \left(\frac{N^{\frac{1}{2}}[d_1,d_2]^{1/6}x^{\delta/6}}{q_0^{\frac{1}{2}}} + \frac{N(c_1,\delta_1')(c_2,\delta_2')}{q_0\delta_1'\delta_2'} \right).$$

Next, use given bounds to variables to get

$$[d_1, d_2] \ll QRUV^2 \prec^o \frac{HQ^2R}{q_0} x^{\delta+2\epsilon}.$$

The proof of the claim is completed by noting as in Step (2.3) that

$$\frac{(c_2, \delta_2')}{\delta_2'} \le \frac{(c_2, \delta_2)}{\delta_2} \le 1 \text{ and } \frac{(c_1, \delta_1')}{\delta_1'} \le \frac{(c_1, \delta_1)}{\delta_1} \le \frac{(c_1, r)}{r},$$

and using (8.65) with the r-component of $\Phi_1(n)\overline{\Phi_2(n)}$ shows a prime $p \mid r$ also divides c_1 when and only when $p \mid m_1 v_2 - m_2 v_1$.

(3.4) Inserting the bound for $T_{l,r}$ derived in Step (3.3), namely

$$|T_{l,r}(\mathbf{p})| \prec^o (q_0, l) \left(\frac{N^{\frac{1}{2}}(Rh Q^2)^{1/6}}{q_0^{\frac{1}{2}}} x^{\delta/3+\epsilon/3} + \frac{N}{q_0 R}(m_1 v_2 - m_2 v_1, r) \right),$$

into the final expression for Υ_2 derived in Step (3.2) and summing the GCD's as in Step (1.3) giving an upper bound for the GCD-sum over m_i and v_i of $hV(hV + R)$, we get

$$\Upsilon_2 \prec^o (q_0, l) h^2 UV^2 \left(\frac{Q}{q_0} \right) \left(N^{\frac{1}{2}}(Rh Q^2)^{1/6} x^{\delta/3+\epsilon/3} + \frac{N}{R} \right) + (q_0, h) hNUV \frac{Q}{q_0^2}.$$

Multiplying this by the upper bound for Υ_1 given in Step (3.2), and using $UV \asymp Q/q_0$, discarding a factor of q_0 to simplify the expression, we get

$$|\Upsilon_{U,V}|^2 \le \Upsilon_1 \times \Upsilon_2$$

$$\prec^o (q_0, l)^2 \frac{NQU}{q_0} \left(\frac{h^2 Q^3 N^{\frac{1}{2}}(h Q^2 R)^{1/6}}{U q_0^3} x^{\delta/3+\epsilon/3} + \frac{h^2 N Q^3}{U R q_0^3} + hN \frac{Q^2}{q_0^3} \right)$$

$$\prec^o (q_0, l)^2 N^2 \left(\frac{Q}{q_0} \right)^4 \left(\frac{h^{13/6} Q^{1/3} R^{1/6}}{N^{\frac{1}{2}}} x^{\delta/3+\epsilon/3} + \frac{h^2}{R} + \frac{h}{V q_0} \right).$$

Consider each summand in the bracketed expression in turn. Using (7.40) from Step (3.1), the definition of h, $QR \prec^o x^{\frac{1}{2}+2\varpi}$ and

$$NQ = \frac{N}{R}(QR) \prec^o x^{\frac{1}{2}+2\varpi+\delta+3\epsilon},$$

we derive

$$\frac{h^{13/6}Q^{1/3}R^{1/6}}{N^{\frac{1}{2}}}x^{\delta/3+\epsilon/3} \prec^o \frac{R^{7/3}Q^{14/3}}{N^{\frac{1}{2}}M^{13/6}}x^{\delta/3+5\epsilon/2} \prec^o \frac{N^{5/3}}{R^{7/3}}x^{1/6+28\varpi/3+\delta/3+5\epsilon/2}$$

$$\prec^o \frac{1}{N^{2/3}}x^{1/6+28\varpi/3+8\delta/3+19\epsilon/2},$$

$$\frac{h^2}{R} \prec^o \frac{RQ^2}{M^2}x^{2\epsilon} \prec^o \frac{(QR)(NQ)^3}{x^2N}x^{2\epsilon} \prec^o \frac{x^{8\varpi+3\delta+11\epsilon}}{N},$$

$$\frac{h}{Vq_0} \prec^o x^{-2\epsilon},$$

where the final estimate comes from the estimates for V of Step (3.1).

Substituting these upper estimates in the estimate we have derived for $|\Upsilon_{U,V}|^2$, using $x^{\frac{1}{2}-\sigma} \prec^o N$ and simplifying we get

$$\Upsilon_{U,V} \prec^o x^{-\epsilon}(q_0,l)Q^2N/q_0^2,$$

that is to say the estimate (8.74) is satisfied for sufficiently small $\epsilon > 0$ provided

$$\frac{28\varpi}{3} + \frac{8\delta}{3} + \frac{1}{6} < \frac{2}{3}\left(\frac{1}{2}-\sigma\right) \text{ and } 8\varpi + 3\delta < \frac{1}{2} - \sigma,$$

which is equivalent to

$$56\varpi + 16\delta + 4\sigma < 1 \text{ and } 16\varpi + 6\delta + 2\sigma < 1 \iff 56\varpi + 16\delta + 4\sigma < 1.$$

We then again use the same approach as used in Step (1.3) to complete the proof of Theorem 8.33. □

Remark: The missing Part (iii) of Theorem 8.33 is very similar in form to Parts (i) and (ii) and gives the same estimate for $\Upsilon_{r,l}$. The conditions are

$$(j,k) = (1,2), \frac{160}{3}\varpi + 16\delta + \frac{34}{9}\sigma < 1, 64\varpi + 18\delta + 2\sigma < 1, \text{ and } N \prec^o x^{\frac{1}{2}-2\varpi-c},$$

so, in particular, it needs $\mathscr{D}(3,x^\delta)$. What is more important, Polymath's proof requires making use of Deligne's form of the Riemann Hypothesis for algebraic varieties over finite fields, not just curves. It might be possible to remove this requirement, for example by varying the dense divisibility concept, which continues to have, at the time of writing, some mysterious components.

8.10 Polymath's Type I and II Estimates

Theorem 8.33 is an essential step in this approach to prime gaps. Its application is to an estimate for the signed discrepancy of $\alpha \star \beta$ for very general coefficient sequences α and β with the range for the moduli $q \leq x^{\frac{1}{2}+2\varpi}$ and with $q \in \mathscr{D}(i, x^\delta)$ with $i = 1$ or 2. We need to wait until Step (17) of Theorem 8.34 to use the $\Upsilon_{r,l}$ estimate, derived with great effort in Theorem 8.33. Note that we continue to have the Type I and II contextual constraints:

$$\frac{1}{6} < \sigma < \frac{1}{2}, \ 0 < \varpi < \frac{1}{4}, \ 0 < \delta < \frac{1}{2} + \varpi, \ \text{and } 2\varpi < \sigma.$$

First recall some definitions. We say a sequence of complex numbers $(c_n)_{n\in\mathbb{N}}$ is a **coefficient sequence** if it is finitely supported for each $x > 0$ (with support generally depending on x) and satisfies the bound for all $n \in \mathbb{N}$

$$|c(n)| \ll \tau(n)^{O(1)}(\log x)^{O(1)}.$$

A complex sequence $(\alpha_n)_{n\in\mathbb{N}}$ is said to obey the **Siegel–Walfiz property** if for any fixed $A > 0$ and any natural numbers q, r and primitive residue class represented by a modulo q, we have as $x \to \infty$,

$$\left| \sum_{\substack{n \leq x \\ n \equiv a \bmod q \\ (n,r)=1}} \alpha(n) - \frac{1}{\varphi(q)} \sum_{\substack{n \leq x \\ (n,q)=1 \\ (n,r)=1}} \alpha(n) \right| \ll_A \tau(qr)^{O(1)} \frac{x}{L_x(A)}. \tag{8.75}$$

We also need the definition of **multiple dense divisibility**. Let $y > 1$ and $i \geq 0$. If $i = 0$, every natural number is said to be 0-tuply y-densely divisible. If $i \geq 1$, we define a natural number n to be i-tuply densely divisible recursively as follows: for every pair of nonnegative integers j, k with $j + k = i - 1$ and real numbers R with $1 \leq R \leq yn$, we have a factorization

$$n = qr \text{ such that } \frac{R}{y} \leq r \leq R$$

such that q is j-tuply y-densely divisible and r is k-tuply y-densely divisible. Let $\mathcal{D}(i, y)$ be the set of all i-tuply y-densely divisible positive integers, and $\mathcal{D}_\Omega(i, y)$ those which are **squarefree** and have all prime factors in a given interval Ω. Examples are given in Section 8.5.

Theorem 8.34 (**Type I and II estimates**) *[174, theorem 7.1] Let ϖ, δ, σ be fixed positive real numbers, $i \geq 1$ a fixed integer, $\Omega \subset \mathbb{R}$ bounded and $a \bmod P_\Omega$ a congruence class with $(a, P_\Omega) = 1$. Furthermore, let M, N be positive integers which are sufficiently large, with $MN \asymp x$ and $x^{\frac{1}{2}-\sigma} \prec^\circ N \prec^\circ x^{\frac{1}{2}}$.*

Furthermore, let α, β be complex coefficient sequences at scales M, N respectively and such that β has the Siegel–Walfisz property. Assume also that one of the following situations regarding the parameters applies:

(i) Type I: $i = 1, j = 0, k = 0,\ 54\varpi + 15\delta + 5\sigma < 1$ and $N \prec^o x^{\frac{1}{2}-2\varpi-c}$ for some fixed $c > 0$, or

(ii) Type I: $i = 2, j = 1, k = 0,\ 56\varpi + 16\delta + 4\sigma < 1$ and $N \prec^o x^{\frac{1}{2}-2\varpi-c}$ for some fixed $c > 0$, or

(iv) Type II: $i = 1, j = 0, k = 0,\ 68\varpi + 14\delta < 1$ and $x^{\frac{1}{2}-2\varpi-c} \prec^o N$ for some sufficiently small fixed $c > 0$.

Then for any fixed $A > 0$, for $x \to \infty$, we have the estimate

$$\sum_{\substack{q \prec^o x^{\frac{1}{2}+2\varpi} \\ q \in \mathscr{D}_\Omega(i,x^\delta)}} |\Delta(\alpha \star \beta, q, a)| \ll \frac{x}{(\log x)^A}, \tag{8.76}$$

where the implied constant depends on all of the fixed parameters.

Proof **(1)** The given coefficient sequence β satisfies the Siegel–Walfisz property. Therefore, by the general Bombieri–Vinogradov Theorem 8.4, for a given $A > 0$ there is a $B > 0$ which depends only on A such that

$$\sum_{q \le x^{\frac{1}{2}}/(\log x)^B} |\Delta(\alpha \star \beta), q, a)| \ll \frac{x}{(\log x)^A}.$$

If for all D satisfying $x^{\frac{1}{2}} \prec^o D \prec^o x^{\frac{1}{2}+2\varpi}$ (assuming the implicit $o(1)$ may be negative to include the entire range we need with x sufficiently large), we had established the estimate

$$\sum_{q \in \mathscr{D}_\Omega(i,x^\delta) \cap [D,2D)} |\Delta(\alpha \star \beta), q, a)| \ll \frac{x}{(\log x)^A},$$

then using Theorem 8.4 and decomposing the region $[x^{\frac{1}{2}}, x^{\frac{1}{2}+2\varpi}]$ into at most $O(\log x)$ intervals of the form $[D, 2D)$, i.e., dyadically, we could conclude

$$\sum_{\substack{q \in \mathscr{D}_\Omega(i,x^\delta) \\ q \prec^o x^{\frac{1}{2}+2\varpi}}} |\Delta(\alpha \star \beta), q, a)| \ll \frac{x}{(\log x)^A}.$$

(2) So fix D in the given range. We first treat moduli q, which have few small prime factors. To do this, we set $D_0 := \exp\left((\log x)^{1/3}\right)$ (so for all $\epsilon > 0$ and $A > 0$, we have $(\log x)^A \ll D_0 \ll x^\epsilon$ for x sufficiently large) and then define

$$\mathscr{E}(D) := \left\{ q \in [D, 2D) : \prod_{\substack{p|q \\ p \le D_0}} p > \exp\left((\log x)^{2/3}\right) \right\}.$$

We next **claim** that $|\mathscr{E}(D)| \ll D/(\log x)^A$. To see this, let $q \in \mathscr{E}(D)$. Then

$$\prod_{\substack{p|q \\ p \leq D_0}} p > \exp\left((\log x)^{2/3}\right) = D_0^{(\log x)^{1/3}}.$$

Thus q has at least $(\log x)^{1/3}$ prime factors so $\tau(q) \geq 2^{(\log x)^{1/3}}$. We also have from Dirichlet's theorem

$$\sum_{D \leq q < 2D} \tau(q) \ll D \log x.$$

But this implies for all $\kappa > 0$,

$$\sum_{\substack{D \leq q < 2D \\ \tau(q) \geq \kappa}} 1 \leq \frac{1}{\kappa} \sum_{D \leq q < 2D} \tau(q) \ll \frac{D \log x}{\kappa},$$

and choosing $\kappa = 2^{(\log x)^{1/3}}$ the claim follows from these estimates.

By the boundedness property of the coefficient sequences α and β and Lemma 8.24(iii) we get the bound for $D \ll q \ll 2D$,

$$|\Delta(\alpha \star \beta, q, a)| \ll \frac{x}{D} \tau(q)^{O(1)} (\log x)^{O(1)}.$$

Then using Cauchy–Schwarz, Lemma 8.24(ii) and the bound for $|\mathscr{E}(D)|$ which we have just demonstrated, we get

$$\sum_{\substack{q \in \mathscr{D}_\Omega(i, x^\delta) \\ q \in \mathscr{E}(D)}} |\Delta(\alpha \star \beta, q, a)| \ll \frac{|\mathscr{E}(D)|^{\frac{1}{2}}}{D} x (\log x)^{O(1)} \left(\sum_{q \in \mathscr{E}(D)} \tau(q)^{O(1)}\right)^{\frac{1}{2}} \ll \frac{x}{(\log x)^A}.$$

Therefore, integers in $\mathscr{E}(D)$ are suitable, and we need only show that

$$\sum_{\substack{q \in \mathscr{D}_\Omega(i, x^\delta) \\ q \in [D, 2D) \setminus \mathscr{E}(D)}} |\Delta(\alpha \star \beta, q, a)| \ll \frac{x}{(\log x)^A}. \tag{8.77}$$

(3) Next, we let $\epsilon > 0$ be small, so that for x sufficiently large, by the all-cases assumption $x^{\frac{1}{2}-\sigma} \prec^o N \prec^o x^{\frac{1}{2}}$ and $x^{\frac{1}{2}} \prec^o D$ from Step (1), we have $1 \leq x^{-3\epsilon} N \leq D$. If $j, k \geq 0$ are such that $i - 1 = j + k$, then, by the properties of multiple dense divisibility, any $n \in \mathscr{D}_\Omega(i, x^\delta)$ can be factored in the form $n = qr$ with $q \in \mathscr{D}_\Omega(j, x^\delta)$ and $r \in \mathscr{D}_\Omega(k, x^\delta)$ and such that

$$x^{-3\epsilon-\delta} \leq r \leq x^{-3\epsilon} N.$$

Next, let $n \in [D, 2D] \setminus \mathcal{E}(D)$ and define

$$s := \prod_{\substack{p \mid n \\ p \leq D_0}} p \leq x^{1/(\log x)^{1/3}} \prec^o 1.$$

If we replace and rename q by $q/(q,s)$ and r by $r(q,s)$, because the definition of multiple dense divisibility sets $\mathcal{D}_\Omega(i, x^\delta)$ ensures elements q are squarefree, we get a factorization $n = qr$ where all of the prime factors of q are greater than D_0 and

$$x^{-3\epsilon-\delta} N \prec^o r \prec^o x^{-3\epsilon} N.$$

Using Lemma 8.5(i), we get

$$q \in \mathcal{D}(j, sx^\delta) = \mathcal{D}\left(j, x^{\delta + o(1)}\right) \text{ and } r \in \mathcal{D}(k, sx^\delta) = \mathcal{D}\left(k, x^{\delta + o(1)}\right).$$

Hence, if we define $\Omega' := \Omega \cap (D_0, \infty)$, this gives $q \in \mathcal{D}_{\Omega'}\left(j, x^{\delta + o(1)}\right)$. Recall, in addition, we have the a priori assumption $n = qr \in \mathcal{D}_\Omega(1, x^\delta)$.

(4) In this step, we use dyadic decomposition, this time for the q and r variables, so that, since we have assumed $MN \asymp x$, the proof of the estimate (8.77) is reduced to the proof of

$$\sum_{\substack{q \in \mathcal{D}_{\Omega'}(j, x^{\delta + o(1)}) \cap [Q, 2Q) \\ r \in \mathcal{D}_\Omega(k, x^{\delta + o(1)}) \cap [R, 2R) \\ qr \in \mathcal{D}_\Omega(1, x^\delta)}} |\Delta(\alpha \star \beta), qr, a)| \ll \frac{MN}{(\log x)^A}$$

for all fixed $A > 0$ and all $Q, R > 0$ which satisfy from Steps (3) and (1) respectively

$$x^{-3\epsilon-\delta} N \prec^o R \prec^o x^{-3\epsilon} N \text{ and } x^{\frac{1}{2}} \prec^o QR \prec^o x^{\frac{1}{2}+2\varpi}.$$

Multiplying the left estimate for R with the upper estimate for QR and cancelling R, we get

$$NQ \prec^o x^{\frac{1}{2}+2\varpi+\delta+3\epsilon}.$$

We also **claim** $RQ^2 \ll x$. To see this, under the assumptions of cases (i) and (ii) of this theorem, we have $\sigma + 4\varpi + \delta < \frac{1}{2}$, so the estimate follows from the stated estimates for R and QR and the theorem assumption

$$x^{\frac{1}{2}-\sigma} \prec^o N \prec^o x^{\frac{1}{2}}.$$

In case (iv), we have $6\varpi + \delta < \frac{1}{2}$ and the assumption $x^{\frac{1}{2}-2\varpi-c} \prec^o N$. The claim follows from this.

(5) In this step, we establish some helpful simplifying notation and a splitting of the discrepancy to reflect the qr factorization. Let Q, R be as in Step (4) and $F(n)$

a summand depending on $n \in \mathbb{N}$. Then set for sums over explicit q (or q_1 or q_2) and explicit r

$$\sum_q F(q) := \sum_{q \in \mathscr{D}_{\Omega'}(j, x^{\delta+o(1)}) \cap [Q, 2Q)} F(q),$$

$$\sum_r F(r) := \sum_{r \in \mathscr{D}_{\Omega}(k, x^{\delta+o(1)}) \cap [R, 2R)} F(r).$$

In addition, we define the split discrepancies

$$\Delta_1(\alpha \star \beta, qr, a) := \sum_{n \equiv a \bmod qr} (\alpha \star \beta)(n) - \frac{1}{\varphi(q)} \sum_{\substack{n \equiv a \bmod r \\ (n,q)=1}} (\alpha \star \beta)(n),$$

$$\Delta_2(\alpha \star \beta, qr, a) := \frac{1}{\varphi(q)} \sum_{\substack{n \equiv a \bmod r \\ (n,q)=1}} (\alpha \star \beta)(n) - \frac{1}{\varphi(qr)} \sum_{(n,qr)=1} (\alpha \star \beta)(n),$$

so $\Delta = \Delta_1 + \Delta_2$. We also need the definition

$$\Delta_0(\alpha \star \beta, a, u, v) := \sum_{\substack{n \equiv a \bmod r \\ n \equiv u \bmod q}} (\alpha \star \beta)(n) - \sum_{\substack{n \equiv a \bmod r \\ n \equiv v \bmod q}} (\alpha \star \beta)(n)$$

for all a, u, v coprime to P_Ω, noting that Δ_0 depends on q and r implicitly.

Note that Δ_0 enables us to use a symmetric form for Bombieri–Vinogradov-style estimates. To see this note that given any finite sequence (x_n) with average $\bar{x}$, then for all u, v and $\epsilon > 0$ we have $|x_u - x_v| \ll \epsilon$ if and only if for all u, $|x_u - \bar{x}| \ll \epsilon$. This provides a simple transformation which should also be useful in other circumstances.

(6) First, we treat the sum for Δ_2. Because $r \leq 2R \ll x^{\frac{1}{2}+o(1)-3\epsilon}$, applying the generalized Bombieri–Vinogradov Theorem 8.4, for each q with α replaced by α_q which takes the same values as α but is set to zero if $(n, q) \neq 1$, and similarly for β_q, we get

$$\sum_{\substack{R \leq r < 2R \\ qr \in \mathscr{D}_{\Omega}(x^\delta)}} \left| \sum_{\substack{(n,q)=1 \\ n \equiv a \bmod r}} (\alpha \star \beta)(n) - \frac{1}{\varphi(r)} \sum_{(n,qr)=1} (\alpha \star \beta)(n) \right| \ll \frac{MN}{(\log x)^A}.$$

Then divide each side by $\varphi(q)$, sum over $q \leq 2Q$ and use (see for example [121, Problem 8.8]) $\sum_{n \leq x} 1/\varphi(n) \ll \log x$, or simply use Merten's theorem with

$$\sum_{n \leq x} \frac{1}{\varphi(n)} = \sum_{n \leq x} \frac{1}{n \prod_{p|n} (1 - 1/p)} \leq \sum_{n \leq x} \frac{1}{n \prod_{p \leq x} (1 - 1/p)}$$

$$\ll (\log x) \sum_{n \leq x} \frac{1}{n} \ll (\log x)^2,$$

to get

$$\sum_{\substack{q,r \\ qr \in \mathscr{D}_\Omega(1,x^\delta)}} |\Delta_2(\alpha \star \beta, qr, a)| \ll \frac{MN}{(\log x)^A},$$

which is all we need.

(7) The sum for Δ_1 is much more difficult than that for Δ_2, and will be first bounded in terms of Δ_0, as defined in Step (5). Using the symmetrization described in Step (5) for the average over q, and the Chinese Remainder Theorem and triangle inequality for each b with $(b, P_\Omega) = 1$, we get

$$\sum_{\substack{q,r \\ qr \in \mathscr{D}_\Omega(1,x^\delta)}} |\Delta_1(\alpha \star \beta, qr, a)| \leq \frac{1}{\varphi(P_\Omega)} \sum_{\substack{b \bmod P_\Omega \\ (b, P_\Omega)=1}} \sum_{\substack{q,r \\ qr \in \mathscr{D}_\Omega(1,x^\delta)}} |\Delta_0(\alpha \star \beta, a, a, b)|.$$

Therefore, it will be sufficient to bound the inner sum for all fixed a, u, v coprime to P_Ω. Indeed, we will show that

$$\sum_{\substack{q,r \\ qr \in \mathscr{D}_\Omega(1,x^\delta)}} |\Delta_0(\alpha \star \beta, a, u, v)| \ll \frac{MN}{(\log x)^A}.$$

(8) To handle Δ_0, we use the **dispersion method** of Linnik, following Zhang and Polymath. (See Appendix G for a simple case.) We introduce complex coefficients $c_{q,r}$ with modulus 1 so that

$$\sum_{\substack{q,r \\ qr \in \mathscr{D}_\Omega(1,x^\delta)}} |\Delta_0(\alpha \star \beta, a, u, v)|$$

$$= \sum_{\substack{q,r \\ qr \in \mathscr{D}_\Omega(1,x^\delta)}} c_{q,r} \left(\sum_{\substack{n \equiv a \bmod r \\ n \equiv u \bmod q}} (\alpha \star \beta)(n) - \sum_{\substack{n \equiv a \bmod r \\ n \equiv v \bmod q}} (\alpha \star \beta)(n) \right).$$

Expanding the Dirichlet products and interchanging the order of summation gives

$$\sum_{\substack{q,r \\ qr \in \mathscr{D}_\Omega(1,x^\delta)}} |\Delta_0(\alpha \star \beta, a, u, v)|$$

$$= \sum_r \sum_m \alpha(m) \left(\sum_{\substack{n:mn \equiv a \bmod r \\ qr \in \mathscr{D}_\Omega(1,x^\delta)}} c_{q,r} \beta(n)(\chi_{mn \equiv u \bmod q} - \chi_{mn \equiv v \bmod q}) \right),$$

where the characteristic functions in the inner sum are evaluated at n and, here and in what follows, recalling α is at scale M, we have $m \ll M$ and β at scale N gives $n \ll N$.

Using the Cauchy–Schwarz inequality for the sums over r and m, the bounds

$$NQ \prec^o x^{\frac{1}{2}+2\varpi+\delta+3\epsilon} \text{ and } RQ^2 \ll x$$

from Step (4) and bounds for the number of divisors function from Lemma 8.20, we get for any smooth real coefficient sequence ψ_m at scale M with $\psi_M(m) \geq 1$, when m belongs to the support of β,

$$\sum_{\substack{q,r \\ qr \in \mathscr{D}_\Omega(1,x^\delta)}} |\Delta_0(\alpha \star \beta, a, u, v)| \leq \sqrt{RM}(\log x)^{O(1)}$$

$$\times \left(\sum_r \sum_m \psi_M(m) \left| \sum_{\substack{n:mn\equiv a \bmod r \\ qr \in \mathscr{D}_\Omega(1,x^\delta)}} c_{qr}\beta(n)(\chi_{mn\equiv u \bmod q} - \chi_{mn\equiv v \bmod q}) \right|^2 \right)^{\frac{1}{2}}.$$

Therefore, all we need is to establish the estimate, with ψ_M any smooth coefficient sequence at scale M,

$$\Sigma := \sum_r \sum_m \psi_M(m) \left| \sum_{\substack{mn\equiv a \bmod r \\ qr \in \mathscr{D}_\Omega(1,x^\delta)}} c_{q,r}\beta(n)(\chi_{n:mn\equiv u \bmod q} - \chi_{mn\equiv v \bmod q}) \right|^2$$

$$\ll \frac{MN^2}{R(\log x)^A}. \tag{8.78}$$

Here the phrase "all we need" is a euphemism, since establishing this will take the next eight steps of the proof!

This will take a veritable family of sigma types to achieve, and we gather them together here so the reader might trace back as needed: $\Sigma(u, v)$ is defined in Step (9); Σ_0, Σ_1 and $\hat{\Sigma}_1$ in Step (13); and Σ'_0 in Step (14).

(9) Next, recall the summary notation from Step (5); we now introduce further summary notation: for integers u, v coprime to P_Ω and $q_i \in \mathbb{N}$ satisfying $q_1, q_2 \in \mathscr{D}_{\Omega'}(j, x^{\delta+o(1)}) \cap [Q, 2Q)$, we define

$$\Sigma(u, v) := \sum_r \sum_m \psi_M(m) \sum_{\substack{q_1,q_2,n_1,n_2 \\ mn_1\equiv mn_2\equiv a \bmod r \\ q_1r, q_2r \in \mathscr{D}_\Omega(1,x^\delta)}} \cdots \sum c_{q_1,r}\overline{c_{q_2,r}}$$

$$\times \beta(n_1)\overline{\beta(n_2)}\chi_{mn_1\equiv u \bmod q_1}\chi_{mn_2\equiv v \bmod q_2},$$

for ψ_M a real smooth coefficient sequence at scale M with $\psi_M(m) \geq 1$ for all m in the support of β. Expanding the square of the left-hand side of the estimate (8.78), we get

$$\Sigma = \Sigma(u, u) - \Sigma(u, v) - \Sigma(v, u) + \Sigma(v, v).$$

(10) The aim is to derive an equation

$$\Sigma(u, v) = X + O\left(\frac{N^2 M}{R(\log x)^A}\right),$$

where the main term X is independent of the values of u and v, so it does not appear in Σ. Here symmetry is working well. Assuming this, the desired estimate follows from the result of Step (8). We will derive this expression for $\Sigma(u, v)$ in Step (17).

Since $(a, qr) = 1$, we have $(n_1 n_2, qr) = 1$, so $n_1 \equiv n_2 \bmod r$ and thus $n_2 = n_1 + lr$. Let $n = n_1$ in the sum of Step (9). Using $(n, q_1 r) = (n + lr, q_2 r) = 1$, we get for some integer l,

$$\Sigma(u, v) = \sum_{\substack{r, 0 \leq |l| \leq N/R}} \sum_{\substack{q_1, q_2 \\ q_1 r, q_2 r \in \mathcal{D}_\Omega(1, x^\delta)}} c_{q_1, r} \overline{c_{q_2, r}} \sum_n \beta(n) \overline{\beta(n + lr)}$$

$$\times \sum_{m \in \mathbb{N}} \psi_M(m) X_{mn \equiv u \bmod q_1} X_{m(n+lr) \equiv v \bmod q_2} X_{mn \equiv a \bmod r}.$$

(11) For the sum from Step (10) over m, we first consider the diagonal terms with $l = 0$, that is to say with $n_1 = n_2$. Let $T(u, v)$ denote this contribution. Then because $RQ^2 \ll x$ and $R \prec^o x^{-3x} N$, we can derive

$$|T(u, v)| \leq \sum_{r \in \mathbb{N}} \sum_{\substack{q_1, q_2 \\ q_1 r, q_2 r \in \mathcal{D}_\Omega(1, x^\delta)}} \sum_{n \in \mathbb{Z}} |\beta(n)|^2$$

$$\times \sum_{m \in \mathbb{Z}} \psi_M(m) X_{mn \equiv u \bmod q_1} X_{mn \equiv v \bmod q_2} X_{mn \equiv a \bmod r}$$

$$\prec^o \sum_{\substack{r \asymp R}} \sum_{q_1, q_2 \asymp Q} \sum_{s \asymp x} \tau(s) X_{s \equiv u \bmod q_1} X_{s \equiv v \bmod q_2} X_{s \equiv a \bmod r}$$

$$\prec^o \sum_{\substack{r \asymp R}} \sum_{q_1, q_2 \asymp Q} \frac{x}{r[q_1, q_2]} \prec^o x \ll \frac{N^2 M}{R(\log x)^A},$$

so the terms with $l = 0$ do not need to contribute to X in the expression for $\Sigma(u, v)$.

(12) Next, consider terms with $l \neq 0$. Because $(n, q_1 r) = (n + lr, q_2 r) = 1$, we get, by the Chinese Remainder Theorem, for a residue class γ,

$$X_{mn \equiv u \bmod q_1} X_{m(n+lr) \equiv v \bmod q_2} X_{mn \equiv a \bmod r} = X_{m \equiv \gamma \bmod [q_1, q_2]r}.$$

Let $q_0 := (q_1, q_2)$. Since q_1, q_2 have no prime factor less than D_0, either $q_0 = 1$ or $q_0 \geq D_0$. Also we restrict the sum over n by ensuring

$$u(n + lr) \equiv nv \bmod q_0,$$

which, for brevity, we denote X^n as its characteristic function, recalling that in each application the function depends on the other variables in the congruence. Because $(q_0, ru) = 1$, we note for future reference that these values of n are in the union of at most

$$(u - v, q_0, lru) \le (q_0, l)$$

congruence classes modulo q_0.

(13) Next, fixing q_1, q_2, r, l and summing over $n \in \mathscr{D}_\Omega(x^\delta)$, if we define $h :=$ $x^\epsilon [q_1, q_2] r / M \ll x^\epsilon Q^2 R / M$, noting that h depends on q_1, q_2, and can be shown to be small relative to x, and we define $\Sigma_0(u, v)$ and $\hat{\Sigma}_1$ by

$$\Sigma_0(u, v) := \left(\sum_{m \in \mathbb{N}} \psi_M(m) \right)$$

$$\times \sum_{r \in \mathbb{N}} \frac{1}{r} \sum_{l \ne 0} \sum_{\substack{q_1, q_2 \\ q_1 r, q_2 r \in \mathscr{D}_\Omega(1, x^\delta)}} \frac{c_{q_1, r} \overline{c_{q_2, r}}}{[q_1, q_2]} \sum_{n \in \mathscr{D}_\Omega(x^\delta)} \beta(n) \overline{\beta(n + lr)} X^n \text{ and}$$

$$\hat{\Sigma}_1 := \sum_r \sum_{l \ne 0} \sum_{\substack{q_1, q_2 \\ q_1 r, q_2 r \in \mathscr{D}_\Omega(1, x^\delta)}} \frac{c_{q_1, r} \overline{c_{q_2, r}}}{h}$$

$$\times \sum_{1 \le |m| \le h} \left| \sum_{n \in \mathscr{D}_\Omega(1, x^\delta)} \beta(n) \overline{\beta(n + lr)} X^n e_{[q_1, q_2] r}(\gamma m) \right|,$$

respectively, then we set $\Sigma_1(u, v)$ to be a function which satisfies $\Sigma_1(u, v) \ll 1 + x^\epsilon \hat{\Sigma}_1(u, v)$, and sum over $n \in \mathscr{D}_\Omega(x^\delta)$ for each q_1, q_2, r, l using Lemma 8.23 with

$$f(m) := \sum_{\substack{n \in \mathscr{D}_\Omega(1, x^\delta) \\ a_i \equiv m \bmod q}} c_i,$$

we can write

$$\Sigma(u, v) = \Sigma_0(u, v) + \Sigma_1(u, v) + O\left(\frac{M N^2}{R (\log x)^A} \right).$$

(14) Next, we consider $\Sigma_0(u, v)$, first summing the terms where $q_0 = (q_1, q_2) = 1$. In that situation, we have $X^n = 1$, and the sum gives the main term for $\Sigma(u, v)$, which we have called X, so

$$X = \left(\sum_{m \in \mathbb{N}} \psi_M(m) \right) \sum_{r \in \mathbb{N}} \frac{1}{r} \sum_{l \ne 0} \sum_{\substack{q_1, q_2 \\ q_1 r, q_2 r \in \mathscr{D}_\Omega(1, x^\delta) \\ (q_1, q_2) = 1}} \frac{c_{q_1, r} \overline{c_{q_2, r}}}{[q_1, q_2]} \sum_{n \in \mathscr{D}_\Omega(x^\delta)} \beta(n) \overline{\beta(n + lr)}.$$

Denote the remaining contribution to the sum for $\Sigma(u, v)$ by $\Sigma'_0(u, v)$. It follows that

$$\Sigma'_0(u, v) \ll \frac{M(\log x)^{O(1)}}{R} \sum_{r \asymp R} \sum_{|l| \leq h} \sum_{\substack{1 \neq q_0 \ll Q \\ q_0 \in S_{\Omega'}}} \frac{1}{q_0} \sum_{q_1, q_2 \asymp Q/q_0} \frac{1}{q_1 q_2}$$

$$\times \sum_{n \in \mathscr{D}_{\Omega}(1, x^\delta)} (\tau(n)\tau(n + lr))^{O(1)} \chi^n.$$

(15) We are now able to obtain an expression for $\sum(u, v)$ in terms of $\hat{\Sigma}_1$. To achieve this, we first bound the sum over l. Because we have $(ru, q_0) = 1$, we can interpret the condition $b_1(n + lr) \equiv b_2 n \bmod q_0$ as a congruence condition for l, which, using the estimate for powers of the number of divisors Lemma 8.24(iii), i.e.,

$$\sum_{\substack{n \leq x \\ n \equiv a \bmod q}} \tau(n)^C \ll \frac{x}{q} \tau(q)^{O(1)} (\log x)^{O(1)},$$

gives

$$\sum_{|l| \ll N/R} \tau(n + lr)^{O(1)} \chi_{u(n+lr) \equiv vn \bmod q_0} \ll \left(1 + \frac{N}{q_0 R}\right) (\log x)^{O(1)}.$$

Because in this sum we have (as shown earlier) $q_0 \neq 1$ so $D_0 \leq q_0 \ll Q$, we can derive using in the final estimate $R \ll x^{-3x} N$ and for all $A > 0$, $(\log x)^A \ll D_0$,

$$\Sigma'_0(u, v) \ll \frac{MN(\log x)^{O(1)}}{R} \sum_{r \asymp R} \sum_{D_0 \leq q_0 \ll Q} \frac{1}{q_0}\left(1 + \frac{N}{q_0 R}\right) \sum_{q_1, q_2 \asymp Q/q_0} \frac{1}{q_1 q_2}$$

$$\ll MN(\log x)^{O(1)} \sum_{D_0 \leq q_0 \ll Q} \frac{1}{q_0}\left(1 + \frac{N}{q_0 R}\right)$$

$$\ll MN(\log x)^{O(1)} + \frac{MN^2}{D_0 R}(\log x)^{O(1)}$$

$$\ll \frac{MN^2}{R(\log x)^A}.$$

Combining this estimate with the results of Steps (13) and (14) gives

$$\Sigma(u, v) = X + O(x^\epsilon |\hat{\Sigma}_1(u, v)|) + O\left(\frac{MN^2}{R(\log x)^A}\right). \qquad (8.79)$$

(16) Next, recalling the definition of $\Upsilon_{r,l}$ (8.66), which is

$$\Upsilon_{l,r}(b_1, b_2, q_0) := $$

$$\sum_{\substack{q_1,q_2 \asymp Q/q_0 \\ (q_1,q_2)=1 \\ q_0,q_1,q_0q_1 \in \mathscr{D}_\Omega(j, x^{\delta+o(1)}) \\ q_0q_1 r, q_0q_2 r \in \mathscr{D}_\Omega(x^\delta)}} \sum_{1 \leq |m| \ll x^\epsilon \frac{RQ^2}{q_0 M}} \left| \sum_{n \in \mathbb{N}} \chi^n \beta(n) \overline{\beta(n+lr)} \Phi_l(m,n,r,q_0,q_1,q_2) \right|,$$

where $|\Phi_1(\mathbf{p})| = 1$, the interpretation of sums over unqualified r from Step (5), and using the expression derived in Step (13), we get

$$|\hat{\Sigma}_1(u,v)| \leq \sum_r \sum_{l \neq 0} \sum_{\substack{q_1,q_2 \\ q_1 r, q_2 r \in \mathscr{D}_\Omega(1,x^\delta)}} \frac{1}{h} \sum_{0 < |m| \leq h} \left| \sum_{n \in \mathscr{D}_\Omega(x^\delta)} \chi^n \beta(n) \overline{\beta(n+lr)} e_{[q_1,q_2]r}(\gamma m) \right|$$

$$\ll x^{-\epsilon} \frac{M}{RQ^2} \sum_{1 \leq |l| \ll N/R} \sum_{q_0 \ll Q} q_0 \sum_r \Upsilon_{l,r}(u,v,q_0). \tag{8.80}$$

Next, replace q_1 by $q_1 q_0$ and q_2 by $q_2 q_0$ so $[q_0, q_1]r$ is to be replaced by $q_0 q_1 q_2 r$. Also note that if $x^\epsilon A^2 R/(q_0 M) \gg 1$ is false, then the sum over m is empty so $\hat{\Sigma}_1(u,v) = 0$.

(17) We can now complete Theorem 8.34. Inserting the estimate of Theorem 8.33, namely (8.67), which gives

$$\sum_{r \in \mathscr{D}_\Omega(k, x^{\delta+o(1)}) \cap [R, 2R]} \Upsilon_{l,r}(b_1, b_2, q_0) \prec^o x^{-\epsilon} \frac{Q^2 R N(q_0, l)}{q_0^2}$$

in the estimate (8.80) from Step (16), the GCD sum given in Lemma 8.25(i), and recalling that we do not need to include terms corresponding to $l = 0$, we get

$$x^\epsilon |\hat{\Sigma}(u,v)| \prec^o x^{-\epsilon} MN \sum_{q_0 \ll Q} \frac{1}{q_0} \sum_{1 \leq |l| \ll N/R} (q_0, l) \prec^o x^{-\epsilon} \frac{MN^2}{R},$$

and then using (8.79) we conclude using the expression for Σ from Step (9), that

$$\Sigma = O\left(\frac{N^2 M}{R(\log x)^A} \right),$$

which is equivalent, as we saw in Step (8), to

$$\sum_{\substack{q \prec^o x^{\frac{1}{2}+2\varpi} \\ q \in \mathscr{D}_\Omega(i, x^\delta)}} |\Delta(\alpha \star \beta, q)| \ll \frac{x}{(\log x)^A}. \tag{8.81}$$

This (at last) completes the proof of Theorem 8.34. $\qquad\square$

8.11 Application to Prime Gaps

Having assembled all these concepts and set out the proofs of the fundamental lemmas, we are now in a position to derive the final results. The hard work has been done and it is only a matter of joining a few rather large dots to reach the goal. First, recall some definitions:

Definition 8.14 (**Densely divisible MPZ property**) Let $i \geq 0$ be an integer. We say $\mathrm{MPZ}^{(i)}[\varpi, \delta]$ is true if for any bounded subset $\Omega \subset \mathbb{R}$ which may depend on x, and $Q \prec^o x^{\frac{1}{2}+2\varpi}$, if $(a, P_\Omega) = 1$ and $A \geq 1$ is a fixed real number, then

$$\sum_{\substack{q \leq Q \\ q \in \mathcal{D}_\Omega(i, x^\delta)}} \sup_{\substack{a \bmod q \\ x \leq n < 2x \\ (a,q)=1}} \left| \sum_{\substack{n \equiv a \bmod q}} \Lambda(n) - \frac{1}{\varphi(q)} \sum_{\substack{x \leq n < 2x \\ (n,q)=1}} \Lambda(n) \right| \ll \tau(qr)^{O(1)} \frac{x}{(\log x)^A}. \tag{8.82}$$

Definition 8.15 (**Type I estimate**) Let $i \geq 1$, ϖ with $0 < \varpi < \frac{1}{4}$, δ with $0 < \delta < \frac{1}{4} + \varpi$, σ with $\frac{1}{6} < \sigma < \frac{1}{2}$ and $\sigma > 2\varpi$ be fixed real numbers, let Ω be a fixed bounded subset of $(0, \infty)$ and define the squarefree number

$$P_\Omega = \prod_{p \in \Omega} p.$$

Let a be an integer with $(a, P_\Omega) = 1$ and $a \bmod P_\Omega$ the corresponding congruence class. We say the estimate $\mathrm{Type}_{\mathrm{I}}^{(i)}[\varpi, \delta, \sigma]$ holds if for all Ω and class $a \bmod P_\Omega$ and any integers $M, N \gg 1$ such that $x \ll MN \ll x$ and for some fixed $c > 0$

$$x^{\frac{1}{2}-\sigma} \prec^o N \prec^o x^{\frac{1}{2}-2\varpi-c},$$

and for any $Q \prec^o x^{\frac{1}{2}+2\varpi}$ and any coefficient sequence α at scale M and β at scale N, with β having the Siegel–Walfiz property, for any fixed $A > 0$, we have

$$\sum_{\substack{q \leq Q \\ q \in \mathcal{D}_\Omega(i, x^\delta)}} \sup_{\substack{a \bmod q \\ (a,q)=1}} \left| \sum_{\substack{n \equiv a \bmod q \\ (a,r)=1}} (\alpha \star \beta)(n) - \frac{1}{\varphi(q)} \sum_{(n,q)=1} (\alpha \star \beta)(n) \right| \ll \frac{x}{(\log x)^A}. \tag{8.83}$$

Definition 8.16 (**Type II estimate**) We say the estimate $\mathrm{Type}_{\mathrm{II}}^{(i)}[\varpi, \delta]$ holds if for all Ω and class $a \bmod P_\Omega$ and any integers $M, N \gg 1$ such that $x \ll MN \ll x$ and for some sufficiently small but fixed $c > 0$,

$$x^{\frac{1}{2}-2\varpi-c} \prec^o N \prec^o x^{\frac{1}{2}}$$

and for any $Q \prec^o x^{\frac{1}{2}+2\varpi}$ and any coefficient sequence α at scale M and β at scale N, with β having the Siegel–Walfiz property, for any fixed $A > 0$ we have the estimate (8.83).

Next, we extract two results from Theorem 8.34:

Lemma 8.35 (**Type I estimate**) *[174, lemma 2.23(ii)] If* $56\varpi + 16\delta + 4\sigma < 1$, *then we have* $\mathrm{Type}_{\mathrm{I}}^{(2)}[\varpi, \delta, \sigma]$.

Proof This follows from Theorem 8.34(ii). ☐

Lemma 8.36 (**Type II estimate**) *[174, lemma 2.23(iv)] If* $68\varpi + 14\delta < 1$, *then we have* $\mathrm{Type}_{\mathrm{II}}^{(1)}[\varpi, \delta]$.

Proof This follows from Theorem 8.34(iv). ☐

Next, we use the main combinatorial Lemma 8.22 to get $\mathrm{MPZ}^{(2)}[\varpi, \delta]$:

Theorem 8.37 (**A Motohashi–Pintz–Zhang style estimate**) *[174, theorem 2.17(ii)] For any fixed pair* $\varpi, \delta > 0$ *with* $168\varpi + 48\delta < 1$, *we have* $\mathrm{MPZ}^{(2)}[\varpi, \delta]$.

Proof Choose σ to be larger and sufficiently close to $1/6$. With this constraint and the other constraints of Lemma 8.22, namely $0 < \varpi < \frac{1}{4}$, $0 < \delta < \frac{1}{4} + \varpi$, $\frac{1}{6} < \sigma < \frac{1}{2}$, and $2\varpi < \sigma$, then

$$168\varpi + 48\delta < 1 \implies 56\varpi + 16\delta + \frac{4}{6} < 1,$$

which by Lemma 8.35 implies we have the estimate $\mathrm{Type}_{\mathrm{I}}^{(2)}[\varpi, \delta, \sigma]$.

Similarly,

$$168\varpi + 48\delta < 1 \implies 68\varpi + 14\delta < 1,$$

which by Lemma 8.36 gives $\mathrm{Type}_{\mathrm{II}}^{(1)}[\varpi, \delta]$. In addition, $\mathrm{Type}_{\mathrm{II}}^{(1)}[\varpi, \delta]$ implies $\mathrm{Type}_{\mathrm{II}}^{(2)}[\varpi, \delta]$. Therefore, using Lemma 8.22 we have $\mathrm{MPZ}^{(2)}[\varpi, \delta]$, which completes the proof. ☐

Now we go way back to Theorem 8.19, and a little computation, to verify that a particular value of k, with given parameter settings, is sufficient to derive $\mathrm{DHL}[k, 2]$:

Theorem 8.38 *[174, theorem 2.3] For* $k = 1783$, *we have* $\mathrm{DHL}[k, 2]$.

Proof By the remarks following the proof of Theorem 8.19, for the given values of the parameters in Table 8.2, ϖ and δ we have $168\varpi + 48\delta < 1$. Therefore, by Theorem 8.37 we have $\mathrm{MPZ}^{(2)}[\varpi, \delta]$. But by Theorem 8.19, this implies we also have $\mathrm{DHL}[k, 2]$ for any k which satisfies the inequality (8.35). Since $k = 1783$ is such an integer, this completes the proof. ☐

Finally we derive the penultimate "large dot", a 1783-tuple that is reasonably narrow, maybe as narrow as possible for that given k:

Theorem 8.39 *[174, theorem 3.1] For $k = 1783$, there exists an admissible k-tuple of width less than or equal to* 14950.

Proof Following Polymath, we commence with all integers in the range $\{1714, \ldots, 16664\}$ and remove all odd integers and all multiples of small odd primes, namely multiples of each $p \leq 211$. We then removed any elements in any of the residue classes, with stated representatives listed as follows, and with moduli which are most of the primes between 223 and 409, found by trial and error:

$$188 \bmod 223, 0 \bmod 227, 222 \bmod 229, 38 \bmod 233, 146 \bmod 239,$$

$$33 \bmod 241, 0 \bmod 251, 229 \bmod 257, 21 \bmod 263, 78 \bmod 269,$$

$$140 \bmod 277, 104 \bmod 277, 106 \bmod 281, 53 \bmod 283, 141 \bmod 293,$$

$$216 \bmod 307, 12 \bmod 311, 17 \bmod 313, 252 \bmod 317, 191 \bmod 337,$$

$$269 \bmod 347, 32 \bmod 353, 142 \bmod 359, 42 \bmod 379, 345 \bmod 383,$$

$$165 \bmod 389, 221 \bmod 409.$$

This gives an admissible 1783-tuple with $h_{1783} - h_1 = 14950$, as may be checked with `AdmissibleTupleQ`. The explicit sequence is stored in PGpack as `NarrowH[1783]`. $\square$

The rest is easy:

Theorem 8.40 **(A bound for H_1)** *[174, theorem 1.3] For an infinite set of natural numbers n, we have $p_{n+1} - p_n \leq 14950$. Thus $H_1 \leq 14950$.*

Proof This follows directly by combining the result of Theorem 8.38 with Theorem 8.39. $\square$

8.12 End Notes

We have presented sufficient details of Polymath8a's approach to derive a significantly smaller prime gap than that of Zhang, namely $H_1 \leq 14950$. This was not their best result. They followed Zhang and joined their method based on dense divisibility with his adoption of techniques based on multidimensional exponential sum estimates. These in turn relied on the work of Deligne and as indicated before, a presentation of the concepts and results would take at least as much again to describe as what has been set out in this chapter. In addition, the result obtained with this significant additional effort, namely $H_1 \leq 4680$ corresponding to $k = 632$, was soon significantly superseded by Maynard, who obtained $H_1 \leq 600$ as we have seen in Chapter 6, and then Polymath8b, refining and enhancing Maynard's approach, with $H_1 \leq 246$, described in Chapter 7. The interested reader could well consult [175] and the references in that work.

Two parts of the development of Polymath8a were omitted in this presentation of the work. These were the proofs of Parts (iii) and (v) of Theorem 8.34, namely

> (iii) If $160\varpi/3 + 16\delta + 34\sigma/9 < 1$ and $64\varpi + 18\delta + 2\sigma < 1$,
>
> then $\text{Type}_\text{I}^{(4)}[\varpi, \delta, \sigma]$ is true, and
>
> (v) If $\sigma > 1/18 + 28\varpi/9 + 16\delta + 2\delta/9$ and $\varpi < 1/12$,
>
> then $\text{Type}_\text{III}^{(1)}[\varpi, \delta, \sigma]$ is true.

We chose not to give the details of these derivations here for the reasons given earlier, namely that they depend on the deep results of Deligne, and Polymath was able to obtain an excellent improvement over Zhang without using them.

Note the view of Motohashi from 2014 [150], reflecting on related issues:

Zhang followed to a large extent the work of Bombieri, Friedlander and Iwaniec [15]. Thus I am unable to confirm his reasoning on my own but have to rely on the affirmative opinion of experts. I have no courage to exploit any result which I do not fully understand; neither have I any other way to trust, with considerable caution, competent authors whose claims depend on works which are far beyond my expertise.

Be that as it may, the energetic reader may wish to pursue these developments, since they are part of a great strand of mathematical thinking. Here we will outline some of the concepts and stepping stones taken up from the works of many, including Weil, Deligne, Fouvry, Iwaniec, Bombieri, Friedlander and others, first by Zhang and then elaborated by Polymath8a. A great example of the evolution of mathematical ideas.

Deligne had already pointed to the significance of his results for multinomial exponential sums – see for example [22, section 9.14], where we find, among other initial applications by Deligne, for a polynomial $f \in \mathbb{F}_p[x_1, \ldots, x_n]$ of degree d which is not a multiple of p and with a leading homogeneous part defining a nonsingular projective hypersurface, then we have the nice estimate

$$\left| \sum_{x_i \in \mathbb{F}_p} \exp\left(\frac{2\pi i}{p} f(x_1, \ldots, x_n) \right) \right| \leq (d-1)^n p^{n/2}.$$

Such results do not come cheaply and are the products of algebraic geometry, in particular the theory of l-adic cohomology of sheaves on the projective line. These terms will not even be defined here. The reader is directed to Polymath's extended summary in the introductory pages of section 8 of [174] and the references given there.

Indeed, Polymath in [174, section 8.1] gives the reader the opportunity to avoid the technical underpinnings by accepting just four main results (A) through (D) set out as follows, and then derived all that is needed completely from these.

First we define the mth **hyper-Kloosterman sum**, i.e. for any finite extension $k = \mathbb{F}_q$ of $\mathbb{F}_p$, point $x \in k$ and additive character ψ of k, we set

$$\mathrm{Kl}_m(x, k) := \frac{1}{q^{(m-1)/2}} \sum_{\substack{y_i \in k \\ y_1 \cdots y_m = x}} \psi(y_1 + \cdots + y_m).$$

Then [173, proposition 8.11] if $m \geq 2$ and ψ an additive character of $\mathbb{F}_p$, we have

$$(A) \quad \left| \sum_{x \in \mathbb{F}_p^*} \mathrm{Kl}_m(x, p) \psi(x) \right| \ll \sqrt{p}, \text{ and}$$

if $a \neq 1$ or ψ is nontrivial, we have

$$(B) \quad \left| \sum_{x \in \mathbb{F}_p^*} \mathrm{Kl}_m(x, p) \overline{\mathrm{Kl}_m(ax, p))} \psi(x) \right| \ll \sqrt{p}.$$

In (A) and (B), the implied constants depend at most on m in a polynomial manner.

Next, consider [173, theorem 8.17]. First, set parameters $a, b, c, d, e \in \mathbb{F}_p$ with $a \neq c$ and define a rational function $f \in \mathbb{F}_p(x, y)$ by

$$f(x, y) := ey + \frac{1}{(y + ax + b)(y + cx + d)}.$$

For any fixed nontrivial additive character ψ of $\mathbb{F}_p$, define the sum

$$K_f(x, p) := -\frac{1}{\sqrt{p}} \sum_{\substack{y \in \mathbb{F}_p \\ y + ax + b \neq 0 \\ y + cx + d \neq 0}} \psi(f(x, y)).$$

Polymath shows that for all $h \in \mathbb{F}_p$, we have the estimate

$$(C) \quad \left| \sum_{x \in \mathbb{F}_p} K_f(x, p) \psi(hx) \right| \ll \sqrt{p},$$

and for any nonzero $h, l \in \mathbb{F}_p$,

$$(D) \quad \left| \sum_{x \in \mathbb{F}_p} K_f(x, p) \overline{K_f(x + l, p)} \psi(hx) \right| \ll \sqrt{p}.$$

Polymath's [173, theorem 2.23(v)=theorem 9.1, page 139] is then proved using these estimates. They acknowledge the sources for their ideas (see [173, page 140]: especially the work of Fouvry–Iwaniec [57], Heath-Brown [97], and Fouvry et al. [58] on the ternary divisor function in arithmetic progressions. For details, the reader is encouraged to consult the original source [173], where the proof of the theorem is set out very fully.

If readers wish to see how (A) through (D) are proved and wish to dig deeper into this aspect of prime gaps, they might want to first scan the concepts which would be needed and go to the references, some of them substantial, noted here and elsewhere, especially [174, section 8]. The main ideas are the projective line, a sheaf on a curve, l-adic sheaves and trace functions on l-adic sheaves. These come from algebraic geometry and the reader is directed for example to [92], then [93], and follow this with [117, section 11.11], and even the whole of [117, chapter 11], which contains references to an abundant set of related sources.

As well as these sources, there are the more immediate developments from analytic number theory. In the introduction to Chapter 5, we gave a list of some of the contributors to results and methods used in Zhang's, and then Polymath8a's, work. Here we give some of the related references, in particular those relating to the variations on the Bombieri–Vinogradov theorem.

In 1982/1983, Deshouillers–Iwaniec [37] modified the Kuznetsov trace formula to derive an estimate for the Fourier coefficients of cuspidal Maass wave forms. The particular estimate arising from this work, theorem 12, page 237, has an error which has been corrected recently [19], and in all cases considered so far found to be either benign or avoidable, leaving the results which use the theorem intact.

Then, in 1983 Fouvry–Iwaniec [55] used that theorem, together with a well-known additive decomposition of $\Lambda(n)$ of Heath-Brown [95], and the concept of a well-factorable arithmetic function of level Q, to break through the $1/2$ barrier for sums over moduli, but weighted by the values of a well-factorable function. In more detail, if the level $Q = x^{9/17-\epsilon}$, then as $x \to \infty$ we have

$$\sum_{q:(a,q)=1} \lambda(q)(\pi(x,q,a) - \mathrm{li}(x)/\varphi(q)) \ll x/(\log x)^A.$$

These two authors had already in 1980 [54] broken the $1/2$ barrier by fixing the residue class in the Bombieri-Vinogradov theorem.

Another dependency came in 1985, when Friedlander–Iwaniec [61] extended the range of moduli for the ternary divisor sum in arithmetic progressions with fixed class to beyond $1/2$. This paper has an appendix by Birch–Bombieri, and it is an exponential sum estimate from this part of the paper which is used by Zhang, namely equation (1.26), page 330. The Birch–Bombieri estimate was used by Zhang only in the derivation of Type III estimates.

Zhang and then Polymath used many advanced techniques from analytic number theory, including the so-called dispersion method first given by Linnik (see Appendix G), but used by Bombieri–Friedlander–Iwaniec [15] and Fouvry–Iwaniec [55], and a Weil bound for Kloosterman sums.

Finally, we give a brief summary of two of the differences between Zhang's and Polymath8a's approaches. Where Zhang used fixed values for parameters, Polymath introduced ranges. Where Zhang used moduli which were smooth, Polymath defined and used the parameterized generalization of smoothness, which they denoted multiple dense divisibility. These choices enabled them, using the fundamentally the same callouts to the methods and results of mathematicians described earlier, to find a gap between prime pairs significantly smaller than that of Zhang.

Here is a list of the PGpack functions which relate to the work discussed in this chapter: DenseDivisibility, DenselyDivisibleQ, EnlargingElementQ, FeasibleKPolymathQ, GetPolymathTuple, LargeSubsetQ, MultipleDenseDivisibleQ, SmallestKPolymath0, SmallestKPolymath1, SmallestKStar, SmallSubsetQ, SmoothIntegerQ, StabilizingElementQ and StabilizingElements. Manual entries may be found in Appendix I.

9

Further Work and the Epilogue

9.1 Introduction

Further results are flowing like a stream. By early 2018, the principal works which have formed the core of this book, namely [72, 138, 173, 175, 215] had collectively received over 210 citations registered on MathSciNet. In this chapter, we summarize some of these results, depending also on the list of Andrew Granville [82, pages 175–176], and the mathematician's ordinary resource, MathSciNet.

The material covered is but an indication of the fundamental nature of the breakthroughs which have been described, and much more is expected in the future. It includes results on gaps between almost primes, clusters of primes in intervals, the set of limit points of the sequence of normalized consecutive prime differences, Artin's primitive root conjecture, arithmetic progressions of primes with a fixed common difference, prime ideals in the ring of integers of number fields, irreducible polynomials, the coefficients of a class of modular forms including Ramanujan's tau-function form, quadratic twists of a class of elliptic curves including the congruent number elliptic curve and results obtained assuming the Elliott–Halberstam and generalized Elliott–Halberstam conjectures.

References are given for each section, and readers are encouraged to follow the citations of these, and of the principal works, for up-to-date developments.

9.2 Assuming Elliott–Halberstam's Conjecture

The expression EH[θ] (Elliott–Halberstam) means that for all fixed $\epsilon > 0$, $A > 0$, and $Q \leq x^{\theta-\epsilon}$, we have as $x \to \infty$,

$$\sum_{q \leq Q} \sup_{\substack{1 \leq a \leq q \\ (a,q)=1}} \left| \sum_{\substack{x \leq n \leq 2x \\ n \equiv a \bmod q}} \Lambda(n) - \frac{1}{\varphi(q)} \sum_{\substack{x \leq n \leq 2x \\ (n,q)=1}} \Lambda(n) \right| \ll \frac{x}{(\log x)^A},$$

where the implied constant depends on the fixed constants.

We have called θ the (rather "a") **level of distribution of primes**. The **Elliott–Halberstam conjecture** is the assertion EH[1]. Note that by the Bombieri–Vinogradov theorem we can assert EH[$\frac{1}{2}$] unconditionally. No value of θ is known in the range $\frac{1}{2} < \theta < 1$ without some restriction on the moduli q. At the upper end of the range for θ, say $1 - \epsilon$, it has been proved by Friedlander and Granville [60] that setting $\epsilon = 0$ is not possible, indeed even the range of moduli up to $Q = x/(\log x)^B$ fails.

The relationship between the value of θ and prime gaps was already known to GPY who showed assuming EH[1] (see [72]) that

$$\liminf_{n\to\infty}(p_{n+1} - p_n) \le 16.$$

Maynard using a particular choice of his function F showed that $M_5 > 2$ – see Lemma 6.11(i). Assuming EH[1], we then get $\theta M_5/2 > 1$ with $\theta = 1$, so with admissible $\mathscr{H} = \{0, 2, 6, 8, 12\}$, and the assumption we could derive

$$\liminf_{n\to\infty}(p_{n+1} - p_n) \le 12.$$

So even with this strong assumption (which has no method of proof anywhere in sight), using these methods we fail to demonstrate the existence of an infinite number of twin primes.

9.3 Assuming the Generalized Elliott–Halberstam Conjecture

The generalized Elliott–Halberstam conjecture says that a Bombieri–Vinogradov estimate should hold for all prime distributions $0 < \theta < 1$ and a very broad class of arithmetic functions, ones which do not grow too quickly with $n \in \mathbb{N}$ and which can be expressible as Dirichlet products of suitably constrained functions. Polymath8b showed that this assumption was strictly stronger than EH[θ] [175, proposition 13], and Motohashi [146] that it was true unconditionally for $0 < \theta < \frac{1}{2}$. Of course, since Elliott–Halberstam is currently well out of reach, its generalized form is even more so.

What is important, though, is that Polymath, assuming this generalized form of EH (called of course GEH), showed that $p_{n+1} - p_n \le 6$ for infinitely many n, [175, theorem 4(xii)]. Looking back to Figure 1.4, prime gaps of size 6 could be in a sense more abundant than other gap sizes, especially up to small values of the range, and could be a goal to aim for using other methods.

9.4 Gaps between Almost Primes

It is natural in the context of this work to study gaps between integers which are close to being prime. We say a positive integer is E_j if in its standard prime factorization it has exactly j prime factors and they are all distinct. It is P_j if it has at most j prime factors, not necessarily distinct. Chen's famous proof (for example, [152, chapter 10]) shows that there are an infinite number of primes p such that $p + 2 = P_j$. Asymptotically (for example, [90, theorem 437]), the number of P_j up to x is for $j \geq 2$,

$$\frac{x (\log\log x)^{j-1}}{(j-1)! \log x}.$$

Positive integers which are either E_j or P_j are called **almost primes**.

Goldson, Graham, Pintz and Yildirim (GGPY, of course) used the methods of GPY to study bounded gaps between almost primes. See [77, 78, 79]. They proved in particular a theorem which has some remarkable easy to derive consequences, their so-called "Basic Theorem":

Let $a_1, a_2, a_3 \in \mathbb{Z}$ be positive and $b_1, b_2, b_3 \in \mathbb{Z}$. Define three affine forms

$$L_1(n) := a_1 n + b_1,$$
$$L_2(n) := a_2 n + b_2,$$
$$L_3(n) := a_3 n + b_3,$$

which are such that for each $p \in \mathbb{P}$ there is at least one integer n_p, dependent on p, such that

$$p \nmid L_1(n_p) L_2(n_p) L_3(n_p).$$

We say such a triple of affine forms is "admissible". See Section 9.5 in this chapter. Let $C > 0$ be a given real constant. Then, as shown by GGPY in their "Basic Theorem" (proved unconditionally in reference [77]), there exist two of the forms which simultaneously take E_2 values, with each of the prime factors larger than C, for infinitely many positive integer values n.

GGPY use a particular admissible triple of forms to prove their theorem 3, namely there are infinitely many positive integers n such that

$$\omega(n) = \omega(n+1) = A$$

in the special case $A = 3$. To show this, let

$$L_1(n) = 6n + 1, \quad L_2(n) = 8n + 1, \quad L_3(n) = 9n + 1.$$

Because we can choose $n = 0$ for all p, this set of forms is admissible. They also satisfy

$$4L_1(n) = 3L_2(n) + 1,$$
$$3L_1(n) = 2L_3(n) + 1,$$
$$9L_2(n) = 8L_3(n) + 1.$$

Let the constant $C = 3$. By the Basic Theorem, two of the forms simultaneously take E_2 values with all prime factors greater than or equal to 5. The two forms could be $\{L_1, L_2\}$, $\{L_1, L_3\}$ or $\{L_2, L_3\}$, so we consider each of the possibilities for a pair to take E_2 values simultaneously separately:

In case the pair are $\{L_1, L_2\}$, let $m = 3L_2(n)$ and then $m + 1 = 4L_1(n)$.

In case the pair are L_1, L_3, let $m = 2L_3(n)$ and then $m + 1 = 3L_1(n)$.

In case the pair are L_2, L_3, let $m = 8L_3(n)$ and then $m + 1 = 9L_2(n)$.

In each case, we have $\omega(m) = \omega(m + 1) = 3$, so this equation is satisfied by an infinite number of positive integers m.

The reader might recall similar startling derivations following from Dickson's conjecture ([39, axiom D]) of 1904.

Closely related, but with a different flavour, are results which give explicit bounds, denoted r_k, for the number of prime factors of products of k admissible affine forms. Here we simply summarize results in Table 9.1 and give references.

Table 9.1 *Almost prime k-tuples prime factor bounds for $2 \le k \le 10$.*

Year	Author	Reference	r_2	r_3	r_4	r_5	r_6	r_7	r_8	r_9	r_{10}
1972	Porter	[176]	–	8	–	–	–	–	–	–	–
1973	Chen	[28]	2	–	–	–	–	–	–	–	–
1997	Diamond et al.	[38]	–	–	12	16	20	–	–	–	–
2006	Ho/Tsang	[107]	–	–	–	–	–	24	28	33	38
2014	Maynard	[137]	–	8	11	15	18	22	26	30	34
2013	Maynard	[136]	–	7	–	–	–	–	–	–	–

Conjecture: It is tempting, following these results and some simple numerics, to make a conjecture to add to those of Erdős (see for example those in [105]) and others: for each $j \ge 2$, there are an infinite number of sets of $j + 1$ consecutive integers which are P_j. The same applies to E_j, and $j + 1$ is the best possible in both cases.

9.5 Affine Forms and Clusters of Primes in Intervals

Maynard in [139] made wide-ranging extensions to his seminal methods outlined in Chapter 6. Instead of k-tuples of integers, he applied them to finite sets of linear forms

$$\mathscr{L} = \{L_1(n) = a_1 n + b_1, \cdots, L_k(n) = a_k n + b_k\},$$

where admissibility in this case means for each prime p there is an integer n such that no $L_j(n) \equiv 0 \bmod p$ (admissibility in Section 9.4 is the case $k = 3$, of course).

He also applied the methods to subsets $\mathscr{A} \subset \mathbb{N}$ and $\mathscr{P} \subset \mathbb{P}$ which satisfied good distribution hypotheses for arithmetic progressions, but such that $\mathscr{A}$ is not too concentrated in any particular progression.

He showed, amongst many other results, given these distribution properties and some mild uniformity restrictions in terms of x for the linear form coefficients and parameters, that

$$\# \left\{ n \in \mathscr{A} : \#\{L_1(n), \ldots, L_k(n)\} \cap \mathscr{P}) \geq \frac{\delta \log k}{C} \right\} \gg \frac{\#\mathscr{A}(x)}{(\log x)^k \exp(Ck)}.$$

The hypotheses apply for example to the primes with even rank,

$$\mathscr{P} = \{p_{2n} : n \in \mathbb{N}\} = \{3, 7, 13, \ldots\},$$

for which the twin primes conjecture, and many other prime pattern conjectures, all fail.

He showed that for all $x, y \geq 1$ there are $\gg x \exp\left(-\sqrt{\log x}\right)$ integers $n \in (x, 2x]$ with more than a positive constant times $\log y$ primes in every interval $(n, n + y]$.

He also showed that if we fix $m \in \mathbb{N}$ and $\epsilon > 0$ then there is a $k_m \leq \exp(Cm)$, with $C > 0$ a fixed constant, such that for any x sufficiently large and y in the range $x^{7/12+\epsilon} \leq y \leq x$, we have for any admissible set of k_m affine forms $L_j(n) = a_j n + b_j$ with $a_j \ll (\log x)^{1/\epsilon}$ and $b_j \ll x$,

$$\#\{n \in [x, x + y] : \text{ at least } m \text{ of the } L_j(n) \in \mathbb{P}\} \gg \frac{y}{(\log x)^{k_m}}.$$

Closely related is the work of Alweiss and Luo [2], who showed that for any $\delta \in [21/40, 1]$ there exist positive integers k, d such that the interval $[x - x^\delta, x]$ contains $\gg_k x^\delta/(\log x)^k$ pairs of consecutive primes differing by at most d.

9.6 Limit Points of Normalized Consecutive Prime Differences

This application has a long history with recent significant advances coming from PMGY/Zhang/Maynard and Polymath's work. Let $(p_n)_{n \in \mathbb{N}}$ be the sequence of primes with $p_1 = 2$ and define a related set

$$\mathbf{L} := \overline{\left\{ \frac{p_{n+1} - p_n}{\log n} : n \in \mathbb{N} \right\}} = \overline{\left\{ \frac{d_n}{\log n} : n \in \mathbb{N} \right\}},$$

where the overbar represents the closure in the usual topology on $\mathbb{R}$.

Erdős conjectured that $\mathbf{L} = [0, \infty]$. Already in 1931, Westzynthius had shown $\infty \in \mathbf{L}$ [213], but it took until 2006 with the work of GMPY [70, 72] to show that $0 \in \mathbf{L}$. Erdős (1955) and Ricci (1956) [46, 183] showed $\mu(\mathbf{L}) > 0$, where μ is Lebesgue measure on $\mathbb{R}$, and in 1988 Hildebrand and Maier [106] showed that there is a positive constant c such that for all T sufficiently large we have $\mu([0,T] \cap \mathbf{L}) \geq cT$. In 2015, Goldston and Ledoan [75], using a method of Erdős, showed that there exists a class of specific subintervals of positive measure in $\mathbf{L}$. For example, $[1/8,2]$ and $[1/40,1]$.

In 2016, Pintz [168] showed that there was an ineffective constant $c > 0$ such that $[0,c] \subset \mathbf{L}$. This is the only known result where an interval with positive measure, which includes 0, is in $\mathbf{L}$.

Finally in this summary, in 2016 Banks, Freiberg and Maynard [6], using Maynard's extension to linear forms of his original bounded gaps results described in Section 9.5, and other results, showed that for any sequence of nine nonnegative real numbers $\beta_1 \leq \beta_2 \leq \cdots \leq \beta_9$, the set of nonnegative reals

$$\{\beta_j - \beta_i : 1 \leq i < j \leq 9\} \cap \mathbf{L} \neq \emptyset.$$

This result is then used to show as a corollary that at least one eighth of the real numbers in $[0,\infty]$ are in $\mathbf{L}$. They also showed that for all $T > 0$ we have $\mu([0,T] \cap \mathbf{L}) > T/22$.

9.7 Artin's Primitive Root Conjecture

This application also has a long history. Artin conjectured in 1927 that any integer g, other than ± 1 or a perfect square, is a primitive element or generator for the multiplicative group of the field $\mathbb{Z}/p\mathbb{Z}$ for infinitely many primes p. Although no explicit value of g has been found for which the conjecture has been verified, there is a great deal of evidence that it is true. For example, Hooley showed in 1967 [108] that it followed if one assumed GRH for the Dedekind zeta function of a particular family of number fields, Gupta and Murty in 1984 [87] that it was true unconditionally for an infinite number of integers g, and in 1986 Heath-Brown [98] that there are at most two exceptional prime values of g for which the conjecture is false. However, he was not able to find these primes explicitly.

Assuming g is an integer for which an infinite number of p's exists, say $(p_n)_{n \in \mathbb{N}}$, Paul Pollack [169], combining ideas of Hooley and those of Maynard and Tao, showed that g is a primitive root for each of infinitely many m-tuples of primes which differ by no more than $B_m := e^{8m+5}$, so $p_{n+m} - p_n \leq B_m$ for an infinite number of the p_n.

Then, Pollack and the late Roger Baker [5], using ideas of Maynard and Tao together with those of Murty, Srinivansan, H. Li and Pan, showed that if $D = \{g_1, \ldots, g_r\}$ is a set of nonzero integers with no nontrivial product being either a square or a square times -3, and $m \ll \log r$, then there are infinitely many sets of consecutive primes

$$p_1 < p_2 < \cdots < p_m \text{ with } p_m - p_1 \ll_D \exp(O(m)),$$

such that each prime has a primitive root g_j in the set D.

9.8 Consecutive Primes in AP with a Fixed Common Difference

In 2008, Ben Green and Terence Tao stunned the mathematical world by showing that for each integer $k \geq 2$ there exists an infinite number of sets of primes
$p_{n+1}, p_{n+2}, \ldots p_{n+k}$ which are in arithmetic progression [86]. In fact, they showed
that any subset of the primes of relative strictly positive density contains at least
one arithmetic progression of length k.

In 2010, János Pintz, combining the methods of [72] and [86], showed that if the
level of distribution θ of primes exceeds $^1/_2$, then there is a positive constant $d \in \mathbb{N}$
with $d \leq C_1(\theta) < \infty$ such that there are arbitrarily long arithmetic progressions
of primes such that the common difference is d and the primes are all consecutive
[166].

In addition, Pintz in 2016 [168] combined that result with Zhang's theorem [215]
to show that we can choose these arithmetic progressions such that the primes are
consecutive and each of the common differences is a fixed $d \leq 7 \times 10^7$.

9.9 Prime Ideals and Irreducible Polynomials

There have also been applications to number fields and function fields. In 2014,
Jesse Thorner [205] showed that given any quadratic form $f(x, y) = ax^2 + bxy +
cy^2$ over $\mathbb{Z}$ with discriminant $D = b^2 - 4ac < 0$, there is a constant C_D such that
infinitely many primes p, q with $p - q = C_D$ are values of f.

Now recall that a number field K over $\mathbb{Q}$ is said to be **Galois** if it is algebraic
and $\mathbb{Q}$ is precisely the field fixed by the group of automorphisms of K fixing $\mathbb{Q}$.
Let $\mathcal{O}_K$ be its ring of integers. An ideal I in $\mathcal{O}_k$ is **prime** if $ab \in I$ implies $a \in I$
or $b \in I$. The norm of an ideal $N(I)$ is the integer $\sqrt{D(I)/\Delta}$, where $D(I)$ is the
discriminant of the ideal and Δ the discriminant of the field.

In his paper [205], Thorner applied Maynard's method from [138] in the setting
of Chebotarev sets of primes. A Chebotarev set $P \subset \mathbb{P}$ has the following definition:
let $K/\mathbb{Q}$ be a Galois extension with discriminant Δ and Galois group G. For a
prime $p \nmid \Delta$, there corresponds a conjugacy class $C \subset G$ consisting of the set of
Frobenius automorphisms attached to the prime ideals of K which lie over p. This
conjugacy class is represented by the so-called Artin symbol

$$\left[\frac{K/\mathbb{Q}}{p} \right] := C.$$

We say that a subset $P \subset \mathbb{P}$ is a **Chebotarev set** of primes, if there exists an extension $K/\mathbb{Q}$ and a union of conjugacy classes $C \subset G$ such that P is a union of sets of the form

$$\left\{ p \in \mathbb{P} : p \nmid \Delta, \ \left[\frac{K/\mathbb{Q}}{p} \right] = C \right\}.$$

Using the Chebotarev density theorem and a theorem of Murty and Murty, together with Maynard's method [138], Thorner showed that there exist bounded gaps between primes in any Chebotarev set.

Using this concept he showed, in the same article [205], in a field with a nonabelian Galois group, there are an infinite number of prime ideals with norms being primes which differ by a constant dependent only on K.

He also showed, if the Galois group of $K/\mathbb{Q}$ is abelian and m is the smallest integer such that $K \subset \mathbb{Q}(e^{2\pi i/m})$, there are infinitely many pairs of distinct primes p, q such that $|p - q| \leq 600m$.

In the same number field K setting, in 2015 Castillo, Hall, Oliver, Pollack and Thompson [27] showed that given a positive integer m and admissible k-tuple $\mathcal{H} = \{h_1, \ldots, h_k\}$ in $\mathcal{O}_K$, there exist an infinite number of $g \in \mathcal{O}_K$ such that if k is sufficiently large depending on K and m, we have at least m of the principal ideals

$$(g + h_1), (g + h_2), \ldots, (g + h_k)$$

being prime.

In a function field $\mathbb{F}_q[t]$ setting, where prime elements are the irreducible polynomials, these five authors showed that given a positive integer m and a k sufficiently large in terms of m (see Maynard's Theorem 6.17, in Chapter 6, where we have $k \gg m^2 \exp(4m)$), given any admissible k-tuple $h_1, \ldots, h_k$, there are infinitely many $g \in \mathbb{F}_q[t]$ such that at least m of the $g + h_1, \ldots, g + h_k$ are irreducible in $\mathbb{F}_q[t]$.

9.10 Coefficients of Modular Forms

Thorner, also in [205], studied applications of bounded gaps to congruence properties of the Fourier coefficients of normalized Hecke eigenforms, and derived the following: given any newform for a congruence subgroup of $SL_2(\mathbb{Z})$

$$f(z) = \sum_{n \in \mathbb{N}} a(n)q^n \in S_k^{\text{new}}(\Gamma_0(N), \chi) \cap \mathbb{Z}[[q]], \ q = e^{2\pi i z},$$

for the group $\Gamma_0(N)$ having even weight $k \geq 2$, and any positive integer d and any prime $p_0 \nmid dN$, there is a constant $C_{f,d}$ such that an infinite number of distinct primes p_1, p_2 satisfy $p_1 - p_2 = C_{f,d}$ and

$$a(p_0) \equiv a(p_1) \equiv a(p_2) \bmod d.$$

Applied to Ramanujan's tau function

$$f(z) = q \prod_{n=1}^{\infty} (1 - q^n)^{24} = \sum_{n \in \mathbb{N}} \tau(n) q^n \in S_{12}^{\text{new}}(\Gamma_0(1), \chi_{\text{trivial}}),$$

this work shows that there are bounded gaps between the set of primes p for which $\tau(p) \equiv 0 \bmod d$ for any $d \in \mathbb{N}$.

9.11 Elliptic Curves

A natural number n is called a **congruent number** if n can be represented as the area of a right-angled triangle with rational length base, height and hypotenuse. For example,

$$5 = \frac{1}{2} \cdot \frac{3}{2} \cdot \frac{20}{3} \text{ and } 6 = \frac{1}{2} \cdot 3 \cdot 4,$$

show that 5 and 6 are congruent numbers. If n is congruent then the elliptic curve $y^2 = x^3 - n^2 x$ has a nontrivial rational point (solution) $P = (x, y) \in \mathbb{Q}^2$. The reverse is not quite true: if the curve has a rational point $P = 2Q$, where Q is a second rational point, then n is congruent.

If $E/\mathbb{Q}$ is an elliptic curve with Weierstrass equation $y^2 = x^3 + ax^2 + bx + c$ and d is a squarefree rational number, then the so-called **quadratic twist** of E, denoted E_d, is the curve $dy^2 = x^3 + ax^2 + bx + c$. This is equivalent over $\mathbb{Q}\left(\sqrt{d}\right)$ to

$$y^2 = x^3 + adx^2 + bd^2x + cd^3.$$

So the congruent number elliptic curve for n is the quadratic twist of $y^2 = x^3 - x$, the so-called **congruent number elliptic curve**.

A conjecture, attributed to Joseph Silverman, is that there are infinitely many primes p for which $\text{rk}(E_p) = 0$ or $\text{rk}(E_{-p}) = 0$, and an infinite number of primes q for which $\text{rk}(E_q) > 0$ or $\text{rk}(E_{-q}) > 0$. This conjecture has been shown to be true for the congruent number elliptic curve.

Thorner, in his groundbreaking paper [205], derived from his fundamental Chebotarev sets of primes result quite general results for a family of "good" elliptic curves – where "good" is a list of elliptic curve properties set out in the paper. These results imply there exist infinitely many pairs of distinct primes p, q such that both elliptic curves $py^2 = x^3 - x$ and $qy^2 = x^3 - x$ have finitely many (ditto infinitely many) rational points.

Nontrivial integer solutions of $py^2 = x^3 - x$ are, apparently, not so easy to find. Indeed, a small amount of searching gave $5 \cdot 12^2 = 9^3 - 9$, but nothing else for $p = 5$.

More generally, Thorner showed in his Theorem 1.4 that if $E/\mathbb{Q}$ is good, there is an $\epsilon = \pm 1$, a constant $C_E > 1$, and infinitely many primes p, q such that $|p - q| \le C_E$ and $\mathrm{rk}(E_{\epsilon p}) = \mathrm{rk}(E_{\epsilon q}) = 0$.

He also showed, in his theorem 1.5, that for $E : y^2 + xy = x^3 - x^2 - 2x - 1$, the minimal Weierstrass equation, there are infinitely many primes p, q such that $|p - q| \le 16800$, the elliptic curve L-functions $L(E_{-p}, s)$ and $L(E_{-q}, s)$ have a simple zero at $s = 1$, $\mathrm{rk}(E_{-p}) = \mathrm{rk}(E_{-q}) = 1$, and both Tate–Shafarevich groups $\mathfrak{W}(E_{-p}/\mathbb{Q})$ and $\mathfrak{W}(E_{-q}/\mathbb{Q})$ are finite and of odd size.

9.12 Epilogue

In this final note, we briefly summarize the achievements of PMGY, Zhang, Maynard, Tao and Polymath regarding prime gaps, and make a short comment on "where to next".

Each contributor to these developments used admissible k-tuples $\mathscr{H} = \{h_1, h_2, \ldots, h_k\}$ and related polynomials $P \in \mathbb{Z}[n]$ defined by $P(n) = (n + h_1) \cdots (n + h_k)$ or an equivalent.

PMGY used sieve weights based on an arbitrary admissible k-tuple and a given parameter $l \in \mathbb{N}$, using a truncated von Mangoldt function to detect prime (powers) with $N < n \le 2N$ and R a small power of N:

$$w(n) = \Lambda_R(n, \mathscr{H}, k + l)^2 := \left(\frac{1}{(k+l)!} \sum_{\substack{d|P(n) \\ d < R}} \mu(d) \left(\log \frac{R}{d} \right)^{k+l} \right)^2 .$$

Zhang used similar weights but restricted divisors differently. If $\varpi = 1/1168$ and $\mathscr{P}_1 = \prod_{p < x^\varpi} p$, then his weights for sums $x < n \le 2x$ are

$$w(n) := \left(\sum_{d|(P(n), \mathscr{P}_1)} \mu(d) \left(\log \frac{R}{d} \right)^{k+l} \right)^2 .$$

Maynard used a multidimensional sieve and based the sieve weights on a given but fixed piecewise differentiable function F supported on a simplex $\mathscr{R}_k$ in $[0, \infty)^k$. He defined

$$\lambda(d_1, \ldots, d_k) := coef \times \sum_{\substack{r_1, \ldots, r_k \\ d_i | r_i}} F\left(\frac{\log r_1}{\log R}, \ldots, \frac{\log r_k}{\log R} \right) .$$

and then if $W = \prod_{p < \log\log\log N} p$ and v_0 is a class chosen such that any n in the class modulo W has no small prime divisors, the weight

$$w(n) = \left(\sum_{\substack{d_i \mid n+h_i \\ n \equiv v_0 \bmod W}} \lambda(d_1, \ldots, d_k) \right)^2 .$$

Polymath8b simplified these weights. Firstly, they used functions $f : [0, \infty) \rightarrow \mathbb{R}$ with compact support defining

$$\lambda_f(n) := \sum_{d \mid n} \mu(d) f \left(\frac{\log d}{\log x} \right).$$

They used the Stone–Weierstrass approximation theorem and algebras based on these functions to approximate compactly supported functions $F : [0, \infty)^k \rightarrow \mathbb{R}$ and defined weights having the form

$$w(n) = \left(\sum_{j=1}^{J} c_j \lambda_{f_{1,j}}(n + h_1) \cdots \lambda_{f_{k,j}}(n + h_k) \right)^2$$

and providing a new proof of "Maynard's lemma" using them.

The majority of the authors truncated divisor sums using an upper bound, denoted R, which was a small power of N, where the range of interest was $N < n \leq 2N$ (see Table 9.2).

Table 9.2 *Values chosen for the truncation parameter* R.

People	R
GPY	$N^{\frac{1}{4}-\epsilon}$
Zhang	$N^{\frac{1}{4}+\varpi/2}, \varpi = 1/1168$
Polymath8a	$N^{\frac{1}{4}+\epsilon}, 0 < \epsilon < \frac{1}{4}$
Maynard	$N^{\frac{1}{2}-\epsilon}$
Polymath8b	–

For the optimization step, GPY used two admissible tuples and the Elliott–Halberstam conjecture. No other contributor, thankfully, used more than one tuple, however.

Zhang in a sense optimized by finding a good value for his function omega which made his expression giving bounded prime gaps positive. But no attempt (and he expressly notes this) was made to find the best value for his parameters, work left to others. This did not detract in any way from the significance of his breakthrough result, since discovering an explicit numerical bound for an infinite number of prime gaps was a great thrill for the mathematics world.

Polymath8a employed many steps which could be regarded as optimizations. For example, replacing Zhang's smooth number condition by multiple dense divisibility (MDD). Replacing Zhang's omega with an inequality involving k, the tuple size, which could be optimized over the parameter space, and then a more refined form including MDD, the clever combinatorics based on Heath-Brown's identity, etc.

Optimization played a very visible and essential role in Maynard's approach, where functionals are used to express the fundamental sums. These are based on a multidimensional Selberg sieve, which itself, like all such sieves, is optimal. The functionals are built into a condition for bounded prime gaps, and then optimized over finite dimensional spaces spanned by families of explicit symmetric polynomials using a maximum eigenvalue calculation. Maynard, in a seminal and quite short paper, did a lot more than that.

Finally Polymath8b, using Tao's approach to the Selberg sieve calculations, and a different basis, were able to significantly reduce the prime gap bound using the standard simplex and optimization. Improvements were made using a similar approach but a perturbed simplex.

The values found are set out in Table 9.3.

Table 9.3 *Successive improvements on the upper bound for H_1 and H_m.*

People	H_1	k	H_m	EH[1]	GEH[1]
GPY	$\epsilon \log p_n$	∞	–	$H_1 \leq 16$	–
Zhang	7×10^7	3.5×10^6	–	–	
Polymath8a	4680	632	–	–	–
Maynard	600	105	$H_m \ll m^3 e^{4m}$	$H_1 \leq 12$	–
Polymath8b	246	50	$H_m \ll m e^{600m/157}$	$H_m \ll m e^{2m}$	$H_1 \leq 6$

We have seen in Chapter 7 how the bound $H_1 \leq 246$ arose and the tight limits to further progress with optimizations based on M_k. However, the known bounds for $M_{k,\epsilon}$ are not tight, and there may be room for improvement using an expanded or varied basis, even with the simply perturbed simplex. The way forward could involve varying the simplex for M_k in alternative ways.

To summarize: Maynard optimized over real linear combinations of symmetric polynomials over the standard simplex in $\mathbb{R}^k$ of the form for given $n \in \mathbb{N}$

$$\sum_{j=1}^{N} a_j (1 - P_1)^{u_j} \times P_2^{v_j}, \ u_j, v_j \geq 0, \ u_j + 2v_j \leq 2n + 1.$$

Recall $P_1 := t_1 + \cdots + t_k$ and $P_2 := t_1^2 + \cdots + t_k^2$. This method uses a simpler basis than Polymath and has a worse outcome. It could be improved, but would not give improved results for prime gaps because of Polymath8b's Corollary 7.11.

In turn, Polymath8b first optimized over the standard simplex, but used symmetric polynomials of the form

$$P_\alpha(t) := \sum_{a = \sigma(\alpha)} t_1^{a_1} \cdots t_k^{a_k},$$

where the sum is over all of the distinct permutations of the signature α. They optimized over a basis $\mathscr{B}_d$ parameterized by $d \in \mathbb{N}$ and defined by

$$\mathscr{B}_d := \{(1 - P_{(1)})^a P_\alpha : a \geq 0, \ \alpha \text{ has all even entries}\}.$$

For a more complete description and examples, see Section 7.5. Because of Polymath8b's Corollary 7.11, there is little point in attempting to improve their method in this case.

In turn, Polymath8b then optimized over perturbations of the standard simplex with a single purturbation parameter $\epsilon > 0$. For the matrix A (see Theorem 7.15), they optimized over $(1 + \epsilon)\mathscr{R}_k$ with a basis

$$\{(1 + \epsilon - P_{(1)})^a P_\alpha : a \geq 0, \ \alpha \text{ has no entries } 1\},$$

and for the numerator B used for $(1 - \epsilon)\mathscr{R}_k$ the polynomials

$$\{(1 - \epsilon - P_{(1)})^a P_\alpha : a \geq 0, \ \alpha \text{ has no entries } 1\}.$$

Polymath used $\epsilon = 1/27$ and obtained their best result, the gap size 246. There is scope for some improvement regarding prime gaps in this case.

Polymath also in their paper [175] included some details and results of the so-called Krylov method of Bogaert. This iterates the symmetric function $F(t) = 1$ using the $\mathscr{L}$ operator, defined by

$$\mathscr{L}(F)(t) := \sum_{j=1}^{k} \int_0^{1 - \left(\sum_{i=1}^{k} t_i\right) + t_j} F(t_1, \ldots, t_{j-1}, s, t_{j+1}, \ldots, t_k) ds,$$

to derive symmetric polynomials with coefficients being rational functions of k, deriving matrices which are functions of k, making this variable explicit in the optimization step. Although Bogaert reached $r = 25$ for the matrix size, we were not able to get anywhere close, finding the algorithm to be time bound. There may be value in experimenting with this approach using alternative initial polynomials which are more natural for the operator $\mathscr{L}$.

There is a great deal of value in the other approaches reported here, and progress might come from the improvements to Bombieri–Vinogradov based on dense divisibility or some other idea, even though this appears to be very difficult. Related to this is the need, at the time of writing, for the sets of integers $\mathscr{D}(i, y)$ to be better understood, undergoing studies reflecting those for smooth integers – see for example [143, section 7.1] or [22, chapter 13].

Appendix A

Bessel Functions of the First Kind

In this appendix, we derive the special properties of Bessel functions of the first kind which are used in Chapter 8.

Definition A.1 **(Bessel functions of the first kind)** If $k \in \mathbb{Z}$ then

$$J_k(x) := \sum_{n=0}^{\infty} \frac{(-1)^n}{n! \, (n+k)!} \left(\frac{x}{2}\right)^{2n+k} . \qquad (\text{A.1})$$

Figure A.1 is a plot of $J_0(x)$ and Figure A.2 of $J_1(x)$ and $J_2(x)$.

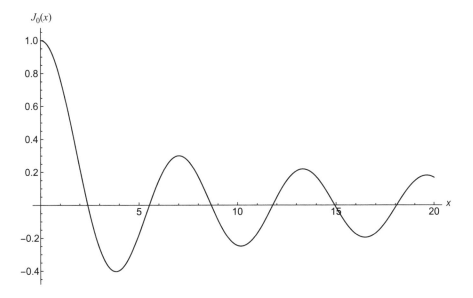

Figure A.1 The Bessel function $J_0(x)$ for $0 \le x \le 20$.

465

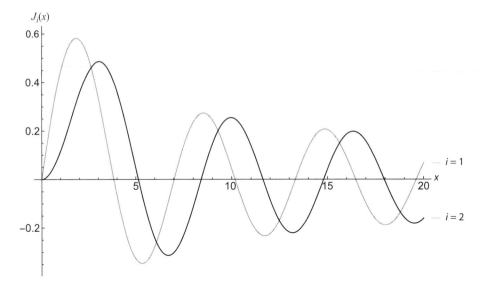

Figure A.2 The Bessel functions $J_1(x)$ and $J_2(x)$ for $0 \le x \le 20$.

Lemma A.1 *For any $k > 0$, $j > 0$ and real $x > 0$, we have*

$$\left(x^{-k/2} J_k \left(j\sqrt{x}\right)\right)' = -\frac{jx^{-(k+1)/2}}{2} J_{k+1}\left(j\sqrt{x}\right).$$

Proof This follows directly by substituting $j\sqrt{x}$ in the series representation Equation (A.1), multiplying by $x^{-k/2}$, and then differentiating term by term.   □

Lemma A.2 *For any $k > 0$, $j > 0$ and real $x > 0$, we have*

$$J_k(x) = \frac{x}{2k}(J_{k-1}(x) + J_{k+1}(x)),$$

$$J_k'(x) = \frac{1}{2}(J_{k-1}(x) - J_{k+1}(x)),$$

and for $k = 0$, we get $J_0'(x) = -J_1(x)$

Proof First, multiply (A.1) by x^k and then differentiate to get for $k \ge 1$,

$$(x^k J_k(x))' = \sum_{n=0}^{\infty} \frac{(-1)^n (2n + 2k)}{2^{2n+k} n! (n + k)!} x^{2n+2k-1}$$

$$= x^k \sum_{n=0}^{\infty} \frac{(-1)^n}{n! (n + k - 1)!} \left(\frac{x}{2}\right)^{2n+k-1}$$

$$= x^k J_{k-1}(x).$$

Similarly, multiplying by x^{-k} for $k \geq 0$, we get

$$\left(x^{-k} J_k(x)\right)' = -x^{-k} J_{k+1}(x).$$

Therefore, for $k = 0$ we get $J_0'(x) = -J_1(x)$ and for $k \geq 1$ we have

$$kx^{k-1} J_k(x) + x^k J_k'(x) = x^k J_{k-1}(x) \text{ and}$$
$$-kx^{-k-1} J_k(x) + x^{-k} J_k'(x) = -x^{-k} J_{k+1}(x).$$

Cancelling x^k and x^{-k} respectively, then adding and subtracting these equations, gives

$$2J_k'(x) = J_{k-1}(x) - J_{k+1}(x) \text{ and } \frac{2k}{x} J_k(x) = J_{k-1}(x) + J_{k+1}(x),$$

which completes the derivations of the identities. ☐

Lemma A.3 (Bessel function square integral) *[49, lemma 5.4] For all $u > 0$ and $k \geq 0$, we have*

$$\int_0^u t J_k(t)^2 \, dt = \frac{u^2}{2} J_k'(u)^2 + \frac{u^2 - k^2}{2} J_k(u).$$

Proof Let $y := y(t) = J_k(t)$. Then y satisfies Bessel's differential equation

$$t^2 y'' + t y' + (t^2 - k^2)y = 0 \implies t(ty')' = (k^2 - t^2)y.$$

Multiply both sides by $2y'$ to get

$$2(ty')'(ty') = (k^2 - t^2)(2y'y) \implies ((ty')^2)' = (k^2 - t^2)(y^2)'.$$

Integrate both sides from 0 to u and use integration by parts to then get

$$(ty')^2 \Big|_0^u = (k^2 - t^2)y^2 \Big|_0^u + \int_0^u 2ty^2 \, dt.$$

The evaluations at $t = 0$ all vanish, so

$$2 \int_0^u ty^2 \, dt = u^2 y'(u)^2 + (u^2 - k^2)y(u)^2,$$

which completes the proof. ☐

Lemma A.4 *For any integer $k > 0$ and real $j, x > 0$, we have*

$$\int_0^x J_k(j\sqrt{t})^2 \, dt = x \left(J_k \left(j\sqrt{x} \right)^2 - J_{k-1} \left(j\sqrt{x} \right) J_{k+1} \left(j\sqrt{x} \right) \right).$$

Proof Make the change of variables $s = j\sqrt{t}$ with $u = j\sqrt{x}$ to show the expression of the lemma is equivalent to

$$\int_0^u s J_k(s)^2 \, ds = (u^2/2)\left(J_k(u)^2 - J_{k-1}(u)J_{k+1}(u)\right).$$

The proof then follows from Lemma A.3, substituting for both Bessel functions on the right, $J_{k-1}(x)$ and $J_{k+1}(x)$, using the two identities of Lemma A.2. □

We also need a good upper bound for the first positive zero, denoted j_k, of the Bessel function $J_k(x)$ for $k > 0$. An asymptotic formula for this zero in terms of negative powers of k has been known for some time, [162, equation 10.21.40]. The upper bound is related to this expansion, and it was conjectured by Lee Lorch in 1993 [128] to take the form given in Lemma A.6. The range $0 < k < 10$ was demonstrated by Lorch and Uberti [129], and the infinite range $k \geq 10$ by Lang and Wong [124], both in 1996. In Table A.1, we compare the values computed by the Mathematica function `BesselJZero` with the values of the Lorch bound. We then give a few details of the proof for the infinite range to give the reader a flavour of what is involved in the proof. Fundamental to this is the work of Olver [158, 159, 160, 161], covering more than 20 years of work, especially the text [161].

There are a number of relationships between the Airy function $Ai(x)$ (see Figure A.3) and $J_k(x)$ and a particular one is exploited in Lang and Wong's proof. The Airy function is a solution to $y'' - xy = 0$ and gives an entire function which is oscillatory along the negative real axis and exponential on the positive axis. It is

Table A.1 *First positive Bessel J zero upper bound for* $1 \leq k \leq 15$.

k	BesselJZero	LorchBound
1	3.831705970	3.888907385
2	5.135622307	5.158119350
3	6.380161896	6.392811265
4	7.588342350	7.596674651
5	8.771483814	8.777489935
6	9.936109183	9.940698931
7	11.08637044	11.09002313
8	12.22509271	12.22808931
9	13.35430048	13.35681715
10	14.47550069	14.47765332
11	15.58984788	15.59171696
12	16.69824993	16.69989309
13	17.80143515	17.80289487
14	18.89999795	18.90130634
15	19.99443063	19.99561244

only the value of its first negative zero which appears in the Lorch bound, but the Airy function's functional properties are used in the proof.

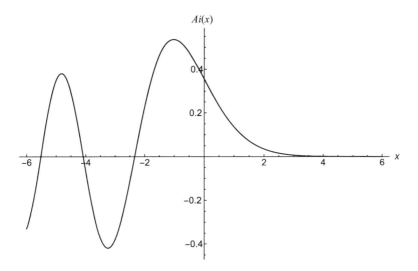

Figure A.3 The Airy function Ai(x) for $-6 \le x \le 6$.

Theorem A.5 (**Hethcote's zero estimate**) *[103, 104] Let a_n be the nth negative zero of* Ai(t), ρ *be sufficiently small such that on the interval* $I := [a_n - \rho, a_n + \rho]$, $m := \min\{|\text{Ai}'(t)| : t \in I\} > 0$ *and* $f(t) := \text{Ai}(t) + \epsilon(t)$, *where* $f(t)$ *is continuous and*

$$E := \max\{|\epsilon(t)| : t \in I\} < \min\{\text{Ai}(a_n - \rho), \text{Ai}(a_n + \rho)\}.$$

Then there is a zero t_n of $f(t)$ in I which satisfies $|t_n - a_n| \le E/m$.

Lemma A.6 (**First Bessel zero upper bound**) *[124] Let the first negative real zero of the Airy function* Ai(x) *be* $a_1 = -2.338107410459767385\ldots x$. *For $k \ge 10$, we have the Lorch upper bound*

$$j_k < k - a_1 \left(\frac{k}{2}\right)^{1/3} + \frac{3a_1^2}{20} \left(\frac{2}{k}\right)^{1/3} \le k + 3.2446076k^{1/3} + \frac{3.15824}{k^{1/3}}.$$

Proof (**1**) The proof starts with an expression for $J_k(kx)$ in terms of a truncated asymptotic expansion and the Airy function. Some details of the derivation can be found in the text [161, chapter 11]. In summary:

$$J_k(kx) = h(k)\frac{\varphi(\zeta)}{k^{1/3}} \left(\text{Ai}(k^{2/3}\zeta)\left(1 + \frac{A_1(k)}{k^2}\right) + \frac{\text{Ai}'(k^{2/3}\zeta)}{k^{4/3}}B_0(k) + \epsilon_3(k, \zeta)\right).$$

$$\text{(A.2)}$$

The $A_1(k)$ and $B_0(k)$ come from the original asymptotic expansion. On $x \in (0, \infty)$, there is a one-to-one relationship between the variables x and ζ given by lines 2 and 3 of Equations A.3.

$$h(k) := \left(1 + \frac{1}{12k} + \frac{1}{288k^2}\right) \frac{e^k \Gamma(k+1)}{\sqrt{2\pi} k^{k+\frac{1}{2}}},$$

$$\zeta := -\left(\frac{3}{2}((x^2 - 1)^{1/2} - \sec^{-1}(x))\right)^{2/3}, \quad x \geq 1,$$

$$\zeta := \left(\frac{3}{2} \log\left(\frac{1 + (1 - x^2)^{1/2}}{x}\right) - \frac{3}{2}(1 - x^2)^{\frac{1}{2}}\right)^{2/3}, \quad 0 < x \leq 1,$$

$$\varphi(\zeta) := \left(\frac{4\zeta}{1 - x^2}\right)^{\frac{1}{4}}.$$

(A.3)

(2) Lang and Wong simplify this expression, introducing a new variable t and error $\epsilon^*(k, \zeta)$, to get the fundamental equation relating the Bessel function to the Airy function, namely

$$f(t) := \frac{k^{1/3}}{h(k)\varphi(k)} J_k(kx) = \text{Ai}(t) + \epsilon^*(k, \zeta),$$

(A.4)

where

$$t := k^{2/3}\zeta + \frac{B_0(\zeta)}{k^{4/3}}.$$

(A.5)

The error ϵ^* is related to the original error $\epsilon_3(k, \zeta)$ by

$$\epsilon^*(k, \zeta) := \text{Ai}(k^{2/3}\zeta)) \frac{A_1(\zeta)}{k^2} - \frac{\text{Ai}''(\theta)}{2!} \frac{B_0^2(\zeta)}{k^{8/3}} + \epsilon_3(k, \zeta),$$

(A.6)

where θ comes from a Taylor expansion.

For ζ in a restricted range, determined by the parameter c, and $k \geq 10$, Lang and Wong derive the bound

$$|\epsilon_3(k, \zeta)| \leq \frac{0.00747}{|c|^{\frac{1}{4}} k^3} e^{0.30/k},$$

(A.7)

which enables them to control the size of $\epsilon^*(k, \zeta)$ also obtaining

$$|\epsilon^*(k, \zeta)| \leq \frac{0.000974}{k^2} \leq 9.74 \times 10^{-6}, \quad k \geq 10.$$

(A.8)

(3) Next, essential use is made of Hethcote's Theorem A.5, using $f(t) = \text{Ai}(t) + \epsilon^*(k, \zeta)$ with

$$f(t) = \frac{k^{1/3} J_k(kx)}{h(k)\varphi(\zeta)}.$$

(A.9)

Then Lang and Wong let t_1 be the zero of $f(t) = f\left(k^{2/3}\zeta + B_0(\zeta)/k^{4/3}\right)$ given by Heathcote's theorem so $|t_1 - a_1| \le E/m$. Let $t_1 = k^{2/3}\zeta_1 + B_0(\zeta_1)/k^{4/3}$, and we can write introducing a new parameter δ,

$$\zeta_1 = \frac{(a_1 + \frac{\delta}{k^{4/3}})}{k^{2/3}}. \tag{A.10}$$

(4) Expanding x as a function of ζ in a Taylor series about 0 gives

$$x(\zeta) = 1 - \frac{\zeta}{2^{1/3}} + \frac{3}{2^{2/3}10}\zeta^2 + \frac{\zeta^3}{700} + \cdots \tag{A.11}$$

for some θ which can be controlled, we can write

$$x_1 := x(\zeta_1) = 1 - \frac{t}{2^{1/3}k^{2/3}} + \frac{3}{2^{2/3}10}\frac{t^2}{k^{4/3}} + \frac{x'''(\theta)}{3!}\frac{t^3}{k^2}. \tag{A.12}$$

Recalling (A.4), we have

$$\frac{k^{1/3}}{h(k)\varphi(k)}J_k(kx) = \mathrm{Ai}(t) + \epsilon^*(k,\zeta) \implies j_k = kx_1,$$

and from (A.12) we then get for some β, which can be controlled, indeed $-0.060804 \le \beta \le -0.000263$:

$$j_k = kx_1 = k - \frac{a_1}{2^{1/3}}k^{1/3} + \frac{3}{20}a_1^2\left(\frac{2}{k}\right)^{1/3} + \frac{\beta}{k}, \tag{A.13}$$

so

$$-\frac{0.060804}{k} \le j_k - \left(k - \frac{a_1}{2^{1/3}}k^{1/3} + \frac{3}{20}a_1^2\left(\frac{2}{k}\right)^{1/3}\right) \le -\frac{0.000263}{k},$$

which in particular implies the Lorch bound but also

$$\left|j_k - \left(k - \frac{a_1}{2^{1/3}}k^{1/3} + \frac{3}{20}a_1^2\left(\frac{2}{k}\right)^{1/3}\right)\right| \le 0.061, \; k \ge 10.$$

□

In addition, we need to know in detail how Bessel functions normally arise in applications, i.e., as solutions to Bessel's differential equation:

$$x^2y''(x) + xy'(x) + (x^2 - k^2)y(x) = 0. \tag{A.14}$$

For us, we will have $k \in \mathbb{Z}$. Solutions with $k \ge 0$ are denoted $J_k(x)$ and are called "Bessel functions of the first kind". A second family of solutions Y_k is given by

$$Y_k(x) := J_k(x)\int_a^x \frac{1}{yJ_k^2(y)}\,dy,$$

where $a > 0$. Then the general solution to (A.14) has the form $y(x) = c_1 J_k(x) + c_2 Y_k(x)$ for arbitrary constants c_1 and c_2 (real or complex, depending on the context).

Fitting alternative second-order linear ordinary differential equations into standard forms, such as Bessel's equation, is an art well practiced by applied mathematicians. The equation we need may be obtained by defining, for some real fixed parameters α, β, γ and with $y(x)$ satisfying (A.14),

$$\varphi(x) := x^\alpha y(\beta x^\gamma).$$

Then $\varphi(x)$ satisfies

$$x^2 u''(x) + x(2\alpha - 1)y'(x) + (\beta^2 \gamma^2 x^{2\gamma} + (\alpha^2 - k^2 \gamma^2))\varphi(x) = 0. \qquad \text{(A.15)}$$

Appendix B

A Type of Compact Symmetric Operator

In this section, we define a compact operator on a Hilbert space and derive the properties which are needed in Appendix C to demonstrate that the functions used in Chapter 8 are optimal.

The setting: the complex Hilbert space $H := L^2([0,1], \mathbb{C})$ with inner product

$$\langle \varphi, \psi \rangle := \int_0^1 \varphi(x)\overline{\psi(x)}\, dx,$$

a Lebesgue square integrable function $K : [0,1]^2 \to \mathbb{R}$ with $0 \le K(x,y) \le 1$ and $K(x,y) = K(y,x)$ for all x, y, and a linear mapping $T : H \to H$ defined by

$$(T\varphi)(x) := \int_0^1 \varphi(y)K(x,y)\, dx. \tag{B.1}$$

In Step (12), we use the special form for K based on a fixed parameter α:

$$K(x,y) := \left(\frac{\min(x,y)}{\max(x,y)} \right)^\alpha, \quad \alpha > 0 \text{ and } K(0,0) := 0.$$

Recall some definitions: An everywhere defined operator on a Hilbert space H is called **self adjoint** if $\langle T\varphi, \varphi \rangle$ is real for all $\varphi \in H$. It is **compact** if the image of every bounded sequence has a convergent subsequence. It is **positive** if $\langle T\varphi, \varphi \rangle \ge 0$ for every φ.

In the following steps, we derive properties of the mapping T, culminating in the proof of the spectral theorem.

(1) The mapping T is symmetric, i.e., $\langle T\varphi, \varphi \rangle = \langle \varphi, T\varphi \rangle$ for all $\varphi \in H$, so $\langle T\varphi, \varphi \rangle$ is real: Using Fubini's theorem and that K is symmetric and real,

$$\langle T\varphi, \varphi \rangle = \int_0^1 \left(\int_0^1 K(x, y)\varphi(y)dy \right) \overline{\varphi(x)}dx$$

$$= \int_0^1 \int_0^1 \overline{K(x, y)\varphi(y)}\varphi(x)dxdy$$

$$= \overline{\langle T\varphi, \varphi \rangle}.$$

Thus T is self-adjoint and all of its eigenvalues are real. Similarly one shows that $\langle T\varphi, \psi \rangle = \langle \varphi, T\psi \rangle$.

(2) The mapping is continuous: using Fubini's theorem, the Cauchy–Schwarz inequality and $|K(x, y)| \leq 1$, we get

$$\langle T\varphi, T\varphi \rangle = \int_0^1 \int_0^1 \int_0^1 K(z, x)\varphi(x)\overline{K(z, y)\varphi(y)}dxdydz$$

$$\leq \int_0^1 \int_0^1 |\varphi(x)||\varphi(y)|dxdy$$

$$\leq \left(\int_0^1 |\varphi(x)|dx \right)^2 \leq \int_0^1 |\varphi(x)|^2 dx = \langle \varphi, \varphi \rangle.$$

(3) All eigenvalues of T are real, and eigenvectors associated with different eigenvalues are orthogonal: let $T\varphi = \lambda\varphi$ with $\lambda \neq 0$. Then

$$\lambda\langle \varphi, \varphi \rangle = \langle T\varphi, \varphi \rangle = \langle \varphi, T\varphi \rangle = \overline{\lambda}\langle \varphi, \varphi \rangle \implies \lambda = \overline{\lambda},$$

so λ is real. If μ is a different nonzero eigenvalue with eigenvector ψ, then

$$(\lambda - \mu)\langle \varphi, \psi \rangle = \langle \lambda\varphi, \psi \rangle - \langle \varphi, \mu\psi \rangle = \langle T\varphi, \psi \rangle - \langle \varphi, T\psi \rangle = 0,$$

so ψ and φ are orthogonal.

(4) We will show that the set of compact operators on H is closed in the operator norm: let (T_n) be a sequence of compact operators on H with $\|T_n - T\| \to 0$, and let u_n have $\|u_n\| \leq 1$ for all $n \in \mathbb{N}$. Since T_1 is compact, there is an increasing sequence $S_1 = (n_{i,1})$ in $\mathbb{N}$ such that $T_1(u_{n_{i,1}})$ converges in H. Inductively, since T_{n+1} is compact, there is an increasing sequence $S_{n+1} \subset S_n$ such that $T_{n+1}(u_{n+1,i})$ converges. By this construction, if we define $v_i := u_{i,i}$, we have $T_n(v_i)$ converges for each $n \in \mathbb{N}$. We **claim** $(T(v_n))$ is Cauchy. To see this,

$$\|T(v_i) - T(v_j)\| \leq \|T(v_i) - T_n(v_i)\| + \|T_n(v_i) - T_n(v_j)\| + \|T_n(v_j) - T(v_j)\|$$

$$\leq \|T - T_n\| \cdot \|v_i\| + \|T_n(v_i) - T_n(v_j)\| + \|T - T_n\| \cdot \|v_j\|,$$

so the sequence is Cauchy. Since H is complete, it converges, showing that T is compact. Therefore, the set of compact operators on H is closed in the operator norm.

(5) The operator T is compact: let (u_n) be an orthonormal basis for H and define $a_{n,m} := \langle Tu_m, u_n \rangle$. Then we can write $\varphi = \sum_{n \in \mathbb{N}} \langle \varphi, u_n \rangle u_n$ so

$$T\varphi = \sum_{n \in \mathbb{N}} \langle T\varphi, u_n \rangle u_n$$

$$= \sum_{n \in \mathbb{N}} \left(\sum_{m \in \mathbb{N}} \langle \varphi, u_m \rangle Tu_m, u_n \right) u_n$$

$$= \sum_{n \in \mathbb{N}} \left(\sum_{m \in \mathbb{N}} a_{m,n} \langle \varphi, u_m \rangle \right) u_n. \tag{B.2}$$

Since $K \in L^2([0,1]^2, \mathbb{C})$, which is a Hilbert space, defining $K_x(y) = K(x,y)$ and setting

$$\alpha_m := \sum_{n \in \mathbb{N}} |a_{mn}|^2 < \infty,$$

which converges because of the symmetry of T and Bessel's inequality, we can derive

$$\|K\|^2 = \int_0^1 \int_0^1 |K(x,y)|^2 \, dx dy$$

$$= \int_0^1 \|K_x\|^2 \, dx = \int_0^1 \sum_{n \in \mathbb{N}} |\langle K_x, u_n \rangle|^2 \, dx$$

$$= \int_0^1 \sum_{n \in \mathbb{N}} \left| \int_0^1 K_x(y) u_n(y) \, dy \right|^2 dx$$

$$= \int_0^1 \sum_{n \in \mathbb{N}} |Tu_n(x)|^2 \, dx$$

$$= \sum_{n \in \mathbb{N}} \int_0^1 |Tu_n(x)|^2 \, dx = \sum_{n \in \mathbb{N}} \|Tu_n\|^2$$

$$= \sum_{n \in \mathbb{N}} \sum_{m \in \mathbb{N}} |\langle Tu_n, u_m \rangle|^2$$

$$= \sum_{n \in \mathbb{N}} \sum_{m \in \mathbb{N}} |a_{m,n}|^2 = \sum_{m \in \mathbb{N}} \alpha_m.$$

Next, we truncate the series representation of T, (B.2), to obtain a sequence (T_n) of operators of finite rank, and thus compact, approximating T. We let

$$T_n \varphi := \sum_{i=1}^n \left(\sum_{j=1}^\infty a_{ij} \langle \varphi, u_j \rangle \right) u_i \implies T\varphi - T_n \varphi = \sum_{i=n+1}^\infty \left(\sum_{j=1}^\infty a_{ij} \langle \varphi, u_j \rangle \right) u_i.$$

We **claim** $T_n \to T$. To see this, use the generalized Pythagorean theorem, Cauchy–Schwarz inequality and Bessel's inequality to derive

$$\|T\varphi - T_n\varphi\|^2 = \sum_{i=n+1}^{\infty} \left| \sum_{j=1}^{\infty} a_{ij}\langle \varphi, u_j \rangle \right|^2$$

$$\leq \sum_{i=n+1}^{\infty} \left(\sum_{j=1}^{\infty} |a_{ij}|^2 \right) \sum_{j \in \mathbb{N}} |\langle \varphi, u_j \rangle|^2$$

$$\leq \|\varphi\|^2 \sum_{i=n+1}^{\infty} \sum_{j=1}^{\infty} |a_{ij}|^2$$

$$= \|\varphi\|^2 \sum_{i=n+1}^{\infty} \alpha_i.$$

Therefore, $\|T_n - T\| \to 0$ so $T_n \to T$. By Step (4), it follows that T is compact.

(6) We have $\|T\| = \sup\{|\langle T\varphi, \varphi \rangle| : \|\varphi\| = 1\}$: To see this, using Cauchy–Schwarz we have

$$N(T) := \sup\{|\langle T\varphi, \varphi \rangle| : \|\varphi\| = 1\} \leq \sup_{\|\varphi\|=1} \|T\varphi\| \cdot \|\varphi\| = \sup_{\|\varphi\|=1} \|T\varphi\| = \|T\|.$$

We **claim** we also have $\|T\| \leq N(T)$ so $N(T) = \|T\|$. To see this, first expand the left-hand side of the following equation and use Step (1) to get

$$\langle T(\varphi + \psi), \varphi + \psi \rangle - \langle T(\varphi - \psi), \varphi - \psi \rangle = 4\Re\langle T\varphi, \psi \rangle.$$

From the definition of $N(T)$, we get for all $\varphi \in H$, $|\langle T\varphi, \varphi \rangle| \leq N(T)\|\varphi\|^2$. Using the parallelogram law in H, we are then able to derive

$$4|\Re\langle T\varphi, \psi \rangle| \leq |\langle T(\varphi + \psi), \varphi + \psi \rangle| + |\langle T(\varphi - \psi), \varphi - \psi \rangle|$$

$$\leq N(T)\|\varphi + \psi\|^2 + N(T)\|\varphi - \psi\|^2$$

$$= 2N(T)(\|\varphi\|^2 + \|\psi\|^2).$$

If $\|\varphi\| = \|\psi\| = 1$, the equation we have derived gives $|\Re\langle T\varphi, \psi \rangle| \leq N(T)$. If $e^{i\theta}\langle T\varphi, \psi \rangle = |\langle T\varphi, \psi \rangle|$, then

$$|\langle T\varphi, \psi \rangle| = \langle T(e^{i\theta}\varphi), \psi \rangle = \Re\langle T(e^{i\theta}\varphi), \psi \rangle \leq |\Re\langle T(e^{i\theta}\varphi), \psi \rangle| \leq N(T),$$

so, choosing $\psi = T\varphi/\|T\varphi\|$ for $T\varphi \neq 0$, we get $N(T) = \|T\|$, as claimed. In addition, for general nonzero φ, ψ, applying the displayed inequality to $\varphi/\|\varphi\|$, $\psi/\|\psi\|$, we get for all φ, ψ,

$$|\langle T\varphi, \psi \rangle| \leq N(T)\|\varphi\| \cdot \|\psi\|.$$

(7) We now **claim** there is an eigenvalue λ_1 such that $|\lambda_1| = \|T\|$: let (φ_n) be such that $\|\varphi_n\| = 1$ and $\lim_{n\to\infty} |\langle T\varphi_n, \varphi_n \rangle| = N(T) = \|T\|$. Because T is a compact operator, there is a convergent subsequence $T\varphi_{n_j}$, say $\psi = \lim_{j\to\infty} T\varphi_{n_j}$. Then since $|\langle T\varphi_{n_j}, \varphi_{n_j} \rangle| \to \|T\| \neq 0$, we must have ψ is not the zero (a.e.) function. Since $\langle T\varphi, \varphi \rangle$ is real for all φ, we can take a further subsequence

$$|a_n| \to a_0 > 0 \implies a_{n_j} \to -a_0 \text{ or } a_0,$$

which we relabel as the original, to deduce that for some $\lambda \neq 0$ we have $\lambda = \lim_{j\to\infty} \langle T\varphi_{n_j}, \varphi_{n_j} \rangle$, and λ is real. Expanding the norm square, we get

$$0 \leq \|T\varphi_{n_j} - \lambda\varphi_{n_j}\|^2$$
$$= \|T\varphi_{n_j}\|^2 - 2\lambda\langle T\varphi_{n_j}, \varphi_{n_j} \rangle + \lambda^2 \|\varphi_{n_j}\|^2.$$

Therefore,

$$0 \leq \lim_{j\to\infty} \|T\varphi_{n_j} - \lambda\varphi_{n_j}\|^2 = \|\psi\|^2 - 2\lambda^2 + \lambda^2 = \|\psi\|^2 - \lambda^2,$$

so $|\lambda| \leq \|\psi\|$.

We can also write from what we have seen

$$\|\psi\| = \lim_{j\to\infty} \|T\varphi_{n_j}\| \leq \lim_{j\to\infty} \|T\| \cdot \|\varphi_{n_j}\| = \|T\| = |\lambda|,$$

so $|\lambda| = \|\psi\|$ and therefore $0 \leq \lim_{j\to\infty} \|T\varphi_{n_j} - \lambda\varphi_{n_j}\|^2 \leq 0$ and, in addition,

$$\|\psi - \lambda\varphi_{n_j}\| \leq \|\psi - T\varphi_{n_j}\| + \|T\varphi_{n_j} - \lambda\varphi_{n_j}\| \to 0.$$

This implies $\varphi_{n_j} \to \psi/\lambda$ and then $T\psi = T(\lim_{j\to\infty} \lambda\varphi_{n_j}) = \lambda \lim_{j\to\infty} T\varphi_{n_j} = \lambda\psi$, so λ is an eigenvalue of T.

(8) **(The spectral theorem for T)** We now **claim** there is a set of eigenvalues λ_n of T, having limit value zero, and associated orthonormal eigenvectors ψ_n such that we have the representation

$$T\varphi = \sum_{n\in\mathbb{N}} \lambda_n \langle \varphi, \psi_n \rangle \psi_n.$$

First, note that T is not the zero operator. Let $H_1 = H$ and λ_1, ψ_1 an eigenvalue and eigenfunction determined by the derivation in Step (7). Let H_2 be the orthogonal complement of ψ_1, i.e., $H_2 = \{\varphi \in H : \langle \varphi, \psi_1 \rangle = 0\}$. Then $H_2 \subset H_1$ is a subspace, and because for $\varphi \in H_2$,

$$\langle T\varphi, \psi_1 \rangle = \langle \varphi, T\psi_1 \rangle = \langle \varphi, \lambda_1\psi_1 \rangle = \lambda_1 \langle \varphi, \psi_1 \rangle = 0,$$

we have $T(H_2) \subset H_2$.

Call the operator T restricted to H_2, T_2. Then if T_2 is nonzero, it is compact and self-adjoint and satisfies $\|T_2\| \leq \|T\|$. Continue inductively. At the n stage, we have a sequence of eigenvalues

$$|\lambda_1| \geq |\lambda_2| \geq \cdots \geq |\lambda_n| > 0$$

and a corresponding orthonormal sequence of eigenfunctions $\psi_1, \ldots, \psi_n$, so $T_n(H_n) \neq \{0\}$ and $T\psi_j = \lambda_j \psi_j$ for $1 \leq j \leq n$. Define H_{n+1} to be the orthogonal complement of the space spanned by the $\psi_1, \ldots, \psi_n$.

If for some n the mapping T restricted to H_{n+1} is zero, then we **claim** the range is spanned by $\psi_1, \ldots, \psi_n$. To see this, since for all φ, $\varphi - \sum_{1 \leq j \leq n} \langle \varphi, \psi_j \rangle \psi_j$ is in the orthogonal complement of the $\psi_1, \ldots, \psi_n$, so it would be mapped by T to zero. But this would give via linearity

$$T\varphi = \sum_{1 \leq j \leq n} \langle \varphi, \psi_j \rangle T\psi_j = \sum_{1 \leq j \leq n} \lambda_j \langle \varphi, \psi_j \rangle \psi_j,$$

which shows the range of T would have finite dimension, which is false, since by Step (10) the kernel is zero.

We next **claim** that $\lim_{n \to \infty} \lambda_n = 0$: in order to derive a contradiction, suppose there is an $\epsilon > 0$ such that $|\lambda_n| \geq \epsilon$ for all $n \in \mathbb{N}$. This implies (ψ_n/λ_n) with $\|\psi_n\| = 1$, is bounded in H so $(T(\psi_n)/\lambda_n)$ contains a convergent subsequence. But for all n, $T(\psi_n/\lambda_n) = \psi_n$ and $\|\psi_n - \psi_m\| = \sqrt{2}$ for all $n \neq m$ in the subsequence of indices. Hence $\lambda_n \to 0$.

Now we can demonstrate the spectral representation for T: define a sequence

$$\tau_n := \varphi - \sum_{j=1}^{n} \langle \varphi, \psi_j \rangle \psi_j. \tag{B.3}$$

Then because for each n we have $\tau_n \perp \sum_{j=1}^{n} \langle \varphi, \psi_j \rangle \psi_j$, we have

$$\|\varphi\|^2 = \|\tau_n + \sum_{j=1}^{n} \langle \varphi, \psi_j \rangle \psi_j\|^2 = \|\tau_n\|^2 + \sum_{j=1}^{n} |\langle \varphi, \psi_j \rangle|^2 \geq \|\tau_n\|^2$$

But T restricted to H_{n+1} has norm $|\lambda_{n+1}|$, so

$$\|T\tau_n\| \leq |\lambda_{n+1}| \|\tau_n\| \leq |\lambda_{n+1}| \|\varphi\| \to 0.$$

Applying T to (B.3) then gives

$$T\tau_n = T\varphi - \sum_{j=1}^{n} \lambda_j \langle \varphi, \psi_j \rangle \psi_j \implies T\varphi = \sum_{j \in \mathbb{N}} \lambda_j \langle \varphi, \psi_j \rangle \psi_j.$$

Finally in this step, every nonzero eigenvalue of T occurs in the sequence $(\lambda_n)_{n\in\mathbb{N}}$: again, to derive a contradiction, let $T\psi = \lambda\psi$ with $\lambda \neq 0$, $\psi \neq 0$ and $\lambda \notin \{\lambda_n : n \in \mathbb{N}\}$. Then by Step (3), we have $\psi \perp \psi_n$ for all n. If that is so, then

$$\lambda\psi = T\psi = \sum_{n\in\mathbb{N}} \lambda_n \langle \psi, \psi_n \rangle \psi_n = 0,$$

which is false. The derivation is complete.

(9) The mapping T is such that $\lambda_1 = \sup_{n\in\mathbb{N}} \lambda_n$ is strictly positive (note an analogous argument shows all the λ_n, in the given situation, are nonnegative). By Steps (6) and (7), we have

$$|\lambda_1| = \|T\| = \sup\{|\langle T\varphi, \varphi \rangle| : \|\varphi\| = 1\}.$$

In addition,

$$\|\varphi\|^2 = \int_0^1 \varphi(x)\overline{\varphi(y)}\, dxdy = \|(|\varphi|)\|^2,$$

and if $\varphi \neq 0$ a.e., we have $\langle T|\varphi|, |\varphi|\rangle > 0$. Therefore,

$$\sup\{|\langle T\varphi, \varphi \rangle| : \|\varphi\| = 1\} = \sup\{|\langle T|\varphi|, |\varphi|\rangle| : \|\varphi\| = 1\}$$

and the right-hand side is positive. Thus, since λ_1 is real, we have $\lambda_1 > 0$.

(10) The kernel of T is $\{0\}$, i.e., T is injective: let $T\varphi = 0$ and assume $\varphi \in C^{(1)}[0,1]$ (this subset of $L^2[0,1]$ is dense). Then for all $x \in [0,1]$, we have $\int_0^1 K(x,y)\varphi(y)\, dy = 0$. Thus

$$0 = \int_0^x \frac{y^\alpha}{x^\alpha}\varphi(y)\, dy + \int_x^1 \frac{x^\alpha}{y^\alpha}\varphi(y)\, dy.$$

Simplifying and differentiating twice, we get $\varphi = 0$.

(11) It follows from Steps (8) and (10) that T has a complete set of positive eigenvalues $\lambda_1 \geq \lambda_2 \geq \lambda_3 \geq \cdots > 0$ and $\lim_{n\to\infty} \lambda_n = 0$. In addition, $\lambda_1 = \|T\|$. This completes the proof of the spectral theorem.

(12) Define for fixed $\varphi \in L^2[0,1]$, a sesquilinear form

$$A(\varphi, \psi) := \int_0^1 \int_0^1 K(x,y)\varphi(x)\overline{\psi(y)}dxdy = \langle T\varphi, \psi \rangle. \tag{B.4}$$

Then, using Step (6),

$$\|T\| = \sup_{\|\varphi\|=1} \|T\varphi\| = \sup_{\|\varphi\|=1} |\langle T\varphi, \varphi \rangle| = \sup_{\|\varphi\|\leq 1} |A(\varphi, \varphi)|.$$

Appendix C

Solving an Optimization Problem

Some definitions: Let $k \geq 4$, $\mathcal{X}$ a suitable function space so all terms are well defined, and define an (extended) positive real number by

$$S(k) := \max_{\substack{q \in \mathcal{X} \\ q(1)=0 \\ q' \neq 0}} \frac{(k-1) \int_0^1 x^{k-2} q(x)^2 \, dt}{\int_0^1 x^{k-1} (q'(x))^2 \, dx}. \tag{C.1}$$

In detail, the spaces employed in the development are

$$\mathcal{X} := \{ q : (0,1] \to \mathbb{R} : q(x) = -\int_x^1 q'(t) \, dt \text{ and } \int_0^1 x^{k-1}(q'(t))^2 \, dt < \infty \}.$$

$$\mathcal{Y} := \{ p \in L^1_{\text{loc}}(0,1] : \int_0^1 t^{(k-1)/2}(p(t))^2 \, dt < \infty \}.$$

Then q is an absolutely continuous function on each compact in $\mathbb{R}$ subinterval of $(0,1]$.

Define a function of two real variables on $[0,1] \times [0,1]$ by $K(0,0) = 0$ and

$$K(x,y) := \left(\frac{\min(x,y)}{\max(x,y)} \right)^{\frac{k-1}{2}}, \quad (x,y) \neq (0,0) \implies 0 \leq K(x,y) \leq 1,$$

where we always assume $k \geq 4$. It is expected that as this theory develops the lower range, $k \in \{2,3\}$ will be important.

Figure C.1 is a plot of $K(\cos(\theta), \sin(\theta))$ for $0 \leq \theta \leq \pi/2$.

Define a continuous, everywhere-defined linear mapping or "operator" T: $L^2([0,1], \mathbb{C}) \to L^2([0,1], \mathbb{C})$ by

$$(T\varphi)(y) := \int_0^1 \varphi(y) K(x,y) \, dx. \tag{C.2}$$

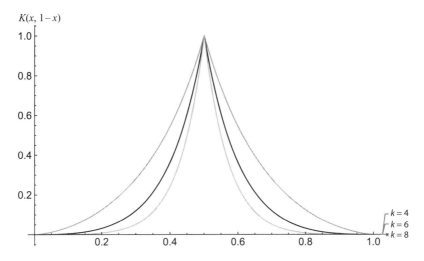

Figure C.1 Diagonal section of $K(x, y)$ for $k \in \{4, 6, 8\}$.

Recall an everywhere defined operator on a Hilbert space H is called **self-adjoint** if $\langle T\varphi, \varphi \rangle$ is real for all $\varphi \in H$. It is **compact** if the image of every bounded sequence has a convergent subsequence. It is **positive** if $\langle T\varphi, \varphi \rangle \geq 0$ for every φ.

The space $C_0(0, 1]$ is the set of all continuous real-valued functions φ on $[0, 1]$ with $\varphi(0) = 0$.

We now give the basic properties of the operator T in Lemma C.1.

Lemma C.1 *The mapping T:*

(1) is compact, positive and self-adjoint and maps real-valued functions to real-valued functions and positive functions to positive functions.

(2) The maximum $S(k)$ is attained by a function $\varphi \in L^2([0, 1], \mathbb{C})$.

(3) There exists a function $q \in \mathscr{X}$ with $q \not\equiv 0$ and $q(1) = 0$, such that

$$S(k) = -\frac{2 \int_0^1 x^{k-1} q'(x) q(x) \, dx}{\int_0^1 x^{k-1} (q'(x))^2 \, dx}.$$

(4) $S(k)$ is the largest eigenvalue, denoted λ_1, of T, and $\lambda_1 = \|T\|$. Any eigenfunction φ of T with $\|\varphi\| = 1$ with eigenvalue λ_1 is an extremum. The only maximizers are nonzero eigenfunctions of T.

Proof **(1)** The given properties of T follow from the development set out in Appendix B. All of the eigenvalues of T are real and nonnegative and can be ordered $\lambda_1 > \lambda_2 > \cdots > \lambda_n > \cdots$ with

$$\lim_{n \to \infty} \lambda_n = 0 \text{ and } \lambda_1 = \|T\|.$$

Because T takes real-valued functions to real-valued functions, taking the real part of an eigenfunction we get a real valued eigenfunction, so we can write $T\varphi = \lambda_1 \varphi$ with $\varphi \in L^2([0,1],\mathbb{R})$ and multiplying by a suitable constant, $\|\varphi\| = 1$.

(3) We now show the optimization problem of Step (2) is equivalent to Equation (C.1). Begin by integrating that equation by parts to derive

$$S(k) := \sup_{\substack{q\in\mathscr{X} \\ q(1)=0 \\ q'\neq 0}} \frac{-2\int_0^1 x^{k-1}q(x)q'(x)\,dt}{\int_0^1 x^{k-1}(q'(x))^2\,dx}. \tag{C.3}$$

Then set $p(x) := q'(x)$ and $\varphi(x) := p(x)x^{(k-1)/2}$ so, using the definition of $K(x,y)$, the optimization problem of Equation (C.3) becomes

$$S(k) := \sup_{\partial\in\mathscr{Y}} \frac{2\int_0^1 x^{(k-1)/2}p(x)\left(\int_x^1 p(y)\,dy\right)dx}{\int_0^1 p(x)^2 x^{k-1}\,dx}$$

$$= \sup_{\varphi\in H\setminus\{0\}} \frac{2\int_0^1 \varphi(x)x^{(k-1)/2}\left(\int_0^1 \varphi(y)y^{(1-k)/2}\,dy\right)dx}{\int_0^1 \varphi(x)^2\,dx}$$

$$= \sup_{\varphi\in H\setminus\{0\}} \frac{2\int_0^1\int_0^1 \varphi(x)\varphi(y)x^{(k-1)/2}y^{(1-k)/2}\chi_{y>x}(x,y)\,dxdy}{\int_0^1 \varphi(x)^2\,dx}$$

$$= \sup_{\varphi\in H\setminus\{0\}} \frac{\int_0^1\int_0^1 K(x,y)\varphi(x)\varphi(y)\,dxdy}{\int_0^1 \varphi(x)^2\,dx}.$$

(4) Finally, suppose that $\varphi \in H = L^2[0,1]$ is not identically zero and not an eigenfunction of T with eigenvalue λ_1. Let (u_n) be an orthonormal basis so we can write, since there is an m with $\lambda_m < \lambda_1$ and $\langle \varphi, u_m \rangle \neq 0$, using Parseval's identity,

$$\varphi = \sum_{n=1}^{\infty} \langle \varphi, u_n \rangle u_n \implies \langle T\varphi, \varphi \rangle = \sum_{n=1}^{\infty} |\langle \varphi, u_n \rangle|^2 < \lambda_1 \|\varphi\|.$$

Therefore, any such φ is not an extremum. This completes the proof. $\qquad\square$

Next, we prove some additional properties of the mapping T which will be needed in the sequel. First some definitions: We say a real function is **absolutely continuous** on an interval Ω if it is continuous on the interval and differentiable almost everywhere with the derivative being Lebesgue integrable and

$$f(x) - f(a) = \int_a^x f'(t)\,dt$$

for all a, x in the interval. Equivalently, for every $\epsilon > 0$ there is a $\delta > 0$ such that for all finite families of disjoint subintervals $([a_n, b_n] : n \in F)$ with

$$\sum_{n \in F} b_n - a_n < \delta, \text{ we have } \sum_{n \in F} |f(a_n) - f(b_n)| < \epsilon.$$

We say a real function $f : [a, b] \rightarrow \mathbb{R}$ has **bounded total variation** if there is a real number M such that for every partition P defined by points $a = x_0 \leq x_1 \leq \cdots \leq x_n = b$, we have

$$V(f, P) := \sum_{j=1}^{n} |f(x_{j-1}) - f(x_j)| \leq M.$$

If this is so, then we define $V(f) = V(f, [a, b]) := \sup_P V(f, P)$, where the supremum is taken over all finite partitions of the interval, and call $V(f)$ the **total variation** of f on $[a, b]$. Note that if $f \in C^{(1)}[a, b]$, we can write, using the mean value theorem,

$$V(f) = \int_a^b |f'(x)| \, dx.$$

Lemma C.2 *In addition, the mapping T:*

(1) satisfies $T(L^2([0, 1], \mathbb{R})) \subset C_0[0, 1]$,
(2) has eigenfunctions continuous and which satisfy $\varphi(0) = 0$,
(3) when restricted to $C[0, 1]$ is compact and maps into $C[0, 1]$, and has the same eigenvalues and eigenfunctions as T on $L^2([0, 1], \mathbb{C})$,
(4) satisfies $T(L^2([0, 1], \mathbb{C})) \subset C^{(1)}(0, 1]$, and if $\varphi \in L^2[0, 1]$ and $x \in (0, 1]$, then we can write

$$(T\varphi)'(x) = \int_0^1 \varphi(y) \frac{\partial}{\partial x} K(x, y) \, dy$$

$$= -\frac{k-1}{2x}(T\varphi)(x) + (k-1)x^{(k-3)/2} \int_x^1 \frac{\varphi(y)}{y^{(k-1)/2}} \, dy, \qquad \text{(C.4)}$$

(5) all functions in the range of T are absolutely continuous with bounded total variation which satisfies

$$V(T\varphi, [0, 1]) = \int_0^1 |(T\varphi)'| \leq \|\varphi\|_1,$$

(6) if ψ is in the range, then $T\psi \in C^{(1)}[0, 1]$ and $(T\psi)'(0) = 0$. Thus if φ is an eigenfunction, then $\varphi \in C^{(1)}[0, 1]$ and $\varphi'(0) = 0$.

Proof **(1)** Since $0 \le K(x, y) \le 1$, we have $\varphi(y)K(x, y) \le |\varphi(y)|$, and since $[0, 1]$ is bounded, we have $|\varphi(y)| \in L^1[0, 1]$. In addition, we have for all $y \in [0, 1]$, if $x \to x_0$, then $\lim_{x \to x_0} K(x, y) = K(x_0, y)$. Therefore, by the Lebesgue bounded convergence theorem, (e.g. [22, theorem D.1, appendix D]),

$$\lim_{x \to x_0} (T\varphi)(x) = \int_0^1 \varphi(y) \lim_{x \to x_0} K(x, y) \, dy = \int_0^1 \varphi(y) K(x_0, y) \, dy = T\varphi(x_0).$$

Thus $T\varphi \in C[0, 1]$. Finally since $K(0, y) = 0$ for all $y \in [0, 1]$, if $\varphi \in L^2[0, 1]$, we get $T\varphi(0) = 0$. Hence the range of T is in $C_0[0, 1]$.

(2) It follows immediately from Step (1) that all nonzero eigenfunctions φ of T are continuous and satisfy $\varphi(0) = 0$. This shows, given the correspondence of Step (2) of Lemma C.2 between the spaces $L^2[0, 1]$ and $\mathscr{Y}$ is given by and $\mathscr{X}$, namely that

$$\varphi(x) = x^{(k-1)/2} p(x) = x^{(k-1)/2} q'(x),$$

if φ is an eigenfunction, we get $x^{(k-1)/2} q'(x) \in C_0(0, 1]$. In addition,

$$\lim_{x \to 0+} x^{(k-1)/2} q'(x) = \lim_{x \to 0+} \varphi(x) = \varphi(0) = 0.$$

On $(0, 1]$, we can write $q(x) = -\int_x^1 q'(t) \, dt$ so $q \in C(0, 1]$. Also

$$\lim_{x \to 0+} x^{(k-3)/2} q(x) = \lim_{x \to 0+} x^{(k-3)/2} \left(-\int_x^1 o(1) t^{(1-k)/2} \, dt \right) = \lim_{x \to 0+} \frac{o(1)}{k-3} = 0.$$

Therefore, $x^{(k-3)/2} q(x) \in C_0(0, 1]$ also.

(3) By Step (1) each of the eigenfunctions of T acting on $L^2[0, 1]$ is in $C_0(0, 1]$, so is an eigenfunction for T considered as a mapping on $C[0, 1]$ with the same eigenvalues. The reverse is also true.

that in $\mathscr{X}$ we have

$$q(x) = -\int_x^1 q'(t) \, dt = -\int_x^1 t^{(1-k)/2} \varphi(t) \, dt.$$

(4) First, define two subsidiary functions κ and ω by $\kappa(0, 0) = \omega(0, 0) = 0$ and for $(x, y) \ne (0, 0)$,

$$\kappa(x, y) := \frac{\min(x, y)}{\max(x, y)},$$

$$\omega(x, y) := \frac{\sum_{n=0}^{k-2} x^{n/2} y^{(k-2-n)/2}}{\sqrt{x} + \sqrt{y}}.$$

Note that we can also write

$$\frac{\partial}{\partial x}\kappa(x,y) = \frac{1}{y}, \; x < y,$$

$$\frac{\partial}{\partial x}\kappa(x,y) = -\frac{y}{x^2}, \; y < x.$$

Then for all $(x, y) \in [0, 1]^2$, we have

$$\frac{k-1}{2}K(x,y) = \omega(\kappa(x,y),\kappa(x,y))\kappa(x,y).$$

We also have

$$0 \le \omega(x,y) \le (k-1)\max(x,y)^{(k-3)/2} \le n$$

and

$$x^{(k-1)/2} - y^{(k-1)/2} = (\sqrt{x} - \sqrt{y})\left(\sum_{n=1}^{k-2} x^{n/2}y^{(k-2-n)/2}\right) = (x-y)\omega(x,y).$$

Thus

$$K(x_1, y) - K(x_0, y) = \omega(\kappa(x_1,y),\kappa(x_0,y))(\kappa(x_1,y) - \kappa(x_0,y)). \tag{C.5}$$

We can also derive a uniform bound for $x_1 > x_0 > 0$, namely with $\Delta := x_1 - x_0$,

$$|\kappa(x_1, y) - \kappa(x_0, y)| \le \frac{\Delta}{x_0}.$$

To see this, first write

$$\kappa(x_1, y) - \kappa(x_0, y) = \begin{cases} \frac{y}{x_1} - \frac{y}{x_0} = -\frac{y\Delta}{x_0 x_1}, & y < x_0 < x_1, \\ \frac{y}{x_1} - \frac{x_0}{y} = \frac{y^2 - x_0 x_1}{y x_1} \le \frac{\Delta}{y} \le \frac{\Delta}{x_0}, & x_0 \le y \le x_1, \\ \frac{x_1}{y} - \frac{x}{y} = \frac{\Delta}{y}, & x_0 < x_1 < y. \end{cases}$$

This implies for $x_0 > 0$

$$|\kappa(x_1, y) - \kappa(x_0, y)| = \begin{cases} \frac{y}{x_1} - \frac{y}{x_0} = -\frac{y\Delta}{x_0 x_1} \le \frac{\Delta}{x_1} \le \frac{\Delta}{x_0}, & y < x_0 < x_1, \\ \left|\frac{y}{x_1} - \frac{x_0}{y}\right| = \frac{|y^2 - x_0 x_1|}{y x_1} \le \frac{\Delta}{y} \le \frac{\Delta}{x_0}, & x_0 \le y \le x_1, \\ \frac{x_1}{y} - \frac{x}{y} = \frac{\Delta}{y} \le \frac{\Delta}{x_1} \le \frac{\Delta}{x_0}, & x_0 < x_1 < y. \end{cases}$$

Using these notations and relationships, we can derive the expression for the derivative of $T\varphi(x_0)$ when $x_0 > 0$:

$$\lim_{x_1 \to x_0+} \frac{T\varphi(x_1) - T\varphi(x_0)}{x_1 - x_0} = \int_0^1 \varphi(y)\omega(\kappa(x_1, y), \kappa(x, y)) \frac{\kappa(x_1, y) - \kappa(x_0, y)}{x_1 - x_0} \, dy$$

$$= \int_0^1 \varphi(y)\omega(\kappa(x_0, y), \kappa(x_0, y)) \frac{\partial}{\partial x} \kappa(x, y)|_{x=x_0} \, dy$$

$$= \int_0^1 \varphi(y)\omega(\kappa(x_0, y), \kappa(x_0, y)) \mathrm{sgn}(y - x_0) \frac{\kappa(x_0, y)}{x_0} \, dy$$

$$= \frac{k-1}{2x_0} \int_0^1 \varphi(y) K(x_0, y) \mathrm{sgn}(y - x_0) \, dy$$

$$= \frac{k-1}{2x_0} \left(\int_{x_0}^1 \varphi(y) K(x_0, y) \, dy \right.$$

$$\left. - \int_0^{x_0} \varphi(y) K(x_0, y) \mathrm{sgn}(y - x_0) \, dy \right)$$

$$= \frac{k-1}{x_0} \left(\int_{x_0}^1 \varphi(y) K(x_0, y) \mathrm{sgn}(y - x_0) \, dy - \frac{1}{2} T\varphi(x_0) \right).$$

A corresponding derivation applies when $x_0 > 0$ to $x_1 \to x_0-$.

(5) Since $T\varphi \in C^{(1)}(0, 1]$, it is absolutely continuous, and we can write $V(T\varphi) = \int_0^1 |(T\varphi)'(x)| \, dx$. Thus, using Equation (C.4) we have

$$V(T\varphi) = \int_0^1 \left| \int_0^1 \varphi(x) \frac{\partial}{\partial x} K(x, y) \, dy \right| dx$$

$$\leq \int_0^1 \int_0^1 |\varphi(y)| \left| \frac{\partial}{\partial x} K(x, y) \right| dx dy$$

$$= \int_0^1 |\varphi(y)| \left(\int_0^1 \left| \frac{\partial}{\partial x} K(x, y) \right| dx \right) dy$$

$$= 2\|\varphi\|_1.$$

(6) Let $\psi = T\varphi$, and note from Steps (1) and (5) that $\psi \in C_0(0, 1]$, so $\psi(0) = 0$ and $\psi \in C^{(1)}(0, 1]$ with $V(\psi) \leq 2\|\varphi\|_1$.

Let $0 < x_1$ and write using integration by parts and the mean value theorem for integrals there is a z with $0 < z, x_1$ such that for $k > 3$,

$$\frac{1}{x_1} \int_0^1 \psi(y) K(x_1, y) \, dy = \frac{1}{x_1} \left(\int_0^{x_1} \psi(y) K(x_1, y) \, dy + \int_{x_1}^1 \psi(y) \left(\frac{x_1}{y} \right)^{(k-1)/2} dy \right)$$

$$= \psi(z) K(x_1, z) + \left(-\left[\psi(y) \frac{x_1^{(k-3)/2}}{(k-3)/2} y^{(3-k)/2} \right]_{x_1}^1 \right.$$

$$\left. + \int_{x_1}^1 \psi'(y) \frac{x_1^{(k-3)/2}}{(k-3)/2} y^{(3-k)/2} \, dy \right).$$

When $x_1 \to 0+$, the first term in this expression tends to $\psi(0) = 0$. The first term in parentheses also tends to zero. For the final term, note that $\psi' \in L^1[0,1]$ and that since for the integral we have $y \geq x_1$ the remaining factors remain bounded. Using the Lebesgue bounded convergence theorem, since the integrand tends to zero when y is fixed we get zero for the limit of the final term as well. This completes the proof. $\square$

The following Lemma C.3 is fundamental. Eigenfunctions of T are shown to satisfy Bessel's differential equation, and this gives us the desired form of solutions. The lemma is then applied in Theorem C.4 to achieve the goal of this section.

Lemma C.3 *Let $\varphi \in L^2[0,1]$ satisfy, for a given $\lambda > 0$, the eigenvalue equation $T\varphi = \lambda\varphi$, $k > 2$, and let*

$$q(x) := -\int_x^1 \varphi(t) t^{-(k-1)/2} \, dt, \; 0 \leq x \leq 1 \implies q(1) = 0.$$

Then:

(1) *The eigenfunction φ is continuous on $[0,1]$ and infinitely differentiable (smooth) on $(0,1]$.*

(2) *The function q satisfies the ordinary differential equation on $(0,1]$*

$$q''(x) + \frac{k-1}{x} q'(x) + \frac{k-1}{\lambda x} q(x) = 0. \tag{C.6}$$

(3) *Conversely, if a function $q(x)$ is a solution to Equation (C.6) in $C^{(2)}(0,1]$, with $q(1) = 0$ and not identically zero, and if we define $\varphi(x) := x^{(k-1)/2} q'(x)$, and φ has a continuous extension to $[0,1]$ with $\varphi(0) = 0$, then $\varphi(x)$ is an eigenfunction of T with eigenvalue λ.*

(4) *Every solution of Equation (C.6) is a linear constant combination of Bessel functions of the first and second kinds having the form*

$$q(x) = c_1 x^{1-k/2} J_{k-2}\left(2\sqrt{\frac{k-1}{\lambda}} \sqrt{x} \right) + c_2 x^{1-k/2} Y_{k-2}\left(2\sqrt{\frac{k-1}{\lambda}} \sqrt{x} \right), \tag{C.7}$$

with arbitrary constants c_1 and c_2.

(5) *Necessarily the constant $c_2 = 0$.*

(6) *The constant $\lambda > 0$ is an eigenvalue of T if and only if $J_{k-2}(2\sqrt{(k-1)/\lambda}) = 0$, so all eigenfunctions corresponding to λ have the form*

$$\varphi(x) = x^{(k-1)/2}q'(x) = c_1 x^{\frac{1}{2}} J_{k-2}\left(2\sqrt{\frac{k-1}{\lambda}}\sqrt{x}\right), \quad c_1 \neq 0.$$

Proof **(1)** Let $\varphi \in L^2[0,1]$ be such that $T\varphi = \lambda\varphi$ with $\lambda > 0$. Then φ is in the range of T so, by Lemma C.2, Step (4), it is continuous on $[0,1]$ and continuously differentiable on $(0,1]$. Substitute $\lambda\varphi$ for $T\varphi$ in Equation (C.4) to get for $x > 0$,

$$\lambda\varphi'(x) = -\frac{k-1}{2x}\lambda\varphi(x) + (k-1)x^{(k-3)/2}\int_x^1 \frac{\varphi(y)}{y^{(k-1)/2}}\,dy.$$

Because the right-hand side of this expression is differentiable it shows $\varphi \in C^{(2)}(0,1]$, and by induction φ is smooth on $(0,1]$.

(2) The next two steps are a little tricky, so we simplify the expressions by letting $n = k - 1$. Making the substitution $\varphi(x) = x^{n/2}q'(x)$ gives, using $q(1) = 0$,

$$\lambda(x^{n/2}q'(x))' = -\frac{n}{2x}\lambda x^{n/2}q'(x) + nx^{n/2-1}\int_x^1 q'(y)\,dy$$

$$= -\frac{\lambda n}{2}x^{n/2-1} - nx^{n/2-1}q(x).$$

Carrying out the differentiation on the right gives and simplifying then gives

$$\lambda x^{n/2}q''(x) + \frac{\lambda n}{2}x^{n/2}q'(x) + nx^{n/2-1}q(x) = 0.$$

Therefore,

$$q''(x) + \frac{n}{x}q'(x) + \frac{n}{x\lambda}q(x) = 0,$$

which completes the derivation of Equation (C.6).

(3) We have $\varphi(x) = x^{n/2}q'(x)$, so under the given conditions $\varphi \in L^2[0,1]$. Set $\psi = T\varphi$. By Lemma C.2(1), we have $\psi \in C_0(0,1]$. Differentiating the definition of φ gives

$$q''(x) = -\frac{n}{2}x^{-n/2-1}\varphi(x) + x^{-n/2}\varphi'(x),$$

and then using Equation (C.6) gives

$$-\frac{n}{x}q'(x) - \frac{n}{x\lambda}q(x) = q''(x) = -\frac{n}{2}x^{-n/2-1}\varphi(x) + x^{-n/2}\varphi'(x),$$

so

$$0 = \frac{n}{2}\varphi(x)x^{-n/2-1} + x^{-n/2}\varphi'(x) - \frac{n}{x\lambda}\int_x^1 \varphi(t)t^{-n/2}\,dt,$$

and thus

$$0 = \frac{\lambda n}{2} x^{-n/2} \varphi(x) + \lambda x^{-n/2+1} \varphi'(x) - n \int_x^1 \varphi(t) t^{-n/2} \, dt.$$

In addition, using $\psi = T\varphi$ with Equation (C.4), we get

$$x^{-n/2} \psi'(x) = -\frac{n}{2} x^{-n/2-1} \psi(x) + \frac{n}{x} \int_x^1 \varphi(t) t^{-n/2} \, dy,$$

so we have

$$x^{-n/2+1} \psi'(x) = -\frac{n}{2} x^{-n/2} \psi(x) + n \int_x^1 \varphi(t) t^{-n/2} \, dt.$$

Therefore,

$$\frac{n}{2} x^{-n/2} \psi(x) + x^{-n/2+1} \psi'(x) = n \int_x^1 \varphi(t) t^{-n/2} \, dt$$

$$= \lambda \left(\frac{n}{2} x^{-n/2} \varphi(x) + x^{-n/2+1} \varphi'(x) \right).$$

Finally in this step, multiply each side by x^{n-1} to get

$$\frac{n}{2} x^{n/2-1} \psi(x) + x^{n/2} \psi'(x) = \lambda \left(\frac{n}{2} x^{n/2-1} \varphi(x) + x^{n/2} \varphi'(x) \right),$$

which implies

$$\frac{d}{dx} \left(x^{n/2} (\lambda \varphi(x) - \psi(x)) \right) = 0$$

for all $x > 0$. Because $\varphi, \psi \in C^{(1)}[0, 1]$ and $x^{(k-1)/2}(\lambda\varphi(x) - \psi(x))$ vanishes at 0, it must vanish on $[0, 1]$. Hence $\psi(x) = \lambda\varphi(x)$ on $[0, 1]$, which completes the proof of the converse.

(4) Pattern matching with the general Bessel equation given in Appendix A, and making the choices

$$v = k - 2, \ \alpha = 1 - k/2, \ \beta = 2\sqrt{\frac{k-1}{\lambda}}, \ \gamma = \frac{1}{2}$$

unifies, via the substitution $u(x) := x^\alpha y(\beta x^\gamma)$, the Bessel equation and Equation (C.4), i.e.,

$$y'' + \frac{1}{x} y' + \left(1 - \frac{v^2}{x^2} \right) y = 0 \text{ and } u'' + \frac{2\alpha - 1}{x} + \left(\beta^2 \gamma^2 x^{2\gamma-2} + \frac{\alpha^2 - v^2 \gamma^2}{x^2} \right) u = 0,$$

gives $q'' + (k-1)q'(x)/x + (k-1)q(x)/\lambda = 0$. Since the Bessel equation has the two linearly independent solutions $J_{k-2}(x)$, $Y_{k-2}(x)$, the equation for $q(x)$ therefore has the general solution

$$q(x) = c_1 x^{1-k/2} J_{k-2}\left(2\sqrt{\frac{k-1}{\lambda}}\sqrt{x}\right) + c_2 x^{1-k/2} Y_{k-2}\left(2\sqrt{\frac{k-1}{\lambda}}\sqrt{x}\right). \quad \text{(C.8)}$$

(5) First note that asymptotically as $x \to 0+$, we have

$$J_k(x) \asymp x^k \text{ and } Y_k(x) \asymp x^{-(k-1)},$$

so $\sqrt{x} J_k(a\sqrt{x}) \in C_0(0,1]$, but $\sqrt{x} Y_k(a\sqrt{x})$ is unbounded, so necessarily $c_2 = 0$.

(6) Finally, returning to Equation (C.7), and using $q(1) = 0$, we obtain the condition

$$J_{k-2}\left(2\sqrt{\frac{k-1}{\lambda}}\right) = 0,$$

so if j_{k-2} is the first positive zero of $J_{k-2}(x)$, we can satisfy this constraint by choosing λ such that

$$2\sqrt{\frac{k-1}{\lambda}} = j_{k-2} \implies q(x) = x^{-(k-2)/2} J_{k-2}\left(j_{k-2}\sqrt{x}\right),$$

completing the proof. $\qquad\qquad\qquad\qquad\qquad\qquad\qquad\qquad\qquad\qquad\square$

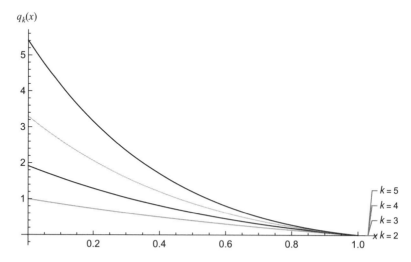

Figure C.2 The normalized Bessel function $q_k(x)$ on $[0,1]$ for $k \in \{2,3,4,5\}$.

Figure C.2 gives a plot of the normalized Bessel functions for four small values of k.

Theorem C.4 **(Prime gaps optimization problem)** *[48, theorem 16] Let $k \geq 4$ and recall the definitions*

$$\mathcal{X} := \{q : (0, 1] \to \mathbb{R} : q(x) = -\int_x^1 q'(t) \, dt \text{ and } \int_0^1 x^{k-1}(q'(t))^2 \, dt < \infty\},$$

and

$$S(k) := \max_{\substack{q \in \mathcal{X} \\ q(1)=0}} \frac{\int_0^1 x^{k-2} q(x)^2 \, dx}{\int_0^1 x^{k-1}(q'(x))^2 \, dx}, \quad q(1) = 0.$$

Then solutions of this optimization problem are constant nonzero multiples of

$$f(x) := x^{1-k/2} J_{k-2}(j_{k-2}\sqrt{x}),$$

and these solutions are unique.

Proof This is a direct application of Lemma C.3. $\qquad\square$

Appendix D

A Brun–Titchmarsh Inequality

The purpose of this appendix is to state and prove the Brun–Titchmarsh–Shiu inequality for a class of multiplicative functions. It gives an upper bound for sums of these functions in arithmetic progressions in intervals. It is very useful and is referred to in Chapters 2, 7 and 8. The proof takes ten steps and is due to Shiu.

Let $P(1) = 1$ and for $n > 1$ let $P(n)$ be the maximum prime number which divides n. Let $p(1) = 1$ and for $n > 1$ let $p(n)$ be the smallest prime dividing n.

The theorem of Shiu uses quite a few different symbols, so in the proof we list most of them in Step (0) with the steps at which they are introduced. The reader should skip this and refer back as may be necessary. Note that in Step (5) the crucial factorization $n = b_n d_n$ is defined. Here we give we trust a clearer specification than the original.

> **Theorem D.1** **(Brun–Titchmarsh inequality for arithmetic progressions)** *[195, theorem 1] Let $f(n)$ be a nonnegative multiplicative function with real values, and which satisfies for fixed constants A and B,*
>
> $f(p^m) \leq A^m$ *for all primes p and $m \in \mathbb{N}$ and $f(n) \leq Bn^\epsilon$ for all $\epsilon > 0$ and $n \in \mathbb{N}$.*
>
> *Let $0 < \alpha, \beta < \frac{1}{2}$ and $0 < a < q$ with $(a, q) = 1$. If $q < y^{1-\alpha}$ and $x^\beta < y \leq x$, then we have uniformly in integers a, q and real y the estimate*
>
> $$\sum_{\substack{x-y < n \leq x \\ n \equiv a \bmod q}} f(n) \ll \frac{y}{\varphi(q) \log x} \exp\left(\sum_{\substack{p \leq x \\ p \nmid q}} \frac{f(p)}{p}\right).$$

Proof **(0)** The following symbols are used in the proof:

$$A, B > 0 \qquad\qquad \text{statement,}$$

$$0 < \alpha, \beta < \frac{1}{2} \qquad\qquad \text{statement,}$$

$$a \in \mathbb{Z}, \; q \in \mathbb{N} \qquad\qquad \text{statement,}$$

$x, y \in \mathbb{R}$	statement,
$\Psi(x, y)$	Step (1),
$\Phi(x, y, z, q, a)$	Step (2),
$C > 0$	Step (9),
$n = b_n d_n$	Step (5),
$z \geq 2$	Steps (1) and (5),
$3/4 \leq \delta \leq 1$	Step (4),
$r_0 \in \mathbb{N}$	Step (9),
$S \subset \mathbb{N}$	Step (5),
$S_1 \subset S$	Steps (5) and (6),
$S_2 \subset S$	Steps (5) and (7),
$S_3 \subset S$	Steps (5) and (8),
$S_4 \subset S$	Steps (5) and (9).

(1) Let

$$\Psi(x, z) := \sum_{\substack{n \leq x \\ P(n) \leq z}} 1.$$

By the prime number theorem and Abel's theorem [4, theorem 4.2], we have for $y > 0$ sufficiently large

$$\sum_{p \leq y} \frac{1}{\log p} \leq \frac{2y}{(\log y)^2}. \tag{D.1}$$

Let $\epsilon > 0$ and again let y be sufficiently large. Using the estimate (D.1) in the final line, we get

$$\Psi(x, y) \leq x^\epsilon \sum_{\substack{n \leq x \\ P(n) \leq y}} \frac{1}{n^\epsilon}$$

$$\leq x^\epsilon \sum_{\substack{n \in \mathbb{N} \\ P(n) \leq y}} \frac{1}{n^\epsilon}$$

$$= x^\epsilon \prod_{p \leq y} \left(1 + \frac{1}{p^\epsilon} + \frac{1}{p^{2\epsilon}} + \cdots\right)$$

$$= x^\epsilon \prod_{p \leq y} \left(1 + \frac{1}{p^\epsilon - 1}\right)$$

$$\leq \exp\left(\epsilon \log x + \sum_{p \leq y} \frac{1}{p^\epsilon - 1}\right)$$

$$\leq \exp\left(\epsilon \log x + \frac{1}{\epsilon} \sum_{p \leq y} \frac{1}{\log p}\right)$$

$$\implies \Psi(x, y) \leq \exp\left(\epsilon \log x + \frac{2y}{\epsilon(\log y)^2}\right).$$

Finally, set $y = \log x \log\log x$ and $\epsilon = 1/\sqrt{\log\log x}$ to show that as $x \to \infty$, we have

$$\Psi(x, \log x \log\log x) \leq \exp\left(\frac{3\log x}{\sqrt{\log\log x}}\right).$$

Note that this is an application of the trick which has become known as "Rankin's method".

(2) Next, define for $x, y, z > 0$, $q \in \mathbb{N}$ and $a \in \mathbb{Z}$,

$$\Phi(x, y, z, q, a) := \sum_{\substack{x-y < n \leq x \\ n \equiv a \bmod q \\ p(n) > z}} 1,$$

and suppose $q < y \leq x$ and $2 \leq z$. Then an application of the Selberg sieve, Theorem 2.25, gives

$$\Phi(x, y, z, q, a) \leq \frac{y}{\varphi(q) \log z} + z^2.$$

(3) Because of the estimates satisfied by $f(n)$ and multiplicativity, we can write

$$\sum_{\substack{n \leq x \\ (n,q)=1}} \frac{f(n)}{n} \leq \prod_{\substack{p \leq x \\ p \nmid q}} \left(1 + \frac{f(p)}{p} + \sum_{j \geq 2} \frac{f(p^j)}{p^j}\right) \leq \exp\left(\sum_{\substack{p \leq x \\ p \nmid q}} \left(\frac{f(p)}{p} + \sum_{j \geq 2} \frac{f(p^j)}{p^j}\right)\right)$$

$$= \exp\left(\sum_{\substack{p \leq x \\ p \nmid q}} \frac{f(p)}{p} + O(1)\right)$$

$$\ll \exp\left(\sum_{\substack{p \leq x \\ p \nmid q}} \frac{f(p)}{p}\right).$$

Therefore, as $x \to \infty$ we have

$$
\sum_{\substack{n \le x \\ (n,q)=1}} \frac{f(n)}{n} \ll \exp\left(\sum_{\substack{p \le x \\ p \nmid q}} \frac{f(p)}{p} \right).
$$

(4) This is the main and most intricate step. It begins replacing ϵ by δ in Rankin's trick, which does not go to zero but is chosen to satisfy $3/4 \le \delta \le 1$. We derive

$$
\sum_{\substack{n \ge x \\ P(n) \le y \\ (n,q)=1}} \frac{f(n)}{n} \le x^{\delta-1} \sum_{\substack{n \ge x \\ P(n) \le y \\ (n,q)=1}} \frac{f(n)}{n^\delta}
$$

$$
\le x^{\delta-1} \sum_{\substack{n \ge 1 \\ P(n) \le y \\ (n,q)=1}} \frac{f(n)}{n^\delta}
$$

$$
\le \exp\left((\delta - 1) \log x + \sum_{\substack{p \le y \\ p \nmid q}} \frac{f(p)}{p^\delta} + O(1) \right). \qquad \text{(D.2)}
$$

Following Shiu, we replace $f(p)/p^\delta$ by $f(p)/p + (p^{1-\delta} - 1)f(p)/p$, use the condition $f(p^l) \le A$ for all $p \in \mathbb{P}$, and use the exponential series and the bound (see for example [4, theorem 4.10])

$$
\sum_{p \le y} \frac{\log p}{p} \le 2 \log y,
$$

to get

$$
\sum_{p \le y} \frac{f(p)}{p} (p^{1-\delta} - 1) \le \sum_{p \le y} \frac{A}{p} \sum_{n=1}^{\infty} \frac{((1 - \delta) \log p)^n}{n!}
$$

$$
\le A \sum_{n=1}^{\infty} \frac{(1 - \delta)^n (\log y)^{n-1}}{n!} \sum_{p \le y} \frac{\log p}{p}
$$

$$
\le 2A \sum_{n=1}^{\infty} \frac{(1 - \delta)^n (\log y)^n}{n!}
$$

$$
\le 2A \exp((1 - \delta) \log y) = 2A y^{1-\delta}.
$$

Inserting this in the estimate (D.2) gives

$$\sum_{\substack{n \geq x \\ P(n) \leq y \\ (n,q)=1}} \frac{f(n)}{n} \ll \exp\left((\delta - 1) \log x + \sum_{\substack{p \leq y \\ p \nmid q}} \frac{\log p}{p} + 2Ay^{1-\delta} \right).$$

Now we make the assignments

$$x = z^{1/2}, \ y = z^{1/r}, \ \delta = 1 - r \log r/(4 \log z),$$

where r in $\mathbb{N}$ is chosen to satisfy

$$1 \leq r \leq \frac{\log z}{\log\log z} \quad \Longrightarrow \quad \frac{3}{4} < \delta \leq 1.$$

In addition, these choices give

$$(1 - \delta) \log x = \frac{r \log r}{4 \log z} \times \frac{\log z}{2} = \frac{r \log r}{8} \quad \text{and}$$

$$y^{1-\delta} = z^{(1-\delta)/r} = z^{\log r/(4 \log z)} = r^{\frac{1}{4}}.$$

Using these choices, we get

$$\sum_{\substack{n \geq z^{\frac{1}{2}} \\ P(n) \leq z^{1/r} \\ (n,q)=1}} \frac{f(n)}{n} \ll \exp\left(\sum_{\substack{p \leq z^{1/r} \\ p \nmid q}} \frac{\log p}{p} - \frac{r \log r}{8} + 2Ar^{\frac{1}{4}} \right)$$

$$\ll \exp\left(\sum_{\substack{p \leq z^{1/r} \\ p \nmid q}} \frac{\log p}{p} - \frac{r \log r}{10} \right).$$

(5) Recall we have assumed that $q < y^{1-\alpha}$ and $x^\beta < y \leq x$. In addition, from now on, let $z = y^{\alpha/10}$. Express each n in the range $x - y < n \leq x$, which satisfies $n \equiv a \bmod q$, in the form $n = b_n d_n$, where for primes $p_1 < p_2 < \cdots < p_m$ we have

$$n = p_1^{\gamma_1} p_2^{\gamma_2} \cdots p_j^{\gamma_j} p_{j+1}^{\gamma_{j+1}} \cdots p_m^{\gamma_m},$$

and b_n is the unique factor of n which satisfies $b_n \leq z < b_n p_{j+1}^{\gamma_{j+1}}$ for some j with $1 \leq j < m$. So, and this is not so clear from Shiu's definition, for $z < p_1^{\gamma_1}$ we set

$b_n = 1$ and for $z \geq n$ we set $b_n = n$, and in every other case b_n is the product of an initial segment of prime powers so $(b_n, d_n) = 1$, i.e.,

$$b_n = p_1^{\gamma_1} p_2^{\gamma_2} \cdots p_j^{\gamma_j} \text{ and } d_n = n/b_n.$$

Next, we divide the integers n in the set

$$S := \{n \in \mathbb{Z} : n \in (x - y, x], \ n \equiv a \bmod q\}$$

into four subsets S_1, S_2, S_3 and S_4, which we consider in turn in the next four steps.

(6) $S_1 := \{n \in S : p(d_n) > z^{1/2}\}$: setting $b\bar{b} \equiv 1 \bmod q$ and $a' \equiv a\bar{b} \bmod q$, we get

$$\sum_{n \in S_1} f(n) = \sum_{n \in S_1} f(b_n) f(d_n)$$

$$\leq \sum_{b \leq z} f(b) \sum_{\substack{n \in S \\ n \equiv a \bmod q \\ n \equiv 0 \bmod b \\ p(n/b) > \sqrt{z}}} f(n/b)$$

$$= \sum_{\substack{b \leq z \\ (b,q)=1}} f(b) \sum_{\substack{bd \in S \\ d \equiv a' \bmod q \\ p(d) > \sqrt{z}}} f(d).$$

Because $z = y^{\alpha/10}$, we get

$$p(d) > z^{1/2} = y^{\alpha/20} > x^{\alpha\beta/20} \implies \Omega(d) \leq \frac{\log x}{\log p(d)} < \frac{20}{\alpha\beta}.$$

Therefore, $f(d) \leq A^{\Omega(d)} \leq A^{20/(\alpha\beta)}$. Thus

$$\sum_{n \in S_1} f(n) \ll \sum_{\substack{b \leq z \\ (b,q)=1}} f(b) \Phi(x/b, y/b, z^{1/2}, q, a').$$

Now $q < y^{1-\alpha}$ and $b \leq z < y^\alpha$ implies $q < y/b$ so, recalling

$$\Phi(x, y, z, q, a) := \sum_{\substack{n \in S \\ p(n) \geq z}} 1,$$

by Step (2), we have

$$\Phi(x/b, y/b, z^{1/2}, q, a') \leq \frac{2y}{\varphi(q) \log z} + z^2.$$

Therefore, using this and the result of Step (3), we get

$$\sum_{n \in S_1} f(n) \ll \left(\frac{y}{\varphi(q)\log z} + z^2\right) \exp\left(\sum_{\substack{p \leq z \\ p \nmid q}} \frac{f(p)}{p}\right).$$

(7) $S_2 := \{n \in S : p(d_n) \leq z^{1/2} \text{ and } b_n \leq z^{1/2}\}$: we proceed to count the elements in S_2. For each $n \in S_2$, here is a prime power p^γ such that $p^\gamma \| n$ and $p \leq \sqrt{z}$, and by the decomposition $n = b_n d_n$, $p^\gamma > \sqrt{z}$. If we let γ_p be the minimum integer γ with $p^\gamma > \sqrt{z}$, then $\gamma_p \geq 2$, and so

$$\frac{1}{p^{\gamma_p}} \leq \min\left(\frac{1}{\sqrt{z}}, \frac{1}{p^2}\right).$$

This implies

$$\sum_{p \leq z^{1/2}} \frac{1}{p^{\gamma_p}} \leq \sum_{p \leq z^{\frac{1}{4}}} \frac{1}{\sqrt{z}} + \sum_{p > z^{\frac{1}{4}}} \frac{1}{p^2} \ll \frac{1}{z^{1/4}}.$$

From this, it follows that

$$\#S_2 \leq \sum_{p \leq z^{1/2}} \sum_{\substack{n \in S \\ n \equiv a \bmod q \\ n \equiv 0 \bmod p^{\gamma_p}}} 1 = \sum_{\substack{p \leq \sqrt{z} \\ p \nmid q}} \left(\frac{y}{qp^{\gamma_p}} + O(1)\right) \ll \frac{y}{qz^{\frac{1}{4}}} + \sqrt{z}.$$

(8) $S_3 := \{n \in S : p(d_n) < \log x \log\log x \text{ and } b_n > z^{1/2}\}$: we will first estimate the number of $n \in S_3$. For each such n, there is $b \mid n$ with $\sqrt{z} < b \leq z$ and $P(b) \leq \log x \log\log x$. Hence, using the result of Step (1),

$$\#S_3 \leq \sum_{\substack{\sqrt{z} < b \leq z \\ P(b) < \log x \log\log x}} \sum_{\substack{n \in S \\ n \equiv a \bmod q \\ n \equiv 0 \bmod b}} 1$$

$$= \sum_{\substack{\sqrt{z} < b \leq z \\ P(b) < \log x \log\log x \\ (b,q)=1}} \left(\frac{y}{qb} + O(1)\right)$$

$$\leq \frac{y}{q\sqrt{z}}\Psi(z, \log x \log\log x) + O(z)$$

$$\ll \frac{y}{qz^{\frac{1}{4}}} + z.$$

Using $q < y^{1-\alpha}$ and $z = y^{\alpha/10}$, we get

$$z < y^\alpha / z^{\frac{1}{4}} < \frac{y}{qz^{\frac{1}{4}}}.$$

Therefore, using the result of Step (7) and the estimate we have derived for $\#S_3$, we get

$$\#S_2 + \#S_3 \ll \frac{y}{qz^{\frac{1}{4}}}.$$

When $y > x^\beta$, using the assumption $f(n) \ll_\epsilon n^\epsilon$ and $z = y^{\alpha/10}$ we get

$$f(n) \ll n^{\alpha\beta/80} \le x^{\alpha\beta/80} < y^{\alpha/80} = z^{1/8} \implies \sum_{n\in S_2} f(n) + \sum_{n\in S_3} f(n) \ll \frac{y}{qz^{1/8}}.$$

(9) $S_4 := \{n \in S : \log x \log\log x < p(d_n) \le z^{1/2} \text{ and } b_n > z^{1/2}\}$: This is the most difficult subset to treat. First we note, with the factorization introduced in Step (5), that

$$\sum_{n\in S_4} f(n) = \sum_{n\in S_4} f(b_n)f(d_n) \le \sum_{\sqrt{z}<b\le z} f(b) \sum_{\substack{n\in S \\ b_n=b \\ P(b)<p(d_n) \\ \log x \log\log x < p(d_n) \le \sqrt{z}}} f(d_n).$$

Following Shiu, define

$$r_0 := \left\lfloor \frac{\log z}{\log(\log x \log\log x)} \right\rfloor.$$

This definition implies $\log x \log\log x > z^{1/(1+r_0)}$.

Take $2 \le r \le r_0$ and $n \in S_4$ such that $z^{1/(1+r)} < p(d_n) \le z^{1/r}$. In this situation, when $b = b_n$ we get

$$P(b_n) = P(b) < p(d_n) < z^{1/r},$$

which enables us to deduce, setting $C := A^{20/(\alpha\beta)}$ and recalling from Step (5) and the hypotheses

$$z^{\frac{1}{2}} = y^{\alpha/20} > x^{\alpha\beta/20}, \text{ and } f(p^m) \le A^m, \ m \ge 1,$$

to get

$$\Omega(d_n) \le \frac{\log x}{\log(p(d_n))} \le \frac{(r+1)\log x}{\log z} < \frac{10(r+1)}{\alpha\beta} < \frac{20r}{\alpha\beta}$$

$$\implies f(d_n) \le A^{\Omega(d_n)} \le C^r.$$

Therefore, if as before we set $b\bar{b} \equiv 1 \bmod q$ and $a' \equiv a\bar{b} \bmod q$, recalling the definition of Φ from Step (2), and using $p(n/b_n) \le z^{1/r}$ implies $P(b_n) < z^{1/r}$, we get

$$\sum_{n \in S_4} f(n) \le \sum_{2 \le r \le r_0} C^r \sum_{\substack{\sqrt{z} < b \le z \\ P(b) < z^{1/r}}} f(b) \sum_{\substack{n \in S \\ b|n \\ z^{1/(r+1)} < p(n/b) \le z^{1/r}}} 1$$

$$\le \sum_{2 \le r \le r_0} C^r \sum_{\substack{\sqrt{z} < b \le z \\ P(b) < z^{1/r} \\ (b,q)=1}} f(b) \Phi\left(x/b, y/b, z^{1/(r+1)}, q, a'\right).$$

The result of Step (2) gives

$$\Phi(x/b, y/b, z^{1/(r+1)}, q, a') \le \frac{(r+1)y}{\varphi(q) b \log z} + z^{2/(r+1)}.$$

Combining these last two estimates then gives

$$\sum_{n \in S_4} f(n) \le \left(\frac{y}{\varphi(q) \log z} + z^2\right) \sum_{2 \le r \le r_0} (r+1) C^r \sum_{\substack{\sqrt{z} < b \le z \\ P(b) < z^{1/r} \\ (b,q)=1}} \frac{f(b)}{b}.$$

Using the definition of r_0 and applying Step (4) to the innermost sum in this expression then gives

$$\sum_{n \in S_4} f(n) \ll \left(\frac{y}{\varphi(q) \log z} + z^2\right) \exp\left(\sum_{\substack{p \le z \\ p \nmid q}} \frac{f(p)}{p}\right) \sum_{2 \le r \le r_0} r C^r \exp(-(r \log r)/10)$$

$$\ll \left(\frac{y}{\varphi(q) \log z} + z^2\right) \exp\left(\sum_{\substack{p \le z \\ p \nmid q}} \frac{f(p)}{p}\right).$$

(10) Having dealt with all of S, we can now complete the proof. Recall the assumptions $0 < \alpha < \frac{1}{2}$ and $z = y^{\alpha/10}$. Then

$$z^2 < \frac{qz^3}{\varphi(q) \log z} < \frac{y^{1-\delta+3\alpha/10}}{\varphi(q) \log z} < \frac{y}{\varphi(q) \log z}.$$

Using this in the estimates from Steps (6) and (9) gives

$$\sum_{n \in S_1} f(n) + \sum_{n \in S_4} f(n) \ll \frac{y}{\varphi(q) \log z} \exp\left(\sum_{\substack{p \leq z \\ p \nmid q}} \frac{f(p)}{p}\right),$$

and this, together with Step (8), gives

$$\sum_{\substack{x-y<n\leq x \\ n \equiv a \bmod q}} f(n) \ll \frac{y}{\varphi(q) \log x} \exp\left(\sum_{\substack{p \leq x \\ p \nmid q}} \frac{f(p)}{p}\right).$$

This completes the proof. $\qquad\qquad\qquad\qquad\qquad\qquad\qquad\qquad\square$

Appendix E

The Weil Exponential Sum Bound

Introduction: The purpose of this appendix is to derive the Weil exponential sum bound used in Chapter 8. It requires techniques based on the theory of finite fields. For background, the reader is directed to the literature, for example, see the very complete account [155] where we have extracted and simplified that approach. A summary background of the properties of finite fields and of the method is provided in section 9.9 of [22].

First some special definitions: p is a fixed prime and $\mathbb{F}_p$ the finite field with p elements and if $q = p^s$, $\mathbb{F}_q$ represents the finite field with q elements, i.e., a vector space over $\mathbb{F}_p$ of dimension s. Then

Φ denotes the set of monic polynomials with coefficients in $\mathbb{F}_p$,

Φ_k denotes the elements of Φ with degree $k \geq 1$,

$\Phi_{\mathscr{I}}$ denotes the irreducible elements of Φ.

The symbols $\sigma_0 = n$, $\sigma_1 = x_1 + \cdots + x_n, \ldots, \sigma_n = x_1 \cdots x_n$, represent the elementary symmetric polynomials, i.e., polynomials in the variables $x_1, \ldots, x_n$ which satisfy

$$(x - x_1) \cdots (x - x_n) = x^n - \sigma_1 x^{n-1} + \cdots + (-1)^n \sigma_n,$$

so

$$\sigma_j(x_1, \ldots, x_n) := \sum_{\substack{U \subset \{1,\ldots,n\} \\ |U|=j}} \prod_{i \in U} x_i, \quad n \geq 1, \ j \geq 0.$$

We also have the definitions:

$$Tr_{\mathbb{F}_{p^s}}(x) := 1 + x^p + x^{p^2} + \cdots + x^{p^{s-1}}, \quad x \in \mathbb{N},$$

$$N(b) := \#\{\gamma \in \mathbb{F}_{p^s} : Tr_{\mathbb{F}_{p^s}}(f(\gamma)) = b\},$$

$$\chi^{(s)}(b) := \chi(Tr_{\mathbb{F}_{p^s}}(b)).$$

Appendix summary:

Figure E.1 shows how the lemmas contribute to the proof of the main Theorem E.9, which is Weil's bound for polynomial sums using a nontrivial additive character modulo a prime p. This applies for example to the character $\mathbf{e}_p$ of Chapter 8:

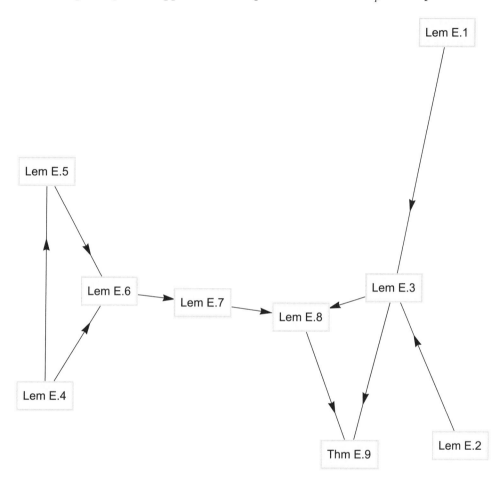

Figure E.1 Dependencies for Appendix E.

We begin with the proof of a formula, attributed to Waring, which gives a formula for the recursive powers $P_k = x_1^k + \cdots = x_n^k$ in terms of the elementary symmetric functions σ_j.

Lemma E.1 (**Waring's formula**) *Let R be a commutative ring with identity and let $\sigma_1, \ldots, \sigma_n$ be the elementary symmetric polynomials in the variables $x_1, \ldots, x_n$. Let $P_0 = n$, and for $k \geq 1$ let $P_k = x_1^k + \cdots x_n^k$, so $P_1 = \sigma_1$. Then for all $k \geq 1$, we have*

$$P_k = \sum_{\substack{\alpha_j \geq 0 \\ \alpha_1 + 2\alpha_2 + \cdots + n\alpha_n = k}} (-1)^{\alpha_2 + \alpha_4 + \cdots} k \frac{(\alpha_1 + \alpha_2 + \cdots + \alpha_n - 1)!}{\alpha_1! \cdots \alpha_n!} \sigma_1^{\alpha_1} \cdots \sigma_n^{\alpha_n}.$$

Proof **(1)** We work in the setting of formal power series in indeterminate z with coefficients in the ring $\mathbb{Q}[x_1, \ldots, x_n]$. Now

$$\prod_{j=1}^{n}(1 - x_j z) = 1 - \sigma_1 z + \cdots + (-1)^n \sigma_n z^n.$$

Taking logarithms of both sides of this equation gives

$$\sum_{m=1}^{n} \log(1 - x_m z) = \log(1 - \sigma_1 z + \cdots + (-1)^n \sigma_n z^n). \tag{E.1}$$

We will compare the coefficients of z^k on both sides of this identity.

(2) The left-hand side of (E.1) is

$$\sum_{m=1}^{n} \sum_{j=1}^{\infty} \frac{x_m^j}{j} z^j = \sum_{j=1}^{\infty} \left(\sum_{m=1}^{n} x_m^j \right) \frac{z^j}{j},$$

so the coefficient of z^k is P_k/k.

(3) The right-hand side of (E.1) is

$$\sum_{j=1}^{\infty} \frac{1}{j} \left(\sigma_1 z - \sigma_2 z^2 + \cdots + (-1)^{n-1} \sigma_n z^n \right)^j.$$

For a given $j \geq 1$, the coefficient of z^k in $\left(\sigma_1 z - \sigma_2 z^2 + \cdots + (-1)^{n-1} \sigma_n z^n \right)^j$ is

$$\sum_{\substack{\alpha_j \geq 0 \\ \alpha_1 + \alpha_2 + \cdots + \alpha_n = j \\ \alpha_1 + 2\alpha_2 + \cdots + n\alpha_n = k}} (-1)^{\alpha_2 + \alpha_4 + \cdots} \frac{j!}{\alpha_1! \cdots \alpha_n!} \sigma_1^{\alpha_1} \cdots \sigma_n^{\alpha_n}.$$

Thus the coefficient of z^k on the right-hand side is

$$\sum_{j=1}^{\infty} \sum_{\substack{\alpha_j \geq 0 \\ \alpha_1 + \alpha_2 + \cdots + \alpha_n = j \\ \alpha_1 + 2\alpha_2 + \cdots + n\alpha_n = k}} (-1)^{\alpha_2 + \alpha_4 + \cdots} \frac{(j-1)!}{\alpha_1! \cdots \alpha_n!} \sigma_1^{\alpha_1} \cdots \sigma_n^{\alpha_n}$$

$$= \sum_{\substack{\alpha_j \geq 0 \\ \alpha_1 + 2\alpha_2 + \cdots + n\alpha_n = k}} (-1)^{\alpha_2 + \alpha_4 + \cdots} \frac{(\alpha_1 + \alpha_2 + \cdots + \alpha_n - 1)!}{\alpha_1! \cdots \alpha_n!} \sigma_1^{\alpha_1} \cdots \sigma_n^{\alpha_n}.$$

This completes the derivation. $\square$

Next we develop a tool to analyze the Φ_k by introducing a type of multiplicative representation λ and assembling them together in a power series. Lemma E.2 gives a useful expression for the derivative of the logarithm of the power series.

Recall the definitions:

Φ denotes the set of monic polynomials with coefficients in $\mathbb{F}_p$,

Φ_k denotes the elements of Φ with degree $k \geq 1$,

$\Phi_\mathscr{I}$ denotes the irreducible elements of Φ.

Lemma E.2 *Let $\lambda \colon \Phi \to \mathbb{C}$ be such that $\lambda(1) = 1$ and $\lambda(gh) = \lambda(g)\lambda(h)$ for all $g, h \in \Phi$. In addition, suppose $|\lambda(g)| \leq 1$ for all $g \in \Phi$. Define a power series, converging absolutely for $|z| < 1/p$ because $|\Phi_k| \leq p^k$, by*

$$L(z) := \sum_{k=0}^{\infty} \left(\sum_{g \in \Phi_k} \lambda(g) \right) z^k.$$

Then

$$z(\log L(z))' := \sum_{s=1}^{\infty} c_s z^s = \sum_{s=1}^{\infty} \left(\sum_{\substack{f \in \Phi_\mathscr{I} \\ \deg(f) | s}} \deg(f) \lambda(f)^{s/\deg(f)} \right) z^s.$$

Furthermore, if for some integer k_0 we have for all $k > k_0$, $\sum_{g \in \Phi_k} \lambda(g) = 0$, then $L(z)$ is a complex polynomial of degree not greater than k_0, we have $L(z) = (1 - \omega_1 z) \cdots (1 - \omega_{k_0} z)$, and $c_s = -\omega_1^s - \cdots - \omega_{k_0}^s$, $s \geq 1$.

Proof **(1)** Using unique factorization into irreducibles in $\mathbb{F}_p[x]$, we can derive

$$L(z) = \sum_{g \in \Phi} \lambda(g) z^{\deg(g)}$$

$$= \prod_{f \in \Phi_\mathscr{I}} \left(1 + \lambda(f) z^{\deg(f)} + \lambda(f^2) z^{\deg(f^2)} + \cdots \right)$$

$$= \prod_{f \in \Phi_\mathscr{I}} \left(1 + \lambda(f) z^{\deg(f)} + \lambda(f)^2 z^{2 \deg(f)} + \cdots \right)$$

$$= \prod_{f \in \Phi_\mathscr{I}} \left(1 - \lambda(f) z^{\deg(f)} \right)^{-1}.$$

(2) Next, use logarithmic differentiation of the result from Step (1), and then a geometric series expansion, to get

$$z(\log L(z))' = \sum_{f \in \Phi_{\mathscr{g}}} \frac{\lambda(f)\deg(f)}{1 - \lambda(f)z^{\deg(f)}} z^{\deg(f)}$$

$$= \sum_{f \in \Phi_{\mathscr{g}}} \left(\lambda(f)\deg(f)z^{\deg(f)} \right)$$

$$\times \left(1 + \lambda(f)z^{\deg(f)} + \lambda(f)^2 z^{2\deg(f)} + \cdots \right)$$

$$= \sum_{f \in \Phi_{\mathscr{g}}} \deg(f) \left(\lambda(f)z^{\deg(f)} + \lambda(f)^2 z^{2\deg(f)} + \cdots \right)$$

$$= \sum_{s=1}^{\infty} \left(\sum_{\substack{f \in \Phi_{\mathscr{g}} \\ \deg(f)|s}} \deg(f)\lambda(f)^{s/\deg(f)} \right) z^s.$$

(3) All that remains is to prove the last part of the lemma statement, so let $\sum_{g \in \Phi_k} \lambda(g) = 0$ for all $k > k_0$. Then returning to the definition of $L(z)$ given in the statement, we see that it is a complex polynomial of degree not greater than k_0. Since $\lambda(1) = 1$, the constant term is 1, so we can write $L(z) = (1 - \omega_1 z) \cdots (1 - \omega_{k_0} z)$ for some complex numbers ω_j, $1 \le j \le k_0$. Using this representation and shifting the index

$$z(\log L(z))' = -\sum_{j=1}^{k_0} \frac{\omega_j z}{1 - \omega_j z}$$

$$= -\sum_{i=0}^{\infty} \left(\sum_{j=1}^{k_0} \omega_j^{i+1} \right) z^{i+1}$$

$$= -\sum_{s=1}^{\infty} \left(\sum_{j=1}^{k_0} \omega_j^s \right) z^s.$$

Comparing this expression with the result of Step (2) completes the proof.   □

The following is the fundamental lemma for the Weil bound derivation. Recall the definition of a lifted character: if $\chi : \mathbb{F}_p \to \mathbb{C}$ is an additive character, then we may define an additive character on the extension field $\mathbb{F}_{p^s}$ by

$$\chi^{(s)}(\beta) := \chi\left(\mathrm{Tr}_{\mathbb{F}_{p^s}}(\beta) \right)$$

whereas before $\mathrm{Tr}_{\mathbb{F}_{p^s}}(\beta) := 1 + \beta^p + \beta^{p^2} + \cdots + \beta^{p^{s-1}} \in \mathbb{F}_p$. The trace is a $\mathbb{F}_p$ linear map from $\mathbb{F}_{p^s}$ to $\mathbb{F}_p$, and maps $\alpha \to s\alpha$ for $\alpha \in \mathbb{F}_p$.

Lemma E.3 *Let $f \in \mathbb{F}_p[x]$ have degree $n \geq 2$ such that $p \nmid n$. Let χ be a nontrivial additive character of $\mathbb{F}_p$. There exist complex numbers $\omega_1, \ldots, \omega_{n-1}$ depending on f and χ, such that for any positive integer s we have*

$$\sum_{x \in \mathbb{F}_{p^s}} \chi^{(s)}(f(x)) = -\omega_1^s - \omega_2^s - \cdots - \omega_{n-1}^s.$$

Proof (1) We use the notation of Lemma E.1. Let $f(x) = b_n x^n + \cdots b_0$ so $b_n \neq 0$. Given $k \in \mathbb{N}$, we can write

$$\sum_{j=1}^{k} f(x_j) = b_n P_n(x_1, \ldots, x_k) + \cdots + b_1 P_1(x_1, \ldots, x_k) + k b_0. \qquad (\text{E.2})$$

Then considering the expressions for the P_j given in Lemma E.1, for $j = 1, 2, \ldots$ we can write for some polynomials G_j, H_j over $\mathbb{F}_p$, splitting off the term in the sum $\alpha_1 + 2\alpha_2 + \cdots = j$ having $\alpha_j = 1$ and the remaining $\alpha_i = 0$:

$$P_1(x_1, \ldots, x_k) = +\sigma_1,$$

$$P_2(x_1, \ldots, x_k) = -2\sigma_2 + G_2(\sigma_1),$$

$$P_3(x_1, \ldots, x_k) = +3\sigma_3 + G_3(\sigma_1, \sigma_2),$$

$$P_j(x_1, \ldots, x_k) = (-1)^{j-1} j\sigma_j + G_j(\sigma_1, \ldots, \sigma_{j-1}), \ 2 \leq j \leq k,$$

$$P_j(x_1, \ldots, x_k) = H_j(\sigma_1, \ldots, \sigma_k), \ j > k.$$

Inserting these evaluations into the right-hand side of (E.2) we can write for polynomials G, H over $\mathbb{F}_p$:

$$\sum_{j=1}^{k} f(x_j) = \begin{cases} (-1)^{n-1} n b_n \sigma_n + G(\sigma_1, \ldots, \sigma_{n-1}), & x \geq n, \\ H(\sigma_1, \ldots, \sigma_k), & 1 \leq k \leq n. \end{cases} \qquad (\text{E.3})$$

(2) Next, we let Φ be the set of monic polynomials in one variable over $\mathbb{F}_p$ and define a function $\lambda \colon \Phi \to \mathbb{C}$ with $|\lambda(g)| = 1$ for all $g \in \Phi$ as follows. Set $\lambda(1) := 1$. If $g(x) = (x - \alpha_1) \cdots (x - \alpha_k)$ is the factorization of g in its splitting field over $\mathbb{F}_p$, because each $\sigma_r(\alpha_1, \ldots, \alpha_k) \in \mathbb{F}_p$ for $1 \leq r \leq k$, by (E.3) we have $f(\alpha_1) + \cdots f(\alpha_k) \in \mathbb{F}_p$ also. Define

$$\lambda(g) := \chi(f(\alpha_1) + \cdots f(\alpha_k)).$$

Given another function $h(x) = (x - \beta_1) \cdots (x - \beta_m) \in \Phi$, since χ is additive,

$$\lambda(gh) = \chi(f(\alpha_1) + \cdots f(\alpha_k) + f(\beta_1) + \cdots f(\beta_m)) = \lambda(g)\lambda(h).$$

(3) Let $\Phi_k \subset \Phi$ be the polynomials of degree k and let

$$S_k := \sum_{g \in \Phi_k} \lambda(g).$$

For the particular

$$g(x) = x^k + \sum_{j=1}^{k} (-1)^j a_j x^{k-j} = (x - \alpha_1) \cdots (x - \alpha_k) \in \Phi_k,$$

where $a_j \in \mathbb{F}_p$, we get for $1 \leq j \leq k$, $\sigma_j(\alpha_1, \ldots, \alpha_k) = a_j$. Inserting these values into (E.3) then gives

$$f(\alpha_1) + \cdots f(\alpha_k) - (-1)^{n-1} n b_n a_n + G(a_1, \ldots, a_{n-1}).$$

Now $p \nmid n \implies b := (-1)^{n-1} n b_n \neq 0$. Thus, since χ is additive, and noting the first sum in the penultimate line is 0,

$$S_k = \sum_{a_j \in \mathbb{F}_p, \ 1 \leq j \leq n} \chi(b a_n + G(a_1, \ldots, a_{n-1}))$$

$$= p^{k-n} \sum_{a_j \in \mathbb{F}_p, \ 1 \leq j \leq n} \chi(b a_n) \chi(G(a_1, \ldots, a_{n-1}))$$

$$= p^{k-n} \left(\sum_{a_n \in \mathbb{F}_p} \chi(b a_n) \right) \left(\sum_{a_j \in \mathbb{F}_p, \ 1 \leq j \leq n-1} \chi(G(a_1, \ldots, a_{n-1})) \right)$$

$$= 0.$$

Thus for all $k \geq n - 1$ we have $S_k = \sum_{g \in \Phi_k} \lambda(g) = 0$.

(4) It follows from Step (3) and Lemma E.2, using the notation in that lemma, for all $s \geq 1$,

$$c_s = -\sum_{j=1}^{n-1} \omega_j^s. \tag{E.4}$$

We can also calculate c_s from Step (2) of the lemma to get

$$c_s = \sum_{\substack{g \in \Phi_{\mathscr{g}} \\ \deg(g) \mid s}} \deg(g) \lambda \left(g^{s/\deg(g)} \right).$$

Let $\gamma \in \mathbb{F}_{p^s}$ be a root of g, so since $\deg g \mid s$, $g(\gamma) = 0$ in $\mathbb{F}_{p^s}$ and $\gamma, \gamma^p, \ldots, \gamma^{p^{s-1}}$ are conjugate in $\mathbb{F}_{p^s}$, we get

$$g(x)^{s/\deg(g)} = (x - \gamma)(x - \gamma^p) \cdots \left(x - \gamma^{p^{s-1}} \right).$$

Therefore, from the definition of λ given in Step (2), we have

$$\lambda(g^{s/\deg(g)}) = \chi\left(f(\gamma) + f(\gamma^p) + \cdots + f\left(\gamma^{p^{s-1}}\right)\right).$$

(5) Now the distinct conjugates of γ are $\gamma^p, \gamma^{p^2}, \ldots, \gamma^{p^{\deg(g)-1}}$ and we can write therefore

$$\deg(g)\lambda(g^{s/\deg(g)}) = \sum_{\substack{\gamma \in \mathbb{F}_{p^s} \\ g(\gamma)=0}} \chi\left(f(\gamma) + f(\gamma^p) + \cdots + f\left(\gamma^{p^{s-1}}\right)\right),$$

and thus

$$
\begin{aligned}
c_s &= \sum_{g \in \Phi} \sum_{\substack{\gamma \in \mathbb{F}_{p^s} \\ g(\gamma)=0}} \chi\left(f(\gamma) + f(\gamma^p) + \cdots + f\left(\gamma^{p^{s-1}}\right)\right) \\
&= \sum_{\gamma \in \mathbb{F}_{p^s}} \chi\left(f(\gamma) + f(\gamma^p) + \cdots + f\left(\gamma^{p^{s-1}}\right)\right) \\
&= \sum_{\gamma \in \mathbb{F}_{p^s}} \chi\left(f(\gamma) + f(\gamma)^p + \cdots + f(\gamma)^{p^{s-1}}\right) \\
&= \sum_{\gamma \in \mathbb{F}_{p^s}} \chi\left(f(\gamma) + f(\gamma)^p + \cdots + f(\gamma)^{p^{s-1}}\right) \\
&= \sum_{\gamma \in \mathbb{F}_{p^s}} \chi\left(\mathrm{Tr}_{\mathbb{F}_{p^s}}(f(\gamma))\right) = \sum_{\gamma \in \mathbb{F}_{p^s}} \chi^{(s)}(f(\gamma)).
\end{aligned}
$$

Combining this expression for c_s with that given by (E.4) completes the proof. $\square$

Next, we define a specialized higher derivative for polynomials: if $n > j$, let the binomial coefficient $\binom{j}{n} := 0$. Then we set for $n \in \mathbb{N} \cup \{0\}$ and a_j in any field K,

$$D^{(n)}\left(\sum_{j=0}^{d} a_j x^j\right) := \sum_{j=0}^{d} a_j \binom{j}{n} x^{j-n},$$

and call this operator the **nth hyperderivative**. Lemma E.4 gives a product rule expansion for the hyperderivative of a product of polynomials.

Lemma E.4 *For polynomials $f_1, \ldots, f_s \in K[x]$, we have the hyperderivative product rule*

$$D^{(n)}(f_1 \cdots f_s) = \sum_{\substack{n_i \geq 0, \ 1 \leq i \leq s \\ n_1 + \cdots + n_s = n}} D^{(n_1)}(f_1) \cdots D^{(n_s)}(f_s).$$

Proof First, note that $D^{(n)}$ is linear. Thus we need only prove the given formula in case each f_j has the form $f_j(x) = x^{e_j}$ for some $e_j \in \mathbb{N}$. Then

$$(x+1)^{e_1 + \cdots + e_s} = (x+1)^{e_1} \cdots (x+1)^{e_s}.$$

Expanding each side and comparing the coefficients of x^n gives

$$\binom{e_1 + \cdots + e_s}{n} = \sum_{\substack{n_i \geq 0,\ 1 \leq i \leq s \\ n_1 + \cdots + n_s = n}} \binom{e_1}{n_1} \cdots \binom{e_s}{n_s},$$

which is equivalent to the given formula in this case. This completes the proof. □

Lemma E.5 *Let $f \in K[x]$ and let γ be a root of $D^{(n)} f(x)$ for $n = 0, 1, 2, \ldots, m - 1$. Then γ is a root of $f(x)$ with multiplicity m or greater.*

Proof Let $f(x) = a_0 + a_1(x - \gamma) + \cdots + a_d(x - \gamma)^d$. By Lemma E.4, choosing each $f_j = (x - \gamma)$ we get

$$D^{(n)}\left((x - \gamma)^s\right) = \binom{s}{n}(x - \gamma)^{s-n}.$$

Thus since $D^{(n)}$ is linear, we get

$$D^{(n)} f(x) = a_n + \binom{n+1}{n} a_{n+1}(x - \gamma) + \cdots + \binom{d}{n} a_d(x - \gamma)^{d-n}.$$

Letting $n = 0, 1, \ldots, m - 1$ and substituting $x = \gamma$, we get $a_n = 0$ in this range. Thus $(x - \gamma)^m \mid f(x)$, which completes the proof. □

Lemma E.6 is a key step in the derivation of Weil's estimate. For each polynomial $f(x) \in \mathbb{F}_{p^s}[x]$, there is an associated polynomial $h(x) \in \mathbb{F}_p[x]$ such that $h(\gamma)$ vanishes, with given multiplicity, for each γ such that the trace of $f(\gamma)$ has a given value b.

Recall the definitions:

$$\operatorname{Tr}_{\mathbb{F}_{p^s}}(x) := 1 + x^p + x^{p^2} + \cdots + x^{p^{s-1}}, \quad x \in \mathbb{N},$$
$$N(b) := \#\{\gamma \in \mathbb{F}_{p^s} : \operatorname{Tr}_{\mathbb{F}_{p^s}}(f(\gamma)) = b\},$$
$$\chi^{(s)}(b) := \chi(\operatorname{Tr}_{\mathbb{F}_{p^s}}(b)).$$

Lemma E.6 *Let $f \in \mathbb{F}_p[x]$ have degree $n \geq 1$ be such that $p \nmid n$, and let $E) = \mathbb{F}_{p^s}$ with $s \geq 3$ and $b \in \mathbb{F}_p$. Let $k = \lfloor s/2 \rfloor$ and m be a positive integer with $p \mid m$ and such that $m \deg(f) \leq p^{s-k-1}$. Then there is a polynomial $h \in \mathbb{F}_p[x]$ such that every $\gamma \in \mathbb{F}_{p^s}$ with $\operatorname{Tr}_{\mathbb{F}_{p^s}}(f(\gamma)) = b$ has $h(\gamma) = 0$ with multiplicity at least m, and*

$$\deg h \leq m p^{s-1} p^s \deg(f).$$

Proof **(1)** Let $d = \deg(f)$, $d_p = \min(d, p)$, and define

$$g(x) := f(x)^{p^k} + \cdots f(x)^{p^{s-1}} = f(x^{p^k}) + \cdots f(x^{p^{s-1}}).$$

We will search for a polynomial h which satisfies the stated degree bound having a particular form as follows. Let $e_{i,j} \in \mathbb{F}_p[x]$ with degree less than $m_p p^{s-2}$ and $u = \lfloor m/p \rfloor$. Then set

$$h(x) = \sum_{i=0}^{p-1} \sum_{j=0}^{u} e_{i,j}(x) g(x)^i x^{jp^s}.$$

Thus we can write $h(x) = k(x, x^{p^k})$, where we define

$$k(x, y) := \sum_{i=0}^{p-1} \sum_{j=0}^{u} e_{i,j}(x) \left(f(y) + f(y^p) + \cdots + f(y^{p^{s-k-1}}) \right)^i y^{jp^{s-k}}.$$

(2) Next, we **claim** that if $h(x) = v(x, x^{p^s})$ for some $s \in \mathbb{N}$ and $v \in \mathbb{F}_p[x, y]$, then for $0 \le p < p^s$, then $D^{(n)}$ is given by the nth hyperderivative of $v(x, y)$ with respect to x, followed by the substitution $y = x^{p^s}$. To see this, note that for $0 < n < p^s$ the binomial coefficient

$$\binom{p^s}{n} = \frac{p^s}{n} \binom{p^s - 1}{n - 1}$$

is divisible by p, so for n in this range $D^{(n)}(x^{p^s}) = 0$. Let $f_1(x) = x^j, \ldots,$ $f_{k+1}(x) = x^{p^s}$ and apply the formula of Lemma E.4 to get

$$D^{(n)}(h) = \binom{j}{n} x^{j-n} x^{kp^s}.$$

Calculating the nth hyperderivative of $v(x, y)$ with respect to x, followed by the substitution $y = x^{p^s}$, gives this same result, completing the proof of the claim.

(3) Now $s \le 2k + 1 \implies m \le p^{s-k-1} \le p^k$, so using the result of Step (2) with $0 \le n \le m - 1$ with $\xi_{i,j,n}(x) := D^{(n)}(e_{i,j}(x))$, we get, setting also

$$\rho_n(x) \sum_{i=0}^{p-1} \sum_{j=0}^{u} \xi_{i,j,n}(\gamma)(b - G(\gamma))^i x^j,$$

$$D^{(n)}(h))(\gamma) = \sum_{i=0}^{p-1} \sum_{j=0}^{u} \xi_{i,j,n}(\gamma)(b - G(\gamma))^i \gamma^j = \rho_n(\gamma).$$

(4) Next, we **claim** that all of the polynomials ρ_n for $0 \le n \le m - 1$ vanish. To see this, first recall $u \le m/p \le p^{k-1}$ so

$$\deg(\rho_n) < d_p p^{s-2} + d(p-1)p^{k-1} + u$$
$$\le d_p p^{s-2} + d(p-1)p^{k-1} + p^{k-1}$$
$$\le d_p p^{s-2} + dp^k.$$

If $n \ge 1$, we have $\deg(\rho_n) < d_p p^{s-2} + mp^k - 1$.... It follows from these bounds that the total number R of coefficients of the ρ_n for $n \in [0, m-1]$ satisfies

$$R < m(d_p p^{s-2} + dp^k) \le md_p p^{s-2} + p^{s-1}.$$

In addition, the number of possible coefficients N of all of the $e_{i,j}$ satisfies

$$N = p(u+1)d_p p^{s-2} = md_p p^{s-2} + d_p p^{s-1} \ge md_p p^{s-2} + p^{s-1} > R.$$

Setting each $\rho_n = 0$ for $0 \le n \le m-1$, we obtain R homogeneous linear equations in the N coefficients of the $e_{i,j}$. Since $N > R$, there are nontrivial solutions. Take one of these and insert them in the polynomials $e_{i,j}$ to get nonzero values for at least one of these. Then set

$$h(x) = \sum_{i=0}^{p-1} \sum_{j=0}^{u} e_{i,j}(x)g(x)^i x^{jp^s}.$$

By Step (3) and Lemma E.5, every γ with $\rho_n(\gamma) = 0$ which satisfies $\mathrm{Tr}_{\mathbb{F}_{p^s}}(f(\gamma)) = b$ is a root of h with multiplicity at least m.

Because $u = m/p$ and $d_p \le d$, we get

$$\deg h < d_p p^{s-2} + d(p-1)p^{s-1} + up^s < mp^{s-1} + mp^s.$$

(5) In this final step, we **claim** $h(x) \ne 0$, i.e., h is not the zero polynomial. To see this, define

$$c_{i,j}(x) := e_{i,j}(x)g(x)^i x^{jp^s} \implies \deg(g_{i,j}) = \deg(e_{i,j}) + imp^{s-1} + jp^s.$$

Thus if $c_{i,j} \ne 0$, we must have

$$\deg(e_{i,j}) < d_p p^{s-2} \le p^{s-1} \implies p^{s-1}(im+jp) \le \deg(c_{i,j}) < p^{s-1}(1+im+jp).$$

If $(i,j) \ne (i',j')$ with $i,i' \in [0, p-1]$ and $j,j' \in [0,u]$, then since $p \nmid m$, we have $im + jq \ne i'm + j'p$. Thus the degrees of the $c_{i,j}$ are distinct so we must have $h \ne 0$. This completes the proof. □

Next, keeping to the notation used in Lemma E.6, we find a good estimate for $N(b)$.

Lemma E.7 *If $f \in \mathbb{F}_p[x]$ has degree $n \in \mathbb{N}$ and $p \nmid n$, then for any finite extension $\mathbb{F}_{p^s}$ of $\mathbb{F}_p$ we have*

$$|N(b) - p^{s-1}| < 2n^2 p^{4+s/2}.$$

Proof **(1)** If n is sufficiently large so $p^s < n^2 p^4$, using $N(b) \le p^s$ the bound of the lemma follows. Thus we can assume that $p^s \ge n^2 p^4$.

(2) Let $k := \lfloor s/2 \rfloor$. This gives

$$n \le p^{(s-4)/2} \le p^{s-k-2}.$$

Thus if we also set $m := \lfloor p^{s-k-2}/n \rfloor p$, then we must have m a positive multiple of p and $mn \le p^{s-k-1}$, and so we are able to use the result of Lemma E.6, i.e., there exists a polynomial $h(x)$ such that every $\gamma \in \mathbb{F}_{p^s}$ with $\mathrm{Tr}_{\mathbb{F}_{p^s}}(f(\gamma)) = b$ has $h(\gamma) = 0$ with multiplicity at least m, so since h can have at most $\deg(h)$ roots, we must have $N(b)m \le \deg(h)$. Using the result of Lemma E.6 and the bound $m \ge p^{s-k-1}/(2n)$, we get for all $b \in \mathbb{F}_p$,

$$N(b) < p^{s-1} + \frac{np^s}{m} < p^{s-1} + 2n^2 p^{k+1}.$$

Therefore,

$$N(b) = p^s - \sum_{\substack{x \in \mathbb{F}_p \\ x \neq b}} N(x)$$

$$> p^s - (p-1)p^{s-1} - 2(p-1)n^2 p^{k+1}$$

$$> p^{s-1} - 2n^2 p^{k+2}.$$

Combining these bounds, we get

$$|N(b) - p^{s-1}| < 2n^2 p^{k+2} \le 2n^2 p^{2+s/2}.$$

This completes the proof. □

In the next step, we use the result of Lemma E.7 to derive an excellent bound for the complex numbers ω_j, which were introduced in Lemma E.3.

Lemma E.8 *The complex numbers ω_j which appear in the statement of Lemma E.3 each satisfy $|\omega_j| \le \sqrt{p}$.*

Proof **(1)** First, we **claim** that if B, C are positive real constants such that for all $s \in \mathbb{N}$, we have

$$|\omega_1^s + \cdots + \omega_n^s| \le CB^s. \tag{E.5}$$

We **claim** for $1 \leq j \leq n$, we have $|\omega_j| \leq B$. To see this, let $z \in \mathbb{C}$ be such that for all j, $|z\omega_j| < 1$. Then

$$-\log((1 - \omega_1 z)(1 - \omega_2 z)(\cdots)(1 - \omega_n z)) = \sum_{s=1}^{\infty} \frac{1}{s}(\omega_1^s + \cdots + \omega_n^s)z^s.$$

By the hypothesis, the series on the right converges when $|z| < 1/B$, so the function on the left is holomorphic on that disk. If for some j we had $|\omega_j| > B$, we could find a z in the disk with $1 = \omega_j z$, which is false. Thus for all j we have $|\omega_j| \leq B$, which completes the proof of the claim.

(2) By the definition of the character lifted to $\mathbb{F}_{p^s}$, we have

$$\sum_{\gamma \in \mathbb{F}_{p^s}} \chi^{(s)}(f(\gamma)) = \sum_{\gamma \in \mathbb{F}_{p^s}} \chi(\mathrm{Tr}_{\mathbb{F}_{p^s}}(f(\gamma))) = \sum_{b \in \mathbb{F}_p} N(b)\chi(b).$$

Write $N(b) - p^{s-1} = R(b)$, so by Lemma E.7, $|R(b)| \leq 2n^2 p^{4+s/2}$. Therefore,

$$\left| \sum_{\gamma \in \mathbb{F}_{p^s}} \chi^{(s)}(f(\gamma)) \right| = \left| \sum_{b \in \mathbb{F}_p} (p^{s-1} + R(b))\chi(b) \right|$$

$$= \left| \sum_{\gamma \in \mathbb{F}_{p^s}} \chi^{(s)}(f(\gamma)) \right| = \left| \sum_{b \in \mathbb{F}_p} R(b)\chi(b) \right|$$

$$\leq \sum_{b \in \mathbb{F}_p} |R(b)| \leq 2n^2 p^{5+s/2}.$$

By Lemma E.3, we have for all $s \in \mathbb{N}$,

$$|\omega_1^s + \cdots + \omega_{n-1}^s| \leq 2n^2 p^{5+s/2}.$$

Therefore, by Step (1) with $B = p^{\frac{1}{2}}$ we have $|\omega_j| \leq \sqrt{p}$ for $1 \leq j \leq n - 1$. This completes the proof. $\square$

Having proved the fundamental lemma and found an upper bound for the $|\omega_j|$, we are now able to derive the Weil bound easily using Lemmas E.3 and E.8:

Theorem E.9 **(Weil bound)** *Let $f \in \mathbb{F}_p[x]$ have $n = \deg f \geq 1$ and suppose also that $p \nmid n$. Let χ be a nontrivial additive character of $\mathbb{F}_p$. Then*

$$\left| \sum_{x \in \mathbb{F}_p} \chi(f(x)) \right| \leq (n-1)\sqrt{p}.$$

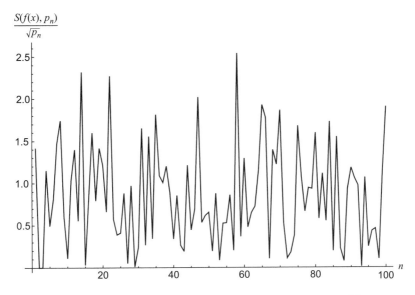

Figure E.2 Weil sums over $\sqrt{p_n}$ for $\mathbf{e}_{p_n}$ and $1 \le n \le 100$.

Proof If $n = 1$, write $f(x) = a + bx$ so

$$\left| \sum_{x \in \mathbb{F}_p} \chi(a + bx)) \right| = |\chi(a)| \left| \sum_{x \in \mathbb{F}_p} \chi(x)) \right| = 0 \le (n-1)\sqrt{p},$$

and we can assume $n \ge 2$. Then by Lemma E.3, we get complex numbers ω_j such that

$$\sum_{x \in \mathbb{F}_p} \chi(f(x)) = -\omega_1 - \cdots - \omega_{n-1},$$

so applying Lemma E.8, we get

$$\left| \sum_{x \in \mathbb{F}_p} \chi(f(x)) \right| \le (n-1)\sqrt{p}.$$

This completes the proof.   □

Figure E.2 is an example of Weil sums divided by $\sqrt{p_n}$ for $f(x) = 1 + x + x^5$ for the first 100 primes p_n.

Appendix F

Complex Function Theory

In Chapter 4, we need an expansion for the logarithmic derivative of $\zeta(s)$ on a disk. This can be derived using some standard techniques of complex function theory, namely the Schwarz lemma, the Borel–Carathéodory lemma, Jensen's formula and the logarithmic derivative estimate in terms of a function's zeros. These results are stated and proved in this appendix. References would include [1, 185].

> **Theorem F.1** **(Schwarz lemma)** *Let $f(s)$ be holomorphic inside the open unit disk $\Omega := B(0, 1)$ with $f(0) = 0$ and such that $|f(s)| < 1$ for all $s \in \Omega$. Then $|f(s)| \leq |s|$ for all $s \in \Omega$ and $|f'(0)| \leq 1$. In addition, if $|f(s)| = |s|$ for some $s \neq 0$ or $|f'(0)| = 1$, then $f(s) = e^{i\theta} \cdot s$ for some fixed θ and all $s \in \Omega$.*

Proof Since $f(0) = 0$, we can write $f(s) = sg(s)$, where $g(s)$ is holomorphic on Ω and, from the Taylor series, for $f(s)$, $g(0) = f'(0)$. Let $0 < r < 1$. By the maximum modulus principle applied to $B(0, r)$, there exists an s_r with $|s_r| = r$ such that for all s with $|s| \leq r$, we have

$$|g(s)| \leq |g(s_r)| = \frac{|f(s_r)|}{|s_r|} < \frac{1}{r}.$$

Take the limit as $r \to 1-$ to get $|g(s)| \leq 1$. If we had either of the additional conditions, then this would give $|g(s)| = 1$ at some point in Ω. By the maximum modulus principle again, this implies $g(s)$ is constant on Ω, and then this constant must have modulus 1, completing the proof. □

> **Theorem F.2** **(Borel–Carathéodory)** *Let $f(s)$ be holomorphic on the closed disk $F := B(0, R]$ and let $0 < r < R$. Then*

$$\max\{|f(s)| : |s| \leq r\} \leq \left(\frac{2r}{R - r}\right) \max\{\Re f(s) : |s| \leq R\} + \left(\frac{R + r}{R - r}\right) |f(0)|.$$

Proof First, assume $f(0) = 0$ and let $M := \max\{\Re f(s) : |s| \leq R\}$. Assume $M > 0$ (else replace f by $-f$) and let $H := \{s : \Re s \leq M\}$ so $f(F) \subset H$. Define

$$g(z) := \frac{Rz}{z - 2M},$$

which is the composite of $z \to z/M - 1$ and $z \to R(z+1)/(z-1)$, so $g(0) = 0$ and g maps H into the disk $B(0, R)$.

Next, apply Theorem F.1 to the composite $f \circ g$ to get, for $|z| \leq r$,

$$\frac{|Rf(z)|}{|f(z) - 2M|} \leq |z| \implies R|f(z)| \leq r|f(z) - 2M|$$

$$\leq r|f(z)| + 2Mr \implies |f(z)| \leq \frac{2Mr}{R - r}.$$

Finally, replace $f(z)$ by $f(z) - f(0)$ to get

$$|f(z)| - |f(0)| \leq |f(z) - f(0)| \leq \frac{2r}{R - r} \max\{\Re(f(z) - f(0)) : |z| \leq R\}$$

$$\leq \frac{2r}{R - r} (\max\{\Re f(z)) : |z| \leq R\} + |f(0)|),$$

and the result follows. □

Assuming $f(0) = 0$, we can also get a useful bound for the first derivative, assuming the real part of $f(z)$ is bounded.

Theorem F.3 *Let $f(z)$ be holomorphic in $B(0, R]$ and satisfy $\Re f(z) \leq M$. Then if $0 < r < R$ on $B(0, r]$, we get*

$$|f'(z)| \leq \frac{2Mr}{(R - r)^2}.$$

Proof First, we have by Cauchy's integral formula

$$\int_0^1 f(Re^{2\pi i u})\, du = \frac{1}{2\pi i} \int_{|z|=R} \frac{f(z)}{z}\, dz = f(0) = 0.$$

If $n > 0$, since $f(z)z^{n-1}$ is holomorphic inside and on the disk, we get

$$\int_0^1 f(Re^{2\pi i u})e^{2\pi i n u}\, du = \frac{1}{2\pi i R^n} \int_{|z|=R} \frac{f(z)}{z} z^{n-1}\, dz = 0.$$

By Cauchy's formula for the nth derivative, we get

$$\int_0^1 f(Re^{2\pi i u})e^{-2\pi i n u}\, du = \frac{R^n}{2\pi i} \int_{|z|=R} \frac{f(z)}{z} z^{-n-1}\, dz = \frac{R^n f^{(n)}(0)}{n!}.$$

Hence, for any real ϕ,

$$\int_0^1 f(Re^{2\pi iu})(1 + \cos(2\pi(nu + \phi)))\, du = \frac{R^n e^{-2\pi i\phi} f^{(n)}(0)}{2(n!)}.$$

Taking the real part of each side, we get

$$\Re\left(\frac{1}{2}\frac{R^n e^{-2\pi i\phi} f^{(n)}(0)}{2(n!)}\right) \le M \int_0^1 (1 + \cos(2\pi(nu + \phi)))\, du = M.$$

Choosing ϕ so that $e^{-2\pi i\phi} f^{(n)}(0) = |f^{(n)}(0)|$, we get $|f^{(n)}(0)|/n! \le 2M/R^n$. Therefore,

$$|f'(z)| \le \sum_{n=1}^{\infty} \frac{|f^{(n)}(0)|nr^{n-1}}{n!} \le \frac{2M}{R}\sum_{n=1}^{\infty} n\left(\frac{r}{R}\right)^{n-1} = \frac{2MR}{(R-r)^2}.$$

This completes the proof. $\qquad\square$

Theorem F.4 **(Jensen's formula)** *Let $f(z)$ be holomorphic on an open neighbourhood of the closed disk $B(0, R]$ with $R > 0$ and let $f(0) \neq 0$. Suppose $f(z)$ has no zero on the boundary $|z| = R$. Then*

$$\frac{1}{2\pi}\int_0^{2\pi} \log|f(Re^{i\theta})|\, d\theta = \log|f(0)| + \sum_{|\alpha|<R,\ f(\alpha)=0} m_\alpha \log\frac{R}{|\alpha|}, \qquad (F.1)$$

where m_α is the multiplicity of the zero α.

Proof For $r > 0$, let C_r be the circle of radius r centre 0. We begin by assuming there are no zeros on C_R. Write

$$I := \log|f(Re^{i\theta})| = \Re\log\left(f\left(Re^{i\theta}\right)\right)$$
$$= \Re\left(\log f(0) + \int_0^R \frac{d}{dr}\left(\log f\left(re^{i\theta}\right)\right) dr\right)$$
$$= \log|f(0)| + \Re\int_0^R \frac{f'\left(re^{i\theta}\right)e^{i\theta}}{f\left(re^{i\theta}\right)}\, dr.$$

Next, integrate both sides with respect to θ to get

$$\frac{1}{2\pi}\int_0^{2\pi} \log|f(re^{i\theta})|\, d\theta = \log|f(0)| + \Re\frac{1}{2\pi}\int_0^{2\pi}\left(\int_0^R \frac{f'\left(re^{i\theta}\right)e^{i\theta}}{f\left(re^{i\theta}\right)}\, dr\right) d\theta$$
$$= \log|f(0)| + \Re\int_0^R \frac{1}{2\pi ir}\left(\int_0^{2\pi} \frac{f'\left(re^{i\theta}\right)re^{i\theta}}{f\left(re^{i\theta}\right)}\, d\theta\right) dr$$
$$= \log|f(0)| + \Re\int_0^R \frac{1}{2\pi ir}\left(\int_{C_r} \frac{f'(z)}{f(z)}\, dz\right) dr.$$

By the principle of the argument,

$$n(r) := \frac{1}{2\pi i} \int_{C_r} \frac{f'(z)}{f(z)} \, dz = \sum_{f(\alpha)=0, \, |\alpha|<r} m_\alpha.$$

Therefore,

$$I = \log |f(0)| + \int_0^R \frac{n(r)}{r} \, dr.$$

Now let the zeros of $f(z)$ in the disk $B(0, R)$ be $\alpha_1, \ldots, \alpha_n$ and set $|\alpha_j| = r_j$, ordered with $0 < r_1 \le r_2 \le \cdots \le r_n$. Also let $r_0 = 0$, $r_{n+1} = R$. Then $n(r)$ is constant on each interval (r_j, r_{j+1}) for $0 \le j \le n$. Let this constant value be denoted n_j, so $n + 0 = 0$. Then

$$I = \log |f(0)| + \sum_{j=0}^n \int_{r_j}^{r_{j+1}} \frac{n(r)}{r} \, dr$$

$$= \log |f(0)| + \sum_{j=0}^n \int_{r_j}^{r_{j+1}} \frac{n_j}{r} \, dr$$

$$= \log |f(0)| + \sum_{j=1}^n n_j (\log(r_{j+1}) - \log(r_j)) - n_1 \log r_1$$

$$= \log |f(0)| - \left(\sum_{|\alpha|<R, \, f(\alpha)=0} m_\alpha \log |\alpha| \right) + n(R) \log R$$

$$= \log |f(0)| + \sum_{|\alpha|<R, \, f(\alpha)=0} m_\alpha \log \frac{R}{|\alpha|}.$$

This completes the proof. $\square$

Note that if $f(z)$ has a zero of order m at $z = 0$, the formula can be applied to $f(z)/z^m$. If the function has zeros on the boundary, the formula remains true with these zeros included – see the argument given in [185, lemma 15.17, theorem 15.18]. However, the formula is insensitive to zeros on the boundary – they contribute zero to both sides. Also, it can easily be extended to include poles, since it uses the principle of the argument. There is nothing special regarding the point zero – simply consider $f(z + z_0)$ on $B(z_0, R]$.

> **Lemma F.5** (Logarithmic derivative estimate) *Let $f(z)$ be holomorphic on the disk $B(0, 1]$ and bounded with $|f(z)| \le M$ on that disk. Let $f(0) \ne 0$ and $0 < r < R < 1$. Then on $B(0, r]$, if the (finite) set of zeros of $f(z)$ in $B(0, R]$ are labelled $z_1, \ldots, z_N$, then we have the estimate*

$$\frac{f'(z)}{f(z)} = \sum_{n=1}^{N} \frac{1}{z - z_n} + O_{r,R}\left(\log \frac{M}{\log |f(0)|}\right),$$

where the implicit constant depends only on the radii r and R.

Proof **(1)** If $f(z)$ has (at most a finite number of) zeros on $|z| = R$, we can slightly increase the value of R and then assume in what follows $f(z) \neq 0$ for $|z| = R$.

First, note from Jensen's formula Theorem F.4 that the number of zeros in the disk N satisfies

$$N \leq \frac{\log M/|f(0)|}{\log 1/R} \ll \log \frac{M}{|f(0)|}.$$

(2) Next, define

$$g(z) := f(z) \prod_{n=1}^{N} \frac{R^2 - z\overline{z_n}}{R(z - z_n)}.$$

On the circle $|z| = R$, the product has modulus 1, so $|g(z)| \leq M$. Since $g(z)$ is holomorphic inside and on the disk $B(0, R]$, by the maximum modulus principle, we have $|g(z)| \leq M$ for all $z \in B(0, R]$.

(3) Now note that

$$|g(0)| = |f(0)| \prod_{n=1}^{N} \frac{R}{|z_n|} \geq |f(0)|$$

and that $g(z)$ has no zeros in $B(0, R]$. Therefore, if we define $h(z):=\log(g(z)/g(0))$, we get $h(0) = 0$ and for $z \in B(0, R]$

$$\Re h(z) = \log|g(z)| - \log|g(0)| \leq \log M - \log|f(0)| = \log\left(\frac{M}{|f(0)|}\right).$$

Using Theorem F.2 applied to $h(z)$, on $B(0, r]$, and then Cauchy's integral theorem for the first derivative, we get the bound $h'(z) \ll \log M/|f(0)|$. Since for $|z| \leq r$, we have $|z - R^2/\overline{z_n}| \geq R - r$, and using the bound from Step (1), we also have

$$h'(z) = \frac{g'(z)}{g(z)} = \frac{f'(z)}{f(z)} - \sum_{n=1}^{N} \frac{1}{z - z_n} - \sum_{n=1}^{N} \frac{1}{z - R^2/\overline{z_n}}$$

$$= \frac{f'(z)}{f(z)} - \sum_{n=1}^{N} \frac{1}{z - z_n} + O\left(\frac{N}{R - r}\right)$$

$$= \frac{f'(z)}{f(z)} - \sum_{n=1}^{N} \frac{1}{z - z_n} + O\left(\log \frac{M}{|f(0)|}\right).$$

Therefore,

$$\frac{f'(z)}{f(z)} = \sum_{n=1}^{N} \frac{1}{z - z_n} + O\left(\log \frac{M}{|f(0)|}\right),$$

which completes the proof. ☐

Appendix G

The Dispersion Method of Linnik

Overview: we want to solve an equation of the form

$$N = a_n + Db,$$

where a_n are given positive integers and there are given restrictions on D and b. We assume there is an approximation to a main term for the number of solutions given N and D, denoted $A(N, D)$. The aim of the method is to show that, in particular circumstances, $A(N, D)$ leads to a correct asymptotic expression as $N \to \infty$ for the number of solutions. It could be arrived at in various ways and is not properly part of the method. We examine the error in the approximation for the number of solutions for given N, namely

$$\sum_{D \in \mathscr{D}} A(N, D),$$

by looking at an expression close to its standard deviation, namely its **dispersion**.

(1) First, we define the dispersion. Let $(a_n)_{n \in \mathbb{N}}$ be a sequence of positive integers. Let E, F, L, M be positive integers, $\mathscr{D} \subset [E, E + F]$ and $\mathscr{B} \subset [L, L + M]$. Let $D \in \mathscr{D}$ and $b \in \mathscr{B}$ and for each $m \in \mathbb{N}$ set

$$R(m) = \#\{n : a_n = m\}.$$

We define the dispersion, or the dispersion of (a_n), for $N > (E + F)(L + M)$, to be

$$\mathrm{Dis}(N) := \sum_{D \in \mathscr{D}} \left(\left(\sum_{b \in \mathscr{B}} R(N - Db) \right) - A(N, D) \right)^2.$$

Now the restrictions on D and b are somewhat constrained – see Step (6) below – and sometimes Linnik uses notation which is confusing (at least for this author). A priori, we would have $Db \leq N$. The method is best understood in association with an example application. We list Linnik's examples from [127] following the description of the simplest possible case. These show that the method is really a template, with much work still to be done in particular cases to ensure its success.

To see how this is intended to work, consider the exact number of solutions to $N = a_n + Db$. Using Cauchy–Schwarz to bound the error this is

$$\sum_{b \in \mathcal{B}} \sum_{D \in \mathcal{D}} R(N - Db) = \sum_{D \in \mathcal{D}} \left(\left(\sum_{b \in \mathcal{B}} R(N - Db) \right) - A(N, D) \right) + |\mathcal{B}| \sum_{D \in \mathcal{D}} A(N, D)$$

$$= |\mathcal{B}| \sum_{D \in \mathcal{D}} A(N, D) + error, \text{ where}$$

$$error \leq \sum_{D \in \mathcal{D}} \left| \left(\sum_{b \in \mathcal{B}} R(N - Db) \right) - A(N, D) \right|$$

$$\leq \sqrt{\text{Dis}(N)} \sqrt{|\mathcal{D}|}.$$

(2) Next, we bound $\text{Dis}(N)$. First, we enlarge the sum and then expand the square to obtain

$$\text{Dis}(N) \leq \sum_{E \leq D \leq E+F} \left(\sum_{b \in \mathcal{B}} R(N - Db) - A(N, D) \right)^2$$

$$= V_3 - 2V_2 + V_1, \text{ where}$$

$$V_1 := \sum_{E \leq D \leq E+F} A(N, D)^2,$$

$$V_2 := \sum_{E \leq D \leq E+F} A(N, D) \sum_{b \in \mathcal{B}} R(N - Db),$$

$$V_3 := \sum_{E \leq D \leq E+F} \left(\sum_{b \in \mathcal{B}} R(N - Db) \right)^2.$$

In the following steps, we find conditions under which asymptotic expressions can be derived for V_1, V_2, V_3.

(3) For the sum V_1, we assume a good approximation can be derived using standard techniques. If not, of course, the method fails. For V_2, we first write

$$V_2 = \sum_{b \in \mathcal{B}} \sum_{E \leq D \leq E+F} A(N, D) R(N - Db).$$

We note that

$$\sum_{E \leq D \leq E+F} R(N - Db)$$

is the number of solutions of the congruence $x \equiv N \bmod b$ with x an element of the sequence (a_n). This sum needs to be modified by the factor $A(N, D)$. To complete the derivation for V_2, we might need to make assumptions such as $L + M \leq N^{\frac{1}{2} - \epsilon}$ and $F > N^{\frac{1}{2} + \epsilon}$.

(4) The derivation for V_3 is often the most difficult step. First, assume we can readily find an upper bound for the diagonal component

$$\sum_{E \leq D \leq E+F} \sum_{b \in \mathcal{B}} R(N - Db)^2.$$

We are left with

$$V_3' := \sum_{E \leq D \leq E+F} \sum_{\substack{b_1 \neq b_2 \\ b_1, b_2 \in \mathcal{B}}} R(N - Db_1) R(N - Db_2).$$

(5) To evaluate the sum from Step (4), we first compute the number of solutions to the system

$$N - Db_1 = x_2,$$
$$N - Db_2 = x_1,$$
$$D = \frac{N - x_2}{b_1} = \frac{N - x_1}{b_2}, \tag{G.1}$$

with x_1, x_2 terms of the sequence (a_n) and $E \leq D \leq E + F$.

Let $d = (b_1, b_2)$ be the GCD and divide the set of pairs $\{b_1, b_2\}$ into subsets labelled by their GCD. Then we **claim** that for a given pair with a given GCD d, the system is equivalent to

$$b_1 x_1 - b_2 x_2 = N(b_1 - b_2) \text{ with } b_2 \mid N - x_1 \text{ and } \frac{N - x_1}{b_2} \in [E, E + F]. \tag{G.2}$$

To see this, if x_1, x_2 satisfy the system (G.1) for some D then, eliminating D, they also satisfy the system (G.2), and the number of all such solutions with x_i in the sequence (a_n) is the same. If x_1, x_2 satisfy (G.2) and we set $b_1 = db_1'$, $b_2 = db_2'$, then $(b_1', b_2') = 1$ and

$$b_1' x_1 - b_2' x_2 = N(b_1' - b_2'). \tag{G.3}$$

Thus

$$(N - x_1)b_1' = (N - x_2)b_2' \implies N \equiv x_1 \bmod b_2' \text{ and } N \equiv x_2 \bmod b_1'.$$

If $d = 1$, this shows $N - x_1)/b_2$ is an integer. If $d > 1$, by (G.3) and

$$\frac{N - x_1}{b_2} = \frac{N - x_2}{b_1},$$

their common value gives a unique integer D, which satisfies (G.1). Thus D is uniquely determined by x_1, x_2, which completes the proof of the claim.

(6) It is useful to rewrite (G.2) in the form

$$b_1(N - x_1) = b_2(N - x_2) \text{ with } b_2 \mid N - x_1 \text{ and } \frac{N - x_1}{b_2} \in [E, E + F], \quad \text{(G.4)}$$

which is regarded as the key underlying equation for the dispersion method. An asymptotic equation of the number of solutions to this system leads to an estimate for V_3. Since we have at best an upper bound V_3', we need the ratio V_3'/V_1 to be sufficiently small, in particular asymptotically smaller than the supposed main term.

(7) Now we consider the variables more closely to describe situations in which the method might be effective. First, we had $D \in \mathscr{D} \subset [E, E + F]$. We restrict $\mathscr{D}$ to be subsets which are quite dense, indeed at least $\#\mathscr{D} \gg F/(\log N)^{C_1}$, where we use the notation C_i for positive constants. The subset $\mathscr{B}$ is allowed to be sparse.

Suppose we have successfully counted the solutions to (G.4) and found that for some $C_2 > 0$ we have

$$V_3' \leq \mathrm{Dis}(N) \leq \frac{V_3}{(\log N)^{C_2}}, \quad \text{(G.5)}$$

with C_2 sufficiently large. Then we **claim** we are able to say that the equation

$$N = a_n + Db, \ D \in \mathscr{D}, \ b \in \mathscr{B}$$

is solvable and we have at hand an asymptotic formula for the number of solutions.

To see this first, consider solvability. If it is false, then $R(N - Db) = 0$ for all $D \in \mathscr{D}$ and $b \in \mathscr{B}$. Then by the definition of discrepancy, we would get

$$\mathrm{Dis}(N) = \sum_{D \in \mathscr{D}} A(N, D)^2.$$

Because $\mathscr{D}$ is sufficiently dense, if $A(N, D)$ changes sufficiently slowly as a function of D, this contradicts Equation (G.5), proving the first part of the claim.

To demonstrate the second part of the claim, Linnik resorts to the argument used by Chebychev in his demonstration of the Law of Large Numbers. We choose those D which occur in the sum for V_3' which satisfy

$$\left| \sum_{b \in \mathscr{B}} R(N - Db) - A(N, D) \right| \geq \frac{A(N, D)}{(\log N)^2}.$$

The estimate (G.5) implies the number of such D is relatively small, so it is sufficient for finding for each of these a weak upper bound for

$$\sum_{L \leq b \leq L+M} R(N - Db).$$

The number of solutions to $N = a_n + Db$, namely

$$\sum_{D \in \mathscr{D}} \sum_{b \in \mathscr{B}} R(n - Db),$$

will differ by a relatively small amount from $\sum_{D \in \mathscr{D}} A(N, D)$, which will be the sought-after asymptotic solution.

Linnik, in his monograph [127], goes on to describe variations and enhancements to the basic simple template that we have just outlined. For example, the situation when D and b are related, $Db \leq N$ say. The monograph is rich with examples and references to problems wherein he and others have used, or attempted to use, dispersion and other methods to verify an asymptotic main term in an approximation. We have summarized a selection of these in the following. In each case, dispersion was successful.

Linnik's applications

(1) An additive divisor problem: [127, theorem 3.2.1] For $k \in \mathbb{N}$, let $\tau_k(n)$ be the number of representations of the positive integer n as k factors with order counting, so in particular $\tau(n) = \tau_2(n)$. For all integers $k \geq 2$, if we define

$$S_k := \lim_{T \to \infty} \frac{1}{(\log T)^{k-1}} \int \cdots \int_{1 \leq t_1 \leq \cdots \leq T/(t_1 \cdots t_{k-1})} \frac{dt_1 \cdots dt_{k-1}}{t_1 \cdots t_{k-1}} \text{ and}$$

$$A_k := \sum_{n \in \mathbb{N}} \prod_{p \mid n} p \left(1 - \left(1 - \frac{1}{p} \right)^{k-1} \right),$$

then using the dispersion method it can be shown that as $N \to \infty$, we have

$$\sum_{n \leq N} \tau(n + 1)\tau_k(n) = k! \, A_k \, S_k \times n(\log n)^k + O\left(n(\log n)^{k-1}(\log\log n)^{k^4} \right).$$

(2) Norms in different algebraic number fields: [127, chapter V, section 1] Let E be an imaginary quadratic number field with discriminant e and F a number field of arbitrary degree and discriminant f. Let n be such that $(n, 2ef) = 1$ and such that it is not divisible by small primes. Then using dispersion, it can be shown that the number of solutions $Q(n)$ to

$$N(a) + N(b) = n,$$

where a and b are elements of given integral ideals in E and F respectively, can be written

$$Q(n) = \mathfrak{S}(E, F, n)n + O\left(\frac{n(\log\log n)^2}{\log n}\right),$$

where $\mathfrak{S}(E, F, n)$ is a singular series associated with E and F.

(3) Quadratic form plus a P_2 almost prime [127, theorem 6.1.1] Let $Q(n)$ be the number of solutions to

$$n = x^2 + y^2 + p_1 p_2 \text{ with } p_1, p_2 > \exp\left((\log\log n)^2\right).$$

Then using dispersion, it can be shown that

$$Q(n) = (1 + o(1))\pi A_0 \times n \frac{\log\log n}{\log n} \prod_{p|n} \frac{(p-1)(p - \chi_4(p))}{p^2 - p + \chi_4(p)}, \text{ where}$$

$$A_0 = \prod_{p}\left(1 + \frac{\chi_4(p)}{p(p-1)}\right), \text{ and}$$

$\chi_4(p) := (-4|p)$ is the Kronecker symbol.

(4) The Titchmarsh divisor problem [127, theorem 8.2.1] Consider the equation

$$p - xy = 1, \ xy \leq N.$$

Using the method, one can show that as $N \to \infty$, we have for $\epsilon = 1/10^4$,

$$\sum_{p \leq N} \tau(p - 1) = \frac{315\zeta(3)}{2\pi^4}n + O\left(\frac{n}{(\log n)^{1-\epsilon}}\right).$$

A conjecture of Euler

We conclude this appendix by stating a not-so-well-known conjecture of Euler which might be accessible using the dispersion method. For every positive integer N which is congruent to 3 modulo 8, there is a square such that

$$N = x^2 + 2p,$$

where p is prime.

Appendix H

One Thousand Admissible Tuples

Some readers may find it useful to consider a set of what we call "inductive tuples". These are a squence of tuples $(I_n)_{n\in\mathbb{N}}$ such that $I_1 = \{0\}$ and for $k \in \mathbb{N}$ and all positive integers $l < k$, I_l is the l term initial segment of I_k. Thus if we have constructed $I_n = \{h_1, \ldots, h_n\}$, to obtain a tuple which is admissible with smallest width and having one more additional element to I_n, all we need do is find the minimum integer $h_{n+1} > h_n$ which is such that for all primes $p \leq n + 1$, there is an $a_p \notin I_n \bmod p$, then $h_{n+1} \not\equiv a_p \bmod p$ also. The PGpack function `InductiveTuple` for an explicit k returns the kth tuple computed using this algorithm.

To save space, we include in this appendix the offsets $a_n = h_n - h_{n-1}$, $n \geq 2$ setting $a_1 = 0$, and include this in PGpack as `InductiveTuplesOffsets1000` or `ito`.

$\{0, 2, 4, 2, 4, 6, 2, 6, 4, 2, 4, 6, 6, 2, 6, 6, 6, 4, 6, 8, 4, 6, 2, 4,$

$8, 6, 4, 8, 4, 6, 2, 6, 6, 4, 2, 4, 6, 8, 6, 4, 2, 10, 2, 10, 2, 4,$

$14, 10, 2, 4, 6, 8, 6, 4, 2, 6, 4, 6, 18, 2, 4, 8, 16, 2, 4, 6, 2,$

$12, 10, 8, 6, 6, 4, 14, 10, 2, 4, 2, 10, 12, 2, 10, 6, 2, 6, 12, 4,$

$6, 8, 4, 2, 4, 14, 6, 12, 4, 6, 14, 6, 4, 8, 6, 4, 6, 2, 10, 2, 10,$

$12, 6, 2, 6, 4, 12, 8, 6, 16, 6, 14, 4, 6, 8, 6, 10, 2, 4, 14, 6, 16,$

$2, 12, 6, 10, 14, 6, 10, 2, 10, 6, 6, 6, 2, 10, 12, 14, 4, 6, 2, 22,$

$6, 2, 4, 8, 6, 6, 4, 2, 10, 12, 2, 4, 26, 4, 2, 4, 6, 2, 6, 6, 4, 24,$

$8, 4, 6, 8, 4, 2, 6, 4, 14, 4, 6, 2, 10, 2, 6, 12, 4, 6, 2, 12, 6,$

$22, 6, 12, 2, 10, 6, 6, 6, 2, 22, 6, 8, 6, 10, 8, 6, 22, 2, 6, 6, 4,$

$12, 2, 6, 6, 16, 20, 10, 12, 2, 16, 14, 6, 4, 18, 6, 6, 8, 10, 12, 6,$

$2, 6, 12, 4, 8, 6, 4, 20, 10, 2, 4, 20, 6, 4, 6, 6, 14, 6, 4, 14, 4,$

8, 12, 6, 4, 6, 6, 6, 14, 4, 12, 2, 12, 4, 2, 10, 2, 12, 4, 2, 4, 14,

6, 4, 20, 12, 22, 2, 4, 2, 12, 10, 18, 8, 10, 6, 8, 6, 4, 2, 4, 2,

10, 2, 10, 2, 6, 4, 18, 6, 12, 12, 2, 6, 10, 12, 2, 10, 8, 10, 6, 2,

12, 6, 6, 4, 6, 18, 6, 26, 6, 4, 20, 10, 6, 2, 6, 16, 6, 8, 4, 8, 4,

18, 12, 8, 6, 6, 4, 2, 10, 8, 10, 12, 20, 4, 20, 4, 14, 6, 4, 2, 10,

6, 18, 2, 4, 14, 12, 10, 14, 10, 8, 6, 10, 2, 12, 12, 6, 4, 8, 6, 6,

10, 14, 4, 2, 6, 22, 6, 2, 22, 6, 6, 6, 2, 18, 4, 18, 2, 6, 10, 2,

10, 18, 8, 4, 2, 4, 14, 6, 6, 24, 6, 4, 6, 18, 6, 6, 14, 6, 6, 12,

12, 4, 12, 12, 18, 8, 4, 12, 6, 2, 10, 12, 14, 4, 2, 18, 16, 6, 26,

4, 6, 12, 6, 12, 2, 12, 4, 12, 2, 6, 10, 6, 6, 14, 6, 4, 20, 10, 18,

8, 12, 6, 6, 4, 6, 6, 12, 2, 28, 8, 10, 8, 10, 14, 6, 4, 2, 16, 12,

6, 6, 14, 12, 4, 8, 6, 6, 4, 2, 10, 2, 28, 12, 2, 4, 2, 4, 20, 4, 24,

2, 16, 12, 6, 2, 18, 10, 6, 8, 6, 10, 2, 10, 2, 10, 2, 16, 18, 12, 2,

12, 4, 6, 6, 2, 6, 6, 6, 4, 18, 8, 4, 18, 6, 2, 18, 10, 12, 2, 6, 10,

2, 10, 2, 12, 10, 20, 10, 6, 2, 6, 4, 8, 4, 14, 6, 4, 2, 4, 8, 6, 4,

6, 6, 6, 14, 16, 2, 10, 32, 6, 4, 6, 14, 4, 6, 6, 8, 4, 2, 28, 2,

4, 24, 26, 6, 6, 18, 10, 2, 10, 12, 8, 4, 6, 2, 12, 18, 22, 6, 8, 6,

6, 4, 8, 6, 30, 10, 6, 6, 8, 4, 2, 28, 6, 6, 6, 12, 2, 4, 6, 18, 8,

10, 8, 24, 6, 4, 20, 6, 6, 4, 6, 6, 14, 6, 12, 10, 6, 14, 10, 18, 8,

12, 10, 8, 6, 18, 6, 4, 6, 2, 10, 20, 4, 8, 12, 6, 10, 18, 2, 4, 6,

8, 6, 4, 14, 6, 6, 10, 14, 6, 10, 24, 6, 6, 14, 6, 4, 6, 8, 6, 6, 30,

10, 2, 6, 6, 4, 24, 8, 22, 6, 2, 4, 8, 10, 6, 2, 4, 6, 8, 6, 6, 10,

8, 4, 2, 16, 14, 4, 2, 18, 4, 6, 12, 2, 6, 4, 2, 10, 6, 8, 6, 10, 18,

8, 12, 6, 6, 6, 4, 8, 10, 2, 4, 2, 24, 4, 2, 4, 8, 10, 42, 14, 10, 6,

24, 20, 10, 2, 10, 2, 6, 6, 6, 10, 2, 10, 2, 4, 6, 14, 4, 20, 22, 6,

14, 4, 2, 4, 8, 6, 4, 6, 2, 12, 6, 12, 4, 6, 14, 4, 2, 22, 2, 10, 8,

6, 24, 6, 6, 10, 6, 12, 2, 28, 6, 20, 6, 4, 2, 10, 8, 4, 6, 12, 2,

12, 4, 6, 8, 6, 6, 4, 12, 2, 10, 18, 18, 20, 4, 6, 6, 20, 6, 4, 18,

2, 6, 10, 2, 10, 8, 4, 6, 2, 12, 16, 8, 12, 10, 8, 4, 2, 4, 2, 28, 2,

28, 14, 6, 4, 8, 10, 2, 16, 2, 10, 2, 12, 4, 6, 6, 8, 6, 6, 4, 14, 6,

18, 6, 30, 6, 6, 10, 12, 30, 6, 6, 8, 10, 12, 14, 4, 8, 4, 6, 8, 4,

6, 24, 12, 6, 2, 16, 14, 6, 10, 2, 10, 2, 10, 8, 12, 6, 4, 2, 10, 14,

4, 2, 6, 4, 20, 4, 2, 18, 16, 6, 18, 2, 4, 8, 6, 4, 20, 10, 6, 2, 18,

6, 4, 6, 6, 12, 2, 16, 8, 10, 6, 24, 2, 10, 8, 16, 6, 2, 10, 6, 2,

22, 2, 28, 2, 4, 6, 8, 10, 2, 6, 4, 20, 18, 6, 10, 2, 4, 6, 2, 12,

16, 12, 2, 4, 30, 12, 12, 2, 16, 2, 6, 4, 2, 24, 6, 4, 20, 12, 10, 2,

12, 10, 20, 4, 14, 12, 10, 6, 8, 4, 18, 2, 10}

Plotting a tally of the offsets produces a figure, Figure H.1 which is remarkably similar to that of the prime gap jumping champions of Section 1.7.

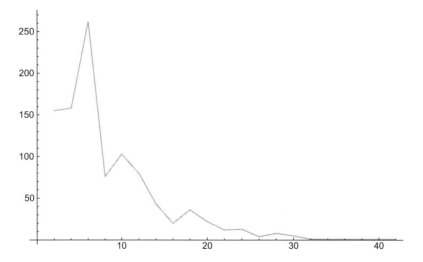

Figure H.1 A plot of the tallied offsets for inductive tuples.

Appendix I

PGpack Minimanual

I.1 Introduction

This appendix is the manual for a set of functions written to assist the reader to reproduce, and potentially extend, the calculations mentioned in the main part of the book. The software for the package is provided at the web page for the book linked to the author's home page:

www.math.waikato.ac.nz/~kab

and is in the form of a standard Mathematica add-on package. To use the functions in the package, you will need to have a version of Mathematica at level 10.0 or higher.

I.1.1 Installation

First, connect to the website given in the preceding paragraph and click on the link for PGpack listed under the heading "Software" to get to the PGpack home page. Instructions on downloading the files for the package will be given on the home page. If you have an earlier version of PGpack, first delete the file **PGpack.m**, and any other associated files related to the old version. The name of the package file is **PGpack.m**. To install the package, if you have access to the file system for programs on your computer, place a copy of the file in the standard repository for Mathematica packages or any other directory, which is listed by evaluating $Path in Mathematica, to which you have access. You can then type ≪**PGpack.m** and then press the Shift and Enter keys to load the package. You may need administrator or super-user status to complete this installation. Alternatively, place the package file **PGpack.m** anywhere in your own file system where it is safe and accessible.

Instructions for Windows systems: The package can be loaded by typing

```
SetDirectory["c:\\your\\directory\\path"]<Shift/
Enter>
```
```
<<PGpack.m;ResetDirectory[]<Shift/Enter>
```

or

```
Get["PGpack.m",Path->{"c:\\your\\directory\\path"}]
<Shift/Enter>
```

where the path-name in quotes should be replaced by the actual pathname of directories and subdirectories which specify where the package has been placed on the computer, including the quotes.

Instructions for Unix/Linux systems: These are the same as for Windows, but the pathname syntax style should be like `/usr/home/your/subdirectory`.

Instructions for Macintosh systems: These are the same as for Windows, but the pathname syntax style should be like `HD:Users:ham:Documents:`.

All systems: The package should load printing a message. The functions of PGpack are then available to any Mathematica notebook you subsequently open.

I.1.2 About This Minimanual

This appendix contains a list of all of the functions available in the package PGpack, in chapter order, followed by a manual entry for each function in alphabetical order. To contain information about bug fixes and updates to PGpack, consult the website for the text linked to the author's home page.

There are many issues to do with computer algebra and mathematical software that will arise in any serious evaluation or use of Mathematica and PGpack. Some notes:

PGpack function arguments are first evaluated and then checked for correct data type. If the user calls a function with an incorrect number of arguments or an argument of incorrect type, rather than issue a warning and proceeding to compute (default Mathematica style), PGpack prints an error message, aborts the evaluation and returns the user to the top level, no matter how deeply nested the function which makes the erroneous call happens to be placed. This is a tool for assisting users to debug programs which include calls to this package.

In Mathematica (and other symbolic computation systems), the mathematical concept of a "variable", such as x, is encapsulated by a "symbol", which also looks like x, but that is just its "print name", the string `"x"`. Symbols are much more multifaceted than variables and need to be used with care. Normally when a PGpack function argument, say $f(x, y)$, calls for a variable x, the variable should be a symbol which evaluates to itself, or very occasionally to another symbol which evaluates to itself. This is best ensured by removing any values attached to the symbol by calling `Clear[x]` before using x as an argument. Such a variable or symbol x will then be termed "inert", but it is just a regular symbol whose value pointer points back to itself.

Functions which include numerical floating point evaluations frequently have an integer argument "d". This represents the precision which is used for internal computations, and not the accuracy of the resulting output.

Here is a list of the functions in **PGpack**. Each also has a 3 letter name.

AdmissibleTupleQ

BesselZeroUpperBound

BombieriDavenportK

BrunsConstant

ComputeI

ComputeJ

ContractTuple

CountResidueClasses

DenseDivisibility

DenselyDivisibleQ

DenseTuple

EnlargeAdmissibleTuple

EnlargingElementQ

EnlargingElements

EratosthenesTuple

ExpandSignature

FeasibleKPolymathQ

GetPolymathTuple

GPYEH

GreedyTuple

HenselyRichardsTuple

HValues

InductiveTuple

InductiveTupleOffsets1000

IntegrateRk

KappaZhang

KrylovF

KrylovNewMonomial

KrylovWs

LargeSubsetQ

MonToPoly

NarrowestTuples

NarrowH

NarrowTuples

NextPrimeGapDist

OmegaZhang

OperatorL
PlotHValues
PlotTwins
PolyMonDegrees
PolyToMon
PrimeTuple
RayleighQuotientBogaert
RayleighQuotientFromWs
RayleighQuotientMaynard
RayleighQuotientPolymath
RhoStar
SchinzelTuple
ShiftedPrimeTuple
SignatureQ
SmallestKPolymath0
SmallestKPolymath1
SmallestKStar
SmallSubsetQ
SmoothIntegerQ
StabilizingElementQ
StabilizingElements
SymmetricPolynomialQ
TupleDiameter
TupleGaps
TupleMinimalCompanions
TwinPrimesConstant
VonMangoldtR
Global variables: verbose$, Verbose$, DB.
Protected variables: HValues, InductiveTupleOffsets1000.

I.2 PGpack Functions

■ **AdmissibleTupleQ (adq)**

This predicate function determines whether a tuple, represented by a list of integers, is admissible. It uses the greedy algorithm checking all primes up to the number of terms in the tuple. See Sections 1.6 and 1.10.

AdmissibleTupleQ[H] $\longrightarrow$ Pred

H is a nonempty list of two or more distinct integers representing a tuple.
Pred is **True** is *H* represents an admissible tuple, otherwise **False**.

■ **BesselZeroUpperBound (bzb)**

This function returns an upper bound for the first positive real zero of the Bessel function $J_n(x)$ – see Appendix A.

$$\textbf{BesselZeroUpperBound[n]} \longrightarrow \textbf{val}$$

n is an integer being the order of the Bessel function with $n \geq 0$,

val is a positive real number being such that the first zero of $J_n(x)$ is in the interval $(0, val]$.

■ **BombieriDavenportK (bdk)**

This function gives a floating point approximation to the value of $K(n)$ as used by Bombieri and Davenport in their work on small gaps between primes defined as

$$K(n) := C_2 \prod_{2 < p | n} \left(\frac{p-1}{p-2} \right) = \prod_{p > 2} \left(1 - \frac{1}{(p-1)^2} \right) \prod_{2 < p | n} \left(\frac{p-1}{p-2} \right),$$

where C_2 is the twin primes constant – see Section 3.4. Note that we use K rather than the corresponding H adopted by Bombieri and Davenport, since H is rather overworked in this field.

$$\textbf{BombieriDavenportK[n, D, d]} \longrightarrow \textbf{val}$$

n is a natural number with $n \geq 2$,

D is a natural number being the length of an internal step such that for those iterations exact integer arithmetic is used,

d is a natural number being the precision for internal floating point computations,

val is a floating point approximation to the constant of Bombieri and Davenport.

■ **BrunsConstant (brc)**

This function computes the value of Brun's constant, which is defined here as

$$B := \left(\frac{1}{3} + \frac{1}{5} \right) + \left(\frac{1}{5} + \frac{1}{7} \right) + \left(\frac{1}{11} + \frac{1}{13} \right) + \left(\frac{1}{p_n} + \frac{1}{p_n + 2} \right) + \cdots,$$

where $(p_n, p_n + 2)$ are the nth pair of twin primes. See Section 2.4.

$$\textbf{BrunsConstant[n, D, d]} \longrightarrow \textbf{val}$$

n is a positive integer being the number of prime twins taken in the evaluation not less than 10,

D is a natural number being the length of an internal step during which iterations exact integer arithmetic is used not greater than one half the first argument,

d is a natural number being the precision for internal floating point computations,

val is a floating point approximation to Brun's constant B.

■ **ComputeI (coi)**

This function computes Maynard's

$$I_k(F) = \int_{\mathscr{R}_k} F(x_1, \dots, x_k)^2 \, dx_1 \cdots dx_k,$$

where $F \in \mathbb{Q}[x_1, \dots, x_n]$ is an explicit polynomial in at least some of the variables **vars**. This function is not suitable for medium or large values of k. See Section 6.6.

$$\text{ComputeI[F, vars]} \longrightarrow \text{val}$$

F is a polynomial in the variables **vars** with rational coefficients,

vars is a list of k inert symbols (see Section I.1.2) being the variables of integration,

val is a rational number being the value of $I_k(F)$.

■ **ComputeJ (coj)**

This function computes Maynard's

$$J_k(F) = J_k^{(1)}(F) = \int_{\mathscr{R}_{k-1}} \left(\int_0^{1-x_2-x_2-\cdots-x_k} F(x_1, \dots, x_k) \, dx_1 \right)^2 dx_2 \cdots dx_k,$$

where F is an explicit symmetric polynomial in $\mathbb{Q}[x_1, \dots, x_k]$. This function is not suitable for medium or large values of k. See Section 6.6.

$$\text{ComputeJ[F, vars]} \longrightarrow \text{val}$$

F is a symmetric polynomial with rational coefficients in the variables **vars**,

vars is a list of k inert symbols (see Section I.1.2) being the variables of integration,

val is a rational number being the value of $J_k(F)$.

■ **ContractTuple (cot)**

This function takes an admissible tuple and returns the tuple obtained by removing the largest or smallest element, which results in a tuple with the smaller diameter. See Sections 1.6 and 1.10.

$$\text{ContractTuple[H}_1] \longrightarrow \text{H}_2$$

H₁ is a nontrivial tuple in standard form of length 3 or more,

H₂ is **H₁** or a narrower subtuple of **H₁**.

■ CountResidueClasses (crc)

This function counts the number of residue classes represented by a finite subset of $\mathbb{Z}$ modulo a prime. If p is the prime and H the subset, this number is often written $v_p(H)$ or $v_H(p)$ – see Section 4.3.

$$\textbf{CountResidueClasses[H, p]} \longrightarrow \textbf{c}$$

H is a nonempty list of distinct integers representing a tuple which is not necessarily admissible,

p is a rational prime number,

c is a count of the number of distinct residue classes modulo **p** represented by the elements of **H**.

■ DenseDivisibility (ddi)

This function computes the largest $i \geq 0$ such that a given natural number is i-tuply y-densely divisible for a given $y > 1$ should such a finite i exist – see Section 8.5 and the manual item for DenselyDivisibleQ.

$$\textbf{DenseDivisibility[n, y]} \longrightarrow \textbf{i}$$

n is a natural number,

y is a real number with $y > 1$,

i is a nonnegative integer, being the smallest i such that n is i-tuply y-densely divisible.

■ DenselyDivisibleQ (ddq)

This function determines whether a given natural number is i-tuply y-densely divisible for a given $i \geq 0$ and $y > 1$ – see Section 8.5. Recall that if $i = 0$, every natural number is 0-tuply y-densely divisible. If $i \geq 1$, Polymath defined a natural number n to be i-tuply **densely divisible** recursively as follows: for every pair of nonnegative integers j, k with $j + k = i - 1$ and real numbers R with $1 \leq R \leq yn$, we have a factorization

$$n = qr \text{ such that } \frac{R}{y} \leq r \leq R$$

such that q is j-tuply y-densely divisible and r is k-tuply y-densely divisible. We sometimes write in this case $n \in \mathscr{D}(i, y)$. See Section 8.5.

$$\textbf{DenselyDivisibleQ[n, i, y]} \longrightarrow \textbf{Pred}$$

n is a natural number,

i is a nonnegative integer,

y is a real number with $y > 1$,

Pred is **True** if n is i-tuply y-densely divisible and **False** otherwise.

■ **DenseTuple (dtu)**

This function determines whether a tuple, represented by a list of integers, is dense, i.e., whether any tuple obtained by removing an endpoint and inserting an interior point is admissible. If so, it returns the first such tuple it finds, otherwise it returns the original tuple. See Sections 1.6 and 1.10.

$$\textbf{DenseTuple}[\textbf{H}_1] \longrightarrow \textbf{H}_2$$

$\textbf{H}_1$ is a nonempty list of two or more distinct integers representing a tuple.

$\textbf{H}_2$ is a tuple equal to $\textbf{H}_1$ or to a tuple of the same length, but smaller diameter, having all elements except one in common with $\textbf{H}_1$.

■ **EnlargeAdmissibleTuple (ent)**

This function transforms an admissible tuple to one with a larger diameter by incrementing the largest element until the resulting tuple is admissible. See Sections 1.6 and 1.10.

$$\textbf{EnlargeAdmissibleTuple}[\textbf{H}_1] \longrightarrow \textbf{H}_2$$

$\textbf{H}_1$ is an admissible tuple of size $k \geq 2$,

$\textbf{H}_2$ is an admissible tuple of strictly larger diameter and the same number of elements.

■ **EnlargingElementQ (eeq)**

This function determines whether a given element when adjoined to some "small" subset produces a "large" subset – see Section 8.8, Lemma 8.20 and the manual entries for `SmallSubsetQ` and `LargeSubsetQ`.

$$\textbf{EnlargingElementQ}[\textbf{j, ts, } \sigma] \longrightarrow \textbf{Pred}$$

$\textbf{j}$ is the given element with $1 \leq j \leq n$,

$\textbf{ts}$ is a nonempty list of nonnegative real numbers $\{t_1, \ldots, t_n\}$ with $n \geq 2$ and

$$\sum_{1 \leq i \leq n} t_i = 1,$$

σ is a real number in the range $1/6 < \sigma < 1/2$,

$\textbf{Pred}$ is **True** if j is enlarging, else **False**.

■ **EnlargingElements (ees)**

This function returns the set of all enlarging elements – see Section 8.8 and Lemma 8.20 and the manual entry for `EnlargingElementQ`.

$$\textbf{EnglargingElements}[\textbf{ts, } \sigma] \longrightarrow \textbf{lis}$$

ts is a nonempty list of nonnegative real numbers $\{t_1, \ldots, t_n\}$ with $n \geq 2$ and

$$\sum_{1 \leq i \leq n} t_i = 1,$$

σ is a real number in the range $1/6 < \sigma < 1/2$,
lis is a list of all of the enlarging elements in the set $\{1, \ldots, n\}$.

■ **EratosthenesTuple (ert)**

This function given an interval $[2, x]$ and integer $k \geq 2$ successively removes all integers divisible by an increasing sequence of primes until the k smallest survivors form an admissible tuple. The size parameter k must be sufficiently small for such a tuple to exist. See Sections 1.6 and 1.10.

EratosthenesTuple[x, k] $\longrightarrow$ **H**

x is a positive real number, sufficiently large in terms of k, namely $x > 4k$,
k is a positive integer 2 or more,
H is an admissible tuple with smallest term 0.

■ **ExpandSignature (esg)**

This function first pads a signature α with zeros so its length is precisely a given explicit k, and then returns a symmetric polynomial in variables $t[j]$, where t is an inert symbol, according to the form

$$P_\alpha(t_1, \ldots, t_k) = \sum_{a = \sigma(\alpha)} t_1^{a_1} \cdots t_k^{a_k},$$

where the sum is over all permutations of α giving distinct values of a – see Section 7.5, where there are examples which could clarify this notation.

ExpandSignature[α, k, t] $\longrightarrow$ **poly**

α is a list of natural numbers of length not greater than k,
k is a natural number,
t is an inert symbol (see Section I.1.2),
poly is a symmetric polynomial in the variables $t[j]$ corresponding to the signature α.

■ **FeasibleKPolymathQ (fkq)**

This function checks a set of constraints to determine whether the inequality of Theorem 8.19, namely

$$(1 + 4\varpi)(1 - 2\kappa_1 - 2\kappa_2 - 2\kappa_3) > \frac{j_{k-2}^2}{k(k-1)},$$

and $\mathrm{MPZ}^{(i)}[\varpi, \delta]$ are satisfied – see Chapter 8.

FeasibleKPolymathQ[k, ϖ, δ, δ', A, $\{c_\delta, c_\varpi\}$, i, d] $\longrightarrow$ **Pred**

k is an integer with $k \geq 2$,

ϖ is a real number with $0 < \varpi < \frac{1}{4}$,

δ is a real number with $0 < \delta < \varpi + \frac{1}{4}$,

δ' is a real number with $\delta \leq \delta' < \varpi + \frac{1}{4}$,

A is a positive real number,

c_δ, c_ϖ are positive real numbers which must satisfy $c_\varpi \varpi + c_\delta \delta < 1$,

i is a nonnegative integer,

d is a positive integer giving the precision for internal computations,

Pred is **True** if the constraints are all satisfied, else **False**.

verbose\$ is a flag that if set to true gives information regarding intermediate results. For information regarding which constraints are not satisfied, the reader can consult the Mathematica source code.

■ **GetPolymathTuple (gpt)**

This function computes and returns the tuple of length 1,783 and width 14,950 used by Polymath to show that Zhang's bound could be reduced to $H_1 \leq 14950$ – see Sections 1.6, 1.10 and 8.7. This function computes the tuple from scratch. It can be checked with AdmissibleTupleQ. It is also available as NarrowH[1783].

$$\text{GetPolymathTuple[]} \longrightarrow \text{H}$$

H is the list {0,4, ...,14950} representing the admissible tuple of size 1783 and diameter 14950 as used by Polymath8a.

■ **GPYEH (gpy)**

This function returns the largest eigenvalue of $b^T B b$ – see Section 4.9, especially Table 4.1.

$$\text{GPYEH[k,}\theta\text{,L]} \longrightarrow \lambda$$

k is a natural number with $k \geq 2$,

θ is a real number with $\frac{1}{2} < \theta < 1$ representing the assumed level of distribution of primes,

L is a natural number,

λ is a floating point number being the maximum eigenvalue (not the maximum absolute value).

■ **GreedyTuple (grt)**

This function takes the integers in $[2, x]$ and for each prime $p \leq k$ removes a minimally occupied residue class modulo p until the remaining integers form an admissible tuple. See Sections 1.6 and 1.10.

$$\text{GreedyTuple[x, k]} \longrightarrow \text{Tuple}$$

x is a real number with $x \geq 2$,
k is a natural number with $k \geq 2$ and $k < \sqrt{x}$, representing the length of the tuple
    to be generated,
Tuple is a list of integers with first term 0 representing an admissible k-tuple, or
    False if no such tuple exists.

■ HensleyRichardsTuple (hrt)
This function takes the integers in $[-x/2, x/2]$ and for each prime $p \leq k$
removes a residue class 0 modulo p until the remaining integers of least
absolute value form an admissible tuple. See Sections 1.6 and 1.10.

$$\textbf{HensleyRichardsTuple[x, k]} \longrightarrow \textbf{Tuple}$$

x is a real number with $x \geq 2$,
k is a natural number, with $k \geq 2$ and $k < \sqrt{x}$, representing the length of the tuple
    to be generated,
Tuple is a list of integers with first term 0 representing an admissible k-tuple, or
    False if no such tuple exists.

■ HValues (hvs)
This global variable has as its value the initial set of values $\{$H[1],H[2],H[3],
..,H[342]$\}$, i.e., for each k with $1 \leq k \leq 342$, the diameter of a k-tuple which
is admissible has k elements, and is such that no other admissible k-tuple has a
smaller diameter. Currently the initial 341 values of Engelsma are stored, together
with $H[1] = 0$. See Sections 1.6 and 1.10. To access the individual values, use the
notation HValues[[k]] with k between 1 and 342.

$$\textbf{HValues} \longrightarrow \textbf{Lis}$$

Lis is a list of 342 values $H[k]$ representing the minimal diameters of k-tuples for
    $k \geq 1$.

■ InductiveTuple (idt)
This function returns an admissible k-tuple which is such that every initial
segment of length $1 \leq l \leq k$ is an admissible l-tuple returned by the function.
Starting with the tuple $\{0\}$, the function adds each element in turn inductively
so that the new tuple is admissible and of minimal diameter. See Sections 1.6
and 1.10, and the PGpack variable InductiveTuplesOffsets 1000
described in Appendix H, which should enable the rapid generation of up to
1,000 "inductive" tuples with some simple programming.

$$\textbf{InductiveTuple[k]} \longrightarrow \textbf{Tuple}$$

k is a natural number with $k \geq 1$ representing the length of the tuple to be
    generated,

Tuple is a list of integers with first term 0 representing an admissible **k**-tuple.

■ **IntegrateRk (irk)**

This function for an explicit value of $k \geq 1$ with $k = Length[vars]$ integrates a polynomial function with real coefficients over the simplex $\mathscr{R}_k$ – see Chapters 6 and 7. The function should be able to be processed by Mathematica's `Integrate`, for example, a polynomial in standard form.

$$\textbf{IntegrateRk[F, vars]} \longrightarrow \textbf{val}$$

F is a polynomial expression with real coefficients and in many variables being the symbols in the second argument,

vars is a list of $k \geq 1$ inert symbols (see Section I.1.2),

val is the value of the integral of **F** over the simplex $\mathscr{R}_k$.

■ **KappaZhang (kaz)**

This function evaluates Zhang's parameters κ_1 and κ_2 numerically, verifying the derivation of Lemma 5.7 – see Section 5.7.

$$\textbf{KappaZhang[i,k,l,}\varpi\textbf{,d]} \longrightarrow \textbf{val}$$

i is an indicator with value 1 for κ_1 and 2 for κ_2,

k is an integer with $k \geq 2$,

l is a positive integer,

ϖ is a real number in the range $0 < \varpi < 1$,

d is a natural number for the precision of internal computations,

val is the logarithm of the value of κ_i which has been computed.

■ **KrylovF (kfm)**

This function, for small values of n, returns the symmetric polynomials obtained through successive applications of the operator $\mathscr{L}$, where

$$\mathscr{L}F(x_1, \ldots, x_k) = \int_0^{1-x_2-\cdots-x_k} F(x_1, \ldots, x_k)\, dx_1$$

$$+ \cdots + \int_0^{1-x_1-\cdots-x_{k-1}} F(x_1, \ldots, x_k)\, dx_k,$$

to the constant function with value 1. These polynomials are given in so-called monomial form, i.e., a list giving rules of the form

$$\{n_1, \cdots, n_L\} \to expr,$$

where the initial list represents the monomial $\prod_{j=1}^{L} P[k, j]^{n_j}$ with $P[k, j] = x_1^j + \cdots x_k^j$, and $expr$ is a function of k representing the coefficient of the monomial. The value returned then represents the sum of each of these coefficients times monomials. See Section 7.7. The first few applications of $\mathscr{L}$ are stored in the protected variable array $FM[n,k]$ for $0 \leq n \leq 4$.

$$\mathbf{KrylovF[n,k]} \longrightarrow \mathbf{mon}$$

n is an integer with $0 \leq n \leq 4$,

k is an inert symbol,

mon is the nth polynomial with coefficients functions of k in monomial form.

■ KrylovNewMon (knm)

This function takes a symmetric polynomial in many variables in monomial form, and returns the value of the operator $\mathscr{L}$ acting on that polynomial, also in monomial form – see Section 7.7. For examples, see values of the protected array FM [n, k], which represent the first five applications of $\mathscr{L}$ to the function with value 1 on $\mathscr{R}_k$ and 0 elsewhere.

Recall

$$\mathscr{L}F(x_1, \ldots, x_k) = \int_0^{1-x_2-\cdots-x_k} F(x_1, \ldots, x_k)\, dx_1$$
$$+ \cdots + \int_0^{1-x_1-\cdots-x_{k-1}} F(x_1, \ldots, x_k)\, dx_k.$$

These polynomials are given in so-called monomial form, i.e., a list giving rules of the form

$$\{n_1, \cdots, n_L\} \rightarrow expr,$$

where the initial list represents the monomial $\prod_{j=1}^{L} P[k,j]^{n_j}$ with $P[k,j] = x_1^j + \cdots x_k^j$, and $expr$ is a function of k representing the coefficient of the monomial. The value returned then represents the sum of each of these coefficients times monomials. See Section 7.7.

$$\mathbf{KrylovNewMon[k, F]} \longrightarrow \mathbf{mon}$$

k is an inert symbol (see Section I.1.2),

F is a polynomial given in monomial form with coefficients being functions of k and variables $P[k,n]$,

mon is the monomial form result of $\mathscr{L}(F)$.

■ KrylovWs (krw)

This function returns expressions in the tuple size k with $1 \leq k$ determined by the $\mathscr{L}$ operator via $\langle \mathscr{L}^j 1, 1 \rangle$, where $\mathscr{L}$ is defined by

$$\mathscr{L} F(x_1, \ldots, F_k) = \int_0^{1-x_2-\cdots-x_k} F(x_1, \ldots, x_k) \, dx_1$$

$$+ \cdots + \int_0^{1-x_1-\cdots-x_{k-1}} F(x_1, \ldots, x_k) \, dx_k,$$

and where integration for the inner product is over the standard simplex in $\mathbb{R}^k$ – see Section 7.7.

$$\mathbf{KrylovWs[k, j]} \longrightarrow \mathbf{expr}$$

$\mathbf{k}$ is an inert symbol (see Section I.1.2),
$\mathbf{j}$ is an integer with $\mathbf{0 \leq j \leq 11}$ representing the number of applications of $\mathscr{L}$,
$\mathbf{expr}$ is a rational function in $\mathbf{k}$ with integer coefficients.

■ LargeSubsetQ (lsq)

This function determines whether or not a finite subset S of $\{1, \ldots, n\}$ is "large" in that

$$\sum_{j \in S} t_j \geq \frac{1}{2} + \sigma.$$

See Section 8.8 and Lemma 8.20.

$$\mathbf{LargeSubsetQ[S, ts, \sigma]} \longrightarrow \mathbf{Pred}$$

$\mathbf{S}$ is a list of integers representing a subset of $\{\mathbf{1, 2, \ldots, n}\}$,
$\mathbf{ts}$ is a list of nonnegative real numbers with $\mathbf{t_1 + \cdots + t_n = 1}$,
$\mathbf{\sigma}$ is a real number in the range $\mathbf{1/6 < \sigma < 1/2}$,
$\mathbf{Pred}$ is $\mathbf{True}$ if the subset is large, otherwise $\mathbf{False}$.

■ MonToPoly (m2p)

This function converts a polynomial in monomial form, with coefficients which are functions of k, to a polynomial in normal form – see Section 7.7.

$$\mathbf{MonToPoly[k, form, vars]} \longrightarrow \mathbf{poly}$$

$\mathbf{k}$ is the symbol $\mathbf{k}$, which should be inert (see Section I.1.2),
$\mathbf{form}$ is a list of rules $\{n_1, \ldots, n_m\} \rightarrow expr$, where $\mathbf{expr}$ is a rational function in $\mathbf{k}$,
$\mathbf{vars}$ is a list of inert symbols representing the variables of $\mathbf{poly}$ where each $\mathbf{m}$ in $\mathbf{form}$ is less than or equal to the length of $\mathbf{vars}$,
$\mathbf{poly}$ is a polynomial in standard form with coefficients being functions of $\mathbf{k}$.

■ NarrowestTuples (nts)

This function returns the set of all tuples of a given size which are admissible and narrowest amongst all tuples of that size. See Sections 1.6 and 1.10. It is suitable only for small values of k

$$\text{NarrowestTuples[k]} \longrightarrow \text{lis}$$

k is a natural number with $k \leq 54$,
lis is a list of narrow admissible tuples of size **k**.

■ NarrowH (nah)

This function gives an admissible tuple of size k for each k in the ranges

$$1 \leq k \leq 54, \ k = 102, \ k = 105, k = 1783,$$

which in all cases, other than the last which may not, has minimal width for tuples of the given size. Note that generally the tuple returned is not unique. See Sections 1.6 and 1.10 and the tables of Engelsma and those of Clark and Javis.

`www.opertach.com/primes/k-tuples.html` and `math.mit .edu/primegaps`.

$$\text{NarrowH[k]} \longrightarrow \text{tuple}$$

k is a natural number,
tuple is a tuple of size **k** and small, possibly minimal, width.

■ NarrowTuples (ntu)

This function returns a list of all admissible tuples of width less than or equal to $k \leq 54$ which are of minimal width among tuples of a given size. See Sections 1.6 and 1.10. This function is suitable only for small values of k.

$$\text{NarrowTuples[k]} \longrightarrow \text{lis}$$

k is a natural number with $k \leq 54$,
lis is a list of admissible narrow tuples.

■ NextPrimeGapDist (pgd)

This function takes the gaps between primes up to a given real $x > 0$ and returns the gap size which appears most frequently and the number of occurrences of that gap size. As a side effect, it gives a plot of the distribution of gap sizes up to the maximum gap size in the given range – see Section 1.7. Note from that section that 6 is certainly not overall the most frequently occurring gap size, but that maximum gap sizes depend on the range and have primorial values. For example, asymptotically, 30 occurs more frequently than 6. In terms of prime gaps and their frequencies, 210, at the time of writing, could be the next gap size goal, given the current best value of Polymath8b, 246.

$$\mathbf{NextPrimeGapDist[x]} \longrightarrow \{\mathbf{gap,frequency}\}$$

x is a positive real number giving an upper bound for the primes which are considered,

gap is the gap size which occurs most frequently in the given range,

verbose\$ if set to **True** gives as a side effect a list of all of the potential gaps and their frequencies which occur in the given range.

■ OmegaZhang (omz)

This function returns the value of the critical parameter ω in Zhang's method:

$$\omega := \frac{2k(1-\kappa_2)(2l+1)}{(l+1)(k+2l+1)} - \frac{4(\kappa_1+1)}{4\varpi+1}$$

– see Theorem 5.10 in Chapter 5. If $\omega > 0$, then Zhang's method succeeds with the given values of k, l, ϖ.

$$\mathbf{OmegaZhang[k, l,\varpi, d]} \longrightarrow \mathbf{val}$$

k is an integer with $k \geq 2$,

l is a positive integer significantly smaller than k (here $l \leq \sqrt{k}$ is required),

ϖ is a real positive number,

d is a positive integer being the internal precision for numerical computations,

val is the value of ω.

■ OperatorL (opl)

This function applies the operator $\mathcal{L}$ to any expression where the Mathematica function `Integrate` is able to be used, where

$$\mathcal{L}F(x_1,\ldots,x_k) = \int_0^{1-x_2-\cdots-x_k} F(x_1,\ldots,x_k)\,dx_1$$
$$+ \cdots + \int_0^{1-x_1-\cdots-x_{k-1}} F(x_1,\ldots,x_k)\,dx_k.$$

See Section 7.7. This function is suitable only for small values of k which must be explicit.

$$\mathbf{OperatorL[F, vars]} \longrightarrow \mathbf{val}$$

F is an expression in the variables *vars*,

vars is a list of k inert symbols being the variables of integration (see Section I.1.2),

val is a real number being the result of applying $\mathcal{L}$ to **F**.

■ PlotHValues (phv)

This function plots the values of $H(k) - \eta k \log k$ for $2 \le k \le b \le 342$, i.e., the difference between the diameter of a narrowest admissible k-tuple in the given range minus an approximation, as a "side effect". See Section 1.10.2.

$$\textbf{PlotHValues[b, eta]} \longrightarrow \textbf{val}$$

b is a natural number with $b \le 342$,

eta is a nonnegative real number,

val is the value returned, namely `HValues[[b]]`.

■ PlotTwins (ptw)

This function evaluates the number of twin primes up n and then plots this times $(\log^2 n)/n$ for $2 \le n \le b$ as a "side effect". See Figure 1.2. It returns the number of twin primes (counting one for each pair) up to b. Asymptotically, the graph should approximate twice the twin primes constant.

$$\textbf{PlotTwins[b]} \longrightarrow \textbf{cnt}$$

b is a real number with $b \ge 3$,

cnt is the number of twin primes up to b.

■ PolyMonDegrees (pmd)

For this function, see Section 7.7.

$$\textbf{PolyMonDegrees[form]} \longrightarrow \textbf{degrees}$$

form is a list representing a polynomial in monomial form,

degrees is a list of the maximum power of each variable that appears in the polynomial in the order corresponding to that of the monomial form.

■ PolyToMon (p2m)

This function converts a polynomial function in standard form to one in so-called monomial form – see Section 7.7.

$$\textbf{PolyToMon[poly, vars]} \longrightarrow \textbf{form}$$

poly is an expression which must be poynomial in at least the variables ***vars***,

vars is a list of inert symbols (see Section I.1.2),

form is the representation of ***poly*** in monomial form with respect to the variables ***vars*** in order.

■ PrimeTuple (prt)

This function returns an admissible k-tuple $\{h_1, \ldots, h_k\}$, each of whose elements are primes in the range

$$\left\{ p_{\pi(k)+1}, \ldots, p_{\pi(k)+k} \right\}$$

shifted left to 0 by subtracting $p_{\pi(k)+1}$ from each element. See Sections 1.6 and 1.10. For a given k, normally significantly narrower admissible tuples may be constructed.

$$\textbf{PrimeTuple[k]} \longrightarrow \textbf{tuple}$$

k is a natural number,

tuple is a k-tuple.

■ RayleighQuotientBogaert (rqb)

For this function, see Section 7.7 and especially Section 7.8. It is not suitable for values of r which are 9 or more.

$$\textbf{RayleighQuotientBogaert[r, \{ka,kb\}, d]} \longrightarrow \textbf{b}$$

r is a natural number giving the size of the Krylov matrix which is determined,

{ka,kb} is a pair of natural numbers with $ka \le k \le kb$ giving a range of k values,

d is a natural number being the floating point precision for internal computations,

b is a list of floating point numbers representing lower bounds for M_k for the values of k in the given list.

■ RayleighQuotientFromWs(rqw)

This function computes a lower bound for M_k using the Krylov method with very small matrices, namely up to 6×6, based on the functions of k stored in `KrylovWs` – see Section 7.7.

$$\textbf{RayleighQuotientFromWs[k, n, d]} \longrightarrow \lambda$$

k is a natural number with $k \ge 2$,

n is an integer with $1 \le n \le 6$,

d is a natural number being the internal precision for numerical computations,

λ is a floating point number being the maximum eigenvalue of $A^{-1}.B$.

■ RayleighQuotientMaynard (rqm)

This function for a given $k \ge 2$ returns a lower bound for M_k – see Chapter 6, especially Section 6.8.

$$\textbf{RayleighQuotientMaynard[k, n, \{d_1,d_2,d_3\}]} \longrightarrow \textbf{b}$$

k is a natural number 2 or more,

n gives the degree bound $2n + 1$ for basis approximating polynomials,

d_1 is a natural number being the precision for the final output,

d_2 is a natural number being internal precision for eigenvector calculations,

d_3 is a natural number being the precision for taking rational approximations to floating point numbers,

b is a floating point number computed to precision d_1, being a lower bound for M_k,

verbose$ is a flag being **True** or **False**. If set to **True**, then, as the algorithm progresses, items of information are printed to the output stream, normally the screen. These are not returned values.

■ RayleighQuotientPolymath (rqp)

This function given $k \geq 2$ and $0 \leq \epsilon < 1$ returns a lower bound for $M_{k,\epsilon}$ – see Chapter 7, especially Section 7.5.

RayleighQuotientPolymath[k, n, ϵ, Mgoal, Mrange, method, sigs, d] $\longrightarrow$ b

k is a natural number 2 or more representing the size of the tuple,

n is the degree bound for basis approximating polynomials,

ϵ is the size of the perturbation parameter $0 \leq \epsilon < 1$,

Mgoal, Mrange are the positive parameters used only when the **"bisection"** method is chosen. The bound computed is expected to be in the range $(Mgoal - Mrange, Mgoal + Mrange)$, and bisection stops when computed values are less than $1/10^d$ apart,

method is a string which is either **"bisection"** corresponding to Polymath's approach to the Rayleigh quotient using the Mathematica function PositiveDefiniteMatrixQ, **"eigenvector"** corresponding to Maynard's use of the generalized form of the function eigenvectors to compute the vector corresponding to the largest eigenvalue and **"eigenvalue"** using the functions Inverse and Eigenvalue to calculate the largest eigenvalue of $M_1^{-1}.M_2$,

sigs is a string which is either "even", "odd" or "full" corresponding to elements of the signatures α which are to be included in the forms $(1 - P_{(1)})^a P_\alpha$,

d is a positive integer, being the precision chosen for internal computations,

b is a floating point number being a lower bound for $M_{k,\epsilon}$,

verbose$ is a flag being **True** or **False**. If set to **True** then, as the algorithm progresses, items of information are printed to the output stream, normally the screen. These are not returned values.

■ RhoStar (rst)

This function returns values of the function $\rho_\star(x)$ for $0 < x \leq 342$, where

$$\rho_\star(x) := \max(k : H(k) \leq x).$$

See Section 1.10.3 and Figure 1.6.

$$\textbf{RhoStar[x]} \longrightarrow \textbf{k}$$

x is a positive real number not greater than 342,

k is a nonnegative integer being the value of $\rho_\star(x)$.

■ **SchinzelTuple (sct)**

This function returns the tuple of Schinzel, namely the result of sieving from the integers in $[2, x]$ all those congruent to 1 modulo primes p for $p \leq x/4$ and to 0 modulo p for those with $p > x/4$, until the least k survivors form an admissible tuple. See Section 1.10.

$$\textbf{SchinzelTuple[x, k]} \longrightarrow \textbf{tuple}$$

x is a positive integer greater than 2 and the square of the second argument,
k is an integer with $k \geq 2$,
tuple is an admissible k-tuple, or **False** if no admissible tuple can be found
        through the given process.

■ **ShiftedPrimeTuple (spt)**

This function constructs an admissible k-tuple consisting of primes $p_{\pi(k)+s+j}$ for $1 \leq j \leq k$ translated to have smallest term 0. See Sections 1.6 and 1.10.

$$\textbf{ShiftedPrimeTuple[k, s]} \longrightarrow \textbf{tuple}$$

k is a natural number,
s is a nonnegative integer,
tuple is a k-tuple.

■ **SignatureQ (siq)**

This function tests to see whether a list of positive integers is a "signature", i.e., is nonempty and in nonincreasing order – see Section 7.5.

$$\textbf{SignatureQ[lis]} \longrightarrow \textbf{pred}$$

lis is a nonempty list of integers,
pred is **True** if *lis* represents a valid signature, else **False**.

■ **SmallestKPolymath0 (sk0)**

This function finds the smallest value of an integer $k \geq 2$ such that

$$\frac{j_{k-2}^2}{k(k-1)} < (1 + 4\varpi)(1 - 2\kappa_1 - 2\kappa_2),$$

where j_{k-2} is the upper bound returned by the PGpack function BesselZero-UpperBound for the first positive zero of the Bessel function $J_{k-2}(x)$ and where κ_1, κ_2 are real functions of k and the parameters ϖ, δ defined by integrals – see Chapter 8 Theorem 8.17 and Appendix A Lemma A.6.

$$\textbf{SmallestKPolymath0[} \varpi, \delta, \textbf{d]} \longrightarrow \textbf{k}$$

ϖ is a real parameter in the range $0 < \varpi < 1/4$,
δ is a real parameter in the range $0 < \delta < \varpi + 1/4$,

d is a positive integer being the precision used for internal computations,

k is the smallest value of the tuple size *k* returned by the algorithm which satisfies the given inequality,

verbose$ if set to **True** prints, as a side effect, information relating to intermediate values and the progress of the algorithm.

■ **SmallestKPolymath1 (sk1)**

This function finds the smallest value of an integer $k \geq 3$ such that

$$\frac{j_{k-2}^2}{k(k-1)} < (1 + 4\varpi)(1 - 2\kappa_1 - 2\kappa_2 - 2\kappa_3),$$

where instead of j_{k-2} we use the value returned by `BesselZeroUpperBound` and where the κ_i are functions of k and the other parameters – see Chapter 8, especially Theorem 8.17 and the remark before its statement and proof. See also Appendix A, especially Lemma A.6.

SmallestKPolymath1[ϖ, δ, δ', A, {c_ϖ, c_δ}, i, d] $\longrightarrow$ **k**

ϖ is a real number being the value of ϖ with $0 < \varpi < \frac{1}{4}$,

δ is a real number with $0 < \delta < \frac{1}{4} + \varpi$,

δ' is a real number with $0 < \delta \leq \delta' < \frac{1}{4} + \varpi$,

A is a positive real number,

c_ϖ a positive integer being the coefficient of ϖ with $c_\varpi \varpi + c_\delta \delta < 1$,

c_δ is a positive integer being the coefficient of δ with $c_\varpi \varpi + c_\delta \delta < 1$,

i is a nonnegative integer being the dense divisibility level for integers in $\mathcal{D}(i, x^\delta)$,

d is a natural number being the precision for internal computations and accuracy goal for the integrals,

k is a positive integer being the smallest value of *k* determined by the algorithm which satisfies the given constraint,

verbose$ if True prints as a side effect the current value of *k*, the values of the κ_i and the values of the individual constraints (which can be determined by consulting the Mathematica code).

■ **SmallestKStar (sks)**

This function finds the smallest value of an integer *k* such that

$$j_{k-2}^2 < (1 + 4\varpi)k(k-1),$$

where instead of j_{k-2} we use the value returned by `BesselZeroUpperBound` – see Appendix A, especially Lemma A.6.

SmallestKStar[ϖ, d] $\longrightarrow$ **k**

ϖ is a real number with $0 < \varpi < \frac{1}{4}$,

d is a natural number being the precision for internal computations,
k is the smallest value of $k \geq 2$ which satisfies the constraint.

■ **SmallSubsetQ (ssq)**

This function determines whether or not a finite subset S of $\{1, \ldots, n\}$ is "small" in that

$$\sum_{j \in S} t_j \leq \frac{1}{2} - \sigma.$$

See Section 8.8 and Lemma 8.20.

$$\textbf{SmallSubsetQ[S, ts, } \sigma \textbf{]} \longrightarrow \textbf{Pred}$$

S is a list of integers representing a subset of $\{1, 2, \ldots, n\}$,
ts is a list of nonnegative real numbers with $t_1 + \cdots + t_n = 1$,
σ is a real number in the range $1/6 < \sigma < 1/2$,
Pred is **True** if the subset is small, otherwise **False**.

■ **SmoothIntegerQ (siq)**

This function tests a natural number to see whether it is smooth of a given order – see Section 8.5.

$$\textbf{SmoothIntegerQ[n, y]} \longrightarrow \textbf{Pred}$$

n is a positive integer,
y is an integer with $y \geq 2$,
Pred is **True** if n is y-smooth, else **False**.

■ **StabilizingElementQ (seq)**

This function checks whether a given element when adjoined to a small subset will never produce a large subset. If this is so it is called "stabilizing" – see Section 8.8, Lemma 8.20 and the manual entries for `SmallSubsetQ` and `LargeSubsetQ`.

$$\textbf{StabilizingElementQ[j, ts, } \sigma \textbf{]} \longrightarrow \textbf{Pred}$$

j is the given element with $1 \leq j \leq n$,
ts is a list of nonnegative real numbers with $t_1 + \cdots + t_n = 1$ and $n \geq 2$,
σ is a real number in the range $1/6 < \sigma < 1/2$,
Pred is **True** if j is stabilizing, else **False**.

■ **StabilizingElements (ses)**

This function returns the set of all stabilizing elements in $\{1, 2, \ldots, n\}$ – see Section 8.8 and Lemma 8.20 and the manual entry for `StabilizingElementQ`.

$$\textbf{StabilizingElements[ts, } \sigma \textbf{]} \longrightarrow \textbf{lis}$$

ts is a list of nonnegative real numbers $t_1 + \cdots + t_n = 1$ and $n \geq 2$,

σ is a real number in the range $1/6 < \sigma < 1/2$,

lis is a list of all of the stabilizing elements in the set $\{1, \ldots, n\}$.

■ SymmetricPolynomialQ (spq)

This function tests to see if a given polynomial is symmetric with respect to permutations of a given set of its variables – see Section 7.7.

$$\textbf{SymmetricPolynomialQ[F, vars]} \longrightarrow \textbf{Pred}$$

F is an expression which should be of polynomial form in the variables *vars*,

vars is a list of inert symbols (see Section I.1.2),

Pred is **True** if *F* is a polynomial in **vars** which is invariant under all permutations of the variables *vars* and **False** otherwise.

■ TupleDiameter (tud)

This function computes the difference between the largest and smallest element of a tuple, sometimes called its width or diameter – see Sections 1.6 and 1.10.

$$\textbf{TupleDiameter[H]} \longrightarrow \textbf{d}$$

H is a tuple being a nonempty list in increasing order which is a subset of the integers,

d is the difference between the maximum element of *H* and the minimum element.

■ TupleGaps (tug)

This function computes the differences between the elements of a tuple when it is in increasing order – see Sections 1.6 and 1.10. Its use is to explore the structure of tuples, especially those which are "narrow".

$$\textbf{TupleGaps[H]} \longrightarrow \textbf{gaps}$$

H is a tuple being a nonempty list with two or more elements in increasing order which is a subset of the integers,

gaps is a list of the sorted gap widths between tuple elements when in increasing order.

■ TupleMinimalCompanions (tmc)

This function, given a narrow tuple (i.e., admissible and minimal diameter for a given size k), computes all of the other minimal tuples with that value of size k – see Sections 1.6 and 1.10. Its use is to explore the structure of narrow tuples and is suitable only for small values of k.

$$\textbf{TupleMinimalCompanions[H]} \longrightarrow \textbf{lis}$$

H is a tuple being a nonempty list representing a minimal (narrow) admissible
    tuple of size $2 \le k \le 54$, which is a subset of the integers with first term 0,
lis is a list of minimal tuples being all of those with the same value of k and first
    term 0.

■ TwinPrimesConstant (tpc)

This function computes the value of the so-called twin primes constant, which
is defined as

$$C_2 := \prod_{p \ge 3} \left(1 - \frac{1}{(p-1)^2} \right) = 0.66016\ldots.$$

See Section 1.6.

$$\textbf{TwinPrimesConstant[n,D,d]} \longrightarrow \textbf{val}$$

n is a positive integer, not less than 100, being the number of prime twins taken in
    the evaluation,
D is a positive integer being the number of products taken in turn to evaluate as
    rational numbers not greater than one quarter of the first argument,
d is a positive integer being the precision for floating point computations,
val is a floating point approximation to the twin primes constant C_2.

■ VonMangoldtR (vmr)

This function computes the value of the truncated von Mangoldt function at a
positive integer n. The definition is

$$\Lambda_R(n) := \sum_{\substack{m|n \\ m < R}} \mu(m) \log \left(\frac{R}{m} \right).$$

See Section 1.8.

$$\textbf{VonMangoldtR[n, R, d]} \longrightarrow \textbf{val}$$

n is a natural number,
R is a positive rational number greater than 1,
d is a natural number being the precision for internal computations or 0 in which
    case the von Mangoldt function value is returned as the logarithm of a
    rational number,
val is a floating point number being an approximation to the value of the truncated
    von Mangoldt function $\Lambda_R(n)$ using floating point precision d.

References

[1] L. Alfors, *Complex Analysis, 2nd ed.*, McGraw-Hill, 1966.

[2] R. Alweiss and S. Luo, *Bounded Gaps Between Primes in Short Intervals* (preprint).

[3] T. M. Apostol, *Modular Functions and Dirichlet Series in Number Theory*, Springer, 1976.

[4] T. M. Apostol, *Introduction to Analytic Number Theory, 2nd ed.*, Springer, 1990.

[5] R. C. Baker and P. Pollack, *Bounded Gaps Between Primes with a Given Primitive Root,* Forum Math. **28** (2016), 675–687.

[6] W. D. Banks, T. Freiberg and J. Maynard, *On Limit Points of the Sequence of Normalized Prime Gaps,* Proc. Lond. Math. Soc. (3) **113** (2016), 515–539.

[7] M. B. Barban, *New Applications of the Great Sieve of Ju. V. Linnik. (Russian) Akad. Nauk Uzbek*. SSR Trudy Inst. Mat. **22** (1961), 1–20.

[8] I. Bogaert (private communication).

[9] E. Bombieri, *On the Large Sieve,* Mathematika **12** (1965), 201–225.

[10] E. Bombieri and H. Davenport, *Small Differences Between Prime Numbers,* Proc. Roy. Soc. Ser. A **293** (1966), 1–18.

[11] E. Bombieri, *Le Grand Crible dans la Théorie Analytique des Nombres,* Astérisque, no 18 (1974) and 2nd ed. no 18 (1987).

[12] E. Bombieri, *On Twin Almost Primes,* Acta Arith. (1975) **28**, 177–193.

[13] E. Bombieri, *The Asymptotic Sieve*, Rend. Accad. Naz. XL **1/2** (1975/76), 243–269.

[14] E. Bombieri, *Corrigendum to My Paper "On Twin Almost Primes" and an Addendum on Selberg's Sieve*, Acta Arith. **28** (1976), 457–461.

[15] E. Bombieri, J. B. Friedlander and H. Iwaniec, *Primes in Arithmetic Progressions to Large Moduli*, Acta Math. **156** (1986), 203–251.

[16] E. Bombieri, J. B. Friedlander and H. Iwaniec, *Primes in Arithmetic Progressions to Large Moduli, II,* Math. Ann. **277** (1987), 361–393.

[17] E. Bombieri, *Selberg's Sieve and Its Applications,* in *Number Theory, Trace Formulas and Discrete Groups: Symposium in Honor of Atle Selberg, Oslo, Norway, July 14–21* 29–49, Academic Press.

[18] E. Bombieri, J. B. Friedlander and H. Iwaniec, *Primes in Arithmetic Progressions to Large Moduli, III,* J. Amer. Math. Soc. **2(2)** (1989), 215–224.

[19] E. Bombieri, J. B. Friedlander and H. Iwaniec, *Some Corrections to an Old Paper,* arXiv:1903.01371v1 [math.NT] 4 March 2019.

[20] K. A. Broughan, *The gcd-sum Function*, Journal of Integer Sequences, **4** (2002) Article 01.2.2, p1–19.

[21] K. A. Broughan, *Equivalents of the Riemann Hypothesis, Volume One: Arithmetic Equivalents*, Cambridge University Press, 2017.

[22] K. A. Broughan, *Equivalents of the Riemann Hypothesis, Volume Two: Analytic Equivalents*, Cambridge University Press, 2017.

[23] V. Brun, *Über das Goldbachsche Gesetz und die Anzahl der Primzahlpaare,* Arch. Mat. Natur. vol. **34**, no. 8 (1915), 1–19. (1915).

[24] V. Brun, *Le crible d'eratosthéne et le theorém de Goldbach*, Videnskapsselskapets Sknfter Kristiania, Mat. Naturv. Klasse., no 3 (1920), 1–36.

[25] B. Casselman, *The Polyface of Polymath*, Notices of the Amer. Math. Soc., June/July 2015, 659.

[26] W. Castryck, E. Fouvry, G. Harcos, et al., *New Equidistribution Estimates of Zhang Type*, Algebra and Number Theory, **8**, (2014), 2067–2199.

[27] A. Castillo, C. Hall, R. J. Lemke Oliver, P. Pollack and L. Thompson, *Bounded Gaps Between Primes in Number Fields and Function Fields,* Proc. Amer. Math. Soc. **143** (2015), 2841–2856.

[28] J. R. Chen, *On the Representation of a Larger Even Integer as the Sum of a Prime and the Product of at Most Two Primes,* Sci. Sinica **16** (1973), 157–176.

[29] J. R. Chen, *On the Representation of a Larger Even Integer as the Sum of a Prime and the Product of at Most Two Primes,* Sci. Sinica **16** (1973), 157–176, in *The Goldbach Conjecture, 2nd ed.*, Y. Wang, Ed. Chapter 20, 275–294, World Scientific, 2002.

[30] B. Cipra, *Proof Promises Progress in Prime Progressions*, Science **304** (2004), 1095.

[31] B. Cipra, *Third Time Proves Charm for Prime-Gap Theorem*, Science **308** (2005), 1238.

[32] T. Cochrane and C. Plinner, *Using Stepanov's Method for Exponential Sums Involving Rational Functions,* J. Number Theory **116** (2006), 270–292.

[33] A. C. Cojocaru and M. Ram Murty, *An Introuction to Sieve Methods and Their Applications*, Cambridge University Press, 2005.

[34] G. Csicsery, *Counting from Infinity: Yitang Zhang and the Twin Primes Conjecture*, Zala films, Oakland, CA, 2015, www.zalafilms.com/films/countingindex.html.

[35] H. Davenport, *Multiplicative Number Theory, 3rd ed.*, Springer, 2000.

[36] P. Deligne, *La conjecture de Weil. II.* Inst. Hautes Études Sci. Publ. Math. **52** (1980), 137–252.

[37] J. -M. Deshouillers and H. Iwaniec, *Kloosterman Sums and Fourier Coefficients of Cusp Forms*, Invent. Math. **70** (1982), 219–288.

[38] H. Diamond and H. Halberstam, *Some Applications of Sieves of Dimensions Exceeding 1*, 101–107, in *Sieve Methods, Exponential Sums, and Their Applications in Number Theory* G. R. H. Greaves, G. Harman, M. N. Huxley, Eds., (Cardiff, 1995), (London Mathematical Society Lecture Note Series **237**), Cambridge University Press, 1997.

[39] L. E. Dickson, *A New Extension of Dirichlet's Theorem on Prime Numbers*, Messenger of Mathematics, **33** (1904), 155–161.

[40] H. M. Edwards, *Riemann's Zeta Function*, Academic Press, 1974. Reprinted by Dover, 2001.

[41] P. D. T. A. Elliott and H. Halberstam, *A Conjecture in Prime Number Theory*, Symposia Mathematica, Vol. IV (INDAM, Rome, 1968/69), Academic Press, 1970, 59–72.

[42] W. and F. Ellison, *Prime Numbers*, Wiley, 1985.

[43] T. Engelsma, *Narrow Admissible Ktuples*, http://math.mit.edu/~primegaps/.

[44] P. Erdős, *On the Difference of Consecutive Primes*, Quart. J. Math. Oxford Ser. **6** (1935), 124–128.

[45] P. Erdős, *The Difference of Consecutive Primes*, Duke Math. J. **6** (1940), 438–441.

[46] P. Erdős, *On Some Problems on the Distribution of Prime Numbers,* C. I. M. E. Teoria dei numeri Math. Congr. Varenna 1954 (1955), 8.

[47] P. Erdős and E. G. Straus, *Remarks on the Differences Between Consecutive Primes,* Elem. Math. **35** (1980), 115–118.

[48] B. Farkas, J. Pintz and S. Révész, *On the Optimal Weight Function in the Goldston–Pintz-Yildirim Method for Finding Small Gaps Between Consecutive Primes,* in 75–104, *Paul Turán Memorial Volume: Number Theory, Analysis, and Combinatorics*, 75–104, J. Pintz, A. Biró, K. Győry, G. Harcos, M. Simonovits, and J. Szabados, Eds., De Gruyter Proc. Math., De Gruyter, Berlin, 2014.

[49] G. Folland, *Fourier Analysis and Its Applications.* Wadsworth and Brooks, 1992.

[50] K. Ford, *A Simple Proof of Gallagher's Singluar Series Sum Estimate*, arXiv:1108.3861v2 [math.NT] 6 December 2016.

[51] K. Ford, *Long gaps between primes,* J. Amer. Math. Soc. **31** (2018), 65–105.

[52] K. Ford, B. Green, S. Konyagin and T. Tao, *Large Gaps Between Consecutive Primes*, arXiv:1408.4505v2 [math.NT] 9 November 2015.

[53] E. Fouvry, *Autour du theoreme de Bombieri–Vinogradov.* Acta Math. **152** (1984), 219–244.

[54] E. Fouvry and H. Iwaniec, *On a Theorem of Bombieri–Vinogradov Type,* Mathematika, **27** (1980), 135–152.

[55] E. Fouvry and H. Iwaniec, *Primes in Arithmetic Progressions*, Acta Arith. **42** (1983), 197–218.

[56] E. Fouvry and F. Grupp, *On the Switching Principle in Sieve Theory*, J. reine angew Math. **370**, (1986), 101–126.

[57] E. Fouvry and H. Iwaniec, *The Divisor Function over Arithmetic Progressions* (with an appendix by N. Katz), Acta Arith. **61** (1992), 271–287.

[58] E. Fouvry, E. Kowalski and P. Michel, *On the Exponent of Distribution of the Ternary Divisor Function,* Mathematika, **61** (2015), 121–144.

[59] J. B. Friedlander, *Prime Numbers: A Much Needed Gap Is Finally Found,* Notices Amer. Math. Soc. **62** (2015), 660–664.

[60] J. B. Friedlander and A. Granville, *Limitations to the Equi-Distribution of Primes I*, Ann. of Math. **129** (1989), 363–382.

[61] J. B. Friedlander and H. Iwaniec, *Incomplete Kloosterman Sums and a Divisor Problem*, with an Appendix by B. J. Birch and E. Bombieri, *On Some Exponential Sums*, Ann. of Math., **121** (1985), 319–350.

[62] J. B. Friedlander and H. Iwaniec, *The Polynomial $x^2 + y^4$ Captures Its Primes*, Ann. of Math. **148** (1998), 945–1040.

[63] J. B. Friedlander and H. Iwaniec, *What Is the Parity Phenomenon?*, Notices Amer. Math. Soc. **56** (2009), 817–818.

[64] J. B. Friedlander and H. Iwaniec, *Opera de Cribro*, American Mathematical Society, 2010.

[65] P. X. Gallagher, *The Large Sieve*, Mathematika **14** (1967), 14–20.

[66] P. X. Gallagher, *A Larger Sieve*, Acta Arith. **53** (1971), 77–81.

[67] P. X. Gallagher, *On the Distribution of Primes in Short Intervals*, Mathematika **23** (1976), 4–9.

[68] P. X. Gallagher, *Corrigendum: "On the Distribution of Primes in Short Intervals"*, Mathematika **28** (1981), 86.

[69] D. A. Goldston, *Are There Infinitely Many Twin Primes?*, (preprint).

[70] D. A. Goldston, Y. Motohashi, J. Pintz and C. Y. Yildirim, *Small Gaps Between Primes Exist*, Proc. Japan Acad. Ser. A Math. Sci. **82** (2006), 61–65.

[71] D. A. Goldston and C. Y. Yildirim, *Higher Correlations of Divisor Sums Related to Primes. III. Small Gaps Between Primes,* Proc. Lond. Math. Soc. (3) **95** (2007), 653–686.

[72] D. A. Goldston, J. Pintz and C. Y. Yildirim, *Primes in Tuples I,* Ann. of Math. (2) **170** (2009), 819–862.

[73] D. A. Goldston and A. H. Ledoan, *Jumping Champions and Gaps Between Consecutive Primes*, Int. J. Number Theory **7** (2011), 1413–1421.

[74] D. A. Goldston and A. H. Ledoan, *On the Differences Between Consecutive Prime Numbers, I*, Integers **12B** (2012/13), #A3, 8 pages.

[75] D. A. Goldston and A. H. Ledoan, *The Jumping Champion Conjecture*, Mathematicka **61** (2015), 719–740.

[76] D. A. Goldston and A. H. Ledoan, *Limit Points of the Sequence of Normalized Differences Between Consecutive Prime Numbers*, 115–125 in *Analytic Number Theory. In Honor of Helmut Maier's 60th Birthday.* Eds. C. Pomerance and M. Th. Rassias, Springer, 2015.

[77] D. A. Goldston, S. W. Graham, J. Pintz and C. Y. Yildirim, *Small Gaps Between Products of Two Primes,* Proc. Lond. Math. Soc, (3) **98** (2009), 741–774.

[78] D. A. Goldston, S. W. Graham, J. Pintz and C. Y. Yildirim, *Small Gaps Between Primes or Almost Primes,* Trans. Amer. Math. Soc., **361** (2010), 5285–5330.

[79] D. A. Goldston, S. W. Graham, J. Pintz and C. Y. Yildirim, *Small Gaps Between Almost Primes, the Parity Problem, and Some Conjectures of Erdős on Consecutive Integers,* Int. Math. Res. Not. IRMN, (2011), 1439–1450.

[80] T. Gowers, (Ed.) *The Princeton Companion to Mathematics*, Princeton University Press, 2008.

[81] A. Granville, *Least Primes in Arithmetic Progressions*, Théorie des Nombres (Quebec, PQ, 1987), Walter de Gruyter, 1989, 306–321.

[82] A. Granville, *Primes in Intervals of Bounded Length*, Bull. Amer. Math. Soc. **52** (2015), 171–222.

[83] A. Granville, *About the Cover: A New Mathematical Celebrity*, Bull. Amer. Math. Soc. **52** (2015), 335–337.

[84] A. Granville and K. Soundararajan, *An Uncertainty Principle for Arithmetic Sequences*, Annals Math. **165** (2007), 593–635.

[85] G. Greaves, *Sieves in Number Theory*, Springer-Verlag, 2001.

[86] B. Green and T. Tao, *The Primes Contain Arbitrarily Long Arithmetic Progressions,* Ann. of Math. (2) **167** (2008), 481–547.

[87] R. Gupta and M. R. Murty, *A Remark on Artin's Conjecture,* Invent. Math. **78** (1984), 127–130.

[88] H. Halberstam and H. -E. Richert, *Sieve Methods,* Academic Press, 1974.

[89] G. H. Hardy and J. E. Littlewood, *Some Problems of "Partitio Numerorum";
III: On the Expression of a Number as a Sum of Primes,* Acta Math. **44**
(1923), 1–70.

[90] G. H. Hardy and J. M. Wright, *An Introduction to the Theory of Numbers,
6th ed.*, Oxford University Press, 2008.

[91] G. Harman, *Prime Detecting Sieves*, Princeton University Press, 2007.

[92] J. Harris, *Algebraic Geometry: A First Course*, Springer, 1992.

[93] R. Hartshorne, *Algebraic Geometry*, 8th printing, Springer, 1997.

[94] J. K. Haugland, *Application of Sieve Methods to Prime Numbers,* Ph.D.
thesis, Oxford University, 1999.

[95] D. R. Heath-Brown, *Prime Numbers in Short Intervals and a Generalized
Vaughan Identity,* Canad. J. Math. **34** (1982), 1365–1377.

[96] D. R. Heath-Brown, *Prime Twins and Siegel Zeros*, Proc. London Math.
Soc. **47** (1983), 193–224.

[97] D. R. Heath-Brown, *The Divisor Function $d_3(n)$ in Arithmetic Progressions,*
Acta Arith. **47** (1986), 1365–1377.

[98] D. R. Heath-Brown, *Artin's Conjecture for Primitive Roots,* Quart. J. Math.
Oxford (2) **37** (1986), 27–38.

[99] D. R. Heath-Brown, *Primes Represented by $x^3 + 2y^3$,* Acta Math. **186**
(2001), 1–84.

[100] D. R. Heath-Brown, *Obituary Atle Selberg*, Bull. London Math. Soc, **42**
(2010), 949–955.

[101] H. Heilbronn, *On the Class Number of Imaginary Quadratic Fields*, Quart.
J. Math., **5** (1934), 150–160.

[102] D. Hensley and I. Richards, *Primes in Intervals*, Acta Arith. **25** (1973/74),
375–391.

[103] H. W. Hethcote, *Asymptotic Approximations with Error Bounds for Zeros of
Airy and Cylindrical Functions,* Ph.D. thesis, University of Michigan, Ann
Arbor, 1968.

[104] H. W. Hethcote, *Error Bounds for Asymptotic Approximations Zeros of
Transcendental Functions*, SIAM J. Math. Anal. **1** (1970), 147–152.

[105] A. Hilderbrand, *Erdős' Problems on Consecutive Integers,* Paul Erdős and
His Mathematics I (Bolyai Society Mathematical Studies, 11) Budapest,
2002, 305–317.

[106] A. Hildebrand and H. Maier, *Gaps Between Prime Numbers,* Proc. Amer.
Math. Soc, **104** (1988), 1–9.

[107] K. -H. Ho and K. -M. Tsang, *On Almost Prime K-tuples*, J. Number Theory
120 (2006), 33–46.

[108] C. Hooley, *On Artin's Conjecture,* J. Reine Angew. Math. **225** (1967),
209–220.

[109] M. N. Huxley, *Small Differences Between Consecutive Primes,* Mathematika **20** (1973), 229–232.

[110] M. N. Huxley, *Small Differences Between Consecutive Primes II,* Mathematika **24** (1977), 142–152.

[111] M. N. Huxley, *An Application of the Fouvry–Iwaniec Theorem,* Acta Arith. **43** (1984), 441–443.

[112] M. N. Huxley and H. Iwaniec, *Bombieri's Theorem in Short Intervals,* Mathematika **22** (1975), 188–194.

[113] A. E. Ingham, *The Distribution of Prime Numbers*, Cambridge University Press, 1932.

[114] H. Iwaniec, *Primes Represented by Quadratic Polynomials in Two Variables,* Acta Arith. **24** (1973/74), 435–459.

[115] H. Iwaniec, *A New Form of the Error Term in the Linear Sieve*, Acta Arith. **37** (1980), 307–320.

[116] H. Iwaniec, *Conversations on the Exceptional Character*, 97–132, in *Analytic Number Theory: Lecture Notes in Mathematics,* vol. 1891, Springer, 2006.

[117] H. Iwaniec and E. Kowalski, *Analytic Number Theory*, American Mathematical Society, 2004.

[118] W. B. Jurkat and H. -E. Richert, *An Improvement of Selberg's Sieve Method*, I, Acta Arith. **11** (1965), 217–240.

[119] M. Jutila, *A Statistical Density Theorem for L-Functions with Applications,* Acta Arith. **16** (1969/70), 207–216.

[120] E. Klarreich, *Unheralded Mathematician Bridges the Prime Gap*, Quanta Magazine, 2013, www.quantamagazine.org/20130519

[121] J. -M. De Koninck and F. Luca, *Analytic Number Theory: Exploring the Anatomy of Integers*, American Mathematical Society, 2012.

[122] E. Landau, *Bemerkungen zu der vorstehenden Abhandlung von Herrn Franel,* Göttinger Nachrichten (1924), 202–206.

[123] E. Landau, *Handbuch der lehre von der Verteilung der Primzahlen, 2nd ed.,* volumes 1 and 2, Chelsea, 1953.

[124] T. Lang and R. Wong, *"Best Possible" Upper Bounds for the First Two Positive Zeros of the Bessel Function $J_v(v)$: The Infinite Case*, J. Comp. and App. Math. **71** (1996), 311–329.

[125] A. M. Legendre, *Théorie des Nombres, 2nd ed.*, Paris, 1808.

[126] U. V. Linnik, *The Large Sieve,* C. R. (Doklady) Acad. Sci. URSS **30** (1941), 292–294.

[127] U. V. Linnik, *The Dispersion Method in Binary Additive Problems, Translated from the Russian.* Amer. Math. Soc., 1963.

[128] L. Lorch, *Some Inequalities for the First Positive Zeros of Bessel Functions,* SIAM J. Math. Anal. **24** (1993), 814–823.

[129] L. Lorch and R. Uberti, *"Best Possible" Upper Bounds for the First Positive Zeros of the Bessel Functions – the Finite Part,* J. Comp. and Appl. Math. **72** (1996), 249–258.

[130] D. Mackenzie, *Prime Proof Helps Mathematicians Mind the Gaps,* Science **300**(2003), 32.

[131] D. Mackenzie, *Prime-Number Proof's Attempt Falls Short,* Science **300** (2003), 1066.

[132] H. Maier, *Small Differences Between Prime Numbers,* Michigan Math. J. **35** (1988), 323–344.

[133] H. Maier and C. Pomerance, *Unusually Large Gaps Between Consecutive Primes,* Trans. Amer. Math. Soc. **322** (1990), 201–237.

[134] Mathematics Genealogy Project, http://genealogy.math.ndsu.nodak.edu/index.php

[135] J. Maynard, *On the Brun–Titchmarsh Theorem,* Acta Arith. **157** (2013), 249–296.

[136] J. Maynard, *3-Tuples Have at Most 7 Prime Factors Infinitely Often,* Math. Proc. Camb. Phil. Soc. **155** (2013), 443–457.

[137] J. Maynard, *Almost Prime K-tuples,* Mathematica **60** (2014), 108–138.

[138] J. Maynard, *Small Gaps Between Primes,* Ann. of Math. (2) **181** (2015), 383–413.

[139] J. Maynard, *Dense Clusters of Primes in Subsets,* Compos. Math. **152** (2016), 1517–1554.

[140] J. Maynard, *Large Gaps Between Primes,* Ann. of Math. (2) **183** (2016), 915–933.

[141] T. T. Moh, *Zhang, Yitang's Life at Purdue (Jan 1985–1991) (Revised in Bold Face. 2018)* www.math.purdue.edu/~ttm/ZhangYt.pdf.

[142] H. L. Montgomery, *Primes in Arithmetic Progression,,* Mich. Math. J. **17** (1970), 33–39.

[143] H. L. Montgomery and R. C. Vaughan, *The Large Sieve,* Mathematika **20** (1973), 119–134.

[144] H. L. Montgomery and R. C. Vaughan, *Multiplicative Number Theory I: Classical Theory,* Cambridge University Press, 2007.

[145] Y. Motohashi, *On Some Improvements of the Brun–Titchmarsh Theorem,* J. Math. Soc. Japan **26** (1974), 306-323.

[146] Y. Motohashi, *An Induction Principle for the Generalization of Bombieri's Prime Number Theorem,* Proc. Japan Acad. **52** (1976), 273–275.

[147] Y. Motohashi, *Sieve Methods and Prime Number Theory,* Tata IFR and Springer-Verlag, 1983.

[148] Y. Motohashi, *The Remainder Term in the Selberg Sieve*, Number Theory in Progress, **2**, de Gruyter (1999), 1053–1064.

[149] Y. Motohashi, *An Overview of the Sieve Method and Its History*, arXiv:math/0505521v2 [math.NT] 27 Dec. 2006.

[150] Y. Motohashi, *The Twin Primes Conjecture*, arXiv:1401.6614v2 [math.NT] (16 March 2014).

[151] Y. Motohashi and J. Pintz, *A Smoothed GPY Sieve,* Bull. Lond. Math. Soc. **40** (2008), 298–310.

[152] M. B. Nathanson, *Additive Number Theory: The Classical Bases*, Springer, 1996.

[153] M. B. Nathanson, *Elementary Methods in Number Theory*, Springer, 2000.

[154] V. Neal, *Closing the Gap: The Quest to Understand Prime Numbers*, Oxford, 2017.

[155] R. L. H. Niederreiter, *Finite Fields*, Addison-Wesley, 1983.

[156] F. Oberhettinger, *Tables of Mellin Transforms*, Springer-Verlag, 1974.

[157] A. Odlyzko, M. Rubinstein and M. Wolf, *Jumping Champions*, Experiment. Math. **8** (1999), 107–118.

[158] F. W. J. Olver, *The Asymptotic Expansion of Bessel Functions of Large Order*, Philos. Trans. Roy. Soc. London Ser A **247** (1954), 328–368.

[159] F. W. J. Olver, *Error Bounds for First Approximations in Turning Point Problems*, J. Soc. Ind. Appl. Math. **11** (1963), 748–772.

[160] F. W. J. Olver, *Error Bounds for Asymptotic Expansions in Turning Point Problems*, J. Soc. Ind. Appl. Math. **12** (1964), 200–214.

[161] F. W. J. Olver, *Asymptotics and Special Functions*, Academic Press, 1974.

[162] F. W. J. Olver, D. W. Lozier, R. F. Boisvert and C. W. Clark (Ed.), *NIST Handbook of Mathematical Functions*, U.S. Department of Commerce, National Institute of Standards and Technology, Washington, DC; Cambridge University Press, 2010.

[163] C. D. Pan, X. X. Ding and Y. Wang, *On the Representation of Every Large Even Integer as a Sum of a Prime an Almost Prime,* Sci. Sinica **18** (1975), 599–610.

[164] G. Z. Pil'tjai, *The Magnitude of the Difference Between Consecutive Primes*, (Russian) Studies in Number Theory, Izdat. Saratov. Univ., Saratov, **4** (1972), 73–79.

[165] J. Pintz, *Very Large Gaps Between Consecutive Primes*, J. Number Theory **63** (1997), 286–301.

[166] J. Pintz, *On the Singlar Series in the Prime K-tuple Conjecture*, arXiv:1004.1084v1 [math.NT] 7 April 2010.

[167] J. Pintz, *A Note on Bounded Gaps Between Primes*, arXiv1303.1497v4 [math.NT] 17 July 2013.

[168] J. Pintz, *Polignac Numbers, Conjectures of Erdős on Gaps Between Primes, Arithmetic Progressions in Primes and the Bounded Gap Conjecture,* 367–384, in J. Sander, J. Steuding and R. Steuding (Eds.), in *From Arithmetic to Zeta-Functions,* Springer, 2016.

[169] P. Pollack, *Bounded Gaps Between Primes with a Given Primitive Root,* Algebra Number Theory **8** (2014), 1769–1786.

[170] A. de Polignac, *Six propositions arithmologiques déduites du crible d'Ératosthène,* Nouvelles annales de mathématiques (1), **8** (1849), 23–429.

[171] D. H. J. Polymath, *A New Proof of the Density Hales–Jewett Theorem,* Ann. of Math. (2) **175** (2012), 1283–1327.

[172] D. H. J. Polymath, *The "Bounded Gaps Between Primes" Polymath Project, a Retrospective Analysis*, EMS Newletter, December 2014, 13–23.

[173] D. H. J. Polymath, *New Equidistribution Estimates of Zhang Type, and Bounded Gaps Between Primes*, Algebra and Number Theory (8) **9** (2014), 2067–2199.

[174] D. H. J. Polymath, *New Equidistribution Estimates of Zhang Type, and Bounded Gaps Between Primes*, axXiv:1402.0811v2 [math.NT] 12 July 2014.

[175] D. H. J. Polymath, *Variants of the Selberg Sieve,and Bounded Intervals Containing Many Primes,* Polymath Research in the Mathematical Sciences, **1** (2014), 1–83.

[176] J. W. Porter, *Some Numerical Results in the Selberg Sieve Method,* Acta Arith. **20** (1972), 417–421.

[177] O. Ramaré with D. S. Ramana, *Arithmetical Aspects of the Large Sieve Inequality,* Hindustan, 2009.

[178] R. A. Rankin, *The Difference Between Consecutive Prime Numbers,* J. London Math. Soc. **13** (1938), 242–247.

[179] R. A. Rankin, *The Difference Between Consecutive Prime Numbers,* II. Proc. Cambridge Philos. Soc. **36** (1940), 255–266.

[180] R. A. Rankin, *The Difference Between Consecutive Prime Numbers, V,* Proc. Edinburgh Math. Soc. (2) **13** (1962/63), 331–332.

[181] A. Rényi, *On the Representation of an Even Number as the Sum of a Prime and of an Almost Prime*, Amer. Math. Soc. Transl. (2) **19** (1962), 299–321. Translated from the Russian Izv. Akad. Nauk SSSR ser. Mat. **12** (1948), 57–78.

[182] G. Ricci, *Sull'andamento della differenza di numeri primi consecutivi,* Riv. Mat. Univ. Parma **5** (1954), 3–54.

[183] G. Ricci, *Recherches sur l'allure de la suite $((p_{n+1} - p_n)/ \log p_n)$,* Colloque sur la Théorie des Nombres, Bruxelles, 1955 (G. Thone, 1956), 93–106.

[184] J. B. Rosser and L. Schoenfeld, *Approximate Formulas for Some Functions of Prime Numbers*, Illinois J. Math. **6** (1962), 64–94.

[185] W. Rudin, *Real and Complex Analysis*, 2nd ed., McGraw-Hill, 1974.

[186] W. Rudin, *Functional Analysis*, 2nd ed., McGraw-Hill, 1991.

[187] A. Schönhage, *Eine Bemerkung zur Konstruktion grosser Primzahllücken*, Arch. Math. **14** (1963), 29–30.

[188] M. Segal, *The Twin Primes Hero*, Nautilus Magazine, September 2013, http://nautil.us/issue/5/fame/the-twin-primes-hero.

[189] A. Selberg, *On an Elementary Method in the Theory of Primes*, Norske Vid. Selsk. Forhdl. **19** (1947), 64–67.

[190] A. Selberg, *The General Sieve Method and Its Place in Prime Number Theory,* Proc. Intern. Cong. Math., **1** (1950), 286–292.

[191] A. Selberg, *Sieve Methods,* Proc. Symp. Pure Math. **20** (1971), 311–351.

[192] A. Selberg, *Remarks on Sieves,* Proc. 1972 Number Theory conf., Boulder 1972, 205–216.

[193] A. Selberg, *Collected Papers. Vol. I. With a Foreword by K. Chandrasekharan*, Springer-Verlag, Berlin, 1989.

[194] A. Selberg, *Collected Papers. Vol. II. With a Foreword by K. Chandrasekharan*, Springer-Verlag, Berlin, 1991.

[195] P. Shiu, *A Brun–Titchmarsh Theorem for Multiplicative Functions,* J. Reine Angew. Math. **313** (1980), 161–170.

[196] R. Sitaramachandrarao, *On an Error Term of Landau, II,* Rocky Mt. J. Math. **15** (1985), 579–588.

[197] K. Soundararajan, *Nonvanishing of Quadratic Dirichlet L-Functions at $s = \frac{1}{2}$*, Ann. of Math. (2) **152** (2000), 447–488.

[198] K. Soundararajan, *Small Gaps Between Prime Numbers: The Work of Goldston–Pintz–Yildirim,* Bull. Amer. Math. Soc. (N.S.) **44** (2007), 1–18.

[199] R. Strichartz, *A Guide to Distribution Theory and Fourier Transforms,* CRC Press, 1994.

[200] T. Tao, *Structure and Randomness: Pages from Year One of a Mathematical Blog,* Amer. Math. Soc, 2008.

[201] T. Tao, *Every Odd Number Greater Than 1 Is the Sum of at Most Five Primes,* Math. Comp. **83** (2012), 997–1038.

[202] T. Tao, *Notes on Zhang's Prime Gaps Paper*, 1 June 2013, Accessible from the Polymath8 home page in section 9, "Recent Papers and Notes".

[203] T. Tao, *Web Based Lecture Notes on the Bombieri–Vinogradov Theorem,* 2016.

[204] G. Tenenbaum, *Introduction to Analytic and Probabilistic Number Theory*, Cambridge University Press, 1995.

[205] J. Thorner, *Bounded Gaps Between Primes in Chebotarev Sets,* Res. Math. Sci. **1** (2014), Art. 4, 16.

[206] E. C. Titchmarsh, *A Divisor Problem,* Rendiconti del Circolo matematico di Palermo, **54** (1930), 414–429.

[207] E. C. Titchmarsh and D. R. Heath-Brown, *The Theory of the Riemann Zeta-Function, 2nd ed.,* Oxford University Press, 1986.

[208] T. S. Trudgian, *A Poor Man's Improvement on Zhang's Result: There Are Infinitely Many Prime Gaps Less Than 60 Million,* arXiv 1305.6369v2 [math.NT] 4 June 2013.

[209] S. Uchiyama, *On the Difference Between Consecutive prime Numbers,* Acta Arith. **27** (1975), 153–157.

[210] R. C. Vaughan, *An Elementary Method in Prime Number Theory,* Acta Arith. **37** (1980), 111–115.

[211] A. I. Vinogradov, *The Density Hypothesis for Dirichlet L-Series,* (Russian) Izv. Akad. Nauk SSSR Ser. Mat. **29** (1965), 903–934.

[212] D. R. Ward, *Some Series Involving Euler's Function,* J. Lond. Math. Soc. **2** (4) (1927), 210–214.

[213] E. Westzynthius, *Über die verteilung der zahlen, die zu den n ersten primzahlen teilerfremd sind,* Commentationes Physico Mathematicae, Societas Scientarium Fennica, Helsingfors **5** (1931), 1–37.

[214] J. Wu, *Sur la suite des nombres premiers jumeaux,* Acta Arith. **55** (1990) 365–394.

[215] Y. Zhang, *Bounded Gaps Between Primes,* Ann. of Math. (2) **1979** (2014), 1121–1174.

Index